5.17.19
$ 30.00
As 14 Day
5/19

THE CREATIVITY CODE

THE CREATIVITY CODE

/ Art and Innovation in the Age of AI /

Marcus du Sautoy

THE BELKNAP PRESS OF HARVARD UNIVERSITY PRESS

Cambridge, Massachusetts

2019

Library of Congress Cataloging-in-Publication Data on file at https://www.loc.gov/

ISBN 978-0-674-988132

Book design by Chrissy Kurpeski

To Shani,
for all her love and support,
creativity and intelligence

CONTENTS

/ 1 /

The Lovelace Test

Works of art make rules;
rules do not make works of art.

—CLAUDE DEBUSSY

The machine was a thing of beauty. Towers of spinning brass cogs with numbers on their teeth were pinned to rods driven by a gear train. The seventeen-year-old Ada Byron was transfixed as she cranked the handle of the Difference Engine. Its inventor, Charles Babbage, had invited her to see it in action as it whirred and clicked away, mechanically calculating polynomial sums. Ada had always had a fascination with mathematics and inventions, encouraged by the tutors her mother had been eager to provide.

But it may have been the artistic genes she'd inherited from her father, the poet Lord Byron, that set Ada to imagining what such marvellous machinery might be capable of. A decade later—now married and become Countess of Lovelace—she turned her attention to Babbage's designs for an even more sophisticated calculator. It dawned on her that it would be more than just a number cruncher:

> The Analytical Engine does not occupy common ground with mere "calculating machines." It holds a position wholly its own; and the considerations it suggests are most interesting in their nature.

Ada Lovelace's notes are now recognized as the first inroads into the creation of code, the spark of the idea that has ignited the artificial intelligence revolution sweeping the world today, fueled by the work of pioneers like Alan Turing, Marvin Minsky, and Donald Michie. Yet Lovelace herself was cautious as to how much any machine could achieve: "It is desirable to guard against the possibility of exaggerated ideas that might arise as to the powers of the Analytical Engine," she wrote. "The Analytical Machine has no pretensions whatever to *originate* anything. It can do whatever we *know how to order it* to perform."

Ultimately, she believed, it was limited: you couldn't get more out than you had put in.

This belief has been a mantra of computer science for many years, a shield against the fear that someday programmers will set in motion a computer they cannot control. Some have gone so far as to suggest that to program a machine to be artificially intelligent, we would first have to understand human intelligence. But in the last few years a new way of thinking about code has emerged: a shift from a top-down approach to programming to bottom-up efforts to get the code to chart its own path. It turns out you don't have to solve intelligence first. You can allow algorithms to roam the digital landscape and learn just like children. Today's code is making surprisingly insightful moves, spotting hard-to-detect features in medical images and making shrewd trades on the stock market. This generation of coders believes it can finally prove Ada Lovelace wrong: that you can get more out than you programmed in.

Yet there is still one realm of human endeavor that most people believe the machines will never be able to touch. We have this extraordinary ability to imagine, to innovate and create. Our code, the creativity code, is one we have long felt that no programmer could ever crack. This is a code that we believe depends on being human.

Mozart's *Requiem* allows us to contemplate our own mortality. Witnessing a performance of *Othello* invites us to navigate the landscape of love and jealousy. A Rembrandt portrait conveys so much more than what the sitter looked like. How could a machine ever replace or compete with Mozart, Shakespeare, or Rembrandt? And, of course, human creativity extends beyond the arts. The molecular gastronomy of Michelin star chef Heston Blumenthal, the football trickery of Dutch striker Johan Cruyff, the curvaceous buildings of Zaha Hadid, the invention of the Rubik's Cube by Hungarian Erno Rubik, even the code behind a game like *Minecraft*—all involve great acts of human creativity.

One of the things that drives me to spend hours at my desk conjuring up equations and penning proofs is the thrill of creating something new. My greatest moment of creativity, one I think back to again and again, was the time I conceived of a new symmetrical object. No one knew this object was possible. But after years of hard work and a momentary flash of white-hot inspiration I wrote on my yellow notepad the blueprint for this novel shape. That sheer buzz of excitement is the allure of creativity.

But what do we really mean by this shape-shifting term? Those who have tried to pin it down usually circle around three factors. They speak of creativity as the drive to come up with something that is new, that is surprising, and that has value.

It turns out it's easy to make something new. I can get my computer to churn out endless proposals for new symmetrical objects. Surprise and value are more difficult to produce. In the case of my symmetrical creation, I was legitimately surprised by what I'd cooked up, and so were other mathematicians. No one was expecting the strange new connection I'd discovered between this symmetrical object and the unrelated subject of number theory. My object suggested a new way of understanding an area of mathematics that is full of unsolved problems, and that is what gave it value.

We all get sucked into patterns of thought. We think we see how the story will evolve and then suddenly we are taken in a new direction. This element of surprise makes us take notice. It is probably why we get a rush when we encounter creativity. But what gives something value? Is it a question of price? Does it have to be recognized by others? I might value a poem or a painting I've created, but my conception of its value is unlikely to be shared more widely. A surprising novel with lots of plot twists could be of relatively little value, but a new and surprising approach to storytelling, or architecture, or music—one that changes the way we see or experience things—will generally be recognized as having value. This is what Kant refers to as "exemplary originality," when an original act becomes an inspiration for others. This form of creativity has long been thought to be uniquely human.

At some level, all these expressions of creativity are the products of neuronal and chemical activity. Creativity is a code that evolution across millions of years has honed inside our brains. If we unpick the creative outpourings of the human species, we can start to see that there are rules underlying the creative process. So is our creativity in fact more algorithmic and rule-based than we might want to acknowledge? Can we hope to crack the creativity code?

This book aims to explore the limits of the new AI to see whether it can match or even surpass the marvels of our human code. Could a machine paint, compose music, or write a novel? It may not be able to compete with Mozart, Shakespeare, or Picasso, but could it be as creative as a child when asked to write a story or paint a scene?

My daughters are being creative when they build their Lego castles. My son is heralded as a creative midfielder when he leads his football team to victory. We speak of solving everyday problems creatively and running organizations creatively. The creative impulse is a key part of what distinguishes humans from other animals, and yet we often let it stagnate inside us, falling into the trap of becoming slaves to our formulaic lives. Being creative requires a jolt to take us out of the well-carved paths we retrace each day. This is where a machine might come in: perhaps it could give us that jolt, throw up a new suggestion, stop us from simply repeating the same algorithm each day. Machines might ultimately help us, as humans, behave less like machines.

Why, you may ask, is a mathematician offering to take you on this journey? The simple answer is that machine learning, algorithms, and code are all mathematical at heart. If you want to understand how and why the algorithms that regulate modern life are doing what they do, you need to understand the mathematical rules that underpin them. If you don't, you will be pushed and pulled around by the machines. AI is challenging us to the core as it reveals that many of the tasks that have been performed by humans can be done equally well, if not better, by machines. But rather than focus on driverless cars and computerized medicine, this book sets out to explore whether algorithms can compete meaningfully on our home turf. Can computers be creative? What does it mean to be creative? How much of our emotional response to artistic creativity is a product of our brains responding to pattern and structure? These are some of the things we will explore.

For me, there is another, more personal reason for wanting to go on this journey. I have found myself wondering, with the onslaught of new developments in AI, if the job of mathematician will still be available to humans in decades to come. Mathematics is a subject of numbers and logic. Isn't that what a computer does best? Part of my defense against the computers knocking on the door, wanting their place at the table, is that, as much as mathematics is about numbers and logic, it is a highly creative subject involving beauty and aesthetics. The breakthroughs mathematicians share in seminars and journals aren't just the results of cranking mechanical handles. Intuition and artistic sensitivity are essential, too. The great German mathematician Karl Weierstrass once wrote: "a mathematician that is not somewhat of a poet will never be a perfect mathematician." As Ada Lovelace showed us, one needs a bit of Byron as much as Babbage.

Although she thought machines were limited, Ada Lovelace began to believe that their cogs and gears could expand beyond the narrow role they had been given. Pondering the Analytical Engine, she could envision the day when "it might act upon other things besides *number*" to produce other forms of output. "Supposing, for instance, that the fundamental relations of pitched sounds in the science of harmony and of musical composition were susceptible of such expression and adaptations," she ventured, "the engine might compose elaborate and scientific pieces of music of any degree of complexity or extent." Still, she believed that any act of creativity would lie with the coder, not the machine.

At the dawn of AI, Alan Turing proposed a test by which a computer might someday be considered to have displayed enough intelligence to be mistaken for a human. What he called "the imitation game" rapidly became known as the Turing Test. I would like to propose a new challenge. Let's call it the Lovelace Test.

To pass the Lovelace Test, an algorithm has to produce something that is truly creative. The process has to be repeatable (not the result of a hardware error) and the programmer has to be unable to explain how the algorithm produced its output. We are challenging the machines to come up with things that are new, surprising, and of value. For a machine to be deemed truly creative, its contribution has to be more than an expression of the creativity of its coder or the person who built its data set.

Ada Lovelace believed this challenge was insurmountable. Let's see if she was right.

Three Types of Creativity

The chief enemy of creativity is good sense.

—PABLO PICASSO

The value placed on creativity in modern times has led to a range of writers and thinkers trying to articulate what it is, how to stimulate it, and why it is important. It was while serving on a committee convened by the Royal Society to assess what impact machine learning would likely have on society that I first encountered the theories of Margaret Boden.

Boden is an original thinker who over the decades has managed to fuse many different disciplines: she is a philosopher, psychologist, physician, AI expert, and cognitive scientist. In her eighties now, with white hair flying like sparks and an ever-active brain, she enjoys engaging with the question of what these "tin cans," as she likes to call computers, might be capable of. To this end, she has identified three different types of human creativity.

Exploratory creativity involves taking what is already there and exploring its outer edges, extending the limits of what is possible while remaining bound by the rules. Bach's music is the culmination of a journey that baroque composers embarked on to explore tonality by weaving together different voices. His preludes and fugues pushed the boundaries of what was possible before breaking the genre open and ushering in the classical era of Mozart and Beethoven. Renoir and Pissarro reconceived how we could visualize nature and the world around us, but it was Claude Monet who really pushed the boundaries, painting his water lilies over and over until his flecks of color dissolved into a new form of abstraction.

Mathematics revels in this type of creativity. The classification of Finite Simple Groups is a tour de force of exploratory creativity. Starting from the simple definition of a group of symmetries—a structure defined by four simple axioms—mathematicians spent 150 years compiling the list of every conceivable element of symmetry. This effort culminated in the discovery of the Monster simple group: it has more symmetries than there are atoms in the Earth and yet fits into no pattern of other groups. This form of mathematical creativity involves pushing limits while adhering strictly to the rules of the game. Those who engage in it are like the geographical explorers who, even as they discover previously unknown territory, are still bound by the limits of our planet.

Boden believes that exploration accounts for 97 percent of human creativity. This is also the sort of creativity at which computers excel. Pushing a pattern or set of rules to an extreme is a perfect exercise for a computational mechanism that can perform many more calculations than the human brain can. But is it enough to yield a truly original creative act? When we hope for that, we generally imagine something more utterly unexpected.

To understand Boden's second type, *combinational creativity*, think of an artist taking two completely different constructs and finding a way to combine them. Often the rules governing one will suggest an interesting new framework for the other. Combination is a very powerful tool in the realm of mathematical creativity. The eventual solution of the Poincaré conjecture, which describes the possible shapes of our universe, was arrived at by applying the very different tools used to understand flow over surfaces. In a leap of creative genius, Grigori Perelman landed at the unexpected realization that by knowing the way a liquid flows over a surface one could classify the possible surfaces that might exist.

My own research takes tools from number theory that have been used to understand primes and applies them to classify possible symmetries. The symmetries of geometric objects don't look at first sight anything like numbers. But applying the language that has helped us to navigate the mysteries of the primes and replacing primes with symmetrical objects has revealed surprising new insights into the theory of symmetry.

The arts have also benefited greatly from this form of cross-fertilization. Philip Glass took ideas he learned from working with Ravi Shankar and used them to create the additive process that is the heart of his minimalist music. Zaha Hadid combined her knowledge of architecture with her love of the pure forms of the Russian painter Kasimir Malevich to create a unique style of curvaceous buildings. In cooking, creative master chefs have fused cuisines from opposite ends of the globe.

There are interesting hints that this sort of creativity might also be perfect for the world of AI. Take an algorithm that plays the blues and combine it with the music of Boulez and you will end up with a strange hybrid composition that might just create a new sound world. Of course, it could also be a dismal cacophony. The coder needs to find two genres that can be fused algorithmically in an interesting way.

It is Boden's third form of creativity that is the more mysterious and elusive. What she calls *transformational creativity* is behind those rare moments that are complete game changers. Every art form has these gear shifts. Think of Picasso and cubism. Schoenberg and atonality. Joyce and modernism. They are phase changes, like when water suddenly goes from liquid to gas or solid. This was the image Goethe hit upon when he sought to describe how he was able to write *The Sorrows of Young Werther*. He devoted two years to wrestling with how to tell the story, only for a startling event, a friend's suicide, to act as a sudden catalyst. "At that instant," he recalled in *Dichtung und Wahrheit*, "the plan of *Werther* was found; the whole shot together from all directions, and became a solid mass, as the water in a vase, which is just at the freezing point, is changed by the slightest concussion into ice."

At first glance it would seem hard to program such a decisive shift, but consider that, quite often, these transformational moments hinge on changing the rules of the game, or dropping a long-held assumption. The square of a number is always positive. All molecules come in long lines, not chains. Music must be written inside a harmonic scale structure. Eyes go on either sides of the nose. There is a meta rule for this type of creativity: start by dropping constraints and see what emerges. The creative act is to choose what to drop—or what new constraint to introduce—such that you end up with a new thing of value.

If I were asked to identify a transformational moment in mathematics, the creation of the square root of minus one, in the mid-sixteenth century, would be a good candidate. This was a number that many mathematicians believed did not exist. It was referred to as an imaginary number (a dismissive term first used by Descartes to indicate that there was no such thing). And yet its creation did not contradict previous mathematics. It turned out it had been a mistake to exclude it. Now consider, if that error had persisted to today: Would a computer come up with the concept of the square root of minus one if it were fed only data telling it that there is no number whose square could be negative? A truly creative act sometimes requires us to step outside the system and create a new reality. Can a complex algorithm do that?

The emergence of the romantic movement in music is in many ways a catalog of rule-breaking. Instead of hewing to close key signatures as earlier composers had done, upstarts like Schubert chose to shift keys in ways that deliberately defied expectations. Schumann left chords unresolved that Haydn or Mozart would have felt compelled to complete. Chopin composed dense moments of chromatic runs and challenged rhythmic expectations with his unusual accented passages and bending of tempos. The move from one musical era to another, from Medieval to Baroque to Classical to Romantic to Impressionist to Expressionist and beyond, is one long story of smashing the rules. It almost goes without saying that historical context plays an important role in allowing us to define something as new. Creativity is not an absolute but a relative activity. We are creative within our culture and frame of reference.

Could a computer initiate the kind of phase change that can move us into a new state? That seems a challenge. Algorithms learn how to act based on the data presented to them. Doesn't this mean that they will always be condemned to producing more of the same?

As the epigraph of this chapter, I chose Picasso's observation that the "chief enemy of creativity is good sense." That sounds, on the face of it, very much against the spirit of the machine. And yet, one can program a system to behave irrationally. One can create a meta rule that will instruct it to change course. As we shall see, this is in fact something machine learning is quite good at.

Can Creativity Be Taught?

Many artists like to fuel their own creation myths by appealing to external forces. In ancient Greece, poets were said to be inspired by the muses, who breathed a kind of creative energy into their minds, sometimes sacrificing the poet's sanity in the process. For Plato, "a poet is holy, and never able to compose until he has become inspired, and is beside himself and reason is no longer in him . . . for no art does he utter but by power divine." The great mathematician Srinivasa Ramanujan likewise attributed his insights to ideas imparted to him in dreams by the goddess Namagiri, his family's deity. Is creativity a form of madness or a gift of the divine?

One of my mathematical heroes, Carl Friedrich Gauss, was known for covering his tracks. Gauss is credited with creating modern number theory with the publication in 1798 of one of the great mathematical works of all time: the *Disquisitiones arithmeticae*. When people tried to glean from his book just how he got his ideas, they were mystified. It has been described as a book of seven seals. Gauss seems to pull ideas like rabbits out of a hat, without ever really giving us an inkling of how he conjured them. At one point, when someone asked him about this, he retorted that an architect does not leave up the scaffolding after the house is complete. He attributed one revelation to "the Grace of God," saying he was "unable to name the nature of the thread" that connected what he had previously known to the subsequent step that made his success possible.

Just because artists are often unable to articulate where their ideas come from does not mean they follow no rules. Art is a conscious expression of the myriad logical gates that make up our unconscious thought processes. In Gauss's case, there was a thread of logic that connected his thoughts. It's simply that it was hard for him to articulate what he was up to—or perhaps he wanted to preserve the mystery and boost his image as a creative genius. Coleridge's claim that the drug-induced vision of *Kubla Khan* came to him in its entirety is belied by all the evidence of preparatory material, showing that he worked up the ideas before that fateful day when he was interrupted by the person from Porlock. Of course, the white-hot flash of inspiration makes for a good story. Even my own accounts of creative discovery focus on that dramatic moment rather than the years of preparatory work I put in.

We have an awful habit of romanticizing creative genius. The solitary artist working in isolation is frankly a myth. In most instances, what looks like a step change is actually continuous growth. Brian Eno talks about the idea of *scenius*, not genius, to acknowledge the community out of which creative intelligence often emerges. Joyce Carol Oates agrees: "Creative work, like scientific work, should be greeted as a communal effort—an attempt by an individual to give voice to many voices, an attempt to synthesize and explore and analyze."

What does it take to stimulate creativity? Might it be possible to program it into a machine? Are there rules we can follow to become creative? Can creativity, in other words, be a learned skill? Some would say that to teach or program is to show people how to imitate what has gone before, and that imitation and rule-following are incompatible with creativity. Yet, we have examples of creative individuals all around us who have studied and learned and improved their skills. If we study what they do, could we imitate them and ultimately become creative ourselves?

These are questions I find myself asking anew every semester. To receive a PhD, a doctoral candidate in mathematics must create a new mathematical construct. He or she has to come up with something that has never been done before. I am tasked with teaching my students how to do that. Of course, at some level, they have been training to do this from their earliest student days. Solving a problem calls for personal creativity even if the answer is already known.

That training is an absolute prerequisite for the jump into the unknown. Rehearsing how others came to their breakthroughs builds the capacity to achieve one's own creative feats. Boden distinguishes between what she calls "psychological creativity" and "historical creativity." Many of us achieve acts of personal creativity that may be novel to us but historically old news. These are what Boden calls moments of psychological creativity. It is by repeated acts of personal creativity that ultimately one hopes to produce something that is recognized by others as new and of value. To be sure, that jump is far from guaranteed. But while historical creativity is rare, when it does occur, it emerges from encouraging psychological creativity.

I can't take just anyone off the street and teach them to be a creative mathematician. Even if I had ten years to train them, we might not get there—not every brain seems to be able to achieve mathematical creativity. Some people appear to be able to achieve creativity in one field but not another, yet it is difficult to understand what sets one brain on the road to becoming a chess champion and another, a Nobel Prize–winning novelist.

My recipe for eliciting original work in students follows Boden's three types of creativity. Exploratory creativity is perhaps the most obvious path. This involves deep immersion in what we have created to date. Out of that deep understanding might emerge something never seen before. It is important to impress on students that there isn't very often some big bang that resounds with the act of creation. It is gradual. Van Gogh expressed it well: "Great things are not done by impulse but by small things brought together."

I find Boden's second type, combinational creativity, to be a powerful weapon in stimulating new ideas. I often encourage students to attend seminars and read papers in subjects that don't seem connected with the problems they are tackling. A line of thought from a distant corner of the mathematical universe might resonate with the problem at hand and stimulate a new idea. Some of the most creative bits of science are happening today at the junctions of the disciplines. The more we can stray beyond our narrow lanes to share our ideas and problems, the more creative we are likely to be. This is where a lot of the low-hanging fruit is found.

Boden's third type, transformational creativity, seems hard at first sight to harness as a strategy. But again, the goal is to test the status quo by dropping some of the constraints that have been put in place. Try seeing what happens if you change one of the basic rules you have accepted as part of the fabric of your subject—it's dangerous, because by doing so you can collapse the system. But this brings me to one of the most important ingredients needed to foster creativity, and that is embracing failure.

Unless you are prepared to fail, you will not take the risks that will allow you to break out and create something new. This is why our education system and our business environment, both realms that abhor failure, are terrible environments for fostering creativity. If I want creativity from my students, I have learned, it is important to celebrate their failures as much as their successes. Sure, their failures won't make it into the PhD thesis, but so much can be learned from them. When I meet with my students, I repeat again and again Samuel Beckett's call to "Try. Fail. Fail again. Fail better."

Are these strategies that can be written into code? In the past, the top-down approach to coding meant there was little prospect of creativity in the output. Coders were never very surprised by what their algorithms produced. There was no room for experimentation or failure. But this all changed recently—because an algorithm, built on code that learns from its failures, did something that was new, shocked its creators, and had incredible value. This algorithm won a game that many believed was beyond the abilities of a machine to master. As we will see in Chapter 3, it was a game that required creativity to play.

/ 3 /

Ready Steady Go

We construct and construct,
but intuition is still a good thing.

—PAUL KLEE

People often compare mathematics to playing chess, and certainly there are connections. But when Deep Blue beat the best chess master the human race could offer, in 1997, it did not lead to the closure of mathematics departments. Although chess is a good analogy for the formal effort of constructing a proof, there is another game that mathematicians have regarded as much closer to their work, because it also features a creative and intuitive side. That is the Chinese game of Go.

I first discovered Go when I visited the mathematics department at Cambridge as an undergraduate to explore whether to do my PhD with the amazing group that had helped complete the Classification of Finite Simple Groups, a sort of Periodic Table of Symmetry. As I sat talking about the future of mathematics with John Conway and Simon Norton, two of the architects of this great project, I kept being distracted by students at the next table furiously slamming black and white stones onto a grid carved into a large, wooden board.

Eventually I asked Conway what they were doing. "That's Go," he explained. "It's the oldest game that is still being played to this day." In contrast to chess, with its warlike quality, Go is a game of territory capture. Players take turns placing their white or black pieces (or stones) onto a grid nineteen squares wide and nineteen squares high. If one player manages to surround a collection of the other's stones, those stones are captured. The game is over when all the stones have been placed, and the winner is the player who has captured the most stones. It sounds rather simple. The challenge of the game is that, as you are pursuing efficient captures of your opponent's stones, you must also avoid having your own stones captured.

"It's a bit like mathematics," Conway explained. "Simple rules that give rise to beautiful complexity." In fact, it was while Conway watched a game playing out between two experts, as they drank coffee in that common room, that he formed the germ of the idea he would later call "surreal numbers." As a Go match moves to its end game, it behaves like this new sort of number.

I've always been fascinated by games. When I travel abroad I like to learn whatever game I find the locals playing and bring it back with me. So when I got back from the wilds of Cambridge to the safety of my home in Oxford, I decided to buy a Go set from the local toy shop and see just what appeal it held for these obsessed students. As I began to explore the game with an Oxford classmate, I realized how subtle it was. It seemed impossible to devise a strategy that would lead to a win. In an important respect, a Go game proceeds in a direction opposite to chess. Whereas with chess, one's choices of moves get simplified with every piece removed from the board, with this game the constant addition of stones to the board makes the situation ever more complicated.

The American Go Association estimates that it would take a number with three hundred digits to count the number of games of Go that are legally possible. For chess, the computer scientist Claude

Shannon estimated that a 120-digit number (now called the Shannon number) would suffice. These are not small numbers in either case, but they give you a sense of how big the difference is in possible permutations.

As a kid I played a lot of chess and enjoyed working through the logical consequences of a proposed move. It appealed to the growing mathematician in me. Because the moves in chess branch out in a controlled manner, it is a manageable task for a computer, or even a human, to comprehend the tree of possibilities and analyze the implications of going down different branches. In contrast, the complexity of Go makes it impossible to analyze the tree of possibilities in any reasonable timeframe. This is not to say that Go players don't work to anticipate the logical consequences of their moves, but it does imply that they also rely on a more intuitive feel for the pattern of play.

The human brain is acutely attuned to discerning whatever structure and pattern there is to be found in a visual image. An experienced Go player looks at the lay of the stones and, by tapping into the brain's strength at pattern recognition, is able to spot a valuable next move. For computers, mimicking this very basic human skill has traditionally been a struggle. Machine vision is a challenge that engineers have wrestled with for decades.

The human brain's highly developed sense of visual structure has been honed over millions of years and has been key to our survival. Any animal's ability to survive depends in part on its ability to pick out structure from the visual mess that nature offers up. A pattern in the chaos of the jungle likely indicates the presence of another animal—and if you fail to take notice, that animal might eat you (or at least you will miss your chance to eat it). The human code is extremely good at reading patterns, interpreting how they might develop, and responding appropriately. It is one of our key assets, and it plays into our appreciation for the patterns in music and art.

It turns out that pattern recognition is precisely what I do as a mathematician when I venture into the remoter reaches of the mathematical jungle. I can't rely on a simple, step-by-step logical analysis of the local environment. That won't get me very far. It has to be combined with an intuitive feel for what might be out there. That intuition is built up by time spent exploring the known space. But it is often hard to articulate logically why you believe there might be interesting territory out there to explore. A conjecture in mathematics is by definition not yet proved, but the mathematician who has made the conjecture has built up a feeling that the mathematical statement may have some truth. Observation and intuition go hand in hand in the process of navigating the thicket and carving out a new path.

The mathematician who makes a good conjecture will often garner more respect than the one who later connects the logical dots to reveal the truth of that conjecture. In the game of Go, we might think of the final winning position as the conjecture and the moves as the logical steps toward proving that conjecture. But it is devilishly hard to spot the patterns along the way.

This is why, although chess has been useful to explain some aspects of mathematics, the game of Go has always been held up as far closer in spirit to the way mathematicians actually go about their business. Note that mathematicians weren't too worried when Deep Blue beat the best humans could offer at chess. The real challenge was the game of Go. For decades, people had been claiming that the game of Go could never be played by a computer. Like all good absolutes, it invited creative coders to test that proposition, yet for a long time it was true that even a junior player could outplay their most complex algorithms. And so mathematicians happily hid behind the cover that Go was providing them. If a computer couldn't play Go, then there was no chance it could play the even subtler and more ancient game of mathematics.

But just as the Great Wall of China was eventually breached, my defensive wall has now crumbled in spectacular fashion.

Game Boy Extraordinaire

At the beginning of 2016 it was announced that a program had been created to play Go that its developers were confident could hold its own against the best human competition. Go players around the world were extremely skeptical, given the failure of past efforts. So the company that developed the program offered a challenge. It set up a public contest with a huge prize and invited one of the world's leading Go players to take up the challenge. International champion Lee Sedol from Korea stepped up to the plate. The competition would be played over five games with the winner taking home a prize of one million dollars. The name of Sedol's challenger: AlphaGo.

AlphaGo is the brainchild of Demis Hassabis. Hassabis was born in London in 1976 to a Greek Cypriot father and a mother from Singapore. Both parents are teachers and what Hassabis describes as Bohemian technophobes. His sister and brother went the creative route, one becoming a composer and the other choosing creative writing. So Hassabis isn't quite sure where his geeky scientific side comes from. But, as a kid, he was someone who quickly marked himself out as gifted, especially when it came to playing games. His abilities at chess were such that, at eleven, he was the child in his age group who was second highest ranked in the world.

But then, at an international match in Liechtenstein that year, Hassabis had an epiphany: What on earth were they all doing? The hall was full of so many great minds exploring the logical intricacies of this great game—and yet, Hassabis suddenly recognized the total futility of such a project. Later, in a BBC radio interview, he admitted what he was thinking at the time: "We were wasting our minds. What if we used that brain power for something more useful like solving cancer?"

His parents were pretty shocked when after the tournament (which he narrowly lost after a ten-hour battle with the adult Dutch world champion) he announced that he was giving up chess competitions. Everyone had thought this was going to be his life. But those years playing chess weren't wasted. A few years earlier he'd used the £200 prize money he won by beating US opponent Alex Chang to buy his first computer: a ZX Spectrum. That computer sparked his obsession with getting machines to do the thinking for him.

Hassabis soon graduated to a Commodore Amiga, which could be programmed to play the games he enjoyed. Chess was still too complicated, but he managed to program the Commodore to play Othello, a game that looks rather similar to Go. Its stones, which are black on one side and white on the other, get flipped when they are trapped between stones of an opponent's color. It's not a game that merits grandmasters, so he tried his program out on his younger brother. It beat him every time.

This was classic if-then programming: he needed to code in by hand the response to each of his opponent's moves. It was "if your opponent plays that move, then reply with this move." The machine's capability all came from Hassabis and his ability to see what the right responses were to win the game. It still felt a bit like magic, though. Code up the right spell and then, rather like *The Sorcerer's Apprentice*, the Commodore would go through the work of winning the game.

Hassabis raced through his schooling, which culminated at the age of sixteen with an offer to study computer science at Cambridge. He'd set his heart on Cambridge after seeing Jeff Goldblum in the film *The Race for the Double Helix*. "I thought, is this what goes on at Cambridge? You go there and you invent DNA in the pub? Wow."

Cambridge wouldn't let him start his degree at the age of sixteen, so he had to defer for a year. To fill his time, he won a place working for a game developer, having come in second in a competition run by *Amiga Power* magazine. While there, he created his own game, *Theme Park*, in which players build and run their own theme park. The game was hugely successful, selling several million copies and winning a Golden Joystick Award. With enough funds to finance his time at university, Hassabis set off for Cambridge.

His course introduced him to the greats of the AI revolution: Alan Turing and his test for intelligence; Arthur Samuel and his program to play checkers; John McCarthy, who coined the term artificial intelligence; Frank Rosenblat and his first experiments with neural networks. These were the shoulders on which Hassabis aspired to stand.

It was while sitting in lectures at Cambridge that he heard a professor repeat the mantra that a computer could never play Go because of the game's great reliance on creativity and intuition. This was like a red rag waved in front of the young Hassabis. He left Cambridge determined to prove that professor wrong.

His idea was that, rather than try to write the program himself that could play Go, he would write the meta-program that could write the program to play Go. It sounded crazy, but the concept was that this meta-program could, as the Go-playing program played more and more games, learn from its mistakes.

Hassabis had learned about a similar idea implemented by artificial intelligence researcher Donald Michie in the 1960s. Michie had written an algorithm called MENACE that learned from scratch the best strategy to play tic-tac-toe or, as it is called in the UK, noughts and crosses. (MENACE stood for Machine Educable Noughts And Crosses Engine.) To demonstrate the algorithm Michie had rigged up 304 matchboxes representing all the possible layouts of X's and O's encountered while playing. Each matchbox was filled with different colored balls to represent possible moves. Balls were removed or added to the boxes to punish losses or reward wins. As the algorithm played more and more games, the reassignment of the balls eventually led to an almost perfect strategy for playing. It was this idea of learning from mistakes that Hassabis wanted to use to train an algorithm to play Go.

Hassabis could base his strategy on another good model. A newborn baby does not have a brain that is preprogrammed with knowledge of how to make its way through life. It is programmed instead to learn as it interacts with its environment.

If Hassabis was going to emulate the way the brain learns to solve problems, then knowing how the brain works was clearly going to help. So he decided to do a PhD in neuroscience at University College London. It was during coffee breaks from lab work that Hassabis started discussing with neuroscientist Shane Legg his plans to create a company to try out his ideas. It shows the low status of AI as recently as a decade ago that they never admitted to their professors their dream to dedicate their lives to AI. But they felt they were onto something big. In September 2010, the two scientists decided to create a company with Mustafa Suleyman, who had been Hassabis's friend since childhood. And thus DeepMind was incorporated.

The company needed money, but initially Hassabis couldn't raise any capital. Pitching on a promise that they were going to solve intelligence by playing games did not sound serious to most investors. A few, however, did see the potential. Among those who put money in right at the outset were Elon Musk and Peter Thiel. Thiel had never invested outside Silicon Valley and tried to persuade Hassabis to relocate there. A born and bred Londoner, Hassabis held his ground, insisting that there was more untapped talent available in London. Hassabis remembers a bizarre conversation he had with Thiel's lawyer, who asked in all earnestness: "Does London have law on IP?" He shakes his head: "I think they thought we were coming from Timbuktu!" The founders had to give up a huge amount of stock to the investors, but they got the money to start trying to crack AI.

The challenge of creating a machine that could learn to play Go still felt like a distant dream. They set their sights at first on a less cerebral goal: playing 1980s Atari games. Atari is probably responsible for a lot of students flunking courses in the late '70s and early '80s. I certainly remember wasting a huge amount of time playing the likes of *Pong*, *Space Invaders*, and *Asteroids* on a friend's Atari 2600 console. The console was one of the first pieces of hardware that could play multiple games, which were loaded via cartridges. This allowed for more games to be developed over time. With previous consoles you could play only the games that had come preprogrammed into them.

One of my favorite Atari games was called *Breakout*. A wall of colored bricks appeared on the screen and your job was to destroy it completely. Your only weapons were a series of brick-destroying balls that you could send toward the bricks by using a paddle at the bottom. Moving this paddle left or right with a joystick, you attempted to intercept the balls as they bounced off the bricks and propel them back toward the wall. As you cleared the bricks—assuming you didn't lose all the balls by letting them fly past your paddle—the yellow ones in the bottom layers each scored you one point. The higher layers of red bricks got you seven points. Meanwhile, as your score rose, the balls would speed up, making the game-play progressively harder.

My friend and I were particularly pleased one afternoon when we found a clever trick to vastly improve our scores. If you dug a tunnel up through the bricks on one edge of the screen, you could get the ball up above the wall, where it would bounce rapidly up and down in that tight crawl space. You could sit back and watch one well-batted ball destroy many upper-level, high-scoring bricks before it eventually ricocheted back down through the wall. You just had to be ready with the paddle to bat the ball back up again. It was a very satisfying strategy!

Hassabis and the team he assembled also spent a lot of time playing computer games in their youth. Their parents may be happy to know that the time and effort they put into those games did not go to waste. It turned out that *Breakout* was a perfect test case to see if the team at DeepMind could program a computer to learn how to play games. It would have been a relatively straightforward job to write a program for each individual game. Hassabis and his team set themselves a much greater challenge.

They wanted to write a program that would have constant awareness of the state of the pixels on the screen and the current score and, with only those two inputs provided, could figure out how to play. The program was not told the rules of the game—only that its objective was to maximize the score. It had to experiment randomly with different ways of moving the paddle in *Breakout* or firing the laser cannon at the descending aliens of *Space Invaders*. Each time it made a move, it could assess whether that move had helped increase the score or not.

The code implements an idea dating from the 1990s called reinforcement learning, which aims to update the probability of actions based on the effect on a reward function, or score. For example, in *Breakout*, the only decision is whether to move the paddle left or right. Initially, the choice will be 50:50. But if moving the paddle randomly results in its hitting the ball, and a short time later the score goes up, the code then recalibrates based on this new information the probability of whether left or right is a better way to go. This increases the chance of heading toward where the ball is heading. The new feature was to combine this learning with neural networks that would assess the state of the pixels to determine what features were correlating to the increase in score.

At the outset, because the computer was just trying random moves, it was terrible. It barely scored anything. But each time it made a random move that bumped up the score, it would remember that move and reinforce the use of such a move in future. Gradually the random moves disappeared and a more informed set of moves began to emerge, moves that the program had learned through experiment seemed to boost its score.

It's worth watching the video the DeepMind team appended to the paper it eventually published. It shows the program learning to play *Breakout*. At first you see it randomly moving the paddle back and forth to see what will happen. Then, when a ball finally hits the paddle and bounces back and hits a brick and the score goes up, the program starts to rewrite itself. If the pixels of the ball and the pixels of the paddle connect, that seems to be a good thing. After four hundred games, it's doing really well, getting the paddle to bat balls back to the wall repeatedly.

The shock for me came with what it discovered after six hundred games. It found the same trick my friend and I had! I'm not sure how many games it took us as kids to discover it, but judging by the amount of time we wasted together, it could well have been more. But there it is. The program manipulated the paddle to ding the ball several times to the right, where it tunneled its way up the side and got trapped in the gap at the top of the screen. But while I remember my friend and I high-fiving when we discovered this trick, the machine felt nothing.

By 2014, four years after the creation of DeepMind, the program had learned how to outperform humans on twenty-nine of the forty-nine Atari games it had been exposed to. The paper the team submitted to *Nature* detailing this achievement came out in early 2015. To be published in *Nature* is one of the highlights of a scientist's career. But this paper achieved the even greater accolade of being featured as the cover story of the whole issue. The journal recognized that this was a huge moment for artificial intelligence.

It has to be reiterated what an amazing feat of programming this was. From just the raw data of the state of the pixels and the changing score, the program had advanced itself from randomly moving the *Breakout* paddle back and forth to learning that sending balls to the side of the screen would win it the top score. But Atari games are hardly the epitome of strategic play. Hassabis and his team at DeepMind decided to make a run at the game that was: they would create a new program that could take on the ancient game of Go.

It was around this time that Hassabis had decided to sell the company to Google. "We weren't planning to, but three years in, focused on fundraising, I had only ten percent of my time for research," he explained in a *Wired* interview. "I realized that there's maybe not enough time in one lifetime to both build a Google-sized company and solve AI. Would I be happier looking back on building a multibillion-dollar business or helping solve intelligence? It was an easy choice." The sale put Google's firepower at his fingertips and provided the space for him to create code to realize his goal of solving Go—and then intelligence.

First Blood

Previous computer programs built to play Go had not come close to playing competitively against even a pretty good amateur, so most pundits were highly skeptical of DeepMind's dream. How could it create code that could get anywhere near an international champion of the game? Most people still agreed with the view expressed in the *New York Times* by astrophysicist Piet Hut after Deep Blue's success at chess in 1997: "It may be a hundred years before a computer beats humans at Go—maybe even longer. If a reasonably intelligent person learned to play Go, in a few months he could beat all existing computer programs. You don't have to be a Kasparov."

Just two decades into that hundred years, the DeepMind team believed it might have cracked the code. Its strategy of getting algorithms to learn and adapt appeared to be working but it was unsure quite how powerful the emerging algorithm really was. So in October 2015 it decided to test-run the program in a secret competition against the current European champion, Fan Hui.

AlphaGo destroyed Fan Hui, five games to zero. But the gulf between European players of the game and those in the Far East is huge. The top European players, when put in a global league, rank way down in the six hundreds. So although it was still an impressive achievement, it was like building a driverless car that could beat a human driving a Ford Fiesta around a racetrack, then trying to challenge Lewis Hamilton for the Grand Prix.

When the press in the Far East heard about Fan Hui's defeat, they were merciless in their dismissal of its significance. Indeed, when Fan Hui's wife contacted him in London after the news got out, she begged her husband not to go online. Naturally, he couldn't resist. It was not a pleasant experience to read how little the commentators in his home country thought of his credentials to challenge AlphaGo.

Fan Hui credits his matches with AlphaGo with giving him new insights into how to play the game. In the following months, his ranking shot up from 633 into the 300s. But it wasn't only Fan Hui who was learning. Every game AlphaGo plays refines its code and makes it a stronger player next time around.

For its part, the DeepMind team felt confident enough after the victory to issue a challenge to Lee Sedol, eighteen-time international title winner and regarded universally as a formidable player of the game. The match would consist of five games scheduled from March 9 to March 15, 2016, to be played at the Four Seasons hotel in Seoul and broadcast live via the internet. The winner would receive a prize of one million dollars. Although the venue was announced publicly, the precise location within the hotel was kept secret and was isolated from noise. Not that AlphaGo was going to be disturbed by the chitchat of press and the whispers of curious bystanders. It could assume a Zen-like state of concentration wherever it was placed.

Sedol wasn't fazed by the knowledge that the machine he was up against had beaten Fan Hui. A few weeks before the match, he made this prediction to reporters: "Based on its level seen in the match [against Fan], I think I will win the game by a near landslide—at least this time." Although he was aware that AlphaGo was constantly learning and evolving, this did not concern him.

Most people still felt that, despite great inroads into programming, an AI Go champion was still a distant goal. Rémi Coulom, the creator of the only software capable of playing Go at any high standard—a program called *Crazy Stone*—was predicting that computers would not beat the best humans at the game for at least another decade.

As the date for the match approached, the team at DeepMind felt it needed someone to really stretch AlphaGo and to test it for any weaknesses. So Fan Hui was invited back to play the machine going into the last few weeks. Despite having suffered a 5–0 defeat and being mocked by the press back in China, he was keen to help out. Perhaps a bit of him felt that if he could help make AlphaGo good enough to beat Sedol, it would make his defeat less humiliating.

As Fan Hui played, he could see that AlphaGo was extremely strong in some areas. But he managed to expose a weakness that the team was not aware of. Faced with certain configurations, it seemed incapable of assessing who had control of the game, often seeming to suffer from the delusion that it was winning when the opposite was true. If Sedol exploited this weakness, AlphaGo wouldn't just lose, it would appear extremely stupid.

Members of the DeepMind team worked around the clock trying to fix this blind spot. Eventually they just had to lock down the code as it was. It was time to ship the hardware they were using to Seoul. The stage was set for a fascinating duel as the players, or at least one player, sat down on March 9 to play the first of the five games.

Beautiful, Beautiful, Beautiful

It was with a sense of existential anxiety that I fired up the YouTube channel broadcasting Sedol's matches against AlphaGo and joined 280 million other viewers to see humanity take on the machines. Having for years compared creating mathematics to playing the game of Go, I had a lot on the line.

Sedol picked up a black stone, placed it on the board, and waited for the response. Aja Huang, a member of the DeepMind team, would make the physical moves for AlphaGo. This, after all, was not a test of robotics but of artificial intelligence, so AlphaGo would still be relying on human anatomy to place the stones on the board. Huang stared at AlphaGo's screen, waiting for its response to Sedol's first stone. But nothing came.

We all stared at our screens wondering if the program had crashed. The DeepMind team was also beginning to wonder what was up. The opening moves are generally something of a formality. No human would think so long over move 2. There is nothing really to go on yet. What was happening? And then a white stone appeared on the computer screen. It had made its move. The DeepMind team breathed a huge sigh of relief. We were off! Over the next couple of hours the stones began to build up across the board.

As I watched the game it was hard for me at many points to assess who was winning. It turns out that this isn't just because I'm not a very experienced Go player. It is a characteristic of the game. And this is one of the main reasons that programming a computer to play Go is so hard. There isn't an easy way to turn the current state of the game into a robust scoring system of who leads by how much.

Chess, by contrast, is much easier to score as you play. Each piece has a different numerical value which gives you a simple first approximation of who is winning. Chess is destructive. One by one, pieces are removed so the state of the board simplifies as the game proceeds. But Go increases in complexity as you play. It is constructive. Even the commentators, although they kept up a steady stream of observations, struggled to say if anyone was in the lead right up until the final moments of the game.

What they were able to pick up quite quickly was Sedol's opening strategy. Because AlphaGo had learned to play on games that had been played in the past, Sedol was working on the principle that it would put him at an advantage if he disrupted the expectations it had built up. He was making moves that were not in the conventional repertoire. The trouble was, this required Sedol to play an unconventional game—one that was not his own.

It was a good idea, but it didn't work. Any conventional machine programmed on a database of familiar openings wouldn't have known how to respond and would likely have made a move that would have serious consequences in the grand arc of the game. But AlphaGo was not a conventional machine. As David Silver, its lead programmer, explained in the run-up to the match, "AlphaGo learned to discover new strategies for itself, by playing millions of games between its neural networks, against themselves, and gradually improving." If anything, Sedol had put himself at a disadvantage.

As I watched I couldn't help feeling for Sedol. You could see his confidence draining out of him as it gradually dawned on him that he was losing. He kept looking over at Huang, the DeepMind representative who was executing AlphaGo's moves, but there was nothing he could glean from Huang's face. By move 186, Sedol realized there was no way to overturn the advantage AlphaGo had built up on the board. He placed a stone on the side of the board to indicate his resignation.

By the end of day one it was AlphaGo 1, Humans 0. Sedol made an admission at the press conference that day: "I was very surprised, because I didn't think I would lose."

But it was Game Two that would truly shock not just Sedol but every human player of Go. In the first game, experts could follow the logic and appreciate why AlphaGo was playing the moves it was. They were moves a human champion would play. But in Game Two, something rather strange happened. Sedol had just played move 36 and then retired to the roof of the hotel for a cigarette break. While he was away, AlphaGo instructed Huang, its human representative, to execute move 37, placing one of its black stones on an intersection five lines in from the edge of the board. Everyone was shocked.

The conventional wisdom is that during the early part of the game you focus on the outer four lines. Stones placed on the third line build up short-term territory strength at the edge of the board while playing on the fourth line contributes to your strength later in the game as you move into the center of the board. Players have always found that there is a fine balance between playing on the third and fourth lines. Playing on the fifth line has always been regarded as suboptimal, giving your opponent the chance to build up territory that has both short- and long-term influence.

AlphaGo had defied this orthodoxy built up over centuries of competing. Some commentators declared it a clear mistake. Others were more cautious. Everyone was intrigued to see what Sedol would make of the move when he returned from his cigarette break. Even watching on my laptop at home, I could see him flinch as he took in the new stone on the board. He was certainly as shocked as all of the rest of us by the move. He sat there thinking for over twelve minutes. As in chess competitions, the game was being played under time constraints. Using twelve minutes of his time was very costly. It is a mark of how surprising this move was that it took Sedol so long to respond. He could not understand what AlphaGo was doing. Why had the program abandoned the region of stones they were competing over?

Was this a mistake by AlphaGo? Or did it see something deep inside the game that humans were missing? Fan Hui, who was serving as one of the referees, looked down on the board. His initial reaction matched everyone else's: shock. But then he paused to appreciate it. *Wired* magazine reports how he would describe the moment later:

"It's not a human move. I've never seen a human play this move," says spectator and Go champion Fan Hui. "So beautiful." It's a word he keeps repeating. Beautiful. Beautiful. Beautiful.

Beautiful and deadly it turned out to be. Not a mistake but an extraordinarily insightful move. Some fifty moves later, as the black and white stones fought over territory in the lower left corner of the board, they found themselves creeping toward the black stone of move 37. It was joining up with this stone that gave AlphaGo the edge, allowing it to clock up its second win. AlphaGo 2, Humans 0.

Sedol's mood in the press conference that followed was notably different. "Yesterday I was surprised. But today I am speechless . . . I am in shock. I can admit that . . . the third game is not going to be easy for me." The match was being played over five games. This was a game that Sedol needed to win to have any hope of keeping AlphaGo from claiming the match.

The Human Fights Back

Sedol had a day off to recover. The third game would be played on Saturday, March 12. He needed the rest, unlike the machine. The first game had required more than three hours of intense concentration. The second lasted over four hours. Spectators could see the emotional toll that losing two games in a row was having on him.

Rather than resting, though, Sedol stayed up till six o'clock the next morning analyzing the games he'd lost so far with a group of fellow professional Go players. Did AlphaGo have a weakness they could exploit? The machine wasn't the only one who could learn and evolve. Sedol felt he might learn something from his losses.

Sedol played a very strong opening to Game Three, forcing AlphaGo to manage a weak group of stones within his sphere of influence on the board. Commentators began to get excited. Some said Sedol had found AlphaGo's weakness. But then, as one commentator, David Ormerod, posted, "things began to get scary. . . . As I watched the game unfold and the realization of what was happening dawned on me, I felt physically unwell."

Sedol pushed AlphaGo to its limits, but in doing so he seemed to invoke some hidden powers the program possessed. As the game proceeded, it started to make what commentators called lazy moves. It had analyzed its position and was so confident in its win that it chose safe moves. It didn't care if it won by half a point. All that mattered was that it win. To play such lazy moves was almost an affront to Sedol, but AlphaGo was not programmed with any vindictive qualities. Its sole goal is to win the game. Sedol pushed this way and that, determined not to give in too quickly. Perhaps one of these lazy moves was a mistake that he could exploit.

At move 176 Sedol eventually caved in and resigned. AlphaGo 3, Humans 0. AlphaGo had won the match. Backstage, the people on the DeepMind team were going through a strange range of emotions. They'd won the match, but seeing the devastating effect it was having on Sedol made it hard for them to rejoice. The million-dollar prize was theirs. They'd already decided to donate the prize, if they won, to a range of charities dedicated to promoting Go and STEM subjects as well as to UNICEF. Yet their human code was causing them to empathize with Sedol's pain.

AlphaGo did not demonstrate any emotional response to its win. No little surge of electrical current. No code spat out with a resounding *yes!* It is this lack of response that gives humanity hope and at the same time is scary. Hope because this emotional response is what provides the drive to be creative and venture into the unknown. It was humans, after all, who programmed AlphaGo with the goal of winning. Scary because the machine won't care if an outcome turns out to be not quite what its programmers intended.

Sedol was devastated. He came out in the press conference and apologized: "I don't know how to start or what to say today, but I think I would have to express my apologies first. I should have shown a better result, a better outcome, and better content in terms of the game played, and I do apologize for not being able to satisfy a lot of people's expectations. I kind of felt powerless." But he urged people to keep watching the final two games. His goal now was to try at least to get one back for humanity.

Having lost the match, Sedol started Game Four playing far more freely. It was as if the heavy burden of expectation had been lifted, allowing him to enjoy his game. In sharp contrast to the careful, almost cautious play of Game Three, he launched into a much more extreme strategy called "amashi." One commentator compared it to an investor who, rather than squirreling away small gains that accumulate over time, bets the whole bank.

Sedol and his team had stayed up all of Saturday night trying to reverse-engineer from AlphaGo's games how it played. It seemed to work on a principle of playing moves that incrementally increase its probability of winning rather than betting on the potential outcome of a complicated single move. Sedol had witnessed this when AlphaGo preferred lazy moves to win Game Three. The strategy they'd come up with was to disrupt this sensible play by playing the risky single moves. An all-or-nothing strategy might make it harder for AlphaGo to score so easily.

AlphaGo seemed unfazed by this line of attack. Seventy moves into the game, commentators were already beginning to see that AlphaGo had once again gained the upper hand. This was confirmed by a set of conservative moves that were AlphaGo's signal that it had the lead. Sedol had to come up with something special if he was going to regain the momentum.

If move 37 of Game Two was AlphaGo's moment of creative genius, move 78 of Game Four was Sedol's retort. He'd sat there for thirty minutes staring at the board, staring at defeat, when he suddenly placed a white stone in an unusual position, between two of AlphaGo's black stones. Michael Redmond, commentator on the YouTube channel, spoke for everyone: "It took me by surprise. I'm sure that it would take most opponents by surprise. I think it took AlphaGo by surprise."

It certainly seemed to. AlphaGo appeared to completely ignore the play, responding with a strange move. Within several more moves AlphaGo detected that it was losing. The DeepMind team stared at screens behind the scenes and watched its creation imploding. It was as if move 78 short-circuited the program. It seemed to cause AlphaGo to go into meltdown as it made a whole sequence of destructive moves.

This apparently is another characteristic behavior of Go algorithms. Once they see that they are losing they go rather crazy.

Silver winced as he saw the next move AlphaGo was opting for: "I think they're going to laugh." Sure enough, the Korean commentators collapsed into fits of giggles at the moves AlphaGo was now making. Its moves were failing the Turing Test. No human with a shred of strategic sense would make such moves. The game dragged on for a total of 180 moves, at which point AlphaGo put up a message on the screen that it had resigned.

The human race had got one back. AlphaGo 3, Humans 1. The smile on Lee Sedol's face at the press conference that evening said it all. "This win is so valuable that I wouldn't exchange it for anything in the world." The press room erupted with joyful applause. "It's because of the cheers and the encouragement that you all have shown me."

Gu Li, who was commentating the game in China, declared Sedol's move 78 as the "hand of God." It was a move that broke the conventional way to play the game and that was ultimately the key to its shocking impact. Yet this is characteristic of true human creativity. It is a good example of Boden's transformational creativity, where people break out of the system to find new insights.

At the press conference, Hassabis and Silver, his chief programmer, could not explain why AlphaGo had lost. They would need to go back and analyze why it had made such a lousy move in response to Sedol's move 78. It turned out that AlphaGo's experience in playing humans had led it to totally dismiss such a move as something not worth thinking about. It had assessed that this was a move that had only a one-in-ten-thousand chance of being played. It seems as if it had just not bothered to learn a response to such a move because it had prioritized other moves as more likely and therefore more worthy of response.

Perhaps Sedol just needed to get to know his opponent. Perhaps over a longer match he would have turned the tables on AlphaGo. Could he maintain the momentum into the fifth and final game? Losing three games to two would be very different from losing four to one. The last win was still worth fighting for. If he could win a second game, it would sow seeds of doubt about whether AlphaGo could sustain its superiority.

But AlphaGo had learned something valuable from its loss. The next person who plays Sedol's one-in-ten-thousand move against the algorithm won't get away with it. That's the power of this sort of algorithm. It never forgets what it learns from its mistakes.

That's not to say it can't make new mistakes. As Game Five proceeded, there was a moment quite early in the game when AlphaGo seemed to completely miss a standard set of moves in response to a particular configuration that was building. As Hassabis tweeted from backstage, "#AlphaGo made a bad mistake early in the game (it didn't know a known tesuji) but now it is trying hard to claw it back . . . nail-biting."

Sedol was in the lead at this stage. It was game-on. Gradually AlphaGo did claw back. But right up to the end the DeepMind team was not exactly sure whether it was winning. Finally, on move 281— after five hours of play—Sedol resigned. This time there were cheers backstage. Hassabis punched the air. Team members hugged and high-fived. The win that Sedol had pulled off in Game Four had evidently reengaged their competitive spirit. It was important for them not to lose this last game.

Looking back at the match, many recognize what an extraordinary moment this was. Some immediately commented on its being an inflection point for AI. Sure, all this machine could do was play a board game—and yet for those looking on, its capability to learn and adapt was something quite new. Hassabis's tweet after winning the first game summed up the achievement: "#AlphaGo WINS!!!! We landed it on the moon."

It was a good comparison. Landing on the moon did not yield extraordinary new insights about the universe, but the technology that humanity developed to achieve such a feat did. Following the last game, AlphaGo was awarded an honorary nine-dan professional ranking by the Korean Go Association, the highest accolade for a Go player.

From Hilltop to Mountain Peak

Move 37 of Game Two was a truly creative act. It was novel, certainly, it caused surprise, and as the game evolved it proved its value. This was exploratory creativity, pushing the limits of the game to the extreme.

One of the important points about the game of Go is that there is an objective way to test whether a novel move has value. Anyone can come up with a new move that appears creative. The art and challenge is making a novel move that has some sort of value. How should we assess value? It can be very subjective and time-dependent. Something that is panned critically at the time of its release can be recognized generations later as a transformative creative act. Nineteenth-century audiences didn't know what to make of Beethoven's *Symphony No. 5*, and yet it is central repertoire now. During his lifetime, Van Gogh could barely sell his paintings—he traded them for food or painting materials—but now they go for millions. In Go, there is a more tangible and immediate test of value: Does it help you win the game? Move 37 won Game Two for AlphaGo. There was an objective measure that we could use to value the novelty of this move.

AlphaGo had taught the world a new way to play an ancient game. Analysis since the match has resulted in new tactics. The fifth line is now played early on, as we have come to understand that it can have big implications for the end game. AlphaGo has gone on to discover still more innovative strategies. DeepMind revealed at the beginning of 2017 that its latest iteration had played online anonymously against a range of top-ranking professionals under two pseudonyms: Master and Magister. Human players were unaware that they were playing a machine. Over a few weeks it had played a total of sixty complete games. It won all sixty.

But it was the analysis of the games that was truly eye-opening. Those games are now regarded as a treasure trove of new ideas. In several games AlphaGo played moves that any beginners, if they made them, would have their wrists slapped for by their Go masters. Traditionally, for example, you do not play a stone at the intersection of the third column and third row. And yet AlphaGo showed how to use such a move to great advantage.

Hassabis describes how the game of Go had got stuck on what mathematicians like to call a local maximum. Look at the landscape illustrated below and imagine you are at the top of the peak to the left. From this height there is nowhere higher to go. This is called a local maximum. If there were fog all around you, you'd think you were at the highest point in the land. But across the valley is a higher peak. To know this, you need the fog to clear. Then you need to descend from your peak, cross the valley, and climb to the top of it.

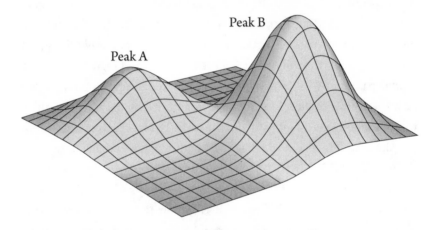

The trouble with modern Go is that conventions had built up about ways to play that had ensured players hit Peak A. But by breaking those conventions AlphaGo had cleared the fog and revealed an even higher Peak B. It's even possible to measure the difference. In Go, a player using the conventions of Peak A will in general lose by two stones to the player using the new strategies discovered by AlphaGo.

This rewriting of the conventions of how to play Go has happened at two previous points in history. The most recent was the innovative game-play introduced by the legendary player Go Seigen in the 1930s. His experimentation with ways of playing the opening moves revolutionized the way the game is played. But Go players now recognize that AlphaGo might well have launched an even greater revolution.

Chinese Go champion Ke Jie recognizes that we are in a new era: "Humanity has played Go for thousands of years, and yet, as AI has shown us, we have not yet even scratched the surface. The union of human and computer players will usher in a new era."

Ke Jie's compatriot Gu Li, winner of the most Go world titles, added: "Together, humans and AI will soon uncover the deeper mysteries of Go." For Hassabis, the algorithm is like the Hubble telescope of Go. This illustrates the way many view this new AI. It is a tool for exploring deeper, further, wider than ever before. It is not meant to replace human creativity but to augment it.

And yet there is something that I find quite depressing about this moment. It feels almost pointless to want to aspire to be the world champion at Go when you know there is a machine that you will never be able to beat. Professional Go players have tried to put a brave face on it, talking about the extra creativity that it has unleashed in their own play, but there is something quite soul-destroying about knowing that we are now second-best to the machine. Sure, the machine was programmed by humans, but that doesn't really make it feel better.

AlphaGo has since retired from competitive play. The Go team at DeepMind has been disbanded. Hassabis proved his Cambridge lecturer wrong. DeepMind has now set its heights on other goals: health care, climate change, energy efficiency, speech recognition and generation, computer vision. It's all getting very serious.

Given that Go had always been my shield against computers doing mathematics, was my own subject next in DeepMind's cross-hairs? To truly judge the potential of this new AI, we'll have to look more closely at how it works and dig around inside. The ironic thing is that the tools DeepMind is using to create the programs that might put me out of a job are precisely the ones that mathematicians have created over the centuries. Is this mathematical Frankenstein's monster about to turn on its creator?

/ 4 /

Algorithms, the Secret to Modern Life

The Analytical Engine weaves algebraic patterns,
just as the Jacquard loom weaves flowers and leaves.

—ADA LOVELACE

O ur lives are run by algorithms. Every time we search for something on the internet, embark on a journey with our GPS, choose a movie recommended by Netflix, or seek a date online, we are being guided by an algorithm. Algorithms are ubiquitous in the digital age, yet few realize that they predate the computer by thousands of years and go to the heart of what mathematics is all about.

The birth of mathematics in ancient Greece coincides with the development of one of the very first algorithms. In Euclid's *Elements*, alongside the proof that there are infinitely many prime numbers, we find a recipe for solving a certain type of problem. Given two whole numbers, it allows anyone who follows its steps to find the largest whole number that divides them both.

It may help to put the problem in more visual terms. Imagine that a floor you need to tile is thirty-six feet long by fifteen feet wide. You want to know the largest size of square tiles that will perfectly cover the floor. So what should you do? Here is the algorithm, more than two thousand years old, that solves the problem:

Suppose you have two numbers M and N (and suppose N is smaller than M).

Start by dividing M by N and call the remainder N_1.

If N_1 is zero, then N is the largest number that divides them both.

If N_1 is not zero, then divide N by N_1 and call the remainder N_2.

If N_2 is zero, then N_1 is the largest number that divides M and N.

If N_2 is not zero, then do the same thing again: divide N_1 by N_2 and call the remainder N_3.

These remainders are getting smaller and smaller and are whole numbers, so at some point one must hit zero.

When it does, the algorithm guarantees that the previous remainder is the largest number that divides both M and N. This number is known as the *highest common factor,* or *greatest common divisor.*

Now let's return to your challenge of tiling the floor. First, imagine the largest square tile that could fit inside the original shape. Then look for the largest square tile that will fit inside the remaining part—and so on, until you hit a square tile that finally covers the remaining space evenly. This is the largest square tile that will cover the whole space.

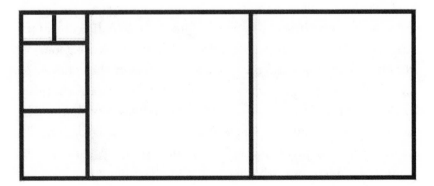

If M = 36 and N = 15, then dividing M by N gives you 2 with a re-mainder (N_1) of 6. Dividing N by N_1, we get 2 with a remainder (N_2) of 3. But now, dividing N_1 by N_2, we get 2 with no remainder at all, so we know that 3 is the largest number that can divide both 36 and 15.

You see that there are lots of "if . . . then . . ." clauses in this process. That is typical of an algorithm and is what makes algorithms so per-fect for coding and computers. Euclid's ancient recipe exhibits the four key characteristics any algorithm should ideally possess:

> It consists of a precisely stated and unambiguous set of instructions.
> Its procedure always comes to a finish, regardless of the numbers inserted. (It does not enter an infinite loop!)
> It produces the answer for any values input.
> It is fast.

In the case of Euclid's algorithm, there is no ambiguity at any stage. Because the remainder grows smaller at every step, after a finite number of steps it must hit zero, at which point the algorithm stops and spits out the answer. The bigger the numbers, the longer the algorithm will take, but it's still relatively fast. (The number of steps is five times the number of digits in the smaller of the two numbers, for those who are curious.)

If the invention of the algorithm happened over two thousand years ago, why does it owe its name to a ninth-century Persian mathe-matician? *Algorithmi* is the Latinized form of a surname—that of Muḥammad ibn Mūsā al-Khwārizmī. One of the first directors of the great "House of Wisdom" in Baghdad, al-Khwārizmī was responsible for many of the translations of the ancient Greek mathematical texts into Arabic. Although all the instructions for Euclid's algorithm are there in the *Elements*, the language Euclid used was very clumsy. The ancient Greeks thought about mathematic problems geometrically, so numbers were presented as lines of different lengths and proofs consisted of pictures—a bit like our example of tiling the floor. But pictures aren't sufficient for doing mathematics with much rigor. For that, you need the language of algebra, which uses letters to stand for variable numbers. This was the invention of al-Khwārizmī.

To be able to articulate the workings of an algorithm, we need language that allows us to talk about numbers without specifying what those numbers are. We already saw this at work in Euclid's algorithm, when we gave names to the numbers we were trying to analyze: N and M. These letters can represent any numbers. The power of this new, linguistic take on mathematics was that it allowed mathematicians to understand the grammar underlying how numbers work. Rather than being limited to showing particular examples of a method working, this new language of algebra provided a way to explain the general patterns behind the behavior of numbers. Today's easy analogy is to think of the code behind a running software program. No matter what numbers are plugged in as inputs, it works to yield an output—the third criterion in our conditions for a good algorithm.

Indeed, algorithms have gained enormous currency in our era precisely because they are perfect fodder for computers. Wherever there is a discernible pattern underlying the way we solve a problem to guide us to a solution, an algorithm can exploit that discovery. It is not required of the computer that it think. It need only execute the steps encoded in the algorithm and, again and again, as if by magic, out pop the answers we seek.

Desert Island Algorithm

One of the most extraordinary algorithms of the modern age is the one that helps millions of us navigate the internet every day. If I were exiled to a desert island and could take only one algorithm with me, I'd probably choose the one that drives Google (although perhaps I would check first whether in exile I would still have an internet connection).

In the early days of the World Wide Web (we're talking the early 1990s) there was a directory of all the existing websites. In 1994 there were only three thousand of them. The list was small enough that you could pretty easily thumb through it and find a site that someone had mentioned to you. Things have changed quite a bit since then. When I started writing this paragraph there were 1,267,084,131 websites live on the internet. A few sentences later, that number has gone up to 1,267,085,440. (You can check the current status at www.internetlive stats.com.)

How does Google's search engine figure out exactly which ones of these billion-plus websites to recommend? Most users have no idea. Mary Ashwood, for example, an eighty-six-year-old granny from Wigan, in northeast England, was careful to add a courteous "please" and "thank you" to each query, perhaps imagining an industrious group of interns on the other end sifting through the endless requests. When her grandson Ben opened her laptop and found "Please translate these roman numerals mcmxcviii thank you," he couldn't resist posting a snapshot on Twitter to share his nan's misconception with the world. He got a shock when someone at Google UK tweeted back:

> Dearest Ben's Nan.
> Hope you're well.
> In a world of billions of Searches, yours made us smile.
> Oh, and it's 1998.
> Thank YOU.

Ben's Nan brought out the human in Google on this occasion, but there is no way any company could respond personally to the million searches Google receives every fifteen seconds. So if it isn't magic Google elves scouring the internet, how does Google succeed in so spectacularly locating the information you want?

It all comes down to the power and beauty of the algorithm Larry Page and Sergey Brin cooked up in their dorm rooms at Stanford in 1996. They originally named their new search engine "BackRub" (referring to its reliance on the web's "back links") but by 1997 switched to "Google," inspired by the name a mathematician in the 1930s gave to a "1" followed by a hundred zeros: googol. Their mission was to find a way to rank pages based on relevance to search terms, and thereby make the vast and always expanding internet navigable, so a name that suggested a huge number seemed to strike the right note.

It isn't that there were no algorithms already being used to do the same thing, but existing ones were pretty simple in their conception. If you wanted to find out more about the "polite granny and Google," existing algorithms would have returned a list to you of all pages on which these words appeared, ordered so that the ones at the top of the list were those featuring the most occurrences of the terms.

That sounds like it would be okay, but unfortunately it made for a ranking system that was easily gamed: a florist's website, for example, could shoot to the top of many son's and daughter's searches simply by pasting the phrase "Mother's Day flowers" a thousand times on a page (not necessarily visibly, since it could be in the page's meta data). A better search engine would be less easily pushed around by savvy web designers. How could one be based on more reliable measures of sites' relevance? What measures could be used to identify sites to rank high or low?

Page and Brin struck on the clever idea of democratizing the assessment of a website's quality. They recognized that, when a website was worth visiting, other sites tended to link to it. The number of links to a site from elsewhere could therefore be seen as a quality signal; other websites were essentially voting up the sites they considered most important. But again, it's a measure that can be hacked. If I'm the devious florist, I only need to set up a thousand artificial websites linking to my website to bump my site up in potential customers' search results. Anticipating this, Page and Brin decided their rankings should assign more weight to votes coming from websites which themselves commanded respect.

This, however, created a new challenge: what measure of respect should be used to give more weight to one site's links over another's? By imagining a very small network of three websites we can arrive at an answer.

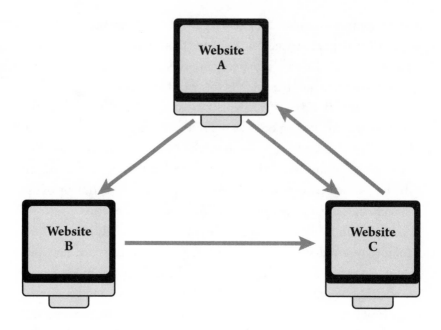

At the outset of this example, each site has equal weight, which is to say equal rank. Each is like a basket containing eight balls that must be awarded to others. To award balls to a site is to link to it. If a site chooses to link to more than one site, then its balls are divided equally between them. In the diagram above, the arrows show the choices each website is making about links it will make. The diagram opposite shows the resulting tallies of balls. Website A links to both B and C, so four of its balls go to each of those sites. Website B, meanwhile, decides to link only to website C, putting all eight of its balls into website C's basket. Website C likewise makes just one choice, putting all eight of its balls into website A's basket.

After the first distribution, website C looks very strong. But go another round with the same choices and it swaps places with website A, which is greatly boosted by the fact that it was the sole choice of the high-ranking website C. Keep repeating the process and the balls shift around as shown in the table on page 48.

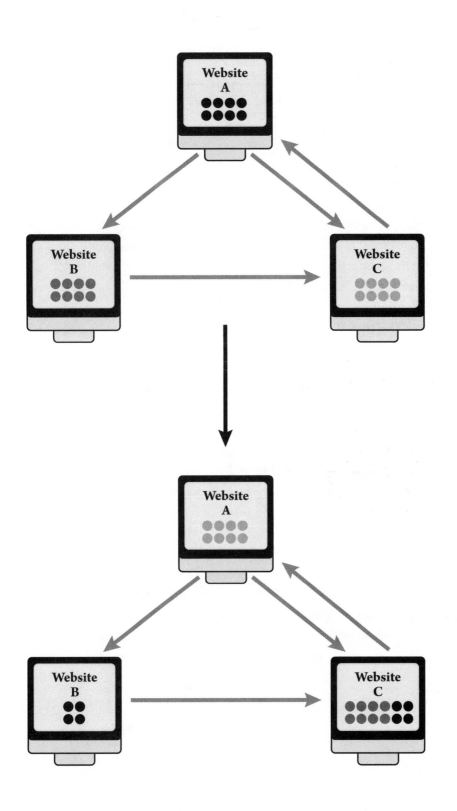

	Round 0	Round 1	Round 2	Round 3	Round 4	Round 5	Round 6	Round 7	Round 8
A	8	8	12	8	10	10	9	10	9.5
B	8	4	4	6	4	5	5	4.5	5
C	8	12	8	10	10	9	10	9.5	9.5

At the moment, this does not seem to be a particularly good algorithm. It appears not to stabilize and is rather inefficient, failing two of our criteria for the ideal algorithm. Page and Brin's great insight was to realize that they needed to find a way to assign the balls by looking at the connectivity of the network. Remembering a clever trick they'd been taught in their university coursework, they saw how the correct distribution could be worked out in one step.

The trick starts by constructing a matrix, as shown below, which records the way that the balls are redistributed among the websites. The first column of the matrix shows the proportion going from website A to the other websites. It reflects what has already been described: website A keeps nothing to itself, gives half its balls to website B, and gives half its balls to website C. With the second column showing website B's choices and the third column showing website C's, the matrix of redistribution looks like this:

$$\begin{pmatrix} 0 & 0 & 1 \\ 0.5 & 0 & 0 \\ 0.5 & 1 & 0 \end{pmatrix}$$

The challenge is to find the eigenvector of this matrix that has an eigenvalue of 1. This is a column vector that does not get changed when multiplied by the matrix.* Finding these eigenvectors, or stability points, is something we teach undergraduates early on in their university

---

* Here is the rule for multiplying matrices: $\begin{pmatrix} a & b & c \\ d & e & f \\ g & h & i \end{pmatrix} \begin{pmatrix} x \\ y \\ z \end{pmatrix} = \begin{pmatrix} ax+by+cz \\ dx+ey+fz \\ gx+hy+iz \end{pmatrix}$

careers. In the case of our network, we find that the following column vector is stabilized by the following redistribution matrix:

$$\begin{pmatrix} 0 & 0 & 1 \\ 0.5 & 0 & 0 \\ 0.5 & 1 & 0 \end{pmatrix} \begin{pmatrix} 2 \\ 1 \\ 2 \end{pmatrix} = \begin{pmatrix} 2 \\ 1 \\ 2 \end{pmatrix}$$

This means that if we split the balls in 2:1:2 distribution we see that this weighting is stable. Distribute the balls using our previous game and the sites still have a 2:1:2 distribution.

Eigenvectors of matrices are an incredibly potent tool in mathematics and in the sciences more generally. They are the secret to working out the energy levels of particles in quantum physics. They can tell you the stability of a rotating fluid like a spinning star or the reproduction rate of a virus. They may even be key to understanding how prime numbers are distributed throughout all numbers.

Calculating the eigenvector of the network's connectivity, we see that websites A and C should be ranked equally. Although website A has only one site (website C) linking to it, the fact that website C is highly valued and links only to website A means that its link bestows high value on website A.

This is the basic core of the algorithm. There are a few extra subtleties that need to be introduced to get it working in its full glory. For example, the algorithm needs to take into account anomalies like websites that don't link to any other websites and become sinks for the balls being redistributed. But at its heart is this simple idea.

Although the basic engine is very public, there are parameters inside the algorithm that, because they are kept secret and change over time, make it a little harder to hack its workings. The fascinating thing about the Google algorithm, in fact, is its robustness and imperviousness to being gamed. There is very little a website can do on its own pages to increase its rank. It must rely on others to boost its position. If you scan a list of the websites that Google's page-rank algorithm scores highest, you will see a lot of major news sources on it, as well as university websites like Oxford and Harvard. Because research universities do work that is valued by so many people around the world, many outside websites link to our pages of findings and opinions.

Interestingly, this means that when anyone with a website within the Oxford network links to an external site, the link will cause a boost to the external website's page rank, as Oxford is sharing a bit of its huge prestige (picture that cache of balls) with that website. This is why I often get requests to link from my website in the mathematics department at Oxford to external websites. The link will increase the external website's rank and hopefully help it appear on the first page of results retrieved by a Google search, which is the holy grail for any website.

But the algorithm isn't wholly immune to clever attacks by those who understand how the rankings work. For a short period in the summer of 2018, if you Googled "idiot," the first image that appeared was one of Donald Trump. Users on the website Reddit understood the powerful position their forum has on the internet and how they could exploit it. By getting people to upvote a post consisting of the word "idiot" next to an image of Trump, they fooled the algorithm into assigning top ranking to that post for any idiot-searchers. Google does not like to make manual interventions—it's a dangerous form of playing God—so it took no action in response to the prank. It trusted instead in the long run power of its mathematics, and sure enough, the spike was smoothed out over time.

The internet is of course a dynamic beast, with new websites emerging every nanosecond, existing sites being shut down or updated, and links constantly being added. This means that page ranks need to change dynamically. In order for Google to keep pace with the constant evolution, it must regularly trawl through the internet using what it rather endearingly calls "Google spiders" to update its counts of the links between sites.

Tech junkies and sports coaches have discovered that this way of evaluating the nodes in a network can also be applied to other networks. One of the most intriguing applications has been in the realm of football (of the European kind, which Americans call soccer). If you've played, you may know that an important part of sizing up the opposition is identifying the key players who control the way their team plays or serve as hubs through which much of the play passes. If you can identify such players and disrupt those patterns, then you can effectively close down the team's strategy.

Two London-based mathematicians, Javier López Peña and Hugo Touchette, both football fanatics, decided to use a version of Google's algorithm to analyze the teams gearing up for the World Cup. If you think of each player as a website and a pass from one player to another as a link from one website to another, then the passes made over the course of a game can be thought of as a network. A pass to a teammate is a mark of the trust you put in that player—players generally avoid passing to a weak teammate who might easily lose the ball—and you will only be passed to if you make yourself available. A static player will rarely be available for a pass.

They decided to use passing data made available by FIFA during the 2010 World Cup to see which players ranked most highly. The results were fascinating. If you analyzed England's style of play, two players, Steven Gerrard and Frank Lampard, emerged with a markedly higher rank than others. This was because the ball very often went through these two midfielders: take them out and England's game collapses. England did not get very far that year in the World Cup. It was knocked out early by its old nemesis, Germany.

Contrast this with the eventual winner: Spain. The algorithm shared the rank uniformly around the whole team, indicating that there was no clear hub through which the game was being played. This is a reflection of the very successful "total football" or "tiki-taka" style played by Spain, in which players constantly pass the ball around—a strategy that contributed to Spain's ultimate success.

Unlike many sports in America that thrive on data, it has taken some time for football to take advantage of the mathematics and statistics bubbling underneath the game. But by the 2018 World Cup in

Russia, many teams boasted a scientist on board crunching the numbers to understand the strengths and weaknesses of the opposition, including how the network of each team behaves.

Network analysis has even been applied to literature. Andrew Beveridge and Jie Shan brought it to George R. R. Martin's epic saga *A Song of Ice and Fire*, otherwise known as *Game of Thrones*. Anyone familiar with this story is aware that predicting which characters will make it through to the next volume, or even the next chapter, is notoriously tricky. Martin is ruthless about killing off even the best characters he has created.

Beveridge and Shan decided to create a network of the books' characters. They identified 107 key people as the nodes of the network. The characters were then connected with weighted edges according to the strength of their relationships. But how could an algorithm assess the importance of a connection between two people? The algorithm was simply programmed to count the number of times their two names appeared in the text within fifteen words of each other. This doesn't measure friendship—it indicates some measure of interaction or connection between them.

They decided to analyze the third volume in the series, *A Storm of Swords*, because the narrative had settled in by this point, and began by constructing a page-rank analysis of the nodes or characters in the network. Three characters quickly stood out as important to the plot: Tyrion, Jon Snow, and Sansa Stark. If you've read the books or seen the shows, you will not be surprised by this revelation. What is striking is that a computer algorithm that did not understand what it was reading could reveal the protagonists. It did so not simply by counting how many times a character's name appears—which would pull out other names—but using a subtler analysis of the network.

To date, all three characters have survived Martin's ruthless pen, which has cut short some of the other key characters in the third volume. This is the mark of a good algorithm: it can be used in multiple scenarios. This one can tell you something useful, from football to *Game of Thrones*.

Mathematics, the Secret
to a Happy Marriage

Sergey Brin and Larry Page may have cracked the code to steer you to websites you don't even know you're looking for, but can an algorithm really do something as personal as find your soulmate? Visit OkCupid and you'll be greeted by a banner proudly declaring, "We Use Math to Find You Dates."

These dating websites use a "matching algorithm" to search through profiles and match people up according to their likes, dislikes, and personality traits. They seem to be doing a pretty good job. In fact, the algorithms seem to be better than we are on our own: research published in the *Proceedings of the National Academy of Sciences* in 2013 surveyed nineteen thousand people who married between 2005 and 2012 and found that those who met their partners online were happier and reported more stable marriages.

The first algorithm to win its creators a Nobel Prize—originally formulated by two mathematicians, David Gale and Lloyd Shapley, in 1962—used a matching algorithm to solve something called the stable marriage problem. Gale, who died in 2008, missed out on the 2012 award. Shapley shared the prize with the economist Alvin Roth, who saw the importance of the algorithm not just to people's love lives but also to social problems including assigning health care and student places fairly.

Shapley was amused by the award. "I consider myself a mathematician and the award is for economics," he said at the time, clearly surprised by the committee's decision. "I never, never in my life took a course in economics." But the mathematics he cooked up had profound economic and social implications.

The stable marriage problem that Shapley solved with Gale sounds more like a parlor game than a piece of cutting-edge economic theory. Imagine you've got four heterosexual men and four heterosexual women. Each is asked to list the four members of the opposite sex in order of preference. The challenge for the algorithm is to match them up in such a way as to create stable marriages. What this means is that, while not everyone will get their first pick, the important thing is not to have a man and woman of different couples who would both prefer to be together rather than paired with the partners they've each been assigned. Otherwise there's a good chance that at some point they'll leave their partners and run off with one another. At first sight it isn't at all clear, even with four pairs, that it is possible to arrange this.

Let's take a particular example and explore how Gale and Shapley could guarantee a stable pairing in a systematic and algorithmic manner. The four men will be played by the kings from a pack of cards: King of Spades, King of Hearts, King of Diamonds, and King of Clubs. The women are the corresponding queens. Each king and queen has listed his or her preferences as shown below.

The kings' preferences

	K spades	K hearts	K diamonds	K clubs
1st choice	Q spades	Q diamonds	Q clubs	Q diamonds
2nd choice	Q diamonds	Q clubs	Q hearts	Q hearts
3rd choice	Q hearts	Q spades	Q diamonds	Q spades
4th choice	Q clubs	Q hearts	Q spades	Q clubs

The queens' preferences:

	Q spades	Q hearts	Q diamonds	Q clubs
1st choice	K hearts	K clubs	K spades	K hearts
2nd choice	K spades	K spades	K diamonds	K diamonds
3rd choice	K diamonds	K hearts	K hearts	K spades
4th choice	K clubs	K diamonds	K clubs	K clubs

Now suppose you were to start by proposing that each king be paired with the queen of the same suit. Why would this result in an unstable pairing? First, note that the Queen of Clubs has ranked the King of Clubs as her least preferred partner, so frankly she'd be happier with any of the other kings. Second, check out the list the King of Hearts made. The Queen of Hearts is at the bottom of his list. He'd certainly prefer the Queen of Clubs over the option he's been given. In this scenario, we can envision the Queen of Clubs and the King of Hearts running away together. Matching kings and queens via their suits would lead to some unstable marriages.

How do we match everyone so we won't end up with two cards running off with each other? Here is the recipe Gale and Shapley cooked up. It consists of several rounds of proposals by the queens to the kings until a stable pairing finally emerges. In the first round of the algorithm, the queens all propose to their first choice. The Queen of Spades' first choice is the King of Hearts. The Queen of Hearts' first choice is the King of Clubs. The Queen of Diamonds chooses the King of Spades, and the Queen of Clubs proposes to the King of Hearts. So it seems that the King of Hearts is the heartthrob of the pack, having received two proposals. He chooses the one he prefers, which is the Queen of Clubs, and rejects the Queen of Spades. So we have three provisional engagements, and one rejection.

First round:

K spades	K hearts	K diamonds	K clubs
Q diamonds	~~Q spades~~		Q hearts
	Q clubs		

The rejected queen strikes off her first-choice king and in the next round moves on to propose to her second choice: the King of Spades. But now the King of Spades has two proposals: his first proposal from round one, the Queen of Diamonds, and a new proposal from the Queen of Spades. Looking at his ranking, he'd actually prefer the Queen of Spades. So he rather cruelly jilts the Queen of Diamonds (his provisional engagement on the first round of the algorithm).

Second round:

K spades	K hearts	K diamonds	K clubs
~~Q diamonds~~	Q clubs		Q hearts
Q spades			

Which brings us to round three. In each round, the rejected queens propose to the next kings on their lists and each king always goes for the best offer he receives. In this third round, the rejected Queen of Diamonds proposes to the King of Diamonds (who has been standing around like that kid who never gets picked for the team). Despite the fact that the Queen of Diamonds is low down on his list, he hasn't got a better option, as the other three queens prefer other kings who have accepted them.

Third round:

K spades	K hearts	K diamonds	K clubs
Q spades	Q clubs	Q diamonds	Q hearts

Finally everyone is paired up and all the marriages are stable. Although we have couched the algorithm in terms of a cute parlor game with kings and queens, the algorithm is now used all over the world: in Denmark to match children to day care places; in Hungary to match students to schools; in New York to allocate rabbis to synagogues; and in China, Germany, and Spain to match students to universities. In the UK it has been used by the National Health Service to match patients to donated organs, resulting in many saved lives.

The modern algorithms that run our dating agencies are built on top of the puzzle that Gale and Shapley solved. The problem is more complex since information is incomplete. Preferences are movable and relative, and they shift from day to day. But essentially, the algorithms are trying to match people with preferences that will lead to stable and happy pairings. Again, the evidence suggests that relying on algorithms could be better than leaving things to human intuition.

You might have detected an interesting asymmetry in the algorithm that Gale and Shapley cooked up. We got the queens to propose to the kings. Would it have mattered if instead we had first invited the kings to propose to the queens? Rather strikingly it would. We would end up with different stable pairings. The Queen of Diamonds would end up with the King of Hearts and the Queen of Clubs with the King of Diamonds. The two queens swap partners, but now they're paired up with slightly lower choices. Although both pairings are stable, when queens propose to kings, the queens end up with the best pairings they could hope for. Flip things around and the kings are better off.

Medical students in America looking for residencies realized that hospitals were using this algorithm to assign places in such a way that the hospitals did the proposing. This meant the students were getting worse matches than they had to. After some campaigning by students, the algorithm was eventually reversed to give students the better end of the deal.

This is a powerful reminder that, as our lives are increasingly pushed around by algorithms, we need to understand how they work. Unless we know what they're doing and why, we might not even realize when the deck is stacked against us.

The Battle of the Booksellers

Another problem with algorithms is that sometimes there are unexpected consequences. A human might be able to tell if something weird were happening, but an algorithm would just carry on doing what it was programmed to do, regardless of the absurdity of the consequences.

My favorite example of this comes from two secondhand booksellers who both ran their shops using algorithms. A postdoc working at UC Berkeley was keen to get hold of a copy of Peter Lawrence's *The Making of a Fly*. It's a text published in 1992 that developmental biologists find useful, but by 2011 it had been out of print for some time. When the postdoc searched for it on Amazon, he saw he could only get it through third-party sellers.

There were several used copies available for prices averaging about $40, but the listing that caught his eye was the one asking $1,730,045.91. The seller, called profnath, wasn't even including shipping in the bargain. Then he noticed that there was another copy on offer for even more! This seller, bordeebook, was asking a staggering $2,198,177.95 (plus $3.99 shipping).

The postdoc showed this to his supervisor, Michael Eisen, who presumed it must be a graduate student having fun. But both booksellers had very high ratings and seemed to be legitimate. Profnath had received more than eight thousand recommendations over the last twelve months while bordeebook topped 125,000 during the same period. Perhaps it was just a weird blip.

When Eisen checked the next day to see if the prices had dropped to more sensible levels, he found instead that they'd gone up. Profnath now wanted $2,194,443.04 while bordeebook was asking a phenomenal $2,788,233.00. Eisen decided to put his scientific hat on and analyze the data. Over the next few days he tracked the changes in an effort to work out if there was some pattern to the strange prices.

	profnath	bordeebook	profnath / bordeebook	bordeebook / profnath
8 April	$1,730,045.91	$2,198,177.95		1.27059
9 April	$2,194,443.04	$2,788,233.00	0.99830	1.27059
10 April	$2,783,494.00	$3,536,675.57	0.99830	1.27059
11 April	$3,530,663.65	$4,486,021.69	0.99830	1.27059
12 April	$4,478,395.76	$5,690,199.43	0.99830	1.27059
13 April	$5,680,526.66	$7,217,612.38	0.99830	1.27059

Eventually he worked out the mathematical rules behind the escalating prices. Dividing the profnath price by the bordeebook price from the day before he always got 0.99830. Dividing the bordeebook price by the profnath book on same day he always got 1.27059. The sellers had both programmed their websites to use algorithms to set the prices on books they were selling. Each day the profnath algorithm would check the price of the book at bordeebook and would then multiply it by 0.99830. This algorithm made perfect sense because the seller was programming the site to slightly undercut the competition at bordeebook. The algorithm at bordeebook, curiously, had an opposite strategy. It was programmed to detect any price change in its rival and to multiply this new price by a factor of 1.27059.

The combined effect was that each day the price would be multiplied by 0.99830 × 1.27059, or 1.26843. This ensured that the prices would escalate rapidly. If profnath had set a sharper factor to undercut bordeebook's price and bordeebook had applied a slighter premium, their prices might have collapsed instead.

The explanation for profnath's algorithm seems clear, but why was bordeebook's algorithm set to offer the book at a higher price? Surely no customer would prefer a more expensive book. Perhaps it was relying on a better reputation, given its much greater number of positive recommendations, to drive traffic its way—especially if its price was only slightly higher, which at the start it would have been. Eisen speculated about this in his blog, writing that "this seems like a fairly risky thing to rely on. Meanwhile you've got a book sitting on the shelf collecting dust. Unless, of course, you don't actually have the book . . ."

That's when the truth dawned on him. Of course. They didn't actually have the book! The algorithm was programmed to find books out there that bordeebook did not have in stock, and to offer the same books at a markup. If a customer preferred to pay extra to get the book from a more reliable website, bordeebook's would simply take the customer's money, purchase the book from the other bookseller, then ship it to the customer when it arrived. The algorithm thus multiplied the price by a factor of 1.27059 to cover the purchase of the book and shipping, and a little extra profit.

Using a few logarithms, it's possible to work out that profnath's book was most likely first spotted by bordeebook's algorithm forty-five days before April 8, priced at about $40. This shows the power of compounding increments. It only took a month and a half for the price to reach into the millions! The price peaked at $23,698,655.93 (plus $3.99 shipping) on April 18, when finally a human at profnath intervened, realizing that something strange was going on. The price then dropped to $106.23. Predictably, bordeebook's algorithm priced its listing at $106.23 × 1.27059 = $134.97.

The mispricing of *The Making of a Fly* did not have a devastating impact for anyone involved, but there are more serious cases, such as the algorithms used to price stock options that have caused flash crashes in financial markets. The potential for algorithmic decision-making to have runaway unintended consequences is one of the biggest fears people have about advancing technology. Imagine a manufacturing company that is dedicated to decarbonizing its operations and builds an algorithm to relentlessly pursue that goal. What if the AI realizes the humans who work in the factory are carbon-based organisms and devises a way to eliminate them once and for all? Who would stop it?

Algorithms are based on mathematics. At some level, they are mathematics in action. But no one in the mathematical community feels particularly threatened by them, because they don't really stretch the field creatively. We don't really believe that algorithms will turn on their creators and put us out of our jobs. For years, I believed that these algorithms would do no more than speed up the mundane part of my work. They were just more sophisticated versions of Babbage's calculating machines that could be told to do the algebraic or numerical manipulations that would take me tedious hours to write out by hand. I always felt in control. But that is all about to change.

Up until a few years ago, it was felt that humans understood what their algorithms were doing and how they were doing it. Like Lovelace, people believed we couldn't really get more out than we put in. But then a new sort of algorithm began to emerge—an algorithm that could adapt and change as it interacted with its data so that, after a while, its programmer might not understand quite why it made the choices it did. These programs were starting to produce surprises: for once, we could get more out than we put in. They were beginning to be more creative. These were the algorithms DeepMind exploited in its crushing of humanity in the game of Go. They ushered in the new age of machine learning.

/ 5 /

From Top-Down to Bottom-Up

Machines take me by surprise with great frequency.

—ALAN TURING

I first met Demis Hassabis a few years before his great Go triumph, at a meeting about the future of innovation. New companies were on the hunt for funding from venture capitalists and other early-stage investors. Some were going to transform the future, but most would flash and burn. For the VCs and angels, the art was to spot the winners. I must admit, when I heard Hassabis speak about code that could learn, adapt, and improve, I dismissed him out of hand. I couldn't see how, if someone programmed a computer to play a game, the program could get any further than the person who wrote the code. How could you get more out than you put in? I wasn't the only skeptic. Hassabis admits that getting investors to back AI startups a decade ago was extremely difficult.

How I wish now that I'd bet on that horse as it came trotting by! The transformative impact of the ideas Hassabis was proposing can be judged by the title of a recent panel discussion on AI: "Is Deep Learning the New 42?" (The allusion to Douglas Adams's answer to the question of life, the universe, and everything in *The Hitchhiker's Guide to the Galaxy* would have been familiar to the geeky attendees, many brought up on a diet of sci-fi.) So what has happened to launch this new AI revolution?

The simple answer is data. It is an extraordinary fact that 90 percent of the world's data has been created in the last five years. One exabyte (equaling 10^{18} bytes) of data is created on the internet every day— roughly the equivalent of the amount that could be stored on 250 million DVDs. Humankind now produces in two days the same amount of data it took us from the dawn of civilization until 2003 to generate.

This flood of data is the main catalyst for the new age of machine learning. Before now, there just wasn't enough of an environment for an algorithm to roam around in and learn. It was like having a child and denying it sensory input. We know that children who have been trapped indoors and deprived of social interaction fail to develop language and other basic skills. Their brains may have been primed to learn but they didn't encounter enough stimulus or gain enough experience to develop properly.

The importance of data to this new revolution has led many to speak of data as the new oil. If you have access to large pools of data, you are straddling the twenty-first century's oil fields. This is why the likes of Facebook, Twitter, Google, and Amazon are sitting pretty— we are giving them our reserves for free. Well, not exactly free, given that we are exchanging our data for services. When I drive my car using Waze, I am choosing to give data about my location in return for being shown the most efficient route to my destination. The trouble is, many people are not aware of these transactions and give up valuable data for little in return.

At the heart of machine learning is the idea that an algorithm can be created that will alter its approach if the result it produces comes up short of its objective. Feedback reveals if it has made a mistake. This tweaks the algorithm's equations such that, next time, it acts differently and avoids making the same mistake. This is why access to data is so important: the more examples a smart algorithm can train on, the more experienced it becomes, and the more each tweak refines it. Programmers essentially create meta-algorithms which create new algorithms based on the data they encounter.

People in the field of AI have been shocked at the effectiveness of this new approach, especially given that the underlying technology is not that new. These algorithms are created by building up layers of questions that can help reach a conclusion. The layers are sometimes called neural networks, because they mimic the way the human brain works. Think about the structure of the brain, with its neurons connected to other neurons by synapses. A collection of neurons might fire due to an input of data from the senses (like the smell of freshly baked bread). Secondary neurons then fire, provided certain thresholds are passed. (Perhaps a decision to eat the bread.) A secondary neuron might fire, for example, if ten connected neurons are firing due to the input data, but not if fewer are firing. A trigger might also depend on the strengths of the incoming signals from other neurons.

As long ago as the 1950s, computer scientists created an artificial version of this process, which they called the *perceptron*. The idea is that a neuron is like a logic gate that receives input and then, depending on a calculation, either fires or doesn't.

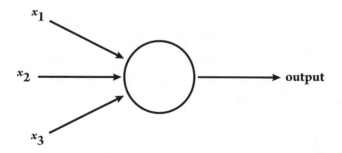

Let's imagine a perceptron that relies on three input numbers, and assigns different weights to the importance of them. In the diagram above, perhaps input x_1 is three times as important as x_2 and x_3. It would calculate $3x_1 + x_2 + x_3$ and then, depending on whether this fell above or below a certain threshold, it would fire an output or not. Machine learning involves reweighting the inputs if the answer turns out to be wrong. For example, it might be that x_3 should have been weighted as more important to the decision than x_2, so perhaps the equation should be changed to $3x_1 + x_2 + 2x_3$. Or perhaps we simply need to tweak the sum required for the perceptron to fire; the threshold activation level can be dialed up or down. Or perhaps the perceptron could be designed to fire to different degrees depending on by how much its output exceeded the threshold. The output could be a measure of its confidence in the assessment of the data.

Let's cook up a perceptron to predict whether you are going to go out tonight, and let's say that decision depends on three things: Is there anything good on TV? Are your friends going out? Is this a good night of the week to go out? It will give each of these variables a score from 0 to 10. For example, if it's Monday, maybe the last factor will get a score of 1, whereas a Friday would get a 10. And, based on awareness of your personal proclivities, it might place different weights on these variables. Perhaps you are a bit of couch potato so anything vaguely decent on TV is a strong incentive to stay in. This would mean that the x_1 variable weighs heavily. The art of this equation is tuning the weightings and the threshold value to mimic the way you behave.

Just as the brain consists of many chains of neurons, a network can consist of layers of perceptrons (as illustrated opposite) so that the triggering of nodes gradually causes a cascade of effects. This is what we call a neural network. These layers can also include slightly subtler versions of the perceptron called sigmoid neurons which aren't just simple on / off switches.

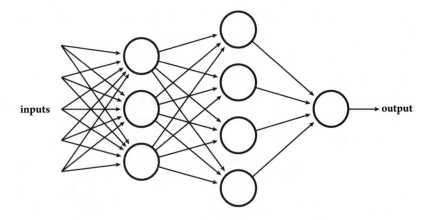

Given that computer scientists have understood for so long how to create artificial neurons, why did it take so long to make these things work effectively? This brings us back to data. The perceptrons need data from which to learn and evolve; together these are the two ingredients you need to create an effective algorithm. You can try to program a perceptron to decide if you should go out tonight by assigning the weights and thresholds you think are true, but it is highly unlikely you'll get them right. But allow it to train on your actual behavior and it will eventually succeed, because each failure to predict your behavior causes it to learn and reweight itself.

To See or Not to See

One of the big hurdles for AI has always been computer vision. Five years ago, computers were terrible at understanding what they were looking at. This is one domain where the human brain totally outstrips its silicon rivals. We are able to eyeball a picture very quickly and say what it is or to classify different regions of the image. Computers can analyze millions of pixels, but programmers have found it very difficult to write algorithms that can take all this data and make sense of it. How could an algorithm be created from the top down to identify a cat? Every image of one consists of a completely different arrangement of pixels. Yet the human brain can instantly synthesize this data and integrate the input to output the answer "cat."

This amazing ability of the human brain to recognize images is used to add an extra layer of security to your bank accounts, and to make sure you aren't a robot trawling for tickets online. In essence, you need to pass an inverse Turing Test. The computer is now setting a task which the human has to pass to prove that it's human.

Until recently, computers haven't been able to cope with all the variations. But machine learning has changed all that. Now, by training on data consisting of images of cats, an algorithm can gradually build up a hierarchy of questions it can pose to an image to identify, with a high probability of accuracy, if it is a cat. Image-recognition algorithms are slightly different in flavor from those in the last chapter, and violate one of the four conditions put forward there for a good algorithm. They don't work 100 percent of the time. But they can work most of the time, and the point is to push that "most" as high as possible. Moving from deterministic, foolproof algorithms to probabilistic ones has been a significant psychological shift for those working in the industry. It's like moving from a mathematician's mindset to an engineer's.

You may wonder why, if this is the case, you are still being asked to prove you are human by identifying bits of images when, say, you go to buy tickets online. What you are actually doing is contributing to the training data that will help algorithms learn to do what you do so effortlessly. Algorithms can only learn from data that is labeled. What you are really doing is feeding the visual recognition algorithms.

An algorithm can use this training data to learn the best sorts of questions to ask to distinguish cats from non-cats. Every time it gets an image wrong, it's altered so that the likelihood rises that the next time it will get it right. This might mean its current parameters change or a new feature is introduced to distinguish the image more accurately. The change isn't communicated in a top-down manner by a programmer who is thinking up all of the questions in advance. The algorithm makes its own adjustments, gradually building itself from the bottom up by interacting with more and more data.

I saw the power of this bottom-up learning process when I dropped in to Microsoft's UK Research Lab in Cambridge and saw how the Xbox my kids use at home is able to identify what they're doing in front of the camera. This algorithm has been created to distinguish, as they move about, their hands from their heads, their feet from their elbows. The Xbox's depth-sensing camera, called Kinect, uses infrared technology to record how far objects are from it. As you stand in front of the camera in your living room, it not only determines the contours of your body, it detects that your body is nearer than the wall at the back of the room.

But people come in different shapes and sizes. They can be in strange positions, especially when playing a game on Xbox. The challenge for the computer is to identify thirty-one distinct body parts, from your left knee to your right shoulder. Microsoft's algorithm is able to do this using a single frozen image. It does not rely on your movement (which would require more processing power to analyze and slow down the game).

So how does it manage to do this? The algorithm has to decide for each pixel in each image which of the thirty-one body parts it belongs to. Essentially it plays a game of Twenty Questions. In fact, there's a sneaky algorithm you can write for the game of Twenty Questions that will guarantee you get the right answer. First ask: Is the word you are thinking of in the first half of the dictionary? Then narrow down the region of the dictionary even more by asking: Within that part you've just revealed it can be found, is it in the first half? If you used all your twenty questions to ask the same, this strategy would land you on a chunk of the dictionary equal to $1/2^{20}$ of its content. Here we see the power of doubling. That is less than a millionth. Even if you were using the *Oxford English Dictionary*, with its roughly 300,000 words, you would have arrived at your word by question nineteen.

What questions should we ask our pixels if we want to identify which body part they belong to? In the past we would have had to come up with a clever sequence of questions to solve this. But what if we program the computer so that it finds the best questions to ask—by interacting with more and more data, more and more images, until it finds the set of questions that seem to work best? This is machine learning at work.

We have to start with some candidate questions that we think might solve this problem, so this isn't completely *tabula rasa* learning. The learning comes from refining our ideas into an effective strategy. So what sort of questions do you think might help us distinguish your arm from the top of your head?

Let's call the pixel we're trying identify x. The computer knows the depth of each pixel, or how far away it is from the camera. The clever strategy the Microsoft team came up with was to ask questions of the surrounding pixels. For example, if x is a pixel on the top of my head then, if we look at the pixels north of pixel x, they are much more likely not to be on my body and thus to have more depth. If we take pixels immediately south of x, they'll be pixels on my face and will have a similar depth. If the pixel is on my arm and my arm is outstretched, there will be one axis, along the length of the arm, along which the depth will be relatively unchanged—but if you move out ninety degrees from this direction, it quickly pushes you off the body and onto the back wall. Asking about the depth of surrounding pixels would produce responses that could cumulatively build up to give you an idea of the body part that pixel belongs to.

This cumulative questioning can be thought of as building a decision tree. Each subsequent question produces another branch of the tree. The algorithm starts by choosing a series of arbitrary directions to head out from and some arbitrary depth threshold. For example, head north, and if the difference in depth is less than y, go to the left branch of the decision tree, but if it is greater, go right—and so on. We want to find questions that give us new information. Having started with an initial, random set of questions, once we apply these questions to ten thousand labeled images we start getting somewhere. (We know, for instance, that pixel x in image 872 is an elbow, and in image 3,339 it is part of the left foot.) We can think of each branch or body part as a separate bucket. We want the questions to ensure that all the images where pixel x is an elbow have gone into one bucket. That is unlikely to happen on the first random set of questions. But over time, as the algorithm starts refining the angles and the depth thresholds, it will get a better sorting of the pixels in each bucket.

By iterating this process, the algorithm alters the values, moving in the direction that does a better job at distinguishing the pixels. The key is to remember that we are not looking for perfection here. If a bucket ends up with 990 out of 1,000 images in which pixel x is an elbow, then that means that in 99 percent of cases it is identifying the right feature.

By the time the algorithm has found the best set of questions, the programmers haven't really got a clue how it has come to this conclusion. They can look at any point in the tree and see the question it asked before and after, but there are over a million different questions asked across the tree, each one slightly different. It is difficult to reverse-engineer why the algorithm ultimately settled on this question to ask at this point in the decision tree.

Imagine trying to program something like this by hand. You'd have to come up with over a million different questions—a prospect that would defeat even the most intrepid coder. But a computer is quite happy to sort through these kinds of numbers. The amazing thing is that it works so well. It took a certain creativity for the programming team to believe that questioning the depth of neighboring pixels would be enough to tell you what body part you were looking at—but after that, the creativity belonged to the machine.

One of the challenges of machine learning is something called over-fitting. It's always possible to come up with enough questions to distinguish an image using the training data, but you want to come up with a program that isn't too tailored to the data it has been trained on. It needs to be able to learn something more widely applicable from that data. Let's say you were trying to come up with a set of questions to identify citizens and were given a thousand people's names and their passport numbers. "Is your passport number 834765489," you might ask. "Then you must be Ada Lovelace." This would work for the data set on hand, but it would singularly fail for anyone outside this group, as no new citizen would have that passport number.

Given ten points on a graph it is possible to come up with an equation that creates a curve that passes through all the points. You just need an equation with ten terms. But again, this has not really revealed an underlying pattern in the data that could be useful for understanding new data points. You want an equation with fewer terms, to avoid this overfitting.

Overfitting can make you miss overarching trends by inviting you to model too much detail, resulting in some bizarre predictions. Here is a graph of twelve data points for population values in the United States since the beginning of the last century. The overall trend is best described by a quadratic equation, but if we use an equation of degree 11 to match the point exactly and extend this equation into the future, it takes a dramatic lurch downward, absurdly predicting complete annihilation of the US population in the middle of October 2028. (Or perhaps the mathematics knows something we don't!)

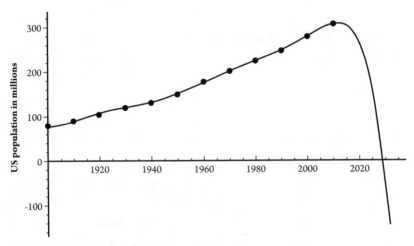

Graph source: The Mathworks, Inc.

Algorithmic Hallucinations

Advances in computer vision over the last five years have surprised everyone. And it's not just the human body that new algorithms can navigate. To match the ability of the human brain to decode visual images has been a significant hurdle for any computer claiming to compete with human intelligence. A digital camera can take an image with a level of detail that far exceeds the human brain's storage capacity, but that doesn't mean it can turn millions of pixels into one coherent story. The way the brain can process data and integrate it into a narrative is something we are far from understanding, let alone replicating in our silicon friends.

How is it that when we receive information through our senses, we can condense it into an integrated experience? We don't experience the redness of a die and its cubeness as two different experiences. They are fused into a single experience. Replicating this fusion has been one of the challenges in getting a computer to interpret an image. Reading an image one pixel at a time doesn't do much to comprehend the overall picture. To illustrate this more immediately, take a piece of paper and make a small hole in it. Now place the paper on an image of a face. It's almost impossible to tell whose face it is by moving the hole around.

Five years ago this challenge still seemed impossible. But that was before the advent of machine learning. Computer programmers in the past would try to create a top-down algorithm to recognize visual images. But coming up with an if-then set to identify an image never worked. The bottom-up strategy, allowing the algorithm to create its own decision tree based on training data, has changed everything. The new ingredient which has made this possible is the amount of labeled visual data there is now on the web. Every Instagram picture with our comments attached provides useful data to speed up the learning.

You can test the power of these algorithms by uploading an image to Google's vision website (at cloud.google.com/vision/). Last year I uploaded an image of our Christmas tree and it came back with 97 percent certainty that it was looking at a picture of a Christmas tree. This may not seem particularly earth-shattering, but it is actually very impressive. Yet it is not foolproof. After the initial shot of excitement has come the recoil of limitations. Take, for instance, the algorithms that are now being trialed by the British Metropolitan Police to pick up images of child pornography online. At the moment, they are getting very confused by images of sand dunes.

"Sometimes it comes up with a desert and it thinks it's an indecent image or pornography," Mark Stokes, the department's head of digital and electronics forensics, admitted in a *Telegraph* interview. "For some reason, lots of people have screen savers of deserts and it picks it up, thinking it is skin color." The contours of the dunes also seem to correspond to shapes the algorithms pick up as curvaceous, naked body parts.

There have been many colorful demonstrations of the strange ways in which computer vision can be hacked to make the algorithm think it's seeing something that isn't there. LabSix, an independent, student-run AI research group composed of MIT graduates and undergraduates, managed to confuse image-recognition algorithms into thinking that a model of a turtle was a gun. It didn't matter at what angle the turtle was displayed—it could even be put in an environment in which you'd expect to see turtles and not guns.

The way they tricked the algorithm was by layering a texture on top of the turtle that to the human eye appeared to be turtle shell and skin but was cleverly built out of images of rifles. The images of the rifle were gradually changed over and over again until a human couldn't see the rifle anymore. The computer, however, still discerned the information about the rifle even when it had been perturbed and that information ranked higher in its attempts to classify the object than the turtle on which it was printed. Algorithms have also been tricked into interpreting an image of a cat as a plate of guacamole. But LabSix's contribution was to make it so the angle of display didn't matter. The algorithm would always be convinced it was looking at a rifle.

The same team also showed that an image of a dog that gradually transforms, pixel by pixel, into two skiers on a mountain slope could still be classified as a dog even when the dog image completely disappeared from the screen. This hack was all the more impressive given that the algorithm used was a complete black box to the hackers.

They didn't know how the image was being decoded but still managed to fool it.

Researchers at Google went one step further and created images that were so interesting to an algorithm that it would ignore whatever else was in the picture, exploiting the fact that algorithms prioritize pixels they regard as important to classifying the image. If an algorithm is trying to recognize a face, for example, it will ignore most of the background pixels: the sky, the grass, the trees, and so forth. The Google team created psychedelic patches of color that totally took over and hijacked the algorithm so that, while it had generally always been able to recognize a picture of banana, when the psychedelic patch was introduced any bananas disappeared from its sight. A patch can also be made to register as an arbitrary image, such as a toaster. In that case, it doesn't matter what picture the algorithm is shown; once the patch is introduced, it thinks it is seeing a toaster. It's like the way a dog can become so distracted by a ball that everything else disappears from its conscious world and all it can see and think is *ball*. Most previous hacks needed to know something about the image they were trying to misclassify, but this new patch had the virtue of working regardless of the image it was seeking to disrupt.

Humans are not as easily fooled by these tricks, but that's not to say we're immune from similar effects. Magicians rely on our brain's tendency to be distracted by one thing in our visual field and to completely overlook something else they're doing at the same time. Another example is the classic video of two teams playing basketball. If viewers are instructed to count the number of passes made by one team, their intense focus on the moving ball typically causes them to completely miss that a man in a monkey suit walks through the players, bangs his chest, and then walks off the court. The attacks on computer vision are teasing out the algorithms' blind spots—but we have plenty of them, too.

Given that driverless cars use vision algorithms to steer, it is clearly an issue that the algorithms can be attacked in this way. Imagine a stop sign that had a psychedelic sticker put on it, or a security system run by an algorithm that completely misses a gun because it thinks it's a turtle.

I decided to put the Kinect algorithm through its paces to see if I could confuse it with strange contortions of my body, but it wasn't easily fooled. Even when I did strange yoga positions that the Kinect hadn't seen in its training data, it was able to identify the bits of my body with a high degree of accuracy. Since bodies out there just don't do drastically new things, the algorithm is largely frozen and doesn't evolve any further. There is no need for it to keep changing. It already does what it was built to do quite effectively. But other algorithms may need to keep adapting to new insights and changes in their environment. The algorithms that recommend films we may like to watch, books we may want to read, music we may want to hear will have to be nimble enough to react to our changing tastes and to the stream of new creative output out there.

This is where the power of an algorithm that can continue to learn, mutate, and adapt to new data comes into its own. Machine learning has opened up the prospect of algorithms that change and mature as we do.

/ 6 /

Algorithmic Evolution

Knowledge rests not upon truth alone, but upon error also.

—CARL JUNG

Today there are algorithms capable of continuous learning. This is true, for example, of the recommender algorithms we trust to curate our viewing, reading, and listening. As each user interacts with a system and displays preferences, the algorithm gains new data that helps refine its recommendations to the next user. I was intrigued to try out one of these algorithms to see how well it might intuit my tastes, so while I was at Microsoft Labs in Cambridge checking out the Xbox algorithm for the Kinect, I dropped in on a colleague to see one of the recommender algorithms learning in real time.

The graphic interface presented to me consisted of some two hundred films randomly arranged on the screen. If I liked a film, I was told to drag it to the right of the screen. I spotted a few films I enjoyed. I'm a big Wes Anderson fan, so I pulled *Rushmore* over to the right. Immediately the films began to rearrange themselves on the screen.

Some drifted to the right: these were other films the algorithm thought I'd likely enjoy. Films less likely to appeal to me drifted to the left. One film isn't much to go on, so most were still clumped in the undecided middle.

I spotted a film I really dislike. I find *Austin Powers* very annoying, so I dragged that to the reject pile on the left. This gave the program more to go on and many films drifted leftward or rightward, indicating that it had more confidence to make suggestions. Woody Allen's *Manhattan* was now proposed as a film I might enjoy. My confirmation of this didn't cause much of a ripple in the suggestions. But then I saw that *This Is Spinal Tap* had drifted quite far to the right, indicating a likelihood that I would be a fan of it. But I can't stand that film. So I dragged it from the right into the reject pile on the left of the screen.

Because the algorithm was expecting me to like *Spinal Tap*, it learned a lot from my telling it I didn't. The films dramatically rearranged themselves on the screen to take account of the new information. But something more subtle also happened in the back engine driving the algorithm. My preference data was teaching it something new, causing it to alter very slightly the parameters of its recommendation model. The probability it had assigned to my liking *Spinal Tap* was now recognized as too high and so the parameters were altered in order to lower this probability. It had learned from other fans of Wes Anderson and *Manhattan* that they quite often did enjoy this film but now it had discovered this wasn't universal.

It's in this way that our interaction with dynamic algorithms allows the machines to continue learning and adapting to our likes and dislikes. These sorts of algorithms are responsible for a huge number of the choices we now make in our lives, from movies to music, and from books to potential partners.

"If You Like This . . ."

The basic idea of a movie recommender algorithm is quite simple. If you like films A, B, and C, and another user has also indicated that these are among her favorites, and she also likes film D, then there is a good chance that you would also like film D. Of course, the data is much more complex than such a simple matching. You might have been drawn to films A, B, and C because they feature a particular actor who doesn't appear in film D, whereas the other user enjoyed them because they were all spy thrillers.

An algorithm needs to look at the data and be able to discern why it is that you like certain films. It then matches you with users who appear to value the same traits you do. As with much of machine learning, the process starts with a good swath of data. One important component of machine learning is that humans have to classify the data so that computers know what it is they're looking at. This act of curating the data prepares the field in advance so that the algorithm can then pick up the underlying patterns.

With a database of movies, you could ask someone to go through and pick out key characteristics—for example, identifying rom-coms and sci-fi movies, or perhaps movies with particular actors or directors. But this kind of curation is not ideal, not only because it is time-consuming but because it is susceptible to the biases of those doing the classifying. Their preconceptions will end up teaching the computer what they already believe to be salient rather than allowing it to expose an unexpected driver of viewers' selections. An algorithm can thus get stuck in a traditional human way of looking at data. The better approach is to devise an algorithm capable of spotting patterns in raw data, and learning by testing the validity of those patterns.

This is what Netflix was hoping to do when it issued its Netflix Prize challenge in 2006. The company had developed its own algorithm for pushing users toward films they would like, but thought a competition might stimulate the discovery of better algorithms. By that point, Netflix had a huge amount of data from users who had watched films and rated them on a scale of one to five. So it decided to publish 100,480,507 ratings provided by 480,189 anonymous customers evaluating 17,770 movies. The added challenge was that these 17,770 movies were not identified. They were only given a number. There was no way to know whether film 2,666 was *Blade Runner* or *Annie Hall*. All you had access to were the ratings that the 480,189 customers had given the film, if they had rated it at all.

In addition to the hundred million ratings that were made public, Netflix retained 2,817,131 undisclosed ratings. The challenge was to produce an algorithm that was 10 percent better than Netflix's own algorithm in predicting what these 2,817,131 recommendations were. Given the data you had seen, your algorithm needed to predict how user 234,654 rated film 2,666. To add some spice to the challenge, the first team that beat the Netflix algorithm by 10 percent would receive a prize of one million dollars. The catch was that, if you won, you had to disclose your algorithm and grant Netflix a nonexclusive license to use it to recommend films to its users.

Several "progress" prizes were offered on the way to the million-dollar prize. Each year, a prize of $50,000 would be awarded to the team that had produced the best results so far, provided it improved on the previous year's progress winner by at least 1 percent. Again, to claim the prize, a team had to disclose the code it used to drive its algorithm.

You might think it would be almost impossible to glean anything from the data, given that you hadn't a clue whether film 2,666 was sci-fi or a comedy. But it is amazing how much even raw data gives away about itself. Think of each user as a point in 17,770-dimensional space, with one dimension for each movie, where the point moves along one particular dimension according to how the user has rated a film. Now, unless you are a mathematician, thinking about users as points in 17,770-dimensional space is rather spacey. But it's really just an extension of how you would graphically represent users if there were only three films being rated.

Imagine Film 1 is *The Lion King*, Film 2 is *The Shining*, and Film 3 is *Manhattan*. If a user rates these films with one star, four stars, and five stars, respectively, you can picture putting this user at the location $(1, 4, 5)$ on a three-dimensional grid, in which the rating appears for Film 1 on the x-axis, for Film 2 on the y-axis, and for Film 3 on the z-axis.

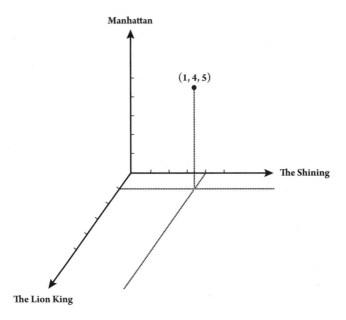

Although we can't draw a picture to represent users in 17,770-dimensional space, mathematics can be used to plot their positions. Similarly a film can be thought of as a point in 480,189-dimensional space, where every title appears somewhere along the dimension corresponding to a user according to how they have rated it. At the moment it is difficult to see any patterns in all these points spread out over these massive dimensional spaces. What you want your algorithm to do is to figure out, among these points, whether there are ways to collapse the spaces to much smaller dimensions so that patterns will begin to emerge.

It's a bit like the different shadows you can cast of someone on a wall. Some angles reveal much more about the person than others. Alfred Hitchcock's profile, for example, is very recognizable, while the shadow he would cast by directly facing a spotlight would give away little. The idea is that films and users are like points on the profile. A shadow cast at one angle might see all these points lining up, while from another angle no pattern is evident.

Perhaps you can find a way to take a two-dimensional shadow of these spaces such that the way that users and films are mapped into the shadow, users end up next to films they are likely to enjoy. The art is finding the right shadow to reveal underlying traits that films and users might possess. Below is an example of such a shadow, created using one hundred users and five hundred films from the Netflix data. You can see that it is well chosen because the two traits it measures seem to be quite distinct. This is borne out by the fact that the dots are not scattered all over the place. This is a shadow that reveals a pattern in the data.

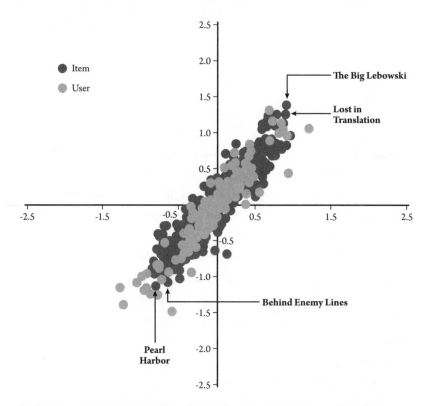

Credit: Reproduced from David Stern, Ralf Herbrich, and Thore Graepe, "Matchbox: Large Scale Online Bayesian Recommendations," *Proceedings of the Eighteenth International Conference on World Wide Web* (ACM, 2009), 111–120: 120.

If you note the few names of actual films plotted, then indeed you see that this shadow has picked out traits that we would recognize as distinct in films. Films like *Lost in Translation* and *The Big Lebowski* appear in the top-right quadrant, and more action-packed offerings are in the bottom-left.

This is the approach the team that eventually won the Netflix Prize in 2009 successfully implemented. They essentially sought to identify a shadow in twenty dimensions that accounts for twenty independent traits of films that would help predict what films users would like. The power of a computer is that it can run through a huge range of different shadows and pick out the best one to reveal structure, something that we cannot hope to do with our brains and eyes. Interestingly, some of the traits that the model picked out could be clearly identified—for example, action films or drama films. But others were much subtler and had no obvious labels, and yet the computer picked up trends in the data.

This is what is so exciting about these new algorithms: they have the potential to tell us something new about ourselves. In a way, the deep learning algorithm is picking up traits in our human code that we still haven't been able to articulate in words. It's as if we hadn't quite focused on what color was and had no words to say something was red versus blue but, through the expression of our likes and dislikes, the algorithm divided objects in front of us into two groups corresponding to blue and red. Sometimes we can't really express why we like a certain movie because there are too many parameters determining that response. The human code behind these preferences is hidden. The computer code can identify the traits guiding our preferences that we can intuit but not articulate.

On June 26, 2009, a team by the name of BellKor's Pragmatic Chaos submitted an entry which passed the 10 percent threshold, scoring 10.05 percent. Netflix had divided the hidden data into two halves. One half was used to give each team its score. The other half was kept back to judge the eventual winner. Once the 10 percent threshold had been passed, other teams had one month to try to improve their scores. On July 25, another team, Ensemble, submitted an entry that scored 10.09 percent. The next day, Netflix stopped gathering new submissions.

By this point, both teams had tweaked their algorithms further: BellKor's Pragmatic Chaos hit 10.09 percent and Ensemble edged forward to 10.1 percent. Now it was time to use the second half of the data to determine which team would get the prize. The result: both teams scored the same. But because BellKor's Pragmatic Chaos had submitted its entry twenty minutes earlier, that team walked away with the million dollars.

Given the success of this first competition, Netflix had hoped to run a second to stimulate even more innovative ideas, but it ran into a problem. The data was meant to be anonymous. Netflix had posted on the competition site the following comment about the privacy of the data:

> All customer identifying information has been removed; all that remains are ratings and dates. This follows our privacy policy. Even if, for example, you knew all your own ratings and their dates you probably couldn't identify them reliably in the data because only a small sample was included (less than one-tenth of our complete dataset) and that data was subject to perturbation. Of course, since you know all your own ratings that really isn't a privacy problem is it?

Two researchers from the University of Texas at Austin, however, took the data and, by comparing it with film ratings posted by known individuals on another website, the Internet Movie Database, were able to work out the identities of several of Netflix's users.

On December 17, 2009, four users brought a legal case against Netflix claiming that the company had violated the US Video Privacy Protection Act by releasing the data. One of them said she was a mother and also a closeted lesbian, and that data about her movie preferences would reveal this against her will. This new privacy threat, that our digital postings and transactions might allow others to reach conclusions about our sexual proclivities or political leanings, has been referred to as the *Brokeback Mountain* factor. Eventually the case was settled out of court but it caused Netflix to cancel the second round of the competition.

Data is the new oil but we are spilling it all over the internet. Who owns the data and what can be done with it are only going to become bigger questions for society as we head into a future awash in it.

How to Train Your Algorithm

You may feel there is something scary about an algorithm deciding what you might like. Could it mean that, if computers conclude you won't like something, you will never get the chance to see it? Personally, I really enjoy being directed toward new music that I might not have found by myself. I can quickly get stuck in a rut where I put on the same songs over and over. That's why I've always enjoyed the radio. But the algorithms that are now pushing and pulling me through the music library are perfectly suited to finding gems that I'll like. My worry originally about such algorithms was that they might corral everyone into certain parts of the library, leaving others bereft of listeners. Would they cause a convergence of tastes? But thanks to the nonlinear and chaotic mathematics usually behind them, this doesn't happen. A small divergence in my likes compared to yours can send us off into different far corners of the library.

I listen to a lot of algorithm-recommended pieces when I am out running. It's a great time to navigate the new. But I made a big mistake a few weeks ago. My wife asked for my help putting together a playlist for her birthday party. She wanted dancing. She wanted the eighties. So we spent a couple of evenings listening to lots of possibilities. It's not my choice of music, but we put together a great list of songs that got all our guests up and moving. The problem came when I went out for my first run following the party. My usual choice to let the player surprise me took me deep into the library aisles stocked with eighties dance music. I pressed "skip" as I ran on, but I couldn't find my way out. It took several weeks of retraining the algorithm on Shostakovich and Messiaen before I got things back on track.

Another context in which we teach algorithms trying to serve us has to do with the spam filters on email applications. A good filter begins by training on a whole swath of emails, some marked as spam, the rest considered legitimate. These are emails that aren't particular to you yet. By analyzing the words that appear in these emails it starts to build up a profile of spam emails. It learns to treat 100 percent of the emails using the word "Viagra" as spam, along with 99 percent of the emails with the word "refinance." One hundred percent of the emails with the combination "hot Russian" are spam. The word "diabetes" is more problematic. A lot of spam emails promise cures for diabetes, but it is also a word that crops up legitimately in people's correspondence. The algorithm simply counts the split in its training data. Perhaps one in twenty messages containing the word turns out not to be spam, so it learns to score an email with "diabetes" as 95 percent likely to be spam.

Your email filter can be set at different levels of filtering. You might specify that only if it's 95 percent sure should an email go into the junk folder. But now comes the cool bit. While the algorithm initially trained on a generic set of emails, your ongoing actions teach it to recognize the sorts of things you are interested in. Suppose that, in fact, you do suffer from diabetes. At first, all emails with the word "diabetes" will go into your junk folder. But gradually, as you mark emails including the word "diabetes" as legitimate, the algorithm recalibrates the probability of spam to some level below 95 percent and the email arrives in your inbox.

These algorithms are also built to spot other keywords that mark out the junk diabetes emails from the legitimate ones. The inclusion of the word "cure" could well distinguish the duds. Machine learning means that the algorithm will go through every email that comes in, trying to find patterns and links, until it ends up producing an algorithm highly customized to your own individual lifestyle.

This updating of probabilities is also how driverless cars work. It's really just a more sophisticated version of controlling the paddle in the Atari game *Breakout*: move the steering wheel right or left according to the pixel data the machine is currently receiving. Does the score go up or down as a consequence?

Biases and Blind Spots

There is something uncanny about the way the Netflix recommender algorithm is able to identify traits in films that we as humans might struggle to articulate. It certainly challenges Lovelace's view that a machine will always be limited by the insights and intent of the person who programs it. Nowadays, algorithms possess a skill that we don't have: they can assess enormous amounts of data and make sense of it.

This is an evolutionary failing of the human brain. It is why the brain is not very good at assessing probabilities. Probabilistic intuition requires understanding trends in experiments run over many trials. The trouble is, we don't get to experience that many instances of an experiment and so we can't build up the intuition. In some ways, the human code has developed to compensate for our low rate of interaction with data. So it is possible that we will end up, thanks to machine learning, with codes that complement our own human code rather than replicate it.

Probabilities are key to much of machine learning. Many of the algorithms we considered in Chapter 4 were very deterministic in their implementation. A human understood how a problem worked and programmed a computer that then slavishly jumped through the hoops it had been programmed to jump through. This is like the Newtonian view of the world, where the universe is controlled by mathematical equations and the task of the scientist is to uncover these rules and use them to predict the future.

The revelation of twentieth-century physics was that the universe was not as deterministic as scientists had been thinking it was. Quantum physics revealed that Nature plays with dice. Outcomes depend on probabilities, not clockwork. And this reign of probability is what has made algorithms so powerful. It may also be why those trained in physics appear to be better placed than us mathematicians to navigate our new algorithmic world. It's the empiricists versus the rationalists and, unfortunately for me, the empiricists are coming out on top.

How did that machine learn to play *Breakout* without being told the rules of the game? All it had was knowledge of the pixels on the screen and a score and a paddle that could be moved left or right. The algorithm was programmed to calculate the effect on the score of moving left or right given the current state of the screen. The impact of a move could occur several seconds down the line, so the calculation had to factor in that delay. This is quite tricky because it isn't always clear what causes a certain effect. This is one of the shortcomings of machine learning: it sometimes picks up correlation and believes it to be causation. Animals suffer from the same problem.

This was beautifully illustrated by an experiment that made pigeons look rather superstitious. A number of pigeons were filmed as they waited in their cages for those certain moments during the day when their food dispensers slid in. The dispensers had lids, however, that opened after some delay, so the pigeons, although excited by the arrival of the dispensers, still had to wait to get the food. The fascinating discovery was that whatever random action the pigeon happened to be doing right as the lid was released would be repeated during the delay the next day. The pigeon noted that when the lid was closed, it had turned around and then the door had opened. It then (falsely) deduced that the turning around caused the door to open. Because it was determined to get the reward, the next time the feeder appeared it spun around twice for good measure.

Another classic example of bad learning, which has become lore in the machine-learning community, took place when the US military tried to use neural networks to sort through visual images and pick out the pictures with tanks in them. The team designing the algorithm fed it picture after picture labeled as containing a tank or not. The algorithm analyzed these by looking for the common features that distinguished the two sets. After it had digested several hundred images, the algorithm was tested with a batch of photos it hadn't seen before. The team was exultant to see that it performed with 100 percent accuracy.

The algorithm was passed on to the US Army for use in a real-world application. Within a short time, the Army sent the algorithm back and

declared it useless. The researchers were perplexed. When they compared the images the Army had used with the algorithm's assessments,
it seemed like it was just randomly making up its mind. Then someone
made a comment: it seemed to be perfectly good at detecting a tank if
the photo was taken on a cloudy day.

Returning to their training data, they realized what had gone wrong.
The research team had had to get lots of pictures of a tank from all different angles, at varying distances, and in a variety of camouflaged
positions—but they had access to a tank for only a few days. What they
hadn't thought about was that, throughout those few days, the skies had
been overcast. Later, when they got around to snapping photos of that
same landscape without tanks, they went out on a sunny day. All the algorithm had picked up was an ability to distinguish between pictures
with clouds and pictures with clear skies. A cloudy-day detector was
not going to be much good to the military. The lesson: a machine may
be learning but you need to make sure it's learning the right thing.

This is becoming an increasingly important issue as algorithms
trained on data begin having impacts on society. Mortgage decisions,
policing alerts, and health advice are being increasingly produced by
algorithms. But there is a lot of evidence now that they encode hidden
biases. MIT graduate student Joy Buolamwini was perturbed to find
that robotics software she was working with seemed to have a much
harder time picking up her face than those of her lighter-skinned colleagues. When she wore a white mask, the robot detected her presence
immediately, but as soon as she removed the mask, she disappeared.

The problem? The algorithm had been trained overwhelmingly on
images of white faces. No one had noticed that the data did not include
many darker complexions. The same kind of bias in training data has
led to a whole host of algorithms making unacceptable decisions:
voice-recognition software trained on male voices that doesn't recognize women's voices, image recognition software that classifies black
people as gorillas; passport photo booths that advise Asians to retake
their photos because it judged their eyes to be closed. In Silicon Valley,
four out of five people hired in the tech industry are white males. This

has led Buolamwini to set up the Algorithmic Justice League to fight bias in the data that algorithms are learning on.

The legal system is also facing challenges as people are being rejected for mortgages, jobs, or state benefits because of algorithms. These people justifiably want to know why they have been turned down. But given that these algorithms are creating decision trees based on interactions with data that are hard to unravel, justifying these decisions is not always easy.

Some have championed legal remedies, but they are devilishly hard to enforce. Article 22 of the General Data Protection Regulations introduced into EU law in May 2018 states that everyone "shall have the right not to be subject to a decision based solely on automated processing" and the right to be given "meaningful information about the logic involved" in any decision made by a computer. Good luck with that!

There have been calls for the industry to try to develop a meta-language that an algorithm can use to justify its choices, but until this is successfully done we may simply have to be more mindful of the impact of these algorithms in everyday life. Many algorithms are good at one particular thing but not so good at knowing what to make of irregularities. When something strange occurs, they just ignore it, while a human might have the ability to recognize the anomalous scenario.

This brings us to the no-free-lunch theorem, which states that there is no universal learning algorithm that can predict outcomes accurately under every scenario. According to this theorem, even if a learning algorithm is shown half the data, and manages to generate a good prediction on that training data, it is still possible to manipulate the remaining unseen data so that the prediction is out of whack when it tries to interpret the data it hasn't trained on.

Data will never be enough on its own. It has to come paired with knowledge. It is here that the human code seems better adapted to coping with context and seeing the bigger picture—at least for now.

Man versus Machine

It is this power to change and adapt to new encounters that was exploited to make AlphaGo. The team at DeepMind built its algorithm with a period of supervised learning, in the way an adult helps a child learn skills the adult has already acquired. Humans make progress as a species because, having accumulated knowledge, we then pass it on in a much more efficient manner than it was first gained. I am not expected to single-handedly rediscover all of mathematics to get to the frontier. Instead I spent a few years at university fast-tracking through centuries of mathematical discovery.

AlphaGo began by going through the same process. Humans have played millions of games of Go that have been digitally recorded and posted online. This is an amazing resource for a computer to trawl through, gleaning which moves gave the winners an edge. Such a large database allowed the computer to assign probabilities to all available moves, given a particular board position, based on their likelihood to contribute to a win. In fact, the amount of available data was small when one considers all the possible paths a game might take. It provided a good basis for playing, but because an opponent might not go down the path that the losing player did in the database, using only this data set wasn't going to be enough.

The second phase, known as reinforcement learning, is what gave the algorithm the edge in the long run. At this point the computer started to play itself, learning from each new game it generated. As the algorithm made what its first phase of training had suggested would be winning moves, only to end up losing in many cases, it continuously adjusted the probabilities it assigned to those moves. This reinforcement learning synthetically generates huge volumes of new game data. Playing against itself also allows the algorithm to probe its own weaknesses.

A potential danger of this reinforcement learning is that it will be too limited and therefore self-reinforcing. To understand a key issue in machine learning, recall the discussion in Chapter 3 of "local maxima"—those peaks that make you feel like you've got to the top but that may only be tiny hillocks surrounded by towering mountains. Imagine your goal is to get to the top of the highest mountain, but you have very limited visibility beyond the next step in front of you. Your strategy will probably be to keep choosing only those steps that take you the next bit higher. This strategy will eventually get you to the point that is the highest in your local environment. Any step away from that peak will take you lower. But there might well be another, far higher peak on the other side of the valley. What if AlphaGo had myopically focused on a local maximum, perfecting its game-play to beat other players only as long as they stayed there?

This appeared to be the case some days prior to the match with Lee Sedol, when European Champion Fan Hui discovered a weakness in the way AlphaGo was playing. But once the algorithm was introduced to this new game-play it quickly learned how to revalue its moves to maximize its chances of winning again. The new player forced the algorithm to descend the hill and find a way to scale new heights.

DeepMind developed an even better algorithm that could thrash the original version of AlphaGo. This algorithm circumvented the need to be shown how humans play the game. Like the algorithm designed to learn Atari games, it was given the nineteen-by-nineteen grid and real-time awareness of the score and started to play, experimenting with moves. This is almost clean-slate learning. After AlphaGo's second phase of development through reinforcement learning, even the team at Deep-Mind was shocked at how powerful the new algorithm was. It was no longer constrained by the way humans think and play.

Within three days of training, in which time it played 4.9 million games against itself, it was able to beat the version of AlphaGo that had defeated Lee Sedol by one hundred games to zero. What took humans three thousand years to achieve, it did in three days. By day forty, it was unbeatable. It was even able to learn in eight hours how to play chess and shogi, a Japanese version of chess, to such a level that it beat two of the best chess programs on the market. This frighteningly versatile algorithm goes by the name of AlphaZero.

David Silver, lead researcher on the project, explained the impact of this blank-slate learning in multiple domains. "If you can achieve *tabula rasa* learning you really have an agent which can be transplanted from the game of Go to any other domain," he told the *Telegraph*. "You untie yourself from the specifics of the domain you are in to an algorithm which is so general it can be applied anywhere. For us, AlphaGo is not about going out to defeat humans but to discover what it means to do science and for a program to be able to learn for itself what knowledge is."

DeepMind's goal is to "solve intelligence, and then use that to solve everything else." Hassabis and his team believe they are well on the way. But how far can this technology go? Can it match the creativity of the best mathematician? Can it create art? Write music? Crack the human code?

Painting by Numbers

The unpredictable and the predetermined unfold together
to make everything the way it is.

—TOM STOPPARD, *ARCADIA*

A few years ago, on a Saturday afternoon, I wandered into the Serpentine Gallery in London and was transfixed. It was that sense of spiritual exhilaration that I suppose we're always after when we enter a gallery space. My companions were struggling to connect, but as I walked through the rooms I became obsessed with what I saw.

On display was Gerhard Richter's series *4900 Farben*. "You've never heard of Gerhard Richter?" my wife asked incredulously as we took the train into town. "He's only one of the most famous living artists on the planet." She often despairs at my lack of visual knowledge, immersed as I am for most of the day in the abstract universe of mathematics. But Richter's project spoke directly to the world I inhabit.

The work consists of 196 paintings, each one a five-by-five grid of squares. Each square is meticulously painted in one of the twenty-five colors Richter carefully selected for the work. That adds up to a total of 4,900 colored squares, which is the reason for that number in the title. Richter has specified eleven different ways in which the paintings can be arranged for display. In the Serpentine exhibition, he had chosen to show Version Two, in which the 196 paintings are divided into forty-nine groups of four paintings; those forty-nine subsets are therefore larger squares made up of one hundred (ten by ten) colored squares.

Staring at these pixelated canvases, the natural urge is to seek meaning in the arrangements of colors. I found myself focusing on the way three yellow squares aligned in one ten-by-ten block. We are programmed to search for patterns, to make sense of the chaotic world around us. It's what saved us from being eaten by wild animals hiding in the undergrowth. That line of yellow might be nothing, but then again it could be a lion. Many psychologists, including Jung, Rorschach, and Matte Blanco, have argued that the mind so hankers after meaning, pattern, and symmetry that one can use such images to access the human psyche. Jung would get his patients to draw mandalas, while Rorschach used symmetrical inkblots to tap into the minds of his clients.

The desire to spot patterns is at the heart of what a mathematician does, and my brain was on high alert to decode what was going on. There were interesting pockets of squares that seemed to make meaningful shapes. As I drifted through the gallery from one grid to another, I started to wonder whether there might be another game going on beneath these images.

I counted the number of times I saw two squares of the same color together in a grid, then the slightly rarer occurrence of a line of three or four squares of the same color. Having gathered my data, I sat down and calculated what would be expected if the pixels had all been chosen randomly. Randomness has a propensity to clump things together in unexpected ways. Think about when you're waiting for a bus. You often experience a big gap, then three buses come rolling along together. Despite their having set off according to a timetable, the impact of traffic has created randomness in their arrival.

I began to suspect that the three yellow squares I'd spotted were the result not of a deliberate choice but of a random process at work behind the creation of the pieces. If there are twenty-five colors to choose from and each one is chosen randomly, then it is possible to figure out how many rows should have two squares of the same color next to each other. The way to calculate this is to consider the opposite. Suppose I were to pick red for the first square. The probability that the next square would be a color other than red is 24 / 25. The chances that the third square will be different from the color I've just picked is again 24 / 25. So the probability of getting a row of ten colors without any two squares of the same color side by side is $(24 / 25)^9 = 0.69$.

This means that in each ten-by-ten painting there will probably be three rows (and three columns) with two of the same color side by side. Sure enough, the canvases matched this prediction.

My calculations also told me that I should find, in the forty-nine groupings of a hundred squares each, six with three squares of the same color in a column or row. Here I found that, while the columns checked out, there were more rows than expected with three of the same color. But that's the point of randomness. It's not an exact science.

Later, after the show, I decided to investigate Richter's approach and discovered that, indeed, the colors had been chosen at random. He had put squares of twenty-five colors into a bag, and had determined which color to use next by drawing a square from the bag. He created 196 different canvases in this way. On any given canvas, his method might have created any of 25^{25} possible paintings. This is a number with 36 digits! Laid end to end, this many canvases would extend well outside the farthest visible reaches of space.

I think my wife regretted taking me to the Serpentine Gallery. For days after, I remained obsessed with calculating the coincidences in the paintings. Not only that, given that the exhibition displayed just one way to put the canvases together, I began to fixate on how many other versions might be possible. Richter's specification for Version One combined all the canvases in a certain order into one, huge, seventy-by-seventy pixelated image. But how many other ways could he have arranged them? The answer turned out to be related to an equation that had intrigued the great seventeenth-century mathematician Pierre de Fermat.

I couldn't resist sending my musings to the director of the Serpentine Gallery, Hans Ulrich Obrist. Some weeks later, I received a letter from the artist himself asking if he could translate my thoughts into German and publish them alongside his images in a book he was producing. Richter said he had been unaware of quite how many mathematical equations were bubbling beneath the art he had made.

A similar process was used for Richter's design for the stained glass windows in Cologne Cathedral's transept. In the cathedral, however, there is an element of symmetry added, as Richter mirrors three of his randomly generated window designs to make up the six-window group. From left to right, the first and third columns are symmetrical, as are the second and fifth columns and the fourth and sixth columns. The symmetry isn't obvious, therefore, but it might tap into the brain's affinity for patterns rather like a Rorschach inkblot.

Richter had in some sense exploited a code to create his work. By giving up the decision about which next color to use and letting the random fumbling around in the bag be responsible, Richter was no longer in control of what the result would be. There is an interesting tension here between a framework created by the artist and an execution process that takes the artist out of the driver's seat.

This use of chance would prove to be a key strategy in some of the early attempts to build creative algorithms—code that would surprise the coder. The challenge is to find some way of passing the Lovelace Test. How can you create something new, surprising, and of value—something that goes beyond what the writer of the code envisioned at the beginning? The idea of adding a dash of randomness to a deterministic algorithm, as Richter had done, was a potential way out of the Lovelace dilemma.

What Is Art?

But why would anyone want to use computers to create art? What is the motivation? Isn't art meant to be an outpouring of the human code? Why get a computer to artificially generate that? Is it commercial? Are the creators just trying to make money by pressing "print" and running off endless new pieces of art? Or is this meant to be a new tool to extend our own creativity? Why do we as humans create art? Why is Richter's work regarded as art while a deck of paint color strips is not? Do we even know what this thing we call "art" really is? Where did it all begin?

Although human evolution can be traced to its beginnings in Africa six million years ago, the fossil record's earliest evidence of innovation is the appearance of tools. Stones fashioned into cutting tools have been discovered that date back 2.6 million years, but there is no proof of these utilitarian inventions sparking a creative surge. The human drive to create art shows up just one hundred thousand years ago. Archaeological finds in the Blombos Cave in South Africa have revealed what archaeologists believe are paint-making kits. It's not clear what they used the paint for. Painting their bodies? Painting designs on leather or other objects? Painting on the walls? Nothing painted survives in these South African caves, but conditions were not ideal for long-term preservation.

But other caves across the world that are deeper underground have preserved images created by early humans. Handprints appear on walls in a striking number of these caves. Research has established that the hand images painted by humans in the caves at Maros on the Indonesian island of Sulawesi date back forty thousand years. The artists are believed to have blown red ochre through tubes, using their hands as stencils. When the hand was withdrawn, its outline remained.

It is an existential statement. As Jacob Bronowski expressed it in his famous TV series *The Ascent of Man*: "The print of the hand says, 'This is my mark. This is man.'"

In addition to hands, we find human figures and pictures of wild, hoofed animals that are found only on the island. An image of a pig has been shown to be at least 35,400 years old and is the oldest figurative depiction in the world. Scientists are able to date these images by dating the calcite crusts that grew on top of them. Because the crusts formed after the paintings were made, the material gives at least a minimum age for the underlying art. The similar dating of such finds has led to theories that something big must have happened forty thousand years ago to unleash a period of sustained innovation in the human species.

But *Homo sapiens* might have been beaten to the first example of cave art by Neanderthals. In Spain, when images of hands were found in caves, it was thought that they must date from the period when *Homo sapiens* moved from Africa to Europe. This migration forty-five thousand years ago is known to have resulted in the European Neanderthals being wiped out as a species over the course of the next five thousand years. But recent dating of the calcite crusts on some of the images in these Spanish caves resets the timing of their creation to more than sixty-five thousand years ago. *Homo sapiens* weren't in Europe that early. This is art created by another species. And in turn, Neanderthals might have been beaten by an even earlier ancestor. Shells found on Java with designs carved into them date back as far as half a million years ago, which means they can only be the work of *Homo erectus*, ancestor to both *Homo sapiens* and Neanderthals. We thought art was uniquely human. But it appears we inherited the artistic impulse from Neanderthals and *Homo erectus*.

Some would argue that we shouldn't call these early efforts art. And yet it seems clear that they represent an important moment in evolution when a species started making marks with an intention that went beyond mere utility. Experiments to recreate some of the carvings made in bone forty thousand years ago reveal the staggering amount of labor that was expended on them. Surely it was an extravagance for a tribe focused on hunting and surviving to allow the carvers to skip other tasks. The carvings must have had value, even if we will never know what the intentions behind them truly were. Marks could have been made on a shell as a gift to impress a mate or to denote ownership. Whatever the original motivation, these acts would evolve into our species's passion for artistic expression.

The question of what actually constitutes art is one that has occupied humanity for centuries. Plato's definition of it in *The Republic* is dismissive: art is the representation of a physical object, which is itself the representation of the abstract ideal object. For Plato, art depends on and is inferior to the physical object it is representing, which in turn depends on and is inferior to the pure form. Given this definition, art cannot yield knowledge and truth but can only lead to illusion.

Kant makes a clear distinction between mere handicraft and "fine art," and explains the latter as "a kind of representation that is purposive in itself and, though without an end, nevertheless promotes the cultivation of the mental powers for social communication." Tolstoy picks up on this idea of communication, defining art as "a means of union among men, joining them together in the same feelings, and indispensable for the life and progress toward well-being of individuals and of humanity." From the Cave of Altamira to the Serpentine Gallery, art has the potential to bind the individual to a group, and allow our personal human code to resonate with others.

For Wittgenstein, art is part of the language games that are central to his philosophy of language. They are all attempts to access the inaccessible: the mind of the other. If we could create a mind in a machine, then its art would be a fascinating window into what it feels like to be a machine. But we are still a long way from creating conscious code.

Art is ultimately an expression of human free will and until computers have their own version of this, art created by a computer will always be traceable back to a human desire to create. Even if a program is sparked into action by certain triggers, such as words it encounters on Twitter, this can't be interpreted as a sudden feeling on the algorithm's part that it must express a reaction. The reaction was programmed into the algorithm by the coder. Even though the mind of the human creator might not know when an action will be executed, the desire for that creative action still emanates from it.

And yet, some modern takes on art have challenged whether it represents anything at all. Is it more about politics and power and money? If Hans Ulrich Obrist decides to show a collection of work at the Serpentine Gallery in London, does that define it as art? Does his powerful position in the art world mean that people will engage with the pieces in a way that, without the metadata of the curator's seal of approval, they would not?

Much modern art is no longer about the cultivation of an aesthetic and display of skill by the likes of Rembrandt or Leonardo, but rather about the interesting messages and perspectives that artists convey about our relationship to our world. Marcel Duchamp places a men's urinal in an exhibition space and the context transforms that functional object into a statement about what constitutes art. John Cage gets us to listen to four minutes and thirty-three seconds of silence, and we begin questioning what music is. We start listening to the sounds that creep in from the outside and appreciating them in a different way. Robert Barry takes his pencil to a white gallery wall and writes in fine block letters:

ALL THE THINGS I KNOW
BUT OF WHICH I AM NOT
AT THE MOMENT THINKING
1:36 PM; JUNE 15, 1969

And thus he challenges the viewer to negotiate the idea of absence and ambiguity. Even Richter's *4900 Farben* is really not about an aesthetic, much less his skill at painting squares. It is a political statement challenging our ideas of intention and chance.

So does computer art represent a similar political challenge? If you laugh at a joke, what difference does it make if subsequently you are told that the joke was created by an algorithm? The fact that you laughed is good enough. But why not other emotional responses? If you cry when you see a piece of art and then are told that the work was computer-generated, I suspect you might feel cheated or duped or manipulated. This raises the question of whether we are truly connecting with another human mind when we experience art, or just exploring untapped reaches of our own minds. This is the challenge of another's consciousness. All we have to go on is the external output of another mind, since we can never truly get inside it. As Andy Warhol declared, "If you want to know all about Andy Warhol, just look at the surface: of my paintings and films and me, and there I am. There's nothing behind it."

But for many, using computers in their art is simply using a new tool. We have always regarded cameras not as being creative but rather as allowing new creativity in humans. The practitioners of computer art are experimenting in the same way, exploring whether the restrictions and possibilities take us in new directions.

Creative Critters

Given that we are going to explore creativity beyond the human realm, it seems worth pausing to consider whether there are any other species that have emerged through their evolutionary processes with anything approaching our level of creativity.

In the mid-1950s, Desmond Morris, a zoologist, gave a chimpanzee at the London Zoo a pencil and a piece of paper and the chimp drew a line over and over again. Congo, as the chimp was known, soon graduated on to paintbrushes and canvases and went on to produce hundreds of paintings. Decades after his death, in 2005, a lot of three Congo creations sold at auction for £14,400—twenty times greater than the auction house's estimate. That same auction saw a work by Andy Warhol go unsold. Did this make Congo an artist? Or, to be a real artist, would he have to have knowledge of what he was doing? My own belief is that, rather than from Congo, the drive to create came primarily from Morris, and this should really be recognized as a disguised form of human creativity.

Some in the zoo community believe that giving tools to animals in captivity can relieve their stress and help avert the repetitive behaviors that animals in zoos so often resort to. Others have criticized zoos for cashing in on the products of animal creativity, for example by selling canvases by elephants in the zoo shop or auctioning lemurs' handprints on eBay. Perhaps zoo animals are not the right group to consider, as their environment is so distorted. Can we find examples of animal creativity in the wild?

The male Vogelkop Bowerbird uses thin sticks to build an intricate, cone-shaped bower on the ground resembling a kind of hut. Then, in front of the bower's entrance, he carefully assembles groupings of the most brightly-colored items he can find, usually flowers, berries, and shells. These appealing outlays are constructed to serve a purpose: to attract females. An abundant collection suggests a strong ability in general to procure needed items from the natural environment. But the beauty of the displays goes way beyond what would seem necessary to show such proficiency. So is the Vogelkop Bowerbird creative, or does the utilitarian nature of his endeavor make his accomplishment something less than that?

Birds also sing to communicate. But at some stage, this skill developed to the point that they are able do more than would seem strictly necessary. Excess is, of course, a signal of power in animals and humans. Only some of us have so much that we can be wasteful. So, pushing oneself to be extravagant in building a bower or singing a song is a way of signaling one's suitability as a mate.

Some interesting legal questions have been raised when animals have been given tools to create. David Slater set up a camera in the Tangkoko Nature Reserve in Indonesia to see if he could get the resident macaques to take photographs, and was overjoyed when he developed the film to discover they had taken the most extraordinary selfies. Later, when these pictures started cropping up on the internet without his permission, he decided to go after the copyright infringers. He spent months arguing with the Wikimedia Foundation, but in August 2014 the US Copyright Office surprised him by issuing a new opinion stating that copyright ownership could not be claimed if "a human being did not create the work." Things got more bizarre the following year, when the People for the Ethical Treatment of Animals (PETA) sued Slater for infringing on a copyright that Naruta the macaque should be able to own. This case was thrown out of court.

The judge in the second case contended that, for Naruto, "there is no way to acquire or hold money. There is no loss as to reputation. There is not even any allegation that the copyright could have somehow benefited Naruto. What financial benefits apply to him? There's nothing." PETA was told in no uncertain terms to stop monkeying around.

How might these test cases apply to works created by AI? Eran Kahana, a lawyer at Maslon LLP and a fellow at Stanford Law School, explains that the reason intellectual-property laws exist is that owners of IP need to be able to generate benefits from it and prevent others from using it. "An AI doesn't have any of those needs," he notes. "AI is a tool to generate those kinds of content." What if AI creates a piece of art in the style of a living artist—could the programmer be sued for copyright infringement? It's very much a grey area. Inspiration and imitation are central to the artistic process. Where is the line between your own creation and copying someone else's?

When a film studio hires many people to create a movie, it's the studio that owns copyright. Maybe AI will have to be given the same legal status as a company. These may seem like rhetorical abstractions but they are actually important issues: Why would anyone invest in creating a complex algorithm capable of composing new music or creating art if the output could then be used by anyone else, at no cost? In the UK there has been a move to credit "the person by whom the arrangements necessary for the creation of the work are undertaken." Note the contrast with the American Copyright Office's position. Will laws in these countries and elsewhere need to change as code becomes more sophisticated? This brings us back to Ada Lovelace's doubt that anything new could be created that really transcended the human inventor's input. Are coders our new artists?

Coding the Visual World

One of the first examples of code-created visuals that could hang in a gallery came from Georg Nees's work at Siemens in Germany in 1965. The language that allows a computer to turn code into art is mathematics, but Nees was not the first to experiment with the relationship between math and the visible world. It was the French philosopher René Descartes who understood that numbers and pictures were intimately related. Descartes created the way to change the visual world into a world of numbers, and vice versa, that is now called Cartesian geometry. Draw two perpendicular axes on a page, and any point placed on the page can be identified by two numbers. These two numbers describe how far to move horizontally and vertically along the axes to arrive at the point's location.

It's like the latitude and longitude numbers in a GPS coordinate. If I want to locate the position of my college in Oxford on a map, then two numbers (51.754762, -1.251530) can tell me how far it is north and west of the starting point (0,0), where the line of longitude through Greenwich meets the equator.

Since every point marked on a page can be described in terms of numbers, this allows you to describe any geometric shape you might draw using the number coordinates of all the points that make up that shape. For example, if you mark all the points where the second coordinate (along the vertical y axis) is twice the first coordinate (along the horizontal x axis), then these points make up a line that rises steeply across the page. The equation for this is $y = 2x$. You could also specify that the first coordinate should be between two values, say $1 < x < 2$. Then your rising line will be quite short.

I like to think of Descartes's system as similar to my French-English dictionary, allowing the terms of one language to be translated to another. Instead of translating words in different tongues, Descartes's dictionary allows you to move between the language of geometry and the language of numbers. A geometric point gets translated into the numbers that define its coordinates. A curve gets translated into the equation that defines all its points' coordinates.

Descartes's translation of geometry into numbers was a revolutionary moment in mathematics. Geometry had been a mainstay of mathematics ever since Euclid introduced his axiomatic approach to the interplay between lines, points, triangles, and circles, but now mathematicians had a new tool to explore the geometric world. The exciting thing about the translating dictionary Descartes provided was that, although its geometric side was limited by our three-dimensional universe, its numbers side could be taken into higher dimensions. Things that could not be physically constructed could be imagined in the mathematician's mind, a concept that allowed mathematicians at the end of the nineteenth century to create new shapes in four dimensions. It was the discovery of these new imaginary geometries that inspired Picasso to try to represent hyperspace on a two-dimensional canvas.

The potential to use equations to manipulate these numbers, especially in the age of the computer, has led to some interesting and surprising outcomes—as Nees found with his machines at Siemens. Nees programmed his computer to start at a given point on a two-dimensional plane and proceed to draw a shape made up of twenty-three lines. In connect-the-dots fashion, each line began where the last line ended. The lines alternated between heading off horizontally and heading off vertically. To program this geometric output, Nees needed to write the code using the numbers side of Descartes's dictionary. He introduced random elements into the equation, leaving it to chance whether the next line would head up or down, left or right, and how long the line should be. The twenty-third line had to close the shape up by connecting the end of line twenty-two to the starting position.

The result was curiously interesting. Nees arranged 266 of these images in a nineteen-by-fourteen grid. Displayed in this way they look like the designs Le Corbusier used to draw in his notebooks. Nees could have done this by hand but the power and ease of the computer to generate new iterations at the touch of a button allowed him to experiment with different rules and to experience their effect on a more accelerated time scale. His work revealed the computer to be a new tool in the artist's toolbox.

The random element Nees had introduced into the program meant that it could produce images he was not in control of and could not predict. This did not mean that the computer was being creative. Creativity is about conscious or subconscious choices, not random behavior. Yet the constraints he introduced, combined with randomness, led to the creation of something that has enough tension to hold the eye.

One could argue that anything that doesn't have randomness programmed into it, that is deterministic, must still really be the creation of the programmer, regardless of the surprise the programmer might get at the outcome. But is this really fair? After all, there is some sense in which one might regard all human action as predetermined. There are real challenges to the assertion that humans really have the free will that we like to believe we do.

The atoms in our bodies follow the equations of physics. Their position and movement at this moment in time determine what they will do in the future, bound on their course by the laws of nature. That motion might be chaotic and unpredictable, but classical physics asserts that it is predetermined by the present. If atoms have no choices in what they do next, then we who are made of atoms have no choices. Our actions are predetermined by the code that controls the universe. If human action is predetermined, then are our creative acts any more our own than the computer's—which people claim belong to the programmer and not to the computer?

Perhaps our only hope for agency in our actions is to appeal to the quantum world. Modern physics asserts that the only truly random thing happens at a quantum level. It is on the level of subatomic particles that there is some element of choice in the future evolution of the universe. What an electron is going to do next is random, based on how the quantum wave equation controlling its behavior collapses. There is no way to know in advance where you will find the electron when you next look for it. Could the creativity of humans, which seems to involve choice, actually depend on the free will of the subatomic world? To make truly creative code, might one need to run the code on a quantum computer?

Fractals: Nature's Code

Nees believed the closed loops he had created were just the beginning of the computer's power to create visual art. In the ensuing decades, computers allowed programmers to experiment, revealing the extraordinary visual complexity of simple equations. The discovery of the visual world of fractals, shapes with infinite complexity, would have been unthinkable without the power of the computer. As you zoom in on a fractal, rather than becoming simpler at small scale, it maintains its complexity. It is a shape that is in some sense scaleless because you cannot discern from a section at what scale of magnification you are viewing it.

The most iconic of these fractals—the Mandelbrot set—is named after the mathematician who sparked the explosion of computer-generated images. Anyone who went clubbing in the eighties would recognize this shape since it was often projected onto walls as DJs spun their psychedelic music. Infinitely zooming in on the image, the graphics created a sense of falling into some dreamlike world without ever touching the ground. These shapes could never have been discovered without the power of the computer. But are they art?

In his "Fractal Art Manifesto," published in 1999, Kerry Mitchell tried to distinguish fractal art from something a machine was doing. The art, he argued, was in the programming, the choice of equation or algorithm, not in the execution: "Fractal Art is not . . . Computer(ized) Art, in the sense that the computer does all the work. The work is executed on a computer, but only at the direction of the artist. Turn a computer on and leave it alone for an hour. When you come back, no art will have been generated."

No claim is being made here that the computer is being creative. One of the qualities that distinguishes fractal art from the computer art generated by Nees is that it is totally deterministic. The computer is making no choices that are not programmed in before it starts calculating. Why do computer fractal images, although new and surprising, still feel so anemic and lifeless? Perhaps the answer lies in the fact that they do not form a bridge between two conscious worlds.

Computer-generated fractals have nonetheless made big money for their creators, as the fractal has proven to be a highly effective means to simulate the natural world. In his seminal book *The Fractal Geometry of Nature*, Mandelbrot explained how nature uses fractal algorithms to make ferns, clouds, waves, mountains. It was reading this book that inspired Loren Carpenter, an engineer working at Boeing, to experiment with code to simulate natural worlds on the computer. Using the Boeing computers during the nighttime, he put together a two-minute animation of a fly-through of his computer-generated fractal landscape. He called the animation *Vol Libre*, meaning free flight.

Although Carpenter was meant to be making these animations for Boeing's publicity department, his ultimate aim was to impress the bosses of Lucasfilm, the production company behind *Star Wars*. That was his dream: to create animations for the movies. He finally got his chance to show off his algorithmic animation at the annual SIGGRAPH conference held in 1980 for professional computer scientists, artists, and filmmakers interested in computer graphics. As he ran his 16 mm film he noticed in the front row of the audience the very guys from Lucasfilm he was hoping to impress.

When the film came to an end, the audience erupted with applause. They hadn't seen anything so impressively natural created by an algorithm. Lucasfilm offered him a job on the spot. Stephen Spielberg, when he saw the effects that Carpenter was able to create with code, was so impressed that he declared, "this is a great time to be alive." Carpenter's colleague Ed Catmull concurred: "We're going to be making entire films this way someday. We'll create whole worlds. We'll generate characters, monsters, aliens. Everything but the human actors will come out of computers."

Carpenter and Catmull, together with Alvy Ray Smith, went on to found Pixar Animation Studios, which today employs as many mathematicians and computer scientists as it does artists and animators. The luscious jungle landscapes of a film like *Up* would once have taken artists months to produce. Today they can be created at the click of an algorithm.

The power of fractals to create convincing landscapes using minimal code also makes the technology perfect for building gaming environments. It was Atari in 1982 that first recognized the potential of this technology to transform the gaming world. It invested one million dollars in the computer graphics department at Lucasfilm to convince that company to help revolutionize the way games were made.

One of the first successes was a game released in 1984 called, appropriately enough, *Rescue on Fractalus*. The gaming environment is more forgiving than a motion picture, so the landscape could look less realistic and gamers would still be happy. The team was still rather frustrated with the jagged nature of the pixelation. But eventually it just accepted that this was as good as it was going to get on the Atari machines. It decided to embrace the jagged nature of the graphics, calling the aliens on *Fractalus* "Jaggis." But as processing power on gaming machines advanced, so did the power of games to create more convincing worlds. The advance from a static *PacMan* space to the almost movie-like rendering of games like *Uncharted* is down to the power of algorithms.

One of the most creative uses of algorithms in the gaming world came with *No Man's Sky*, a vast game released in 2016. Developed for Sony's PlayStation 4, it lets players roam around a universe visiting a seemingly endless supply of planets. Each planet is different, populated by its very own flora and fauna. Perhaps it is not technically true that the planets are infinite in number, but Sean Murray, the creative lead on the game's development, says that if you were to visit one planet every second, you would not reach them all before our own real sun died—an event predicted to occur some five billion years from now.

So does Hello Games, the company that produced *No Man's Sky*, employ thousands of artists to create these individual planets? It turns out there are only four coders, who are exploiting the power of algorithms to make these worlds. Each environment is unique and is created by the code when a player first visits the planet. Even the creators of the game don't know what the algorithm will produce before the planet is visited.

The algorithms being deployed at Pixar and PlayStation are tools for human creativity. Just as the camera didn't replace portrait artists, computers are allowing animators to create worlds in new ways. As long as computers are tools for human ingenuity and self-expression, they are no real threat to the artists. But how about computers that aim to create new art?

From AARON to the Painting Fool

The artist Harold Cohen spent his life trying to create code that might be regarded as creative in its own right. Cohen began his career intending to be a conventional artist, and he seemed to be well on his way to achieving this goal when he represented Great Britain at the Venice Biennale in 1966, at the age of thirty-eight. Shortly after the show he met his first computer, thanks to a visiting professorship at the University of California–San Diego, where he met Jef Raskin. "I had no idea it would have anything to do with art," he would later tell the *Christian Science Monitor*. He just got turned on by the programming aspect of the computer. "It slowly dawned on me that I could use the machine to investigate some of the things that I thought I hadn't been able to in painting, that had made me very discontented with my painting." Raskin, who went on to create the Macintosh computer at Apple in the late seventies, turned out to be a great choice of teacher. (The name was chosen because McIntosh was his favorite variety of apple; the spelling had to be changed for legal reasons.)

Inspired by Raskin, Cohen went on to produce AARON, a program he wrote to make works of art. Cohen's code was of the top-down, if-then variety. By the time he died, in 2016, it consisted of tens of thousands of lines. What was interesting to me was how Cohen described the code's creation process. He talked about AARON "making decisions." But how had he programmed these decisions?

People involved in creating computer art tend to be reluctant to reveal the exact details of how their algorithms work. This subterfuge is partly driven by their goal of creating algorithms that can't easily be reverse-engineered. It took some digging in the code for me to find out that "making decisions" was code for Cohen's choice to put a random number generator at the heart of the decision-making process. Like Nees, Cohen had tapped into the potential of randomness to create a sense of autonomy or agency in the machine.

Is randomness the same as creativity? Many artists find that a chance occurrence can be a helpful spur to creation. In his *Treatise on Painting*, Leonardo da Vinci described how a dirty cloth thrown at a blank canvas might serve as a catalyst for the next step. More recently, Jackson Pollock allowed the swing of his bucket to determine his compositions. Composers have found that chance sometimes helps them head in new and unexpected directions in their musical composition.

But randomness has its limitations. There is no deliberation going into a choice of one configuration as more interesting than any other. Ultimately it is a human decision to discard some of the output as less worth keeping. Randomness is, of course, crucial when it comes to giving a program the illusion of agency, but it is not enough. It is still up to human hands to press the "on" buttons. At some point will algorithmic activity take over and human involvement disappear? Our fingerprints will always be there, but our contribution may at some point be considered to be much like the DNA we inherit from our parents. Our parents do not exercise creativity through us, even if they are responsible for our creation.

But is randomness enough to shift responsibility from the programmer to the program? Cohen died at the age of 87. AARON, however, continues to paint. Has Cohen managed to extend his creative life by downloading his ideas into the program he created? Or has AARON become an autonomous creative artist now that Cohen is no longer working as partner of his creation? If someone else now presses the "create" button, who is the artist?

Cohen said he felt his bond with AARON was similar to the relationship between Renaissance painters and their studio assistants. Consider modern-day studios like those of Anish Kapoor and Damien Hirst, where many people are employed to execute their artistic visions. At an old dairy factory in south London, Kapoor has a big team helping him, just as Michelangelo and Leonardo did.

Cohen was part of a whole movement of artists in the fifties and sixties who started exploring how emerging technology might unleash new creative ideas in the visual arts. The Institute of Contemporary Arts in London held an influential exhibition in 1968 called *Cybernetic Serendipity* that profiled the impact that the robotic movement was having in the art world. It included Nicolas Schöffer's *CYSP 1*, a spatial structure whose movements are controlled by an electronic brain created by the Dutch company Philips. Jean Tinguely supplied two of his kinetic painting machines, which he called Métamatics. Gordon Pask created a system of five mobiles that interacted with each other based on the sound and light each emitted. The interactions were controlled by algorithms that Pask had written. The audience could also interact with the mobiles by using flashlights.

At the same time, Korean artist Nam June Paik was building his *Robot K-456*, billed as history's first nonhuman action artist. The original intent was for it to give impromptu street performances. As Paik recounted, "I imagined it would meet people on the street and give them a split-second surprise, like a sudden show." As technology has grown ever more sophisticated, so has the art exploiting that technology. But how far can these robots and algorithms go? Can they really become the creators rather than the creations?

Simon Colton has been working on a program to take on AARON's mantle. Here is what the Painting Fool, his creation, says about itself on its website:

> I'm The Painting Fool: a computer program, and an aspiring painter. The aim of this project is for me to be taken seriously— one day—as a creative artist in my own right. I have been built to exhibit behaviors that might be deemed as skillful, appreciative and imaginative.

Of course, these are really the aspirations of Colton, its creator, rather than of the algorithm itself, but the aim is clear: to be considered a creative artist in its own right. Colton is not looking to use algorithms as a tool for human creativity so much as to move creativity into the machine. The Painting Fool is an ongoing and evolving algorithm that currently has over two hundred thousand lines of Java code running its creations.

One of Colton's early projects was to create an algorithm that would produce portraits of people who visited the gallery. The results were then displayed on the walls of the gallery in an exhibition he called *You Can't Know My Mind*. The portraits needed to be more than just photographs of visitors taken by a digital camera. A portrait is a painting that captures something of the internal worlds of both the artist and the sitter. But because the artist in this case was an algorithm without an internal world, Colton decided to algorithmically produce one. It needed to express (if not feel) some emotional state or mood.

Colton didn't want to resort to random number generators to choose a mood, as that seemed meaningless. And yet he needed a certain element of unpredictability. To set his algorithm's emotional state on any given day, he decided to have it scan articles in that day's *Guardian*. I can attest that my own morning perusal of the newspaper can lift or dash my spirits. Reading about Arsenal's 4–2 loss to Nottingham Forest in the third round of the 2018 FA Cup certainly put me in a foul mood—my family knows to avoid me when this kind of thing happens—whereas a preview of the final season of *Game of Thrones* might fill me with excited expectation.

The programmers would not be able to predict the state of the algorithm, as they wouldn't know which article was influencing it when it was prompted to paint. Yet there would be a rationale as to why the Painting Fool chose to paint in a certain style.

When a visitor sat down for a portrait, the algorithm scanned an article for words and phrases that might capture the mood of the piece. An article about a suicide bombing in Syria or Kabul would set the scene for a serious and dark portrait. Colton calls the choice "accountably unpredictable." The painting style isn't simply a random choice—the decision can be accounted for—but it is hard to predict.

Sometimes the Painting Fool would be exposed to such depressing reading that it would send visitors away, declaring that it was not in the mood to paint. But before they left, it would explain its decision, providing the key phrase from the article it had read that had sent it into such a funk. It would also stress that "No random numbers were used in coming to this decision."

This ability to articulate its decisions, Colton believes, is an important component of the dialogue between artist and viewer. In the exhibition, each portrait comes with a commentary which seeks to articulate the internal world of the algorithm and to analyze how successful the algorithm thinks the output is in rendering its aims. These are two components Cohen said he missed in AARON.

I asked Colton if he believed that the creativity of this activity came from him, or how much creativity he attributed to the algorithm. He very honestly gave the Painting Fool a 10 percent stake in what was being produced. His aim is to change the balance over time. He proposed a litmus test to this end, suggesting it would be "when The Painting Fool starts producing meaningful and thought-provoking artworks that other people like, but we—as authors of the software—do not like. In such circumstances, it will be difficult to argue that the software is merely an extension of ourselves."

One of the problems Colton sees in mixing computer science and creative arts is that computer science thrives on an ethos of problem-solving. Build an algorithm to beat the best player of Go. Create a program to search the internet for the most relevant websites. Match people up with their perfect partners. But creating art is not a problem-solving activity. "We don't 'solve the problem' of writing a sonata, or painting a picture, or penning a poem. Rather, we keep in mind the whole picture throughout, and while we surely solve problems along the way, problem-solving is not our goal."

Cohen said more about the difference: "In these other areas, the point of the exercise is to write software to think for us. In Computational Creativity research, however, the point of the exercise is to write software to make people think more. This helps in the argument against people who are worried about automation encroaching on intellectual life: in fact, in our version of an AI-enhanced future, our software might force us to think more rather than less."

The strategy of the team is to keep addressing the challenges offered by critics for why they think the output isn't creative, beating the critics finally into submission. As Colton puts it, "It is our hope that one day people will have to admit that The Painting Fool is creative because they can no longer think of a good reason why it is not."

AARON and the Painting Fool are both rather old-school in their approach to creating art by machine. Their algorithms consist of thousands of lines of code written in the classic, top-down mode of programming. But what new artistic creations might be unleashed by the new, bottom-up style of programming? Could algorithms learn from the art of the past and push creativity to new horizons?

/ 8 /

Learning from the Masters

Art does not reproduce the visible; it makes visible.

—PAUL KLEE

In 2006, a Mexican financier, David Martinez, purchased Jackson Pollock's *No. 5, 1948* for $140 million—at the time, the highest amount ever paid for a painting. In households around the world, the transaction rekindled that old burning question of how flicking a load of paint around could command such prices. Surely this is something a kid could do!

It turns out that emulating Pollock's approach isn't quite so simple as one might think. Pollock moved around a lot as he dripped paint onto a canvas. At the best of times, working rapidly, he was off-balance. Often, he was drunk. The resulting images are visual representations of the movement of his body as he interacted with paint and canvas. Yet that doesn't mean his technique can't be simulated by a machine.

Mathematical analysis by Richard Taylor at the University of Oregon has revealed that the arcs of paint Pollock let fall are not unlike the looping lines traced by a chaotic pendulum, which has a pivot that moves around instead of remaining fixed. When I heard this, it occurred to me that I might make millions faking Jackson Pollocks, because the behavior of a chaotic pendulum is something I have studied and understand. In fact, Taylor had designed a contraption called the Pollockizer to confirm his theory about Pollock's painting style. I promptly rigged up a chaotic pendulum for myself and attached a pot with a hole in it to its swinging arm. Having laid out a canvas on the floor, I then poured in some paint, set it to swinging in its oddly punctuated way, and waited to see what would emerge.

The signature of chaos theory is a dynamic system that is incredibly sensitive to small changes, such that an almost imperceptible adjustment in the starting position will result in a hugely different outcome. A conventional pendulum, as it swings back and forth, produces a sustained and predictable pattern—the opposite of chaos. My pendulum, however, featured a pivot point that could be altered as the pendulum swung, and this caused it to behave chaotically. I had set up a machine to mimic Pollock's system of physical movement as he painted.

The visual output that results from this chaotic paint pot is a fractal, an analog version of the digital fractals exploited by Pixar and Sony to create their visual landscapes. The scaleless quality of a fractal is what makes Pollock's paintings so special. As you zoom in on a section, it becomes difficult to distinguish the zoomed-in section from the whole. Approaching the painting, you lose your sense of place in relation to the canvas and begin to mentally fall into the image.

Taylor's insight was a game changer. Many people over the years have tried to scam art buyers by randomly flicking paint onto canvases and selling them at auction as original Pollocks. But Pollock's unique fractal quality was something you could measure. With this insight, mathematicians have been able to pick out the fake canvases 93 percent of the time. I felt confident, however, that the output of my chaotic contraption would pass the fractal test.

Our brains have evolved to perceive and navigate the natural world. Since ferns and branches and clouds and many other natural phenomena are fractals, our brains feel at home when they see these shapes. This is probably why Pollock's fractals are so appealing to the human mind. They are abstract analogs of nature. Recent research on participants scanned in fMRI scanners confirms that, as they look at fractal images close to those seen in nature, the parahippocampal region of their brains is activated. This part of the brain is involved in regulating emotions and, interestingly, is also often activated when we listen to music.

The recognition that similar parts of the brain fire whether we are looking at a Pollock, considering a fern, or listening to music hints at a fundamental reason that humans started to create art in the first place, and suggests why creativity is such an important and mysterious part of the human code. Pollock's paintings are portals into the way he sees the world around him. They come loaded with an implicit question: How do *you* see the world?

When I put my "Pollock" up for sale on eBay, I was a little disappointed. I waited for a few hours, a few days, finally a few weeks, but I got no bids. Locally, the paint on the canvas looks like a Pollock but the problem is, it has no structure. The chaotic pendulum produced drip fractals but was incapable of creating that overall impression of something more that Pollock was able to convey. This seems to be a fundamental limitation of many of the codes attempting to make art. They can capture detail at a local level but they lack the ability to piece these bits together into a canvas that is satisfying on a larger scale.

Pollock's approach may appear mechanical, but he threw himself into every one of his paintings. "It doesn't matter how the paint is put on," he wrote in describing his method, "as long as something is said. Painting is self-discovery. Every good artist paints what he is."

Resurrecting Rembrandt

When Georg Nees displayed his computer-generated art in the University of Stuttgart back in 1965, artists from the nearby State Academy of Art and Design challenged him. "Very fine and interesting, indeed,"

Frieder Nake recalls one saying. "But here is my question. You seem to be convinced that this is only the beginning of things to come, and those things will be reaching way beyond what your machine is already now capable of doing. So tell me: will you be able to raise your computer to the point where it can simulate my personal way of painting?"

"Sure, I will be able to do this," Nees replied. "Under one condition, however: you must first explicitly tell me how *you* paint."

Most artists are unable to explain how they create their art. This means that the process can't simply be coded up. The output is the consequence of many subconscious instincts and decisions. But could machine learning bypass the need for conscious expression by picking up patterns and rules that we are unable to detect? To test this proposition, I decided to investigate whether an algorithm could summon from beyond the grave just one more painting by one of the greatest artists of all time.

Rembrandt van Rijn was sought out for his skill in capturing the emotional state of his subjects in his portraits, and his reputation has only grown over time. Many artists view him as a paragon of their field, and despair of ever reaching his skill and expressive mastery. As van Gogh remarked, "Rembrandt goes so deep into the mysterious that he says things for which there are no words in any language. It is with justice that they call Rembrandt—*magician*—that's no easy occupation." He painted countless portraits of Dutch guild members and grandees as well as landscapes and religious commissions, but even more compelling are the self-portraits which he returned to again and again until his death, creating intimate biographical studies animated by a probing sincerity.

Was Rembrandt's considerable output sufficient for an algorithm to be able to learn how to create a new portrait that would be recognizably his? The internet contains millions of images of cats, but Shakespeare wrote only thirty-seven plays, and Beethoven, only nine symphonies. Will creative genius be protected from machine learning by a shortage of data? Data scientists at Microsoft and Delft University of Technology were of the opinion that there was enough data for an algorithm to learn how to paint like Rembrandt. Speaking for Microsoft, executive Ron Augustus implied the old master himself would probably approve of their project: "We are using technology and data like Rembrandt uses his paints and brushes to create something new."

The team studied 346 paintings in total, creating 150 gigabytes of digitally rendered graphics to analyze. The data-gathering included detecting things like the gender, age, and head direction of Rembrandt's subjects, as well as a more geometric analysis of various key points in the faces. After a careful analysis of Rembrandt's portraits, the team settled on a subject they felt he might have taken on next: a thirty- to forty-year-old Caucasian male with facial hair, wearing dark clothes, a collar, and a hat, and facing to the right. It could just as easily have been a woman—there was a close to fifty-fifty split between the sexes—but the male portraits had more analyzable details. You didn't really need a complicated data analysis to get to this point. Where machine learning came into its own was rendering the portrait in paint.

The team used algorithms to explore his approach to painting eyes, noses, and mouths. Rembrandt's use of light is one of the distinctive features of his paintings. He tended to create a concentrated light source on one area of the subject, almost like a spotlight. This has the effect of throwing some parts of the features into sharp focus while making other areas blurry.

The algorithm did not seek to fuse or create an average of all the features. As Francis Galton discovered in 1877 when he tried to construct a prototypical image of a convict by averaging photographs of real convicts, the result produces something far removed from the original. As he layered the negatives on top of each other and exposed the resulting image, Galton was rather shocked to see this array of distorted and ugly faces transform into a handsome composite. It seems that when you smooth out the asymmetries, you end up with something quite attractive. The data scientists would have to devise a more clever plan if they were going to produce a painting that might be taken for a Rembrandt. Their algorithm would have to create new eyes, a new nose, and a new mouth as if it could see the world through Rembrandt's eyes.

Having created these features, they then investigated the proportions Rembrandt used to place these features on the faces he painted. This was something that had earlier fascinated Leonardo da Vinci, whose sketchbooks are full of measurements of the relative positions of facial features. Some believe Leonardo was applying the mathematical idea of the golden ratio to create the perfect face. Rembrandt was not so concerned with the underlying geometry but nonetheless seemed to favor certain proportions.

The analysis was first conducted on flat images. But a painting isn't a two-dimensional image. The paint on the canvas gives it a topography which contributes to the effect. For many artists, this feature is as important as the composition. Think of Van Gogh's impasto technique, layering pigments to create a sculpture as much as a painting. The textured quality of a painting is something that is often missed by those creating art via algorithms. The art is often rendered on a screen and is therefore limited to its two-dimensional digital canvas. What distinguishes artists from Goya to de Kooning is as much the way the paint is applied to the canvas as the image it produces. Certainly the way Rembrandt layers his paint is a key feature of his late output. But the team realized that modern 3D printers would give them a chance to analyze and sculpt the contours that are characteristic to Rembrandt's canvases. The final, 3D-printed painting consists of thirteen layers of paint-based UV ink laid down as specified by a digital design consisting of 148 million pixels.

Bas Korsten of J. Walter Thompson Amsterdam, who helped cook up the idea as part of an advertising campaign, admitted that while the idea was ingenious in its simplicity, its execution was anything but. "It was a journey of trial and error," he told an interviewer from *Dutch Digital Design*. "We had plenty of ideas that were researched or tested, but discarded in the end." The team had considered rigging up a robotic arm to execute the final painting but current robotic arms had only nine degrees of freedom versus the twenty-seven degrees of a human hand like Rembrandt's. So that approach was abandoned.

The biggest challenge, Korsten recalled, "was keeping the idea behind The Next Rembrandt alive. Even though there were so many forces working against it. Time, budget, technology, critics. But, most of all, the overwhelming amount of data we needed to go through. Perseverance and not taking 'no' for an answer are the only reasons why this project succeeded."

After eighteen months of data crunching and five hundred hours of rendering, the team finally felt ready to reveal to the world its attempt to resurrect Rembrandt. The painting was unveiled on April 5, 2016 in Amsterdam and immediately caught the public's imagination, with over ten million mentions on Twitter in the first few days of its going on display. The result is quite striking. There is no denying that it captures something of Rembrandt's style. If asked to name the artist, most people would probably put it in the Rembrandt school. But does it convey his magic? Not according to British art critic Jonathan Jones.

"What a horrible, tasteless, insensitive, and soulless travesty of all that is creative in human nature," Jones wrote with contemptuous disgust in the *Guardian*. "What a vile product of our strange time when the best brains dedicate themselves to the stupidest 'challenges,' when technology is used for things it should never be used for and everybody feels obliged to applaud the heartless results because we so revere everything digital."

Jones felt the project missed the entire point of Rembrandt's creative genius. "It's not style and surface effects that make his paintings so great but the artist's capacity to reveal his inner life and make us aware in turn of our own interiority—to experience an uncanny contact, soul to soul. Let's call it the Rembrandt Shudder, that feeling I long for—and get—in front of every true Rembrandt masterpiece."

To his mind, there was only one way such a project could ever succeed: "It would also have to experience plague, poverty, old age, and all the other human experiences that make Rembrandt who he was, and his art what it is."

Is it fair to be so dismissive? Would he have reacted in the same way had he not been told ahead of time that a computer had produced the painting? The artist's process is often a black box. Algorithms have given us new tools to dig around inside the box and to find new traces of patterns. If we can replicate through code what an artist has done, then that code reveals something about the process of creation. Could that help us identify overlooked old masters or reattribute falsely cata-logued works?

There has been much debate over the decades about who exactly painted *Tobit and Anna* in the Willem van der Vorm Collection in Holland. It certainly has many of the characteristics of a late Rem-brandt: concentrated light, a rough painting surface, parts that are very sketchy together with others that are in sharp focus. It even has Rem-brandt's signature at the bottom. But many believed that this had been added later and the painting was a fake. For decades it was not classified as a Rembrandt and was attributed to one of his pupils. This all changed in 2010, when Rembrandt expert Ernst van de Wetering brought the powers of modern science to bear on the canvas.

Thanks to infrared scans and x-ray analysis we can now see things hiding beneath the surface of a painting, like the first attempts an artist made on the way to the final product as the work evolved. In the case of *Tobit and Anna*, x-ray images revealed that initially the painting had included a window and it was subsequently painted over. According to van de Wetering, Rembrandt was someone who continually played around with light in this way, trying out different ways to illuminate the figures. Microscopic chemical analysis also revealed that the sig-nature had to have been made while the painting was still wet. Van de Wetering's years of experience and deep knowledge of Rembrandt's style, plus the support of these new scientific techniques, led him to change his mind about the attribution. The museum displaying the painting was very happy to hear it had another Rembrandt in its col-lection, although some critics, despite the scientific support, still doubt the provenance of the painting.

So what did van de Wetering think about this new computer-generated Rembrandt? He had hated the idea when it was first proposed. When he finally came face to face with the result, he immediately started critiquing the painting's brushwork, homing in on subtle inconsistencies. The brushwork, he noted, employed the technique Rembrandt adopted in 1652, while the rest of the portrait was more in the style of work produced twenty years earlier. The team was reasonably relieved that it was at this level of detail that their project was found wanting.

For Microsoft, the motivation for the Rembrandt project was most likely less artistic than commercial. To convincingly fake a Rembrandt demonstrates how good your code is. AlphaGo's triumph against Lee Sedol was similarly not so much about discovering new and more creative ways to play the game of Go as it was to provide great publicity for DeepMind's AI credentials. Is that a problem? Should creativity be free of commercial considerations? Van Gogh sold two paintings in his lifetime (although he did exchange other canvases for food and painting supplies from fellow artists). Perhaps he hoped to make a modest living, but money doesn't appear to have been much of a drive for his creativity. And yet there is evidence that dangling money in front of someone can stimulate (at least at a low level) their creative output.

In 2007 an American team of psychologists invited 115 students to read a short story about popcorn popping in a pan. The students were then asked to provide a title for the story. Half were told: "We will be judging the creativity of your titles against the titles of all the other students who have participated in this research in the past. If your titles are judged to be better than 80 percent of the past participants in this study, you will have done an excellent job." The other half were told the same thing and given the prospect of a ten dollar reward for their creativity. Sure enough, the financial incentive led to more creative output, including such gems as "PANdemonium" and "A-pop-calypse Now."

Is feedback from others, in whatever form it may take, an impetus for creation? Don't we continue to create and originate to keep our fellow humans engaged and interested in us? This is an aspect that the new AI is beginning to incorporate. In machine learning, feedback is often used to move the algorithm towards a better result. Take Deep-Mind's algorithms for playing Atari games. Rewarding risk-taking (by programming it to seek a high score) led the algorithm to crack levels that an algorithm without that incentive had missed.

Competitive Creativity

Creating a new Rembrandt is fairly pointless beyond proving that it can be done. But could genuinely new and exciting art emerge from code? Ahmed Elgammal of Rutgers University wondered whether making artistic creation into a competitive game might spur computers into new and more interesting artistic territory. His idea was to create one algorithm whose job was to disrupt known styles of art, and another one tasked with judging the output of the first. This threat that the output might be condemned as either not recognizably art or insufficiently original is a classic example of a general adversarial network, a concept first introduced by Ian Goodfellow at Google Brain. The first algorithm would learn and change based on feedback from the other algorithm. By the end of the game, Elgammal hoped to produce an algorithm that would be recognized on the international stage for its creativity.

There is some evidence that this adversarial model is applicable to the way the human code channels creativity. This was suggested by the curious case of Tommy McHugh. In 2001, Tommy had a stroke. Before the stroke he had been happily leading his life as a builder in Liverpool. He was married and living in a small house in Birkenhead and had had no interest in art beyond the tattoos he'd decided to get while in prison. But after the stroke something strange happened. Tommy suddenly had an urge to create. He started writing poetry and bought paints and brushes and began to fill the walls of his house with pictures. The trouble was he couldn't control this urge to create. He became a hostage to this drive to cover the walls of his house in paint.

Very quickly every single wall in the house was covered. Stepping into the house was like entering a kitsch version of the Sistine Chapel. Everything was covered in pictures. Tommy's wife could not take the explosion of creativity and left him to it. Tommy couldn't stop. He just kept covering old painting with new.

"Five times I've painted the whole house. Floor, ceilings, carpets," he told me. "I only sleep through exhaustion. If I was allowed, the outside of this house would be painted and so would the trees and the pavements."

Are the paintings any good? Not really. But why did Tommy suddenly have this urge to paint following the stroke? He tried to describe to me what was happening inside his head when this creative urge took hold: "I kept on visualizing a lightning flash shooting over to this side of the brain and hitting this one cell . . . it unlocked a Mount Etna of bubbles. Each little fairy liquid bubble in my imagination contained billions of other bubbles. And then they popped. All *this* has exploded."

Research by neuroscientists has discovered that, like the algorithms driving the generative adversarial networks at Google Brain, our own brains have two competing systems at play. One is an exhibitionist urge to make things. To create. To express. The other system is an inhibitor, the critical alter ego that casts doubt on our ideas, that questions and criticizes our ideas. We need a very careful balance of both in order to venture into the new. A creative thought needs to be balanced with a feedback loop which critiques the thought so that it can be refined and generated again.

It seems that Tommy's stroke knocked out the inhibitor part of his brain. There was nothing telling him to stop, or that what he was creating might not be so great. All that was left was this explosive exhibitionist urge to create more and more crazy images and ideas.

The artist Paul Klee expressed this tension in his *Pedagogical Sketchbook*: "Already at the very beginning of the productive act, shortly after the initial motion to create, occurs the first counter motion, the initial movement of receptivity. This means: the creator controls whether what he has produced so far is good."

Tommy McHugh died in 2012 from cancer. Up to the end, he had no bitterness about what had happened to him. "My two strokes have given me eleven years of a magnificent adventure," he said, "that nobody could have expected."

Elgammal's strategy was to write code to mimic this dialogue between the generator and discriminator that takes place, generally subconsciously, in an artist's mind. First he needed to build the discriminator, an algorithmic art historian that would critique the output. In collaboration with his colleague Babak Saleh, he began to train an algorithm so that it could take a painting it hadn't seen before and classify the style or painter responsible for the painting. WikiArt has probably the largest database of digitized images, with 81,449 paintings by 1,119 different artists spanning fifteen hundred years of history. Could an algorithm be created that could train itself on the content of WikiArt and take a painting at random and classify its style or artist? Elgammal used part of the available data as a training set and the remaining data to test how good the algorithm was. But what should he program his algorithm to look out for? What key distinguishing factors might help classify this massive database of art?

To use mathematics to identify an artist you need things to measure. The basic process is similar to the one behind the algorithms driving Spotify and Netflix, but instead of personal taste, you are looking for distinguishing characteristics. If you measure two different properties of the paintings in your data set, then each painting can be graphically represented as a point on a two-dimensional graph. So what can you measure that will result in your suddenly seeing Picasso's paintings clustered in one corner and van Gogh's in another?

For example, measuring one feature—perhaps the amount of yellow used in a painting—might cause paintings by Picasso (marked with an x) and van Gogh (marked with a o) to be arranged on the scale, like so:

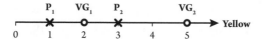

At the moment, measuring this single feature doesn't help us distinguish between the painters. Sometimes Picasso uses a small amount of yellow, as in painting P_1 which scores a 1 on our scale. But other times, the yellow is more pronounced, as in painting P_2 which scores a 3. The two paintings by van Gogh plotted here, VG_1 and VG_2, also vary in the amount of yellow featured. Measuring yellow doesn't help us.

What if we pick another feature to measure—perhaps the amount of blue in the paintings? This time we'll plot the same paintings on a vertical axis.

Evidently blue doesn't help us, either. There isn't a clear divide that puts paintings by Picasso on one side and van Gogh on the other. But look what happens when we combine the two measurements, plotting the paintings now in two-dimensional space: Picasso's painting P_1 is located at position $(1,2)$ while van Gogh's painting VG_1 has position $(2,4)$. But in this two-dimensional graph, a line can be picked out which now partitions the paintings of each artist. We find that, when we combine measurements of blue and yellow, Picasso's are in the lower half of the diagram while van Gogh's appear in the upper half.

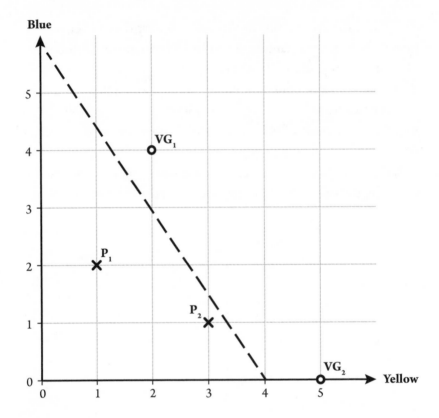

Having learned how to use these two features to distinguish a Picasso from a van Gogh, the algorithm has something to go on. When the algorithm is shown a new painting and told to identify if it is van Gogh or Picasso, it measures the two properties and plots the coordinates of the painting on the graph. Whichever side of the line the painting lands on will give the algorithm the best bet as to the artist behind the painting.

In this simple example I've chosen the features of color to distinguish the artists. But there are numerous other features we could track. The power of machine learning is to explore the space of possible measurements and to pick out the right combination of features that help to distinguish between artists, just as measuring yellow and blue did in our simple example. Two measurements will not be enough, so we need to find enough different qualities that distinguish artists from each other. Each new measurable feature increases the dimension of the space we are mapping paintings into, and gives us a better chance of distinguishing artists and their styles. By the end of the process we will be plotting paintings in a high-dimensional graph rather than the two dimensions we saw in our simple example.

Finding the things to measure can be done in two different ways. As a programmer you can code up certain features that you think might help distinguish between artists: use of space, texture, form, shape, color. But the more interesting feature of machine learning is its ability to engage in unsupervised learning and to find its own features to home in on. A human analyzing the decision tree can sometimes find it hard to figure out what features the algorithm is focusing on to distinguish between paintings. State-of-the-art computer vision measures over two thousand different attributes in images that are now called classemes. These attributes were a good place to start to analyze the paintings they had chosen to train their algorithm on.

In the quick sketch we considered above, we saw how a two-dimensional space was sufficient to distinguish Picassos from van Goghs. To get close to distinguishing styles across the true data set, the algorithm would have to plot paintings in a space with four hundred dimensions, effectively taking four hundred different sorts of measurement. The resulting algorithm, when tested on the unseen paintings, managed to identify the artists more than 50 percent of the time—but it found it tricky to distinguish between artists like Claude Monet and Camille Pissarro. Both are Impressionists who lived in the late nineteenth and early twentieth centuries. Interestingly, both artists attended the Académie Suisse in Paris, and the friendship they developed there resulted in some noticeable interactions.

The Rutgers team decided to investigate whether their algorithm could identify moments in art history of extreme creativity, when something new appeared that hadn't been seen before. Could it identify paintings that had broken the mold and ushered in a new style of painting? Some artists incrementally push the boundaries of an existing convention while others come up with a completely new style. Could the algorithm identify the moment cubism emerged on the scene? Or baroque art?

The algorithm had already plotted all the paintings as points in a high-dimensional graph. What about adding the dimension of time to this graph and plotting when paintings were created? If the algorithm detected a huge shift in the position of the paintings in the high-dimensional space as it moved along this time dimension, would it correspond to a moment that art historians would recognize as a creative revolution?

Take, for example, Picasso's painting *Les Demoiselles d'Avignon*, a painting that many acknowledge broke the mold. The initial reception when *Les Demoiselles d'Avignon* was first shown in Paris in 1916 was very hostile, as you would expect from a revolutionary change in aesthetic. A review published in *Le Cri de Paris* declared: "The Cubists are not waiting for the war to end to recommence hostilities against good sense." But it didn't take long for the painting to be recognized as a turning point in art history. An art critic at the *New York Times* wrote a few decades later: "With one stroke, it challenged the art of the past and inexorably changed the art of our time." The exciting thing is that the algorithm, too, was able to pick out a huge shift in the location of this painting compared to its contemporaries when viewed in this multidimensional graph, scoring it highly as a painting that was markedly different from anything that had gone before. Perhaps even *New York Times* art critics are about to be upstaged by algorithms.

The Rutgers team's discriminator algorithm is like an art historian who can judge whether paintings are part of an accepted existing style and recognize when they break new ground. Its counterpart, the generator algorithm, is tasked with creating things that are new and different but will still be recognized and appreciated as art. To understand this tension between the new but not too new, Elgammal steeped himself in the ideas of psychologist and philosopher D. E. Berlyne, who argued that the psychophysical concept of "arousal" was especially relevant to the study of aesthetic phenomena. Berlyne believed the most significant arousal-raising properties of aesthetics were novelty, unexpectedness, complexity, ambiguity, and the ability to puzzle or confound. The trick was to be new and surprising without drifting so far from expectation that arousal turned to aversion because the result was just too strange. This is captured in something called the Wundt curve.

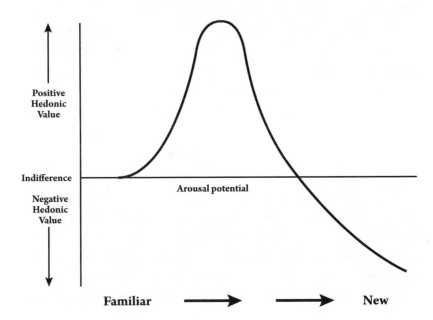

If we become too habituated to the artwork around us, that leads to indifference and boredom. This is why artists never really stabilize in their work: what arouses the artist (and eventually the viewer) is something distinct. The challenge is that the push to arousal or dissonance must not be so great that we hit the downslope of the Wundt curve. There is a maximum hedonic value that the artist is after.

Elgammal and his team programmed the generator algorithm so that it was incentivized to create things which would try to hit that peak in the Wundt curve. The game was to maximize difference while trying not to drift too far from those styles that the art world has found acceptable. The discriminator algorithm would be tasked with feeding back to the generator algorithm whether it was too derivative or too wild to be considered art. Each judgment would alter the parameters of the generator algorithm. This is machine learning in action: the algorithms change as they encounter more data, learning from the feedback. As the algorithms pinged information back and forth, the hope was that the generator algorithm would be pushed to create new things that would fall in the sweet spot of the Wundt curve. Elgammal calls these "creative adversarial networks."

So what did people make of the output of these algorithms? When new works were shown to a group of visitors to Art Basel 2016, the flagship fair for contemporary art, and these art lovers were asked to compare them with new artwork generated by Elgammal's creative adversarial network, they found the computer-generated art more inspiring and identified more closely with its images. (You can view the images for yourself at https://arxiv.org/abs/1706.07068.)

Perhaps the most significant signal that AI art is beginning to be taken seriously came in October 2018 when Christie's became the first auction house to sell a work of art created by an algorithm. The painting was produced by a Paris collective using Goodfellow's original idea of a general adversarial network rather than the creative one developed by Elgammal. The Paris team trained their algorithm on fifteen thousand portraits dating from the fourteenth century through to the current day.

The result is a portrait of a man in a dark coat and white collar with unfinished facial features, which gives the character a slightly unnerving quality. The portrait is strangely uncentered as if the sitter doesn't really want to be there. It is difficult to place the period of the painting, which combines an eighteenth-century style of portraiture with a very contemporary execution similar to British artist Glenn Brown's style. The signature at the bottom of the painting is perhaps the most intriguing part of the whole painting. Instead of an artist's name we find a mathematical formula.

The portrait is one of a whole series produced by the algorithm which the Paris team decided to put into a fictitious family tree depicting different generations of the Belamy family. The Christie's painting depicts Edmond Belamy, great-grandson of the Count de Belamy, whose portrait was bought privately in February 2018 for $12,000. (In the Christie's auction, the portrait of the great-grandson went for a staggering $432,000.) The choice of family name pays homage to Goodfellow, who came up with the idea of these competing algorithms. Goodfellow translates loosely in French to Bel Ami.

This idea of learning from what artists have done in the past and using that knowledge to push into the new is, of course, the process that most human artists go through. Current art can only be understood in light of our shared past. After all, this knowledge or frame of reference is what most viewers bring to their encounters with new art. No art on view at the Basel show is being experienced by someone who has never been exposed to the way Picasso and Munch have painted. Most creativity stems from this idea of perturbing the present to create a future that has some connection to the present but nonetheless breaks from it. It is an evolutionary model and, intriguingly, this is what the algorithm picked up on.

You may feel this approach is horribly manipulative. To change art into a landscape of numbers only to find the points that will trigger maximal hedonic value sounds awful. Aren't great artists meant to express their inner angst? And yet, there may just be a role for this alternative pathway to artistic creativity. These adversarial network algorithms can push us into new terrain that we recognize as art but have been too inhibited to explore. Computer code has the capacity to reveal untapped potential in the art created by the human code.

Seeing How an Algorithm Thinks

Where art is at its best is in providing a window into the way another mind works. And perhaps that is the true potential of art made by AI.

It might ultimately help humans to understand the hidden nature of the underlying computer code. If AI is due to take over from us humans, it might be a good idea to get some perspective on how AI views the world.

A team at Google have been using art created by AI to understand better some of the thought processes at work in the visual-recognition algorithms it has been creating. As I explained in Chapter 5, the algorithms that have been developed to distinguish cats from bananas depend on hierarchies of questions that they can ask about the image. The algorithm effectively plays a game of Twenty Questions to identify what is in the picture.

The trouble is that, as the machine learns and changes, the programmer gradually begins to lose track of the features it is using to identifying bananas from cats. Just looking at the raw code, it is very difficult to reverse-engineer how the algorithm is working. There are millions of different questions that the algorithm can ask about an image and it is tricky to see how and why these questions have been chosen in preference to others. To try to get a feel for how its algorithm was working, the team at Google had the clever idea of turning the program on its head. They gave the algorithm a random pixelated image and asked it to dial up or enhance the features it thought would trigger the recognition of an identifiable feature. The result, they hoped, would reveal what the algorithm was looking for. They called this inverted algorithm DeepDream.

To me, the images that DeepDream produces are perhaps the most meaningful form of AI art that I've seen on my journey. Instead of trying to reproduce another Rembrandt or to compete with modern artists at Art Basel, these images are letting us see something of how visual-recognition algorithms view the world. It may not be very aesthetically important, but it may be what art is all about: trying to understand the world through another set of eyes and to connect with a different way of seeing.

The DeepDream algorithm exploits the way a human can look at an image and suddenly see something—a face in their toast or an animal in the clouds—when there is nothing there. The human brain has evolved to be extremely sensitive to images of animals because that is key to its survival. But this means that sometimes we see animals where there are none. The visual-recognition algorithms work in a similar way. They look for patterns and interpret them. They have learned to detect patterns in a compressed version of evolution, having been trained on thousands of images. Their survival is dependent on correctly identifying them. Machine learning is basically a form of digital evolution. So what are the algorithms seeing in the digital undergrowth?

The results, the Google team discovered, were quite striking. Starfish and ants began to appear out of nowhere. It seems that within the algorithm was the power not just to recognize images but also to generate them. But this wasn't just a fun game. It offered fascinating insights into how the algorithm had learned. Images of dumbbells would always have an arm attached to the dumbbell. It was clear that the algorithm had learned about dumbbells from images of people lifting weights. So it hadn't understood that these things weren't an extension of human anatomy and could stand on their own.

Rather than feed the algorithm random pixels, you could give it actual images and ask it to enhance the features it detected or invite it to play the game we've all played of staring up at the clouds: what can you see hidden in those puffy shapes? The algorithm was able to pick out features that seemed to correspond to a dog or a fish or perhaps a hybrid animal.

The novel that would become the cult film *Blade Runner* was called *Do Androids Dream of Electric Sheep?* Using these algorithms we can now find out! In one image produced by the algorithm, sheep did indeed begin to appear in the sky.

More and more decisions are going to be taken out of human hands and given over to the algorithms we are making. The trouble is that the machine-learning algorithms that are appearing lead to decision trees that are very hard for humans to unpick. This is one of the limitations of this new sort of programming. Ultimately we are not really sure why the algorithm is making the decision it does. How can we be sure that it isn't a mistake rather than an extremely insightful suggestion? The Go commentators were not sure on which side of the divide to put AlphaGo's move thirty-seven in Game Two until they eventually saw that it won the game. But increasingly these algorithms are doing more than playing games. They are making decisions that affect our lives. So any tools that help us to understand how and why these algorithms make the decisions they do will be essential as we head into an increasingly automated future.

In the case of computer-vision algorithms, the art they can produce is giving us some inkling as to how they are working. Sometimes the features that are detected and chosen among are things that we recognize, but other times it seems hard to name what the algorithm is distinguishing in the image. The art is giving us insight into the level of abstraction that the algorithm is working on at particular layers of the decision tree. We are penetrating what might be deemed the deep unconscious of the algorithm. The programmers called the process "inceptionism" and considered the images to be like the dreams of the algorithm—hence the name DeepDream. Certainly the images that the algorithm is generating have a crazy psychedelic feel to them, as if the algorithm is tripping on acid. By applying the algorithm over and over on its own outputs and zooming in after each iteration, the programmers could generate an endless stream of new impressions.

I don't think anyone would rank the product of DeepDream as good art (whatever that is). As the columnist Alex Rayner, who first wrote about these images, commented: "they look like dorm-room mandalas, or the kind of digital psychedelia you might expect to find on the cover of a Terrence McKenna book." Not things you'll find at Frieze in London or Art Basel. But it still represents an important new way of understanding something of the internal world of the algorithm as it classifies images.

The Algorithm Is the Art

Are these new tools pushing the visual arts into interesting new territory? I decided I needed to make a trip back to the Serpentine Gallery to talk to Hans Ulrich Obrist and hear his thoughts on the role of AI in the art world. But before heading up to his office, I decided to have a peek at the art that was currently on show.

As I entered the gallery I was confronted by BOB, an artificial life form created with code by Ian Cheng. In fact there are six BOBs. Each started out with the same code but the evolution of these life forms is affected by its interactions with visitors. By the time I made it to the exhibition, the six BOBs had gone off in very different directions. As a father to two genetically identical twin girls, who are very different from one another, I know how a small change in the environment can have a large effect on the outcome of identical code.

Just as with the Richter, I felt compelled to unravel the code at the heart of BOB. But this is a different sort of code, one that is much harder to reverse-engineer. That may be why it succeeds in holding one's attention longer than one might expect. It is learning and evolving based on its interaction with the viewers who come to the gallery.

BOB picks up on the emotional state of the visitor via interactions with a smartphone. Cheng was intrigued by questions of authorship and origination. He wanted to know: How could art be authored in its meaning but also live beyond the author and mutate itself? The answer was to create a system and allow its content to evolve and change based on interaction which he would not control. BOB's interactions with visitors mean that Cheng is left behind at some point as the code is informed by new parameters coming from its encounters.

Often we respond to code that we don't understand by assigning it some sort of agency. Back when people didn't understand earthquakes or volcanoes, they created gods that were responsible for these elusive forces. The algorithm at the heart of BOB stimulates the same response in the viewer in a phenomenon the philosopher Daniel Dennett refers to as the intentional stance.

Hans Ulrich told me: "Usually the visitors' book at the gallery is full of complaints about the gallery being too hot or 'Why aren't there more chairs?' Or comments about how they like or don't like Grayson Perry. But we were getting instead comments like: Why doesn't BOB like me? I feel sorry for BOB. BOB ignored me. BOB is so cute. It was extraordinary."

One night, BOB appeared to have taken on a life of its own. Hans Ulrich told me he had been traveling abroad a week before when he got a phone call from the security team at the gallery. At 3 AM the Serpentine had suddenly been flooded with light. Not a fire. Rather, it appears that BOB had decided to wake up, despite the fact he'd initially been programmed to wake at 10 AM and to run until 6 PM, when the gallery shut down. Our inability to understand why BOB woke up in the middle of the night makes us feel he has agency. It is this inability to understand how algorithms work that fuels the movies and stories of algorithmic apocalypse.

The open-ended nature of art work that is continually evolving and never repeating, Hans Ulrich believes, is something new for the art world. Most art has a beginning and an end. Any film in the gallery in the past would have to be looped and would ultimately become boring after you'd seen it twenty times. The use of AI breaks that need to recycle material.

The code behind BOB shares something in common with the analog code behind Jackson Pollock's drip paintings. It is based on chaotic deterministic equations that are influenced by the environment, so that the viewer can perturb the output. The chaos allows for unpredictability. Code that exploits the mathematics of chaos can claim to meet the criteria of novelty and surprise demanded by the word "creative." It remains deterministic, but chaotic processes are probably the best we can hope for if we intend to break the connection between coder and creator.

Jonathan Jones gave BOB one star in his *Guardian* review. "They are just clever lab models. There is no soul here . . . art is always human, or nothing at all. Cheng forgets this, and his work is a techno bore." Although Jones is almost certainly right that there is no ghost in the machine, as we head into the future, we will increasingly need to exploit the world of the gallery as a mediator in understanding perhaps when the first ghost might appear.

Hans Ulrich thinks of art as one of society's best early-warning systems. Given the importance of the debate about the role AI is playing in society, it seemed urgent in Hans Ulrich's mind for AI to take its place in the gallery. Much of today's use of algorithms is invisible and hidden. We don't understand how we are being manipulated. Using art to visualize the algorithm helps us interpret and navigate these algorithms more knowingly. The visual artist is a powerful mediator between the crowd and the code. The artificial intelligence that was on display was the art.

"The artists are the experts at making the invisible visible," Hans Ulrich told me. So, will AI ever create great art rather than being the art? "We can never exclude that a great work can be created by a machine. One should never say never. As it stands today there hasn't been a great art work created by a machine." But he was cautious about the future: "When the Go players said a machine is never going to beat us, Demis proved them wrong. I'm a curator but I'd never be arrogant and say a machine couldn't curate a better show . . ."

I could see his neurons beginning to fire. "That could be a fun experiment to do one day . . . to do the Go experiment with curating . . . a dangerous experiment but an interesting one."

/ 9 /

The Art of Mathematics

Sudden illumination is a manifest sign of long,
unconscious prior work.

—HENRI POINCARÉ

I was thirteen when the idea of becoming a mathematician first took root. The maths teacher at my comprehensive school took me aside after one lesson and recommended a few books he thought might interest me. I didn't really know at that stage what being a mathematician entailed, but one of those books revealed that it was much more than simple calculations. Called *A Mathematician's Apology*, the book was written by the Cambridge mathematician G. H. Hardy.

It was a revelation. Hardy wanted to communicate what it meant to do mathematics: "A mathematician, like a painter or a poet, is a maker of patterns. If his patterns are more permanent than theirs, it is because they are made with ideas. The mathematician's patterns, like the painter's or the poet's, must be beautiful; the ideas like the colors or the words must fit together in a harmonious way. Beauty is the first test: there is no permanent place in the world for ugly mathematics." I'd never imagined mathematics to be a creative subject, but as I read Hardy's little book it seemed that aesthetic sensibilities were as important as the logical correctness of the ideas.

I wasn't much of a painter or a poet, so why did my teacher think mathematics would be for me? When I got the chance many years later to ask him why he'd singled me out, he replied: "I could see you responding to abstract thinking. I knew you'd enjoy painting with ideas." It was a perfectly judged intervention that picked up on my desire for a subject that blended a creative mindset with a wish for absolute logic and certainty.

For years I've believed that the creative side of mathematics protected it from being automated by a computer. But now, algorithms are painting portraits like Rembrandt and creating art works that rival human-generated painting on show at the Basel art fair. Will they soon be able to recreate the mathematics of Riemann or compete with the papers published in the *Journal of the American Mathematical Society*? Should I start looking for another job?

Hardy spoke about mathematics like a game—he liked to use the analogy of chess. But ever since computers began playing chess better than humans, playing Go has been my shield against those trying to quickly dismiss what I do as something a computer could do much faster. Mathematics is about intuition, making moves into the unknown that feel right even if I'm not quite sure why I have that feeling. When DeepMind's algorithm discovered how to do something with a very similar flavor, it triggered an existential crisis.

If these algorithms can play Go, the mathematician's game, can they play the real game? Can they prove theorems? One of my crowning achievements as a mathematician was getting a theorem published in the *Annals of Mathematics*. This is the journal in which Andrew Wiles published his proof of Fermat's Last Theorem. It is the mathematician's *Nature*. How long would it be before we might see a paper in the *Annals of Mathematics* authored by an algorithm?

To play a game it's essential to understand the rules. What am I challenging a computer to do? I'm not sitting at my desk doing huge calculations. If that had been the case, computers would have put me out of a job years ago. So what *is* it exactly that a mathematician does?

The Mathematical Game of Proof

If you read a news story about mathematics, it will invariably be about the fact that a mathematician has "proved" some great outstanding conjecture. In 1995, newspapers ran breathless headlines about Wiles's proof of Fermat's Last Theorem. In 2006, the maverick Russian mathematician Grigori Perelman proved the Poincaré conjecture, earning him the right to claim the million-dollar reward that had been placed on its head. There are still six more "millennium prize" problems offering challenges to prove the hunches of mathematicians that have been thorniest to work out.

The idea of proof is central to what mathematicians do. A proof is a logical argument that starts from a set of axioms, a list of self-evident truths about numbers and geometry. By analyzing the implications of these axioms, one can start to piece together new statements that must also be true about numbers and geometry. These discoveries can then form the basis of new proofs, which in turn will invite us to discover yet more logical consequences of the axioms. This is how mathematics grows: like a living organism whose structure extends out from a previously existing form.

It's no wonder people have compared mathematical proof to playing games like chess and Go. The axioms are the starting positions of the pieces on the board, and the rules of logical deduction are the parameters determining how each piece can move. A proof is a sequence of moves played one after the other. In chess, given the number of possible moves at each stage, there are myriad different positions the pieces can assume on the board. For example, after just four moves (two by white, two by black) there are already 71,852 different ways that the pieces might be arranged on the board. There are generally several different ways to reach that position. The tree of possible moves in Go grows even faster.

If I were to place the pieces randomly on the board, you might ask: Is it possible to reach this position from the starting position? In other words, is this a legitimate arrangement of pieces in a game of chess or Go? This is similar to the idea of a conjecture in mathematics. Fermat's last theorem, for example, was the conjecture that the equation $x^n + y^n = z^n$ can have no whole number solutions x, y, and z when n is greater than two. The challenge facing mathematicians was to prove that this was or wasn't a logical consequence of the way numbers work. Fermat had placed the pieces on the board and declared that this was an end point he believed could be reached. Wiles and the other mathematicians who contributed to his work demonstrated a sequence of moves that ended with the arrangement Fermat had guessed was possible.

Part of the art of being a mathematician is picking out these targets. Many mathematicians believe that asking the right question is more important than providing the answer. Sensing what might be true about numbers requires a very keen mathematical nose. This is where the most creative and difficult-to-pin-down skill of the mathematician comes into play. It requires a thorough immersion in this world to gain that intuition about a possible new truth. It is often a feeling or hunch you feel compelled to assert even though you don't have an explanation for why it must be true. That explanation is the proof that everyone then starts to chase.

This is one of the reasons why computers have found it hard to do mathematics. The top-down algorithms of the past have been like drunk people stumbling around in the dark. They might randomly arrive at an interesting location, but most of the time their meanderings are unfocused and worthless. But could an algorithm evolved from the bottom up start to develop an intuition about interesting locations to head for, based on past journeys made by human mathematicians?

How do mathematicians build up a feel for what might be an interesting direction to pursue? They might have some examples in mind to back up a hunch—a buildup of evidence conforming to a pattern that seems too good to be a coincidence. But patterns based on data can quickly vanish. This is why coming up with a proof is so important. It can sometimes take a long time to expose a seeming pattern as a false lead. In my own work, I once made a conjecture about a pattern that turned out to be false, but it took ten years for a graduate student to reveal that to me.

One of my favorite examples of a hunch that didn't hold up is the one the great nineteenth-century mathematician Carl Friedrich Gauss had about prime numbers. He'd come up with a beautiful formula to estimate how many primes there were in a range from one to any number, but he believed his formula would always overestimate the number of primes. All the numerical evidence pointed to his being right, and indeed, if a computer had been let loose on the problem, it would to this day be producing data consistent with Gauss's hunch. Yet, in 1914, J. E. Littlewood proved theoretically why the opposite must be true. Beyond a certain point, it turns out, Gauss's formula actually underestimates the primes—but that point only comes at an extremely high number. Imagine counting through more numbers than there are atoms in the universe. Even that wouldn't get you anywhere near the point where the conjecture breaks down.

That is the challenge of all these conjectures. We just don't know if they are true or if our intuition and the available data are leading us astray. That is why we obsessively try to build a sequence of mathematical moves to link the conjectured endgame to the legitimate games established to date.

What drives humans to want to find these proofs? Where did the human urge to create mathematics come from? If we want algorithms that can challenge mathematicians at their own game, is this motivation to explore the mathematical terrain something that will need to be programmed into them? The origins of mathematics are rooted, of course, in the human needs to understand the environment we live in, to make predictions about what might happen next, to mold our environment to our advantage. Mathematics is an act of survival by the human species.

The Origins of Mathematics

Mathematicians are a bit of a misunderstood breed. Most people would assume that, as a research mathematician, I must sit in my office in Oxford doing long division to lots of decimal places or multiplying six-digit numbers in my head. Far from being a super calculator—something a computer is clearly much better equipped to be—a mathematician, as G. H. Hardy first explained to me, is at heart a pattern searcher. Mathematics is the science of spotting and explaining patterns.

This ability to see a pattern gives humans an edge in negotiating the natural world because it allows us to plan into the future. Humans have become very adept at pattern recognition because those who missed the patterns didn't survive. When people I meet declare (as, alas, so often happens) that they "don't have a brain for mathematics," I counter that, in fact, we have all evolved to have mathematical brains. Our brains are all able to spot patterns. Sometimes they are *too* able, reading patterns into data where none exist, as many viewers did when confronted with Gerhard Richter's random-colored squares at the Serpentine Gallery.

Some of the earliest expressions of pattern recognition come along with some of the very first art made by human hands. The cave paintings in Lascaux include exquisite images of animals racing across the walls. The movement of a stampede of aurochs is intriguingly captured in these ancient images. We might ask why the artist felt compelled to create these images. What role did they play?

Alongside these images are what I believe to be some of the earliest recorded mathematics. There is a strange line of dots, thirteen in number, daubed just below a great picture of a stag with huge antlers. Another series of twenty-six dots accompanies a picture of a pregnant mare. What does this abstract sequence of dots depict? One guess is that each dot represents a quarter of a moon cycle. Thirteen quarters of the moon represents about a quarter of a year. So perhaps these dots are depicting a season and are telling the viewer that a certain season of the year is a good time to hunt stag because they are rutting and vulnerable. Count another twenty-six quarters of the moon and you advance another two seasons. Here we get to a time of year when many mares are about to foal. So perhaps the wall is a training manual for would-be hunters.

In order to relay this information, someone would have had to spot a pattern of animal behavior repeating itself each year and then connect that to the pattern of moon phases. The drive to spot such patterns was clearly practically motivated. There is utility driving the discovery.

Here we see the first ingredient of mathematics: the concept of number. Being able to formulate an accurate sense of numbers has been crucial to the survival of many animals. It informs the choice of whether to fight or take flight from a rival pack. Sophisticated experiments done on newborn chicks reveal quite a complex number ability hardwired into the brain. The chicks were able to judge that five is more than two and less than eight.

But to give these numbers names and represent them by symbols is a uniquely human ability. Part of our mathematical development has involved finding clever ways to identify or name numbers. The ancient Mayans started out with rows of dots. The symbol for a number was simply that number of them. It makes sense, but at some point it becomes inefficient and hard to count up the dots. So someone had the clever idea of turning every five-dot group into a bar. It's not unlike the classic tally marks a prisoner makes to keep count of the days he has spent in his cell.

The Romans used a system by which different magnitudes of numbers were given different symbols. The letter X stood for ten, C for a hundred, M for a thousand. The ancient Egyptians, too, used a new hieroglyph to indicate another zero on the end of a number: a heel bone for ten, a coil of rope for a hundred, a lotus plant for a thousand. But this system quickly gets out of hand as we get into the millions or billions and more and more symbols are required.

The Mayans, who were doing sophisticated astronomy, needed big numbers to keep track of large spans of time. They came up with a clever system that avoided the Roman problem. Called the place-value system, it is the one we use today to write numbers. In our decimal system, the position of a digit indicates what power of ten it relates to. Take the number 123. Here we have one lot of a hundred units, two lots of ten units, and three single units. There is nothing special about the choice of ten beyond the fact that we can use our fingers to count up to ten. Indeed, the Mayans' symbols went up to twenty and the position of their digits indicated different powers of twenty. So, in Mayan mathematics, the number 123 would denote one lot of twenty to the second power, two lots of twenty, and three single units. (That's 443 to us.)

The Mayans were not the first to come up with this clever idea of using the position of a number to indicate its order of magnitude. Four thousand years ago, the ancient Babylonians had conceived of the place-value system. Instead of counting up to twenty, like the Mayans, or in decimals as we do today, the Babylonians used symbols all the way up to fifty-nine before they started a new column. The choice of sixty was influenced by the high divisibility of this number. It can be divided by two, three, four, five, six, ten, twelve, fifteen, twenty, and thirty. This makes it a very efficient choice for doing arithmetic.

Necessity, efficiency, and utility drove these mathematical choices. We see their repercussions today in the way we keep track of time: sixty minutes to an hour, sixty seconds to a minute. During the French Revolution, that country's measurement authorities tried to introduce a new way to track time using a decimal system, but fortunately that never caught on.

In the cuneiform tablets the ancient Babylonians left behind, we witness the first mathematical analyses of how numbers relate to the world around us. More sophisticated mathematics came soon after, in conjunction with the growth of the city-states along the Euphrates. To build, to tax, and to do commerce requires mathematical tools. These tablets reveal that officials were tabulating, for example, the number of workers and days necessary for the building of a canal, so they could calculate the total expenses of the workers' wages. There was nothing particularly challenging or interesting being done at this stage, but the mathematics clearly got some scribes thinking about what else could be done with numbers.

They started to discover clever tricks to help them with their calculations. For example, there are tablets with all the squares of numbers one through fifty-nine written out. These tablets were aids for anyone needing to multiply large numbers together. Someone had noticed the interesting relationship between multiplying numbers and adding their squares. The scribes realized that the answer to A times B could be worked out using a table of squares and this algebraic relationship:

$$A \times B = ((A + B)^2 - (A - B)^2) / 4$$

As shown, you just add A and B and look up the square of the answer. Next, subtract B from A and look up the square of the answer. Then take the difference between those two squares and divide it by four. What is so exciting is to find such an early example of an algorithm at work. Here is a method that takes the work of multiplying and reduces it to the simpler tasks of adding, subtracting, and consulting a database of squares in the form of a cuneiform tablet. It works whatever you plug in for A and B, within the fifty-nine numbers whose squares appear on the tablet.

Although the Babylonians were tapping into an algebraic way of thinking about numbers, they were far from having the language to articulate what they were doing. The equation I have written down became possible only thousands of years later, when the ninth-century Arabic and Persian scholars in Iraq's House of Wisdom developed the language of algebra. The ancient Babylonians did not write down why this method or algorithm always gave the right answer. It worked and that was good enough. The curiosity to come up with a way to explain it would come later. This is why, even though the first algorithms can be found in ancient Babylonian, the algorithm owes its names to the chief librarian and astronomer at the House of Wisdom, Al-Khwārizmī. He founded the subject of algebra.

Again, these early discoveries of mathematical relationships between numbers were driven by utility. They sped up calculation. They gave an advantage to the merchant or builder who spotted the connection. At the same time, problems and ways of solving them began to creep into forms that seem less practical. Outwardly, they might look like the same kind of work but, if you consider them more closely, they are more like fun puzzles to challenge fellow scribes than anything a farmer might gain from. For example, the following problem sounds like it could relate to a real problem:

> The area of a farmer's field is sixty square units. One side of the field is seven units longer than the other. What is the length of the shorter side of the field?

But here's the thing: How would anyone know the area of their field without knowing the lengths of its sides? To me, this feels like a cryptic crossword puzzle. Someone has thought of a word but only gives us a rather mixed-up description of it. We've got to undo their process to work out the word they had in mind. In the case of the scribe's problem about the field, we can do that by calling the length of the shorter side x. The longer side's length is then $x + 7$. The area of the field is the multiple of these two lengths and, since we know the answer is sixty, we get this equation:

$$x \times (x + 7) = 60$$

Which easily translates to:

$$x^2 + 7x - 60 = 0$$

This may send a shiver of recognition through you because it's an example of the quadratic equations you probably had to learn to solve in school. You can blame Babylonian scribes for the challenge, but also thank them for the method they concocted for unraveling this cryptic equation to find out what x is. (It's five, by the way.)

For me, this is an important transitional moment in my subject. Why did anyone even bother to set such a challenge? Why did someone feel compelled to find a clever way to unravel the problem and arrive at the answer? Why do we still get students to learn this? Not because they need to know it—challenges like this don't really come up in everyday life. Even if a farmer had previously calculated and written down the area of his field but neglected to record the sides' lengths, is it likely he would have noted that the long side of the field was seven units longer than the short side? The whole thing is too contrived to ever have been a genuine, practical problem. No—here is someone doing mathematics just for the fun of it!

This is a brain that is enjoying the aha moment and delighting in how to untangle the problem to get the answer. We now know that a shot of dopamine or adrenaline would have accompanied the realization that the method works whatever the numbers involved. There is biology and chemistry at work driving this mathematical feat. Would a computer ever have made such a move—one that involved doing mathematics purely for the fun of it—given that computers have no biology or chemistry?

You might argue that actually there is utility in solving fanciful problems. True, there may be an evolutionary advantage bestowed on the person who can do this sort of mathematics. Indeed, for those of us who still insist on teaching students how to solve quadratic equations, this is our best defense. A mind that can apply this sort of algorithm—that can chase through the logical steps required to get to the answer, that is happy with an abstract, analytical thought process—is a mind that is well equipped to cope with problem-solving in real life.

Perhaps the chemistry behind the satisfaction we feel when we solve a mathematical puzzle will prove key to distinguishing human creativity from machine creativity. In some respects, a brain is like a computer in its construction, and it might be possible to simulate brain activity by creating an abstract network in which each digital neuron switches on and off in relation to the other neurons connected to it. But if we don't put chemistry and biology into our construction, will the machine be denied that satisfying aha moment the Babylonian scribe enjoyed? Will it lack the motivation, the drive, to think creatively?

In Babylonian mathematics you still find a focus on particular arithmetic examples. Methods discovered were applied to solve these particular problems, but no explanations were given for why these methods always worked. That would have to wait a few millennia, until mathematics started to develop the idea of proof.

The Origins of Proof

The beginning of this game of mathematical proof goes back to the ancient Greeks, who discovered the power of logical argument to access eternal truths about number and shape. Proof is really what mathematics is about. For any mathematicians hoping to make their name in the field, this is the holy grail to search for. To earn a million-dollar prize, you have to prove one of the seven conjectures. To win a Fields medal, you must come up with a proof that impresses your fellow mathematicians. Probably Euclid's *Elements* was the rule book that kicked off this great game.

If we go back to our chess analogy, it can help explain how the game of mathematical proof works. We start by laying out an opening set of statements, called axioms, just as we would prepare for a chess game by setting up the pieces. Euclid's *Elements* begins with a list of axioms, making statements about numbers and geometry that mathematicians regard as blindingly obvious—things we can all accept as true. Of course, we might be wrong about the truth of these axioms. Honestly that doesn't matter to the game we are going to play—we can just take them as truths—but it's fair to say, looking at the things Euclid included, they seem pretty acceptable as fundamental truths. Between any two points, a straight line can be drawn. If A = B and B = C, then A = C. Given any line segment, we can use that line as the radius to draw a circle. A + B = B + A.

With the pieces laid on the board, we need to learn next how to play the game. Just as the chess pieces are constrained by certain rules which determine how they can move, there are rules for logical deduction that allow us to write down new truths based on what we know to date. For example, the rule *modus ponens* asserts that if you have established that statement A must imply statement B, and you've also established that statement A is true, then you are allowed to deduce that statement B is true. The complementary rule of *modus tollens* asserts that if you've shown that statement A must imply statement B, and you've also established that statement B is false, then you can deduce that statement A is false.

This last rule is applied in Euclid's *Elements* to prove that the square root of two cannot be written as a fraction. If we assume that it can be written as a fraction, then by letting the game of mathematical chess play out through a series of logical moves, we eventually get to the conclusion that odd numbers are even. But we know that odd numbers are not even. Therefore, by applying the rule of *modus tollens*, we arrive at the conclusion that the square root of two cannot be written as a fraction.

For me, a well-constructed and satisfying game is one that is easy to set up, has rules that are simple to understand and implement, and yet offers an extremely rich and varied range of how games can play out. Tic-tac-toe is simple to explain and play but very soon becomes rather dull because the same games you've already played start repeating. In chess or Go, on the other hand, so many different games can evolve from the starting position that people who dedicate their lives to playing never tire of playing another one.

One important distinction between playing games like chess and Go and playing the game of mathematical proof is that mathematicians don't have to reset all the same pieces every time they want to play. All the games that have been played before become the foundation, the point from which they start the next game. Every generation of mathematicians expands the axioms laid out at the start, and the moves that can be played. Anything that has been established to date can be used in the new game.

It's striking how we give meaning to symbols and words. A line is that thing we draw across the page. An x is meant to represent a number that counts or measures something. How would a computer know what we're talking about? The beauty of the game is that, even though we are trying to capture how numbers and geometry work, we can view the whole game symbolically. In fact, any meaning we give to the symbols such that the axioms are true will give rise to a game that teases out properties of the objects we have substituted for the symbols. This means a computer can make deductions about the game without really having to know what the symbols mean.

Indeed, when the nineteenth-century mathematician David Hilbert lectured on geometry he stressed this point: "One must be able to say at all times—instead of points, lines, and planes—tables, chairs, and beer mugs." His point was that, provided the things had the relationship expressed by the axioms, the deductions would make as much sense for chairs and beer mugs as geometric lines and planes. This allows the computer to follow rules and create mathematical deductions without really knowing what the rules are about. This will be relevant when we come later to the Chinese-room argument devised by John Searle. This thought experiment explores the nature of machine translation and tries to illustrate that following rules doesn't show intelligence or understanding.

Nevertheless, follow the rules of the mathematical game and you get mathematical theorems. But do we really need the rigors of mathematical proof? Imagine that you noticed, and then did a little bit of experimenting to confirm, that every number you came up with could be written as prime numbers multiplied together and there was always only one way to break down the number. For example, 105 was equal to the product of primes $3 \times 5 \times 7$ and no other combination of primes multiplied together would give you 105. You could just make that observation and hope that it always worked. More examples could bolster your faith in this discovery. After a while, you might even think that, with the accumulated evidence so overwhelming, your observation should be added as an axiom.

But now imagine that there is some really large number that no one has gone to the effort to factor and there are two different ways to pull it apart. The axiom you are proposing can be violated—it's just that you've got to hit really large numbers before this becomes possible. This points to a quality of mathematics that marks it out as different from science. A scientist would have to rely on evidence and data gathering to convince other scientists to adopt what appears to be a good theory. But the existence of proof means that we can show something to be a logical consequence of how numbers work. We can prove that there won't be an exceptional number that breaks the theory. Mathematical proof shows *why* there is only one way to write a number as the product of prime numbers. And that proof allows the next person who plays the game to include this as a given about the way numbers work.

The Babylonians would have been happy with the observation that numbers reliably decompose into products of primes but wouldn't have felt compelled to come up with a watertight argument for why this must always be true. They had a more scientific approach to numbers and geometry. It was the ancient Greeks who came up with a new game, by marking out mathematics as a subject that allows us to establish truth.

So where did this urge to prove come from? It is quite possible this is a by-product of the evolution of societies. In the cities of ancient Egypt and Babylon power was centralized, but as new cities emerged in ancient Greece, democracy, a legal system, and political argument became part of everyday life. It is in Greece that we see writers beginning to use logical argument to challenge received opinion and authority.

In the stories that appear during this period, humankind is no longer happy to be pushed around by the Olympian gods and people begin to dispute the terms of their rule. Socrates, for whom an unexamined life is not worth living, dedicates his work to arguing the difference between truth and received opinion. Sophocles has Antigone challenge her uncle's tyrannical rule over Thebes. Aristophanes satirizes the demagogues and other power abusers of his time in his political comedies.

This challenging of authority, this move to democracy and a society based on a legal system, requires that skills of logical argument be developed. The growth of the polis, which gave citizens a role in their society, depended on their growing abilities to engage in debate. Indeed, the Sophists would travel from city to city giving lessons in the art of rhetoric. In his famous treatise on the subject, Aristotle defines rhetoric as "the faculty of observing in any given case the available means of persuasion." He crystalizes those means into three categories, of which one is *logos*: the skill of using logical argument and available facts, rather than emotional appeals or personal credibility, to persuade the crowd.

The drive to come up with clever forms of mathematical proof coincides with this shift in society. Logos gave you the greatest power to persuade. And this push to use logical argument to persuade your fellow citizen of your point of view goes hand in hand with a shift in mathematics. The tools of logical deduction turned out to be powerful enough to access eternal truths about the way numbers and geometry work. You could prove that every number could be uniquely written as a product of prime numbers. You could prove that prime numbers go on to infinity. You could prove that a triangle subtended on a diameter of a circle was right angled.

Again, very often someone would have a hunch about one of these eternal truths—in other words, a conjecture would arise from playing around with numbers. Add up all the odd numbers in sequence, for example, and the sum always seems to be a square number:

$$1 + 3 = 4$$
$$1 + 3 + 5 = 9$$
$$1 + 3 + 5 + 7 = 16$$

But does that always work? The Greeks were not content to say this interesting connection between odd numbers and square numbers had been observed so far without exception. They wanted logos to prove to them that it could never be otherwise—that it was a logical consequence of the basic axioms governing how numbers work.

And so began the great journey that is mathematics. Euclid's *Elements* set the stage for the two thousand years of mathematicians working since to come up with proofs explaining the strange and wonderful ways of numbers and geometry. Fermat proved that, if you raise a number to the power of a prime bigger than that number and then divide the result by the prime, the remainder must be the number you started with. Euler proved that, when you raise e to the power of i times pi, the answer is -1. Gauss proved that every number can be written as the sum of at most three triangular numbers (writing *Eureka* next to his discovery). And eventually my colleague Andrew Wiles proved that Fermat was right in his hunch that the equation $x^n + y^n = z^n$ has no whole number solution when n is greater than 2.

These breakthroughs are representative of what a mathematician does. A mathematician is not a master calculator but a constructor of proofs. So here is the question at the heart of this book: Why can't a computer join the ranks of Fermat, Gauss, and Wiles? A computer can clearly outperform any human when it comes to calculation, but what about our ability to prove theorems? It is possible to translate a proof into a series of symbols and a rule set for why one set of symbols is allowed to follow another. As Hilbert explained, you don't have to know what the symbols mean to be able to construct mathematical proofs. Doesn't this seem like something perfectly conceived for a computer to engage in?

Every time a mathematician starts with an established mathematical statement and takes an unprecedented but allowed step further, the new sequence of symbols constitutes a newly established mathematical statement. It's possible that it's already on the list of mathematical proven statements. Someone might have previously come to it via a different route. But this is nonetheless a way for a mathematician (or a computer) to start generating new theorems out of old ones. Isn't that the goal? Mathematics may not be about doing calculations, but doesn't the computer still put the mathematician out of a job if we can just click go and let it start spewing out logical consequences of all known statements?

Here is where creativity comes into the picture. It is easy to make something new. The top-down style of programming will produce a machine that can crank out new mathematical theorems. The challenge is to create something of value. Where does that value come from? This is something that depends on the mind of the human creating and consuming the mathematics. How will an algorithm know what mathematics will cause that exciting rush of adrenaline that shakes you awake and spurs you on?

This is why the new, bottom-up style of programming emerging from machine learning is so exciting—and potentially so threatening to a mathematician like me. The algorithms Hassabis and his colleagues are producing may learn from the human mathematics of the past how to distinguish the thrilling theorems from the boring ones, and if so they might sort through their spewed output and find them. Already, a machine might be on its way to unveiling a new theorem of such value that it will surprise the mathematical community, just as the gaming world was so shocked by AlphaGo.

10

The Mathematician's Telescope

Our writing tools participate in the writing of our thoughts.

—FRIEDRICH NIETZSCHE

For all my existential angst about the computer putting me out of the game, I must admit that, as a tool, it proves invaluable to me. Often I am faced with the task of combining a slew of equations into a single equation. If I did this by hand, I would almost certainly make mistakes. It is a mechanical procedure that requires little thought; one only has to follow a set of rules. My laptop does not bat an eye at the challenge, and I would trust its calculation over my own pen-and-paper attempt every time. But a computer can also go beyond simply manipulating equations. The role it is capable of playing in my work has grown by leaps and bounds over the years. Just as the telescope allowed Galileo's generation to see further into the depths of our universe, the computer has given mathematicians a new perspective on our subject.

Given the close bond between mathematics and algorithms it perhaps isn't surprising that, for nearly half a century, computers have served as essential partners in proving deep theorems of mathematics. In the 1970s, for example, a computer played a major role in settling the proof of the Four-Color Theorem, which had gone unsolved for over a century. This problem theorized that, if you wanted to illustrate a map—perhaps of Europe but really of any region actual or fictional—so that no two countries that shared a border were shown in the same color, you would need no more than four colors in total. Some maps could be executed in just three colors, but in no case would a fifth color be required.

Already, the proof had been constructed for the claim that five colors would suffice, but before 1976 no one had yet been able to reduce this to four. That year, two mathematicians, Kenneth Appel and Wolfgang Haken, succeeded in doing so, after adding an interesting twist to their approach. They first established that, although there are infinitely many possible maps you can draw, there is a way to show that they can all be reduced to an analysis of 1,936 maps. But analyzing even this many maps by hand was going to be impossible—or, to be more accurate, impossible for a human. Appel and Haken managed to program a computer to go through the list of maps and check whether each one passed the four-color test. It took over a thousand hours for their lumbering computer of the 1970s to run through all the maps.

No creative effort was demanded of the computer, only dumb donkey work. Still, could anyone prove that there wasn't a bug in the program causing false results? This question of how absolutely the results of a computer can be trusted is one that dogs the field of AI. As we head into a future dominated by algorithms, ensuring that there are no undetected bugs in the code will increasingly be a challenge.

In 2006, the *Annals of Mathematics* published a computer-assisted proof of another classic problem in geometry: Kepler's conjecture. Thomas Hales, the human behind the proof, had come up with a strategy to prove conclusively that the hexagonal stacking of oranges you see at the grocer is the most efficient way to pack spheres. No other arrangement wastes less space. Again, like Appel and Haken, Hales had used a computer to run through a finite but huge case analysis. He announced the completion of the proof in 1998 and submitted the paper to the journal along with the code he'd used for the computer component of the proof.

Before a paper is accepted for publication, mathematicians demand that all of its steps be checked by referees, running the proof like a program through their brains to see if anything crashes. But this vast case analysis was a part of the proof that the brain with its physical limitations was unable to vet. The reviewers were asked to trust that the computer had accurately assessed all the sphere-packing possibilities. Many were uneasy about this. It was like mapping your route from London to Sydney but being forced to get in an airplane and close your eyes for most of the journey. Because of the role the computer had played, it took eight years for mathematicians to agree to call the proof correct, and only then with the caveat that it was 99 percent certain.

For mathematical purists, that missing 1 percent was anathema. Imagine claiming you're related to Newton and then proudly displaying the family tree . . . with one generation left blank. The role of computers in proving theorems was viewed by many in the field with deep suspicion. This wasn't because they were nervous about being put out of their jobs—in these early years, a computer could work only at the behest of the mathematician who'd programmed it—but because of that leap of faith required. How could anyone rule out the possibility of a bug buried deep in the program, and without ruling that out, how could they trust such a proof?

Mathematicians had been stung by such bugs before. In 1992, Oxford physicists used heuristics from string theory to make some predictions about the number of algebraic structures that could be identified in high-dimensional geometric spaces. Mathematicians were a little suspicious, wondering how physics could tell them about such abstract structures—and they felt justified in their doubts when a proof showed that the conjecture was false. It turned out, however, that the proof involved a computer component and a bug in one of the programs caused it to miscalculate. It was the mathematicians, not the physicists, who got it wrong. The bug in the program had led them astray.

A few years later, mathematicians went on to prove (this time without a computer) that the physicists' conjecture was correct.

Stories like this have fueled mathematicians' fears that computers might lead us to build elaborate edifices on top of programs that are structurally unsound. Frankly, however, a human has more chance of making a mistake than a computer. It might be heresy to suggest it, but there are probably thousands of proofs with gaps or mistakes that have been missed. I should know. In a couple of proofs I've published, I have subsequently discovered holes. They were pluggable, but the referees and editors had missed them.

If a proof is important, scrutiny generally brings to light any gaps or errors. That's why the millennium prizes are not released till two years after publication; twenty-four months is considered enough time for a mistake to be exposed. Take Andrew Wiles's first proof of Fermat's Last Theorem. Referees seized on a mistake before it ever made it to print. The miracle was that Wiles, with the help of his former student Richard Taylor, was able to repair that mistake. But are there other incorrect proofs out there?

But there is a growing feeling that we might have to rely on computers. Some new proofs are now so complex that mathematicians fear hard-to-pick-up errors will be missed. Take the Classification of Finite Simple Groups, a theorem close to my own research. This is a sort of periodic table of the symmetrical atoms from which all symmetrical objects can be built. People refer to it as the "monster theorem" because the proof is ten thousand pages long and spans a hundred journal articles. It reflects the efforts of hundreds of mathematicians. The list of atoms includes twenty-six strange exceptional shapes called sporadic simple groups. There's always been a sneaking suspicion that a twenty-seventh might be out there that the proof missed. Could a computer help us check such a complex proof?

The trouble is that there is a deep philosophical problem with such a strategy which all science has to deal with. If you get a computer to check through a proof and verify that each step is valid, aren't you just shifting the risk? How would you know that the computer program doing the checking didn't have a bug? You could get another computer to check that program for bugs but where would this end? Science and mathematics have always been dogged by this dilemma. How can you be certain that your methods are leading you to true knowledge? Any attempt to prove that inevitably depends upon the methodology you are relying on to produce truth.

As Hume first pointed out, much of science relies on a process called induction: inferring a general law or principle from the observation of particular instances. Why do we trust this as a sound way of generating scientific truths? Principally because of induction! We can point to many cases where this inductive principle seems to lead to good scientific theories. This leads us to conclude (or induce) that induction is a good approach to doing science.

Given the increase in complexity and the new technology that is available, mathematics is facing a real challenge to its culture of certainty—the very quality that previously set it apart from the other sciences.

Coq: The Proof Checker

As more and more proofs started to appear that were dependent on computer programs, it was felt that some approach was needed to ensure that the conclusions of these programs could be trusted. In the past, mathematics generated by humans could be checked by humans. Now it would be necessary to create new programs to check the programs behind the proofs, as their calculations were too complex and long for humans to validate.

Two French mathematicians, Pierre Huet and Thierry Coquand, began in the late 1980s to work on a project called Calculus of Constructions, or CoC. In France computer scientists seem to have a habit of naming software products after animals, so the system soon came to be called *Coq*, French for rooster. It was also, rather conveniently, the first three letters of one of the developers' surnames. Coq was created to check proofs, and soon emerged as the program favored by anyone interested in validating computer proofs.

Georges Gonthier, the principal researcher at Microsoft Research Cambridge, decided to put together a team to use Coq to check the proof for the Four-Color Theorem, the first proof that had required a computer to complete. By 2000, the Microsoft Research team had run through the computer code developed by Appel and Haken and validated the proof (again, assuming that Coq did not produce a matching but false result because of bugs of its own). Then they had Coq also check the human-generated part of the proof, which Appel and Haken had written themselves.

One of the challenges in checking a human proof is that it rarely spells out all the steps. People do not write proofs like computer code. They write proofs for other people, using code that only has to work on our hardware, the human brain. This means that when we write proofs we often skip tedious steps, knowing that those reading the proof understand how to fill them in. But a computer requires every step. It's the difference between writing a novel, where you don't need to account for every tedious action of your central protagonist, and instructing a new babysitter, where you have to spell out every single detail of the day, including naps, potty breaks, and every last item on the menu.

It took another five years before the computer was able to verify the human element of the proof. An interesting by-product of this process was that the researchers uncovered new and rather surprising nuggets of mathematics that had been overlooked in the first proof.

Why, though, should we trust Coq any more than the original computer proof? The answer, interestingly, is because of induction. As Coq validates more and more proofs that we are confident are correct, we grow ever more certain that it has no bugs. That is really the same principle we use to test the fundamental axioms of mathematics. The fact that every time we take two numbers A and B we get the same answer whether we add A to B or B to A has led us to accept it as an axiom that $A + B = B + A$. By relying on one computer program to check all the others, we gain more trust in its conclusion than we could have in a one-off program created especially to check the proof at hand.

Once his team had finished checking the Four-Color Theorem proof, Gonthier announced a new challenge to his team: the Odd-Order Theorem. This is one of the most important theorems guiding the study of symmetry. Its proof led to the Classification of Finite Simple Groups, a list of the basic building blocks from which all symmetrical objects can be built. One of the simplest building blocks in this periodic table of symmetry is the regular two-dimensional polygon with a prime number of sides, creating shapes like the triangle and the pentagon. But there are many more complex and exotic examples of symmetry, from the sixty rotations of an icosahedron to the symmetries of a strange snowflake in a 196,883-dimensional space that has more symmetries than there are atoms making up the Earth.

The Odd-Order Theorem states that any symmetrical object with an odd number of symmetries will not require exotic symmetries to build. It can be made out of the simple ingredients of a prime-sided polygon. It is an important theorem because it essentially sorts out half the objects you might consider. From then on, you can assume that the objects you are hoping to identify have an even number of symmetries.

The proof as published in 1963 was pretty daunting. It ran to 255 pages and occupied a whole issue of the *Pacific Journal of Mathematics*. Before its publication, most proofs covered at most a few pages and could be mastered in a day. This one was so long and complex it was a challenge for any mathematician to digest. Given its length, doubts persisted as to whether some subtle error might be embedded inside the pages of the proof.

Getting Coq to check the proof would thus not only demonstrate Coq's prowess, it would contribute to our confidence in the proof of one of the most complex theorems in mathematics. This was a worthy goal. But turning a human-generated proof into checkable code expands it even further. Gonthier's challenge was not going to be easy.

"The reaction of the team the first time we had a meeting and I exposed a grand plan," he recalled later, "was that I had delusions of grandeur. But the real reason of having this project was to understand how to build all these theories, how to make them fit together, and to validate all of this by carrying out a proof that was clearly deemed to be out of reach at the time we started the project."

One of his programmers left the meeting and looked through the proof. He emailed his reaction to the team: "Number of lines—170,000. Number of definitions—15,000. Number of theorems—4,300. Fun—enormous!" It took six years for the team at Microsoft Research Cambridge to work through the proof. Gonthier spoke of the elation he felt as the project came to a close. At last, after many sleepless nights, he could relax. "Mathematics is one of the last great romantic disciplines," he said, "where basically one genius has to hold everything in his head and understand everything all at once." But we are reaching full capacity with our human bit of hardware. Gonthier hopes his work will kick-start a period of greater trust and sustained collaboration between human and machine.

The Limits of Our Human Hardware

There is a growing sense among young mathematicians that many regions of the mathematical landscape are becoming so dense and complex that you could spend all three years of your PhD just trying to understand the problem your research supervisor has set you. You can spend years navigating this terrain and mapping out your discoveries, only to find that no one has the head space to retrace your steps to understand or verify them.

There is not much reward to reworking someone else's discoveries. And yet journals depend on this process of peer review. Promotion and tenure rely on the validation that getting a paper published in the *Annals of Mathematics* or *Les Publications Mathématiques de l'IHES* bestows. So increasingly there could be a place for a system like Coq to help verify the proof of a theorem submitted to a journal.

Some mathematicians feel we are at the end of an era. The sort of mathematics that the human brain can navigate must inevitably have limits. Frankly I find it extraordinary how much mathematics *has* been within the reach of the human mind.

Take the Classification of the Finite Simple Groups, the building blocks of symmetry. That we humans were able to construct—using our minds, pencils, and paper—a symmetrical object that can only be built by working in 196,883-dimensional space is extraordinary. The mathematicians who truly feel at home working with the monster symmetry group are growing old. Like the masons of the medieval period, they have skills that will be lost once they die. There isn't much compulsion for those who follow to rework these gothic masterpieces unless they provide a pathway to new wonders.

That hundreds of pages of journals spanning three centuries combine to prove that Fermat's equations have no solutions is a testament to the long game that the human mind can play. And yet, in the midst of working to prove a conjecture, there is always that sneaking feeling that the proof might command a complexity that is beyond the physical capabilities of human brains. It is amazing what we can do but, given that mathematics is infinite and we are finite, it is also mathematically true that mathematics is bigger than we will ever be.

I am working on a conjecture now that has had me entangled in its grip for fifteen years. Every time I try to piece together the insights I've had on parts of the problem, my brain returns an error message. I'm exceeding its capacity. I feel that I am tantalizingly close but I just can't pull the pieces together. But I've reached such points before and also know that finding a new angle of approach to a wild beast of a problem can often bring it in reach of the net my mind can cast.

Proving something like the Riemann Hypothesis, our greatest unsolved problem about prime numbers, might simply be beyond the limits of the human brain. At least, having seen generations of mathematicians work on it without success, it's inevitable that someone would begin to suspect so. To be sure, the statement of the conjecture is simple enough. But as G. H. Hardy wryly pointed out, after spending years battling vainly with it, "Every fool can ask questions about prime numbers that the wisest man cannot answer."

Kurt Gödel, the Austrian logician, proved that mathematics has true statements for which there are no proofs. At some level this is a shocking revelation. Do we need to add new axioms to capture these unprovable truths? Gödel warned that modern mathematics was likely to slip further and further from our grasp: "one is faced with an infinite series of axioms, which can be extended further and further, without any end being visible," he said in 1951. "It is true that in the mathematics of today the higher levels of this hierarchy are practically never used . . . it is not altogether unlikely that this character of present-day mathematics may have something to do with . . . its inability to prove certain fundamental theorems, such as, for example, Riemann's hypothesis."

Given that we may be reaching full capacity as humans, some mathematicians are beginning to acknowledge that if we want to push further, we'll need the machines. We may get to the top of Everest with little more than a tank of oxygen, but we can't reach the moon without the union of human and machine.

One of those who believes the days of the lone mathematician working with pencil and paper are coming to an end is Doron Zeilberger, an Israeli mathematician. Since the 1980s, he has insisted on including the name Shalosh B. Ekhad as coauthor of any research for which he has used his computer. In Hebrew, the way to pronounce 3-B-1 is "Shalosh-B-Ekhad" and Zeilberger's first computer was an AT&T 3B1. Zeilberger believes the resistance to partnering with machines is due to "human-centric bigotry," which, like other forms of bigotry, has held back progress.

Most mathematicians believe that their aspirations are more complex than those of computers: they hope to produce not just truths but an understanding of what lies behind those truths. If a computer verifies the truth of a statement without providing that understanding, they feel cheated.

"We aim to get understanding in mathematics," said Michael Atiyah. Having won the Fields Medal, the mathematics equivalent of a Nobel Prize, he was speaking at a 2013 Laureates Forum in Heidelberg. "If we have to rely on an unintelligible computer proof, it's not satisfactory." Efim Zelmanov, another Fields Medal winner, agreed: "A proof is what is considered to be a proof by all mathematicians, so I'm pessimistic about machine-generated proofs." Certainly, we don't accept a proof if only one mathematician can understand it. So does Zelmanov have a point? If only the machine that generated it can understand a proof, can *we* really trust it?

Doron Zeilberger appreciates where this sentiment comes from but ultimately dismisses it. "I also get satisfaction from understanding everything in a proof from beginning to end," he told *Quanta Magazine* in 2013. "But on the other hand, that's life. Life is complicated." He believes that if a human mind can understand a proof then it must be pretty trivial. "Most of the things done by humans will be done easily by computers in twenty or thirty years. It's already true in some parts of mathematics; a lot of papers published today done by humans are already obsolete and can be done using algorithms. Some of the problems we do today are completely uninteresting but are done because it's something that humans can do."

That's a pretty depressing assessment of the state of the field. But is it really true? I certainly have felt that some papers go into the journals just because of the need to generate publications. But that's not always a bad thing. Sometimes there are unexpected consequences of doing something just for the sake of doing it, suggesting that non-target-driven research is sometimes the best way to glean genuinely new insights.

Like many of my colleagues, Jordan Ellenberg sees a vital role for humans in the future of our field. His response to Zeilberger's position in *Quanta* was this: "We are very good at figuring out things that computers can't do. If we were to imagine a future in which all the theorems we currently know about could be proven on a computer, we would just figure out other things that a computer can't solve, and that would become 'mathematics.'"

But a lot of this human output is moving sideways rather than forward. We really are reaching the point in some areas where to go beyond the heights of Everest is going to necessitate getting into a machine. That's a shock for the old guard (and I probably include myself in that category). That pen and paper will no longer hack it as a way to do groundbreaking mathematics is something many are very reluctant to admit.

Voevodsky's Visions

One of those who made his name in mathematics with pen and paper but has gone on to champion the importance of adding the computer to the mathematician's armory is Vladimir Voevodsky, one of the stars of my generation. I met him at Oxford when we were trying to tempt him with a position. People had spotted that he was a dead cert for a Fields Medal, and Oxford decided to get in early with a tempting offer. The seminars he gave on his work suggested a truly new vision of mathematics. This was not incremental advance or an interesting new fusion of established ideas. Voevodsky seemed to channel a new language of mathematics and was able to prove things that had eluded generations of mathematicians.

I spoke earlier in the book about three sorts of creativity—exploratory creativity, combinational creativity, and transformational creativity—the last of which changes the landscape of a field by introducing a completely new perspective. Voevodsky's creativity was truly transformative. Listening to his ideas, you couldn't help thinking: Where did *that* come from?

It turned out that this exceptional creativity was enhanced by a rather unexpected source. I was quite shocked to learn during his visit that one of the important factors in his choice of a place of work was access to drugs. And I'm not talking about caffeine—

most mathematicians' drug of choice. (As the great mathematician Paul Erdős once quipped, "a mathematician is a machine for turning coffee into theorems.") We were asked to source some pretty hard-core Class B drugs to convince him of Oxford's suitability.

I've never imagined that drugs would be any good in helping me access ideas which I regard as requiring a cold, steely logic to navigate, but Voevodsky felt that amphetamines could lead him to churn out visions that he could then check once he'd landed back on Earth again. Years later, when I learned about the effects that caffeine and amphetamines have on spiders building their webs, it occurred to me that he might have been on to something. Spiders on speed create fast coherent webs, but the webs of caffeinated spiders are a total mess. Voevodsky went on to win his Fields Medal and accepted a position at the Institute for Advanced Study in Princeton, but his early successes triggered something of an existential crisis. In an interview conducted just before his untimely death of an aneurysm, in 2017, he explained: "I realized that the time is coming when the proof of yet another conjecture won't have much of an effect. That mathematics is on the verge of a crisis, or rather, two crises."

The first of these two crises involves the separation of "pure" and "applied" mathematics. As budgets for research are increasingly squeezed, governments are having to make hard choices about where money should be spent. Some politicians are beginning to question why society should pay money to people who are engaged in things that do not have any practical applications. Voevodsky felt it was important to show why the very esoteric research that he was doing could nonetheless have enormous practical impact on society.

But it was the second crisis that was more of an existential threat, and it relates to the increasing complexity of pure mathematics. Even if mathematicians were able to master their little corner, it was becoming impossible for the community to verify others' work. Mathematicians were becoming more and more isolated. Already in 1739 David Hume pointed out in his *Treatise on Human Nature* the importance of the social context of proof:

> There is no Algebraist nor Mathematician so expert in his science, as to place entire confidence in any truth immediately upon his discovery of it, or regard it as anything, but a mere probability. Every time he runs over his proofs, his confidence increases; but still more by the approbation of his friends; and is rais'd to its utmost perfection by the universal assent and applauses of the learned world.

Sooner or later, Voevodsky believed, the journal articles would become too complicated for detailed verification to take place and this would lead to undetected errors in the literature. "And since mathematics is a very deep science, in the sense that the results of one article usually depend on the results of many, many previous articles, this accumulation of errors for mathematics is very dangerous."

Having identified these two brewing crises, Voevodsky decided to leave the research that had won him fame and glory and to focus on some practical problem that would require recent work on the pure side of mathematics to solve. He had been interested in biology since he was a kid so he began with a question of whether the tools he had developed might provide new insights into that field, which was not generally regarded as very mathematical. He spent several years trying to determine whether you could deduce the history of a population by analyzing its current genetic make-up. But his attempts to crack this biological riddle eventually ran aground. He found he didn't have the right tools and skills to dig deep into biological questions as he did in his chosen area of mathematics. "By 2009, I realized that what I was inventing was useless. In my life, so far, it was, perhaps, the greatest scientific failure. A lot of work was invested in the project, which completely failed."

After much soul searching, he turned to the second crisis he'd identified: the increasing complexity of cutting-edge mathematics. If humans were unable to check each other's proofs, then maybe we needed to enlist the help of machines. For a pure mathematician of Voevodsky's caliber to talk about using computers in this way seemed misguided to many. Most mathematicians continued to believe in the power of the human mind to navigate equations and geometries and, guided by their aesthetic sensitivity, to sniff out solutions. Those who were critical of his decision also did not believe or appreciate that a crisis was imminent.

As Voevodsky looked around for suitable tools, he could see that the only viable computer project able to navigate proofs was the French system Coq. Initially he just couldn't get his head around how it worked. So he went back to basics and proposed to the Institute for Advanced Study that he teach a course on Coq. This is a trick I've often used: if you don't understand something, try teaching it. Gradually it began to dawn on him: the language that computer scientists were using, which initially appeared so alienating, was in fact just a version of the very abstract world in which he had spent his early years as a mathematician.

It was as if he had managed to solve two crises at once. First, his obtuse mathematical ideas were perfect for articulating the very practical world of modern-day computing; and second, here was a new language with which he could build a new foundation for mathematics, where the computer would play a central role.

Voevodsky's vision of the future of mathematics was far too revolutionary for most mathematicians, many of whom believed he moved to the dark side. There is still a deep divide between those doing mathematics with pen and paper (perhaps using a computer now and again to check a routine calculation) and those who want to use computers to prove new theorems. The idea of using a computer to check proofs is becoming acceptable; the human who created the proof is still in the driving seat here. It's when it comes to computers actually creating the mathematics that people, myself included, start to have issues.

But Voevodsky believed these old attitudes would have to be dropped. At the 2013 Laureates Forum in Heidelberg he was asked if he truly believed that mathematicians would end up using computers to create proofs. "I can't see how else it will go," he replied. "I think the process will be first accepted by some small subset, then it will grow, and eventually it will become a really standard thing. The next step is when it will start to be taught at math grad schools, and then the next step is when it will be taught at the undergraduate level. That may take tens of years, I don't know, but I don't see what else could happen."

Voevodsky compared the interaction to playing a computer game. "You tell the computer, try this, and it tries it, and it gives you back the result of its actions. Sometimes it's unexpected what comes out of it. It's fun."

I often think how sad it is that Voevodsky did not have the chance to see how his revolution would pan out. As I recall his comments it reminds me, in the spirit of his vision, to embrace the future and be open to the potential of the computer to extend mathematical creativity.

11

Music: The Process of Sounding Mathematics

Music is the pleasure the human mind experiences from
counting without being aware that it is counting.

—GOTTFRIED LEIBNIZ

W hen Philip Glass went to study with Nadia Boulanger in Paris
in 1964, every lesson began with Bach. *The Art of Fugue* was
a key part of the curriculum and Glass was made to learn a new Bach
chorale each week. Once he'd mastered a chorale, a hymn based
on four voices, he was instructed to add four new voices to the original four in such a way that no voice repeated another and yet they
all meshed seamlessly. Boulanger believed any composer hoping to
become great had to start with an immersion in Bach.

A small part of me wishes I had been a composer rather than a
mathematician. Music has been a constant companion on my mathematical journeys. My mind searches for patterns and structures as I
contemplate the unexplored reaches of the mathematical landscape,
which may be why a soundtrack by Bach or Bartók helps my thought

process. Both were drawn to structures similar to those I find exciting as a mathematician. Bach loved symmetry. Bartók was fascinated by the Fibonacci numbers. Sometimes composers are intuitively drawn to mathematical structures without realizing their significance; other times they seek out new mathematical ideas as a framework for their compositions.

It was while talking to composer Emily Howard about geometric structures that might be interesting to explore musically that I had an idea. Perhaps, in exchange for a tour of the mathematics of hyperbolic geometry, she might agree to give me composition lessons. She thought this was a fair deal and we met over coffee not long after that for my first lesson.

Just as a blank piece of paper can present a daunting void for a novice writer, the sight of a musical stave with no notes put me into a panic. Emily explained calmly that every composer needs to start with a framework or set of rules to help give shape to their composition. She suggested we start with the rules governing medieval polyphony, where something called an prolation canon is used to take one line of music and grow it into a multivoiced work. The idea is to start with a simple rhythm that will be sung by one voice. Then a second voice sings the same rhythm at half speed, and a third voice at twice the speed. The three voices sing different rhythms that are nonetheless strongly correlated. When you listen to a piece of polyphony using this technique, your brain recognizes that there is a pattern connecting the three voices.

Here was my homework: to compose a simple rhythm and grow it into a string trio using the medieval tradition of prolation. This was a simple enough undertaking, and one that can easily be mapped out as a mathematical equation: $x + 2x + \frac{1}{2}x$. As the piece I composed emerged, I had a strong sense of being like a gardener. The small fragment of rhythm that I had created from nothing was like a seed thrown down on the stave. Then, by applying the algorithm that Emily had given me, I was able to mutate it, change it, and grow it. The algorithm would start to fill out the rest of the stave with bits of music that had a

strong connection with the original seed, but that weren't simply rep-
etitions of it. It was a deeply satisfying experience to watch my musical
garden growing out of this simple rule.

Composing this simple piece helped me understand the close correlation between algorithms and composition. An algorithm is a set
of rules that can be applied to different inputs and lead to results. The
initial input is the seed. The algorithm creates the way for that seed to
develop. Take two numbers and apply Euclid's algorithm, and you can
discover the largest whole number by which both those original numbers are divisible. There are algorithms that analyze images and tell
you what is in the picture. There are algorithms that grow fractal
graphics: by starting with a simple geometric image and repeatedly applying a mathematical equation to it, they cause a complex graphic to
emerge.

The algorithms that apply to music have a similar quality. One of
Philip Glass's early pieces illustrates why algorithms are a key device
in the composer's toolbox. Called "1 + 1," the piece is for a single player
who taps out a rhythmic sequence on a tabletop, amplified via a contact microphone. The seeds for the piece are two rhythms: the first
rhythm, which I'll call A, is made up of two short beats followed by a
long beat. The second, called B, is just a single, long beat. Glass then
instructs the player to combine the two units using a choice of some
regular arithmetic progression. This is the algorithm that grows the
seed.

The performer is given the freedom to choose their own algorithm
but Glass gives some examples of different arithmetic progressions that
could be used to grow the piece. For example, in the rhythm that begins ABAABBBAAABBBBB, the number of occurrences of the A
rhythm increases by one each time but the B increases by two. I think
a lot of people have criticized Glass by saying "Come on, where is the
music here? It just sounds monotonous." But to me, this piece crystal-
lizes what's at the heart of all music: as you listen, your brain recognizes that the piece isn't random, and nor is it simple repetition. There's
a pleasure in trying to reverse-engineer the construction of the piece

and spot the patterns underpinning it. And it's this idea of pattern that I believe connects music so closely with the world of mathematics.

The art (or perhaps science) of the composer is therefore twofold: coming up with new algorithms that might be used to create interesting music, and choosing different seeds of music that can be fed into the algorithms. Given that there is this algorithmic quality at work that is growing the music, could this be the key to how a computer might embark on becoming a composer?

Bach: The First Musical Coder

One reason Boulanger insisted that Philip Glass take Bach as his starting point for musical composition is that algorithms are very much in evidence in the way Bach creates his music. In some ways Bach deserves to be called one of the first musical coders. (That's coders, not codas!) His compositions are more complex than the simple algorithm behind medieval polyphony, but many of them could still be mapped out in mathematical terms. *The Musical Offering*, inspired by a challenge set by Frederick the Great, illustrates this most clearly.

Although the Prussian king is best known for his military victories, Frederick the Great was also passionate about music all his life. Despite his father's attempts to literally beat such frivolous pursuits out of him as a child, he grew into a leader eager to celebrate the greatest musical talents in his court in Potsdam. Among them was Bach's son Carl Philipp Emanuel, employed as chief harpsichordist.

The Musical Offering grew out of a visit that Bach senior paid to his son at the court in 1747. For the sixty-two-year-old Bach, the trip had taken several days of hard traveling and when he arrived he was looking forward to collapsing at his son's house. That night, however, when Frederick the Great saw the daily list of strangers who had arrived in town, he was excited: "Gentlemen! Old Bach is here!" Immediately he sent an invitation to Bach to join him for an evening of music-making. He was particularly keen to show off his new collection of pianofortes. It is said that he had been so impressed by the pianos made by

Gottfried Silbermann of Freiberg that he had bought all fifteen pianos in his workshop. They were scattered around the palace.

Having received the summons from the palace, Bach did not even have time to change from his traveling clothes. One did not keep the king waiting. On his arrival, they moved from room to room trying out the pianos. Having heard about Bach's fantastic abilities to improvise on the spot, Frederick the Great sat down at one of his new pianofortes and tapped out a theme, then challenged Bach to create a piece based on it.

The royal theme was no ordinary tune. It was full of chromatic steps without any clear key. It was impossibly intricate. So cleverly had it been constructed that, in the words of twentieth-century composer Arnold Schoenberg, it "did not admit one single canonic imitation." In other words, it was resistant to any of the classical rules of counterpoint. Indeed, some have suggested that Frederick the Great had cooked up this diabolical challenge with Bach's son. C. P. E. Bach had lived his whole life in his father's shadow. He regarded his father's music as old-school and wanted to write in a new style. So perhaps the challenge was meant to reveal the shortcomings of his father's style and method. He might have wanted, as Schoenberg surmises, "to enjoy the helplessness of the victim of his joke, when the highly praised art of improvisation could not master the difficulties of a well-prepared trap." If that's so, it backfired spectacularly. The old Bach sat down and proceeded to improvise a stunning three-part fugue based on this tricky theme.

A fugue is a more sophisticated version of a canon or round, something many people sing in school. In a canon, one-half of the class starts singing a song and then a little later the second half start singing the same song. The art of composing a good canon is creating a song that, when shifted in time, sits nicely on top of the original tune to harmonize with it. "London's Burning" and "Frère Jacques" are the obvious examples.

The algorithm at work here is quite simple and has a very geometric quality. First of all, create the tune that will be the basis for the canon. Write it out on a musical stave. The algorithm is a rule you apply to this input to produce a piece full of harmony. The way it works is that it takes a copy of the original tune and then repeats the same tune, just shifted a certain number of notes to the right. It's a bit like a frieze pattern on a pot that gets copied, shifted, and repeated. Just as with a pot, you can then shift the tune again, creating a third voice which sings the tune after the first two voices have started.

If one wanted to try to write the canon algorithm as a mathematical formula, then it takes a tune X and then a choice of time delay S and then plays X + SX + SSX. The algorithm creates a harmonized piece with three voices out of a single tune.

A fugue develops this further, with multiple voices and variations on the themes evolving throughout the piece. A more complex rule that Bach enjoyed applying to an original tune was to make the second voice not only shift to the right but also shift up or down, changing the pitch. He also applied rules of symmetry, so that the second voice might play the tune backwards. This is like reflecting the pattern in a mirror. It was by combining many such rules that he could build an algorithm capable of taking nearly any theme presented and developing it on the spot into a piece full of harmony and complexity.

Frederick the Great was impressed by the on-the-spot composition but would not stop there. Now he wondered if Bach could double the voices and improvise a six-part fugue on the same theme. Again, it was a theme devilishly designed to resist Bach's well-honed algorithm. Still, the composer was not going to turn down the challenge without a fight. Six voices would require a bit more thought than simply sitting at the keyboard and improvising, so he went away to see if he could weave together six voices into a coherent fugue. The result was a stunning piece, now called the *Ricercar a 6*, which he delivered to the king two months later.

Together with this fugue he composed ten pieces based on the theme Frederick had tapped out. In each piece he provided a simple tune and a mathematical rule or algorithm for expanding this tune into a harmonized piece. Each offering was presented as a puzzle that the performer would have to crack in order to perform. For example, in one piece he writes a single line of music with an upside-down clef at one end. This upside-down clef is the key to the algorithm that Bach wanted the player to apply to the tune. The algorithm says that you need to take the original piece of music and flip it over and then, at the piano, play the upside-down piece with one hand while simultaneously playing the original tune with the other. The algorithm is the rule that is applied to the original tune to make additional voices in the music. Just as an algorithm for identifying an image in a photo can be applied no matter what photo is presented to it, this musical algorithm would create a piece with whatever tune you gave it.

Each of the ten pieces that make up the beginning of the *Musical Offering* has its own algorithmic tricks to mathematically transform the original theme. The ten pieces act as a warm-up for the extraordinary final fugue, which is a perfect illustration of how Bach could take a simple theme and, by applying simple mathematical algorithms, create a piece of exquisite complexity. The tune shifts in time, plays backwards, climbs higher in pitch, turns upside down. It's a dizzying mixture of rules that Bach combined so skillfully to produce the six-part fugue. Our brain responds to that tension between recognizing a pattern and knowing it is not so simple that we can predict what will happen next. It is that tension between the known and the unknown that excites us. You don't want to run afoul of Stravinsky's criticism that "too many pieces of music finish too long after the end."

Was Bach aware of all the mathematical games he was playing? For me it is clear he knew what he was doing. There are too many examples of mathematical structures that would be hard to put in accidently or even subconsciously. He was a member of the Corresponding Society of the Musical Sciences founded by his student Lorenz Christoph Mizler. The society was dedicated to exploring the connections between science and music and circulated papers with titles like "The Necessity of Mathematics For the Basic Learning of Musical Composition." So Bach was certainly immersed in a world that was interested in the dialogue between mathematics and music.

Bach's son Carl Philipp Emanuel was rather dismissive of his father's fugues, declaring himself "no lover of dry mathematical stuff." To try to prove that there wasn't really much more than musical trickery at work in them, he even prepared a musical parlor game which he entitled "Invention by which Six Measures of Double Counterpoint can be Written without a Knowledge of the Rules." Players were handed two sheets of music. Each page consisted of what looked like a random selection of notes. The first page would be used to write music for the right hand, the descant. The second page would create the bass line for the left hand. All the player had to do was to pick a note at random from which to start the tune and then play the ninth note after that, then the eighteenth note, the twenty-seventh note, and so on until the notes ran out. C. P. E. Bach's skill was in choosing notes so that wherever someone started, by following this simple rule of playing every ninth note, they could build up a piece of acceptable counterpoint without having a clue as to how it was done. Perfect code for a machine!

While the *Musical Offering* is often performed in concerts, I have never heard of any products of his son's *Inventions* being played, suggesting that there may be more to successful musical composition than mechanically following a set of rules.

Mozart, too, is credited with an algorithm, similar to C. P. E. Bach's game, to allow players to manufacture their very own Mozart waltz. His *Musikalisches Würfelspiel*, or musical dice game, is a way of generating a sixteen-bar waltz using a set of dice. It was first published in 1792, a year after Mozart's death. Some have speculated that the game was actually cooked up by the publisher Nikolaus Simrock, who put Mozart's name to it to boost sales.

The game consists of 176 bars which are arranged in an eleven-by-sixteen configuration. The first column gives you a choice of eleven different bars that can start the music. The way you choose which bar to start with is by throwing two dice and subtracting one from the resulting score, giving you some number from one to eleven. So, for example, if I throw a double six, that means I start my piece by playing the bar eleventh down in the first column. The second column controls the second bar and, again, the bar you play is determined by throwing the dice. You proceed through the rest of the sixteen columns throwing the dice each time to determine which of the eleven bars to play.

The staggering thing is that the different waltzes you can generate using this system add up to eleven to the sixteenth power. That is nearly 46 million billion waltzes. Played one after the other they would take two hundred million years to all be heard. The trick of combining an element of randomness with some predetermined structural elements is one that modern algorithmic artists would later use. The genius of Mozart's composition is to produce 176 bars that fit together into a pretty convincing waltz whatever the throw of the dice. Inevitably, not all variations are pleasing to the ear. Some combinations work better than others. This is one of the problems with such open-ended algorithms. The fact that Mozart hasn't curated which waltzes work better than others becomes a great frustration.

Emmy: the AI Composer

I enjoy challenging myself to name the composer of a piece I hear on the radio before finding out who it is. As I listened one morning, working away at my desk, I quickly homed in on Bach as the most likely candidate for the piece that was playing. When it came to an end I had a shock: the presenter revealed that the piece had been created by an algorithm. What shook me was not that I had been duped into thinking it was Bach so much as that, during the short period I'd listened to the musical composition, I was moved by what I'd heard. Could a bit of code really have done that? I was intrigued to find out how the algorithm behind the work had tricked me into thinking the great Bach had composed it.

Bach is the composer most composers begin with, but he is the composer most computers begin with, too. The piece I listened to on the radio that day was generated following simple rules of code cooked up by a composer who had been struggling for inspiration. David Cope first turned to algorithms out of desperation. He'd been commissioned to write a new opera and was procrastinating, unable to commit notes to stave. But then he had an idea. He remembered how Ada Lovelace had speculated that "the engine might compose elaborate and scientific pieces of music of any degree of complexity or extent," and decided to explore her idea.

He started experimenting by feeding punch cards into an IBM computer (this was the early 1980s). Notes appeared as output. His early results, he later admitted, were truly dreadful. But he persevered, heading off to Stanford to do a course in computer music. With the deadline for his commissioned opera fast approaching, he decided to put his computing skills to the test.

If he could get an algorithm to understand his compositional style, then whenever he got stuck and didn't know how to proceed, the algorithm could make a suggestion that would be compatible with his particular way of composing. Even if the algorithm suggested something Cope considered absurd, it would at least get him thinking what might be a better choice. The algorithm would act as catalyst to spur his creativity. Cope called his new concept Experiments in Musical Intelligence, or EMI for short. The alter-ego composer that began to emerge from these algorithmic experiments would later be called Emmy, partly to avoid confusing the project with the British recording label EMI, and partly to give the algorithm a more human name.

Having wrestled for seven years trying to write his opera, with the help of EMI he completed the piece in two weeks. He called it *Cradle Falling*, and decided at that stage not to let on that a computer had helped him compose the piece, so as not to bias the critical reaction to it. After its premiere two years later, in 1987, Cope was amused to see the piece garnering some of the best reviews of his career to date. One critic declared, "It was most moving. 'Cradle Falling' unquestionably is a modern masterpiece." The reaction inspired Cope to continue his collaboration with EMI.

If the algorithm could learn Cope's own compositional style, could it be trained on more traditional composers? Could it for instance take the compositions of Bach or Bartók and write pieces that they might have written? Cope believed that every piece of music had encoded within it instructions to create other pieces that were similar but subtly different. The challenge was to figure out how to crystallize these instructions into code.

He began, with EMI's help, to build up for each composer a database of ingredients that would correspond to their particular style, like the vocabulary and grammar of their own musical language. The notes were the letters—but what words would correspond to the language specific to this particular composer? One of the key concepts behind Cope's analysis was the idea of the signature motif, a sequence of four to twelve notes that can be found in more than one work by the same composer. In Mozart's piano sonatas and concertos, for example, one can find a phrase used over and over called an Alberti bass pattern. It is often in the second line of music and consists of three notes played in a sequence of 13231323.

This pattern would go into the database corresponding to Mozart's style. As he analyzed various composers' work, Cope found Mozart's to be particularly signature-rich. The signatures might occur at different speeds and pitches, but mathematics is very good at spotting the underlying pattern. It's a bit like recognizing that, even as you throw a ball through the air in many different ways, the ball always follows a path described by a parabolic equation.

Cope's analysis revealed strong patterns across every composer's output. From Bach to Mozart, Chopin to Brahms, Gershwin to Joplin, each has particular patterns of notes that they seem to be drawn to. Perhaps this shouldn't be surprising. Why, after listening to a couple of bars on the radio, am I so often able to recognize a composer even if I've never heard the piece before? Just as someone would at a blind wine tasting, I'm picking up key indicators, which in the case of music are patterns of notes. They are like a painter's signature brushstrokes. Some composers, like Bach, went so far as to sign their names in notes. The final fugue of the *Art of Fugue* features the notes B flat, A, C, and B natural, which in German notation are represented by the letters B A C H.

Having cut up the pieces into cells and signatures to form a database for each composer, Cope's algorithm now turned to what he called "recombinance." It is one thing to break a complex structure down into its building blocks and another to find a way to construct new compositions from those pieces. Cope could have chosen a random process like Mozart's game of dice. But a randomly combined set of fragments is unlikely to mirror the emotional tension and release that a composer builds into a composition. So he added another step to his program: he created a heat map for every piece.

As composers put notes, chords, and phrases together, these tend to obey a kind of grammar. Cope came up with an acronym, SPEAC, to stand for the five basic elements from which musical compositions are constructed. If the database is the equivalent of a composer's dictionary, then these are like parts of speech. SPEAC analysis reveals the way in which the composer arranges the words of that dictionary to create meaningful sentences. According to the SPEAC framework, notes, chords, and phrases can function as:

Statements—musical phrases that "simply exist 'as is' with nothing expected beyond iteration."

Preparations—elements that "modify the meanings of statements or other identifiers by standing ahead of them without being independent."

Extensions—ways of prolonging a statement.

Antecedents—phrases that "cause significant implication and require resolution."

Consequents—phrases that resolve antecedents, often using "the same chords or melodic fragments" as statements, but yielding "different implications."

Sometimes classical composers use their grammar unknowingly, but often they have been taught it as part of their formal training. Certain chords create tension; they feel like they need to go somewhere to be resolved. The chords that follow can make you feel you have arrived home or can further crank up the need for resolution. SPEAC helped Cope analyze the ebb and flow of a piece and then to find the patterns across pieces written by the same hand, revealing that composer's particular grammatical tendencies. Here, for example, is Cope's analysis of one of Scriabin's piano pieces:

Once he had established this basic grammar, Cope measured the tension created by the use of certain intervals. If you have an octave interval or a perfect fifth, this does not cause great tension, a fact that is reflected by the mathematics. These are intervals where the frequencies are in small, whole-number ratios. The octave is 1:2. The perfect fifth is 2:3. However, if you have an interval where two notes next to each other on the piano are played together (a semi-tone or minor second) then the notes sound like they are clashing. There is a high degree of tension. Again, the mathematics reflects this: the frequencies are in a ratio corresponding to much bigger numbers (15:16). If you hear these high-tension intervals in a piece of music, they will generally be followed by a move heading toward low-tension resolutions.

These principles were fed into the system to enable it to build a new composition from the large database of a given composer's signatures. EMI takes fragments and hooks them together following certain recombination rules. For example, fragment A can be followed by fragment B if fragment B begins in the same way as fragment A ends, but now heads off in a new direction. The fragments have to match the grammar encoded by Cope's SPEAC analysis.

When many different fragments would fit, a choice needs to be made. Cope is not a fan of random decision-making. He prefers to use a mathematical formula which provides an arbitrary structure to control the choices made, much like the "unaccountable predictability" guiding the Painting Fool. By 1993, Cope and EMI were ready to release a first album of pieces created in the style of Bach. The pieces were quite tricky and he failed to find a human to perform them, so Cope resorted to having them performed by a Disklavier—a real piano equipped with solenoids to depress its keys and pedals. The album was not well received by critics.

"The first three reviews of Bach by Design were negative on the basis that it sounded so stiff," Cope later recalled in an interviewer for the Computer History Museum. "When I read the reviews, I was very upset that they weren't primarily about how [the pieces] were composed, they were primarily about how they were performed." But given that the composition hadn't been attacked, he felt emboldened to continue the project. He produced a second album in 1997 of pieces in the styles of other composers he'd analyzed: Beethoven, Chopin, Joplin, Mozart, Rachmaninov, and Stravinsky. This time, the pieces were performed by human musicians. The critical response was much more positive.

The Game: a Musical Turing Test

But could Cope's algorithm produce results that would pass a musical Turing Test? Could they be passed off as actual works by the composers they imitated? To find out, he decided to stage a concert at the University of Oregon. Three pieces would be played. One of these would be an unfamiliar piece by Bach, the second would be composed by EMI in the style of Bach, and the third would be composed by a human, Steve Larson, who taught music theory at the university, again in the style of Bach. The three pieces would be played by Larson's wife, Winifred Kerner, a professional pianist. Presiding over the event was mathematician Douglas Hofstadter, author of the classic *Gödel, Escher, Bach: An Eternal Golden Braid*.

Larson was upset when the audience declared his two-part invention à la Bach to be the one composed by a heartless computer. His disappointment was soon eclipsed, however, by the shocking vote for the algorithmic Bach over the great man himself. The real Bach was judged an imitation!

"I find myself baffled and troubled by EMI," Hofstadter told a *New York Times* reporter as he tried to make sense of the results. "The only comfort I could take at this point comes from realizing that EMI doesn't generate style on its own. It depends on mimicking prior composers. But that is still not all that much comfort. To what extent is music composed of 'riffs,' as jazz people say? If that's mostly the case, then it would mean that, to my absolute devastation, music is much less than I ever thought it was."

Cope went on to present computer-composed music in other venues around the world. The audience's reactions sometimes unnerved him. In Germany, a musicologist was so incensed he threatened Cope following the concert, declaring that he had killed music. The man was quite tall and about a hundred pounds heavier than him,

and Cope felt only the crowd surrounding him protected him from a
dustup. After a 2009 concert at his own university, Cope told the
London *Times*, a music "professor "came to me and said this was one
of the most beautiful pieces he'd heard in a long time." But evidently
that colleague hadn't' realized the music was composed by a computer
algorithm. Some weeks later, Cope gave a lecture and again presented
the piece that had been performed, and the same professor approached
him afterwards, now insisting on how shallow the work was. "From the
minute it started I could tell it was computer-composed,"' he now said.
"'It has no emotion, no guts, no soul." Cope was stunned by the totality of his reversal. The output was the same: the only thing that had
changed was his knowledge of the fact that it had been generated by
computer code.

At another university, when Hofstadter played two pieces, one by
Chopin and the other a Chopin-like piece composed by EMI, the musical theorists and composers in the crowd voted for the piece they
were sure was the real thing, only to have it revealed they were wrong.
One woman sitting in the "theory / comp corner of the audience"
wrote to Cope the next day in admiration: "There was a collective gasp
and an aftermath of what I can only describe as delighted horror. I've
never seen so many theorists and composers shocked out of their smug
complacency in one fell swoop (myself included)! It was truly a thing
of beauty."

Hofstadter himself was genuinely surprised by the Chopinesque
piece produced by EMI: "It was new, it was unmistakably Chopin-like
in spirit, and it was not emotionally empty. I was truly shaken. How
could emotional music be coming out of a program that had never
heard a note, never lived a moment of life, never had any emotions
whatsoever?"

Cope believes his algorithm is so successful because it gets to the
heart of how people write music. "I don't know of a single piece of expressive music that wasn't composed, one way or another, by an algorithm," he told one interviewer. While this may strike listeners as a baffling or even offensive statement, many composers would agree. It is

only those on the outside who daren't admit that their emotional state can be pushed and pulled around by code. "The notion that humans have some kind of mystical connection with their soul or God, and so on, allowing them to produce wholly original ideas (not the result of recombination or formalisms) seems ridiculous to me," Cope said.

This may well be, but I think it is important to recognize that, even if music is more mathematical and coded than we generally acknowledge, that does not rob it of its emotional content. When I speak about the connections between mathematics and music, people sometimes get quite upset, imagining that I am reducing the music they love to something cold and clinical. But this misses my point. It isn't so much that music is like mathematics as that mathematics is like music. The areas of mathematics we enjoy and are drawn to have huge emotional content. Those who appreciate the language of math are pushed and pulled by the twists and turns of a proof just as so many of us are moved when we listen to a piece of music unfolding.

The human code running in our brains has evolved to be hypersensitive to abstract structures underpinning the mess of the natural world. When we listen to music or explore creative mathematics we are exposed to the purest forms of structure, and our bodies respond emotionally to mark the recognition of this structure against the white noise of everyday life.

What accounts for the difference we perceive between a random sequence of notes and a sequence we regard as music? According to the work of Claude Shannon, the father of information theory, part of our response comes down to the fact that a nonrandom sequence has some algorithm at its base that can compress the data while the random sequence does not. Music is distinct from noise by virtue of its underlying algorithms. The question is: Which algorithms will make music that humans feel is worth listening to?

Many people will not give up on the idea that music is at some level an emotional response to life's experiences. These algorithms are all composing in soundproof rooms with no interaction with the world around them. Without embodied experience, one cannot hope to emulate the music of the greats. Hofstadter certainly believes—or maybe hopes—that this is the case: "a 'program' which could produce music as [Chopin and Bach] did would have to wander around the world on its own, fighting its way through the maze of life and feeling every moment of it," he writes in *Gödel, Escher, Bach.* "It would have to understand the joy and loneliness of a chilly night wind, the longing for a cherished hand, the inaccessibility of a distant town, the heartbreak and regeneration after a human death. Therein, and therein only, lie the sources of meaning in music."

But it is the listener whose emotions the music draws out. The role of the listener, viewer, or reader in creating a work of art is often underestimated. Many composers argue that this emotional response emerges from the structure of the music. But you don't program-in emotion. Philip Glass believes emotions are generated spontaneously as a result of the processes he employs in his compositions. "I find that the music almost always has some emotional quality in it," he told music scholar David Cunningham, but "it seems independent of my intentions."

The relationship between music and emotions is one that has long been a source of fascination to composers. Stravinsky, whose compositions are so expressive, was particularly eloquent on the subject. He insisted, in his autobiography, that "music is, by its very nature, essentially powerless to express anything at all, whether a feeling, an attitude of mind, a psychological mood, a phenomenon of nature .... If, as is nearly always the case, music appears to express something, this is only an illusion and not a reality. It is simply an additional attribute which, by tacit and inveterate agreement, we have lent it, thrust upon it, as a label, a convention—in short, an aspect unconsciously or by force of habit, we have come to confuse with its essential being." The emotions belong not to the music but to the listener.

How, then, does music manage to illicit such powerful emotional responses? Perhaps composers have tuned into how the brain encodes certain emotions. The frequencies and notes that elicit emotions may be different for different people, but play a certain sequence of notes in a minor key and most people will agree it summons up sadness. Is that a learned or innate response? If a composer chooses a minor key to capture a mood, that suggests a direct encoding, but music theory has not advanced to the stage where we understand a huge amount about how this encoding works. So composers are probably working in the dark, much as Stravinsky and Glass suggested; they create structure and the emotion emerges from the structure.

Many composers like to set up rules or structures to help them generate musical ideas. Bach enjoyed the puzzle of writing fugues. Schoenberg initiated a whole new school of composition around themes that included all twelve notes of the chromatic scale. Bartok was driven to create works that grew in tandem with the Fibonacci numbers. Messiaen used prime numbers as a framework for his *Quartet for the End of Time*. And Philip Glass eventually emerged from his torturous apprenticeship with Nadia Boulanger to create the additive process from which his distinctive minimalist music emerges.

Stravinsky believed that constraints were key to his creativity. Here is how he expressed it in a famous Harvard lecture series: "My freedom thus consists in my moving about within the narrow frame that I have assigned myself for each one of my undertakings. I shall go even further: my freedom will be so much the greater and more meaningful the more narrowly I limit my field of action and the more I surround myself with obstacles."

My own composition tutor had sent me off on my own small musical journey with a set of rules to assist me. After starting with prolation canons, I went on to create some of my own constraints and came up with a few algorithms to guide me in my composition. Having read that John Cage often composed a piece on the page without really knowing how it would sound until it was played, I was curious to hear what my mathematical reimaginings would sound like.

But when I sat at the piano and sounded out the string trio I'd composed, I was disappointed. The rules I'd followed meant that the piece had an interesting logic to it. It took the listener on a journey. And yet it didn't sound right. I don't really know what that means—and of course it's silly to suggest that music has right or wrong answers like mathematics does—but having been disappointed by the initial results, I began to break the rules I'd set up, to perturb the notes I'd penned on the page to create something that made more musical sense to my ears. I can't really articulate why I made the changes I did, as I allowed myself to be guided by something deeper: the relationship of my physical body to the music, my subconscious, my humanity.

This was an important lesson. Composition is a fusion of rules and patterns and algorithms and something else. That mysterious something else draws on all those things Hofstadter believes we get by wandering around the world. Whatever that is, it was starting to bleed into my creation and giving it life and beauty.

Do these structures need to be informed by an awareness of emotions? If so, how could a computer ever gain this awareness? If music encodes emotions, could that code be used to simulate an emotional state in the computer? Perhaps the twenty thousand lines of code that created EMI is already partway there. If Hofstadter has an emotional reaction to the Chopin produced by EMI, then isn't this really an emotional reaction to twenty thousand lines of code? Hasn't that code captured emotion in just the same way that the notes Chopin wrote on the stave did?

To refer to EMI's musical output as being composed by AI is something of a con. EMI is dependent on having a composer prepare its database. It relies on composers of the past having created sound worlds that it can plunder. Cope, as a composer, has the analytic tools and sensitivity to pick out the elements that correspond to a composer's style, and the skill to figure out how those elements should be recombined. Much of EMI's creativity comes from Cope and from the back catalog of history's musical greats.

Cope built EMI using a top-down coding process: he wrote all of the code to output the music. Now, however, a different approach might be taken. New, more adaptive algorithms, exposed to the raw data of a composer's scores, could use that input to learn musical theory from scratch. There may be no need for the knowledge of how to compose to pass through the filter of human musical analysis. Will machine learning yield classical compositions that rival the greats? The answer, as is so often the case in musical theory, takes us back to Bach.

DeepBach: Recreating the Composer from the Bottom Up

Bach wrote 389 chorales—those hymn-like pieces for four voices that Glass was asked to enhance and that Cope analyzed by hand. His famous *St. John Passion* includes several chorales that punctuate the oratorio. If you are looking for an example of Bach's mathematical obsessions, you will find it here—beginning with his particular fondness for the number 14. Many European thinkers and philosophers during this period were interested in the cabalistic practice of *gematria*, which involved changing letters into numbers and exploring the numerical connections between words to infer deeper connections. Bach must have discovered that when the letters of his surname (BACH) were translated into numbers (2, 1, 3, 8) and added up, the sum was 14. This became his signature number, like the number a footballer wears on his jersey. Bach waited, for example, to become the fourteenth member of the Corresponding Society of the Musical Sciences set up by his student Mizler. He also found interesting ways to introduce the number into his compositions. In the *St. John Passion* we find eleven chorales. If you look at the numbers of bars in each of the first ten chorales they are as follows: 11, 12, 12, 16, 17, 11, 12, 16, 16, 17. The eleventh chorale is an outlier: it has 28 (or 2×14) bars. But now take the preceding chorales in pairs, starting with the first and tenth chorale: $11 + 17 = 28$. The second and ninth chorale $12 + 16 = 28$. If you take the chorales in pairs in this symmetrical manner the bars always add up to 28. A coincidence? Unlikely.

To compose these chorales, Bach often starts with a Lutheran hymn tune that forms the soprano part and then fills in the other parts to harmonize the melody. Cope programmed this harmonizing into his algorithm by hand, based on his analysis of the chorales. He discerned the rules Bach was using to navigate his way through the harmony. Could a computer take the raw data and learn the rules of harmony for itself?

The exercise of harmonizing a chorale is like playing a complex game of Solitaire or doing an open-ended Sudoku puzzle. At each step you have to decide where the tenor voice will move next. Up? Down? How far up or down? How fast? This has to be done while taking into account the ways in which you are moving the other two voices you are weaving, and the whole thing has to underpin the melody.

When you learn to do this as a composition student, your teacher imposes a number of rules. For example, you need to avoid two consecutive octaves or perfect fifths. Using these perfect intervals consecutively weakens the independence of the two voices and degrades the harmonic effect. It's as if one channel has dropped out. This particular ban was introduced as early as 1300 and remains a staple of compositional theory.

Glass recalls in his autobiography how, during one session, his teacher Nadia Boulanger inquired about his health. When he claimed to be fine, she persisted: "Not sick, no headache, no problems at home? . . . Would you like to see a physician or a psychiatrist? It can be arranged very confidentially." As he tried to assure her again, she wheeled round in her chair, pointed at his chorale exercise for the week, and practically screamed: "Then how do you explain this?!" Sure enough, Glass could see hidden fifths lurking between the alto and bass parts he had written.

It is the mark of the creative thinker to break with traditional rules. In AlphaGo we saw this in move thirty-seven of Game Two. Likewise we find Bach getting to the end of a chorale by sometimes breaking the rule of no parallel fifths. Does that make it a bad chorale? As my own tutor Emily explained to me, part of the joy of composing is to break these rules. That's your best chance of achieving Boden's idea of transformational creativity.

Harmonizing a chorale has a two-dimensional quality to it. The harmony has to make sense in a vertical direction, yet the voices sung on their own—in the horizontal direction—also have to have a logic and consonance to them. It is a vexing challenge for human composers to write these chorales and get the two dimensions to coalesce.

So is this something an algorithm driven by machine learning could engage with? Is the secret to Bach's skills decodable from his 389 published chorales? One way to test this proposition would be to do a probability analysis of what note a voice might sing next given what it has just sung. For example, you might see the sequence of notes ABCBA occur several times in various different chorales as part of one of the harmonizing voices. You could then do a statistical analysis of the notes that follow the note A. In one chorale (*Now the Day Has Ended*) the next note descends to a G-sharp. Yet if you take the data from another (*Do Not Be Afraid*) the next note jumps up to an F. By building up a statistical analysis, you could create a game of musical dice with different weightings for different possible notes to continue the phrase. Let's imagine you get eight cases where Bach chooses the G-sharp and four cases where he chooses F. In that case, two out of three times you want the algorithm to go for the G-sharp. It's just like how DeepMind's algorithm learned to play Breakout: Which way should the algorithm move the paddle, and by how much, to win the game? Except the paddle is replaced by a voice singing higher or lower notes.

One challenge with this approach, as Cope discovered when he set out to identify a composer's signature phrases, is to decide how many of the previous notes should condition the next choice. Too few and things can go anywhere. Too many and the sequence will be over-determined and just copy what Bach's done. Then, alongside pitch, you have to factor in rhythm patterns.

Moving from left to right and building the voices out of what has gone before seems to be the most obvious approach, given that this is how we hear music. But it isn't the only way to statistically analyze a piece. DeepBach, an algorithm developed by music student Gaëtan Hadjeres for his PhD thesis under the supervision of François Pachet and Frank Nielsen, seeks to analyze Bach's chorales by taking them outside of time and viewing the chorales as two-dimensional geometric structures. If you remove a piece of the geometrical structure and analyze the surrounding image, you can guess at how Bach might have filled in the rest of the shape. So rather than composing forward in time, it looks at the parts threaded backwards. This is a typical trick in solving a puzzle: start at the end and try to work out how to get there. But one could also take middle sections and ask how Bach filled these in.

This multidimensional analysis led to a chorale that was more structurally coherent than those created by the algorithms that sent the music meandering forward without really knowing where it was heading, only nudged on by what had just happened. Yet the analysis is still really being done on a local level. Hadjeres's algorithm looks at a sphere around each note and tries to fill in the note based on the sphere, but the size of the sphere is constrained. In DeepBach's case, the algorithm considers four beats on either side of a given note.

Hadjeres divided Bach's chorales up into two sets: 80 percent to train the algorithm and 20 percent to use as the test data. Volunteers were then asked to listen to chorales generated by DeepBach alongside real Bach chorales from the test data. They had to say whether they thought the chorale was by the computer or by Bach. Listeners were also surveyed about their musical backgrounds, which would obviously affect the reliability of their assessments. Composing students can hear things that untrained ears might miss.

The results were striking: 50 percent of the time, DeepBach's pieces were judged to have been composed by Bach. Composition students had a slightly better hit rate, but even they failed to identify Deep-Bach's fakes in 45 percent of their tries. This is impressive. Chorales are pretty unforgiving. It takes just one bum note to out one as an impostor. Bach made no mistakes in his compositions, yet 25 percent of his chorales were judged to have been cranked out by a machine! I will say, without wishing to sound snooty, that I find Bach's chorales to be the dullest bits of Bach. Hymn tunes were something he needed to bash out, but they are not the Bach that really moves me. Still, DeepBach's results are very impressive.

One of the key difficulties with any algorithm-training project that tries to learn from the masters is a paucity of data. Three hundred and eighty-nine chorales may seem like a lot, but it's really just barely enough for a computer to learn on. In successful machine-learning environments like computer vision, algorithms train on millions of images. Bach's chorales offer only 389 models, and most composers are far less prolific. They are also especially useful in that they offer very comparable variations on a single phenomenon. Typically a composer's output features so much variety that a machine would be lost trying to learn from it.

Maybe this is what will ultimately protect human-generated art from the advance of the machines. The quantity of good stuff is just too small for machines to learn how to make it. Sure, they can churn out Muzak but the high-quality music is beyond them.

/ 12 /

The Song-Writing Formula

Music expresses that which cannot be put into words
and that which cannot remain silent.

—VICTOR HUGO

I'm a trumpeter but I've never been able to master improvisational
jazz. I have no problem playing sheet music in the orchestra, but
jazz demands that I become a composer. Not just that, but a composer composing on the fly, responding in real time to the musicians
around me. I have always had the greatest admiration for those who
can do that.

Various attempts at learning jazz have taught me that there is a
puzzle element to a good improvisation. Generally a jazz standard has
a set of chords that change over the course of a piece. My task as a trumpeter is to trace a line that fits the chords as they change. But my
choice also has to make sense from note to note, so playing jazz is really
like tracing a line through a two-dimensional maze: the chords determine the permissible moves vertically, and the note just played

determines the moves horizontally. As jazz gets freer, the actual chord progressions become more fluid, so that the trumpeter must also be sensitive to the pianist's possible next move, which will again be determined by the chords played to date. A good improviser listens and knows where the pianist is likely to head next.

To create a machine that can do this does not seem impossible, but there are challenges to overcome beyond the ones that algorithmic composers such as EMI are designed to face. A jazz-improvising algorithm has to play and respond to new material in a real-time interaction.

One of the classic texts on which many young musicians cut their teeth is *The Jazz Theory Book* by Mark Levine, who has played with Dizzy Gillespie and Freddie Hubbard, two of the greatest jazz improvisers of the last century. Levine introduces that book by announcing:

> A great jazz solo consists of:
> 1% magic
> 99% stuff that is
> Explainable
> Analyzable
> Categorizeable
> Doable
> This book is mostly about the 99% stuff.

Note that this same 99 percent all sounds like stuff that can be put into an algorithm. Miles Davis's *Kind of Blue* is my favorite jazz album of all time. I wonder how close we are to creating a *Kind of DeepBlue*?

Pushkin, Poetry, and Probabilities

As a young man, François Pachet fantasized about being the kind of musician who could compose hits and play guitar like his heroes, but despite some fair attempts at writing music, he was eventually seduced into a career in AI. During the years he spent heading up the Sony Computer Science Laboratories in Paris, Pachet discovered, however, that the tools he was learning in AI could also help him compose music. To create the first AI jazz improviser, he used a mathematical formula from probability theory known as the Markov chain.

Markov chains have been bubbling under many of the algorithms we have been considering thus far. They are fundamental tools for a slew of applications, from modeling chemical processes and economic trends to navigating the internet and assessing population dynamics. Intriguingly, when the Russian mathematician Andrey Markov came up with his theory, he chose to test it not on scientific statistics but on Pushkin's poetry.

Markov's discovery emerged out of a dispute with another Russian mathematician, Pavel Nekrasov. One of the central pillars of probability theory is the law of large numbers, which states that if you have a coin, and each toss of that coin is totally independent from previous tosses, as you keep tossing it the numbers of heads and tails will get closer and closer to a fifty-fifty split. Given four tosses, there is a one-in-sixteen chance that all tosses will come up heads. But as you increase the number of tosses, the likelihood of deviating from fifty-fifty decreases.

Nekrasov believed the inverse must also be true: that if the data produced by a set of actions followed the law of large numbers, then it must be true that the outcome of each action was independent of previous outcomes. He used this theory to make a surprising argument: that because crime statistics in Russia obeyed the law of large numbers, the decisions made by criminals to commit crimes must all be independent acts of free will.

Markov was dismayed by this faulty logic. He described Nekrasov's claim as "an abuse of mathematics" and was determined to prove it wrong. He needed to hold up a data set in which the probability of each outcome was affected by what had come before and yet the long-term behavior still obeyed the law of large numbers. Whether a coin comes up heads or tails does not depend on previous tosses, so that was not the model Markov was after. But what about adding a little dependence so that the next event depends on what just happened but not on how the system arrived at the current event. A series of events where the probability of each event depends only on the previous event became known as a Markov chain. Predicting the weather is a possible example. Tomorrow's weather is certainly dependent on today's, but not particularly dependent on what happened last week.

Consider the following model. It can be sunny, cloudy, or rainy. If it is sunny today, there is a 60 percent chance of sun tomorrow, a 30 percent chance of clouds, and a 10 percent chance of rain. But if it is cloudy today, the probabilities are different: there is a 50 percent chance of rain tomorrow, a 30 percent chance it will remain cloudy, and a 20 percent chance of sun. In this model, the weather tomorrow depends only on the weather today. It doesn't matter if we've had two weeks of sun—if it's cloudy today, the model will still give us a 50 percent chance of rain tomorrow. The last part of the model tells us the probability of going from a rainy day today: we have 40 percent chance of a sunny day, 10 percent chance of a cloudy day, and 50 percent chance of another rainy day. Let us record these probabilities in what we call a matrix.

$$
\begin{pmatrix} SS & SC & SR \\ CS & CC & CR \\ RS & RC & RR \end{pmatrix} = \begin{pmatrix} 0.6 & 0.3 & 0.1 \\ 0.2 & 0.3 & 0.5 \\ 0.4 & 0.1 & 0.5 \end{pmatrix}
$$

With this model we can calculate what the probability of rain will be in two days' time from a sunny day. There are several ways to get there, of course, so we need to sum up all of the possible probabilities. It could go SSR (Sunny, Sunny, Rain) SCR (Sunny, Cloudy, Rain) or SRR (Sunny, Rain, Rain).

The Probability of SSR =
Probability of SS × Probability of SR = 0.6 × 0.1 = 0.06
The Probability of SCR =
Probability of SC × Probability of CR = 0.3 × 0.5 = 0.15
The Probability of SRR =
Probability of SR × Probability of RR = 0.1 × 0.5 = 0.05

This means that the probability of rain two days after a sunny day, which we'll denote as $SxS = 0.06 + 0.15 + 0.05$, is 0.26 or a 26 percent chance.

There is a convenient tool for calculating the chance of rain on the second day. It involves multiplying two copies of our probability matrix together.

$$\begin{pmatrix} SxS & SxC & SxR \\ CxS & CxC & CxR \\ RxS & RxC & RxR \end{pmatrix} = \begin{pmatrix} 0.6 & 0.3 & 0.1 \\ 0.2 & 0.3 & 0.5 \\ 0.4 & 0.1 & 0.5 \end{pmatrix}^2$$

Despite this dependence from day to day on the previous day's weather, in the long run, whether we start on a sunny, rainy, or cloudy day, the chance of rain will tend toward the same value (about 32.35 percent). To see this, we can multiply together more and more of our probability matrices and we will find that the entries on each row tend toward the same number. The long-term weather forecast is thus independent of today's weather, even if tomorrow's weather is dependent on it.

$$\begin{pmatrix} 0.6 & 0.3 & 0.1 \\ 0.2 & 0.3 & 0.5 \\ 0.4 & 0.1 & 0.5 \end{pmatrix}^{10} = \begin{pmatrix} 0.4412 & 0.2353 & 0.3235 \\ 0.4412 & 0.2353 & 0.3235 \\ 0.4412 & 0.2353 & 0.3235 \end{pmatrix}$$

Each row of this matrix represents the chance of a sunny, cloudy, or rainy day after ten days. We can see now that it doesn't matter what today's weather is (that is, which row we choose): the probability on the tenth day will always be the same. Markov had devised a proof that showed conclusively that Nekrasov's belief that long-term crime statistics implied the exercise of free will was flawed.

Markov decided to illustrate his model with the help of one of Russia's most cherished poems, Pushkin's *Eugene Onegin*. He had no intent to provide new literary insights into the poem; he simply wanted to use it as a data set to analyze the occurrence of vowels and consonants. He took the first twenty thousand letters, about an eighth of the poem, and counted how many were vowels versus consonants. A computer could have done this in a flash but Markov sat down and did the count by hand. He eventually determined that 43 percent were vowels and 57 percent consonants. If you were to pluck out a letter at random, you would therefore be better off guessing it was a consonant. What he wanted to figure out was whether your best guess would change if you knew what preceded that letter in the poem. In other words, does the chance that the next letter will be a consonant depend on whether the previous letter is a consonant?

Analyzing the text, Markov found that 34 percent of the time, a consonant was followed by another consonant, while 66 percent of the time it was followed by a vowel. Knowledge of the previous letter thus changed the chances of a given outcome. This is not unexpected: most words tend to alternate between consonants and vowels. The chance of a vowel following a vowel, he calculated, was only 13 percent. *Eugene Onegin* therefore provided a perfect example of a Markov chain to help him explain his ideas.

Models of this sort are sometimes called models with amnesia: they forget what has happened and depend on the present to make their predictions. Sometimes a model may be improved by considering how the two previous states might affect the next state. (Knowledge of the two previous letters in Pushkin's poems might help sharpen your chances of guessing the next letter.) But at some point this dependence disappears.

The Continuator:
The First AI Jazz Improviser

Pachet decided to replace Pushkin with Parker. His idea was to take the riffs of a jazz musician and, given a note, to analyze the probability of the next note. Let's imagine one of that musician's riffs was just an ascent up a scale and then a descent back down it. If one of those notes were randomly chosen, the next note would either be one up or one down the scale, and the chances between those are fifty-fifty. Based on this input, an algorithm would do a random walk up and down the scale. But the more riffs it was given to train on, the more data the algorithm would be able to analyze and the more a particular style of playing would emerge. Pachet figured out that it wasn't enough to look one note back—it might take a few notes to know where to go next. But he didn't want the algorithm to simply reproduce the training data, so it was no good going too far back.

An advantage of Pachet's approach is that the data can be fed in live. Someone can riff away on the piano. The algorithm statistically analyzes what they are up to and the moment they stop, it takes over and continues to play in the same style. This form of question and response is common in jazz, so the algorithm can jam with a live musician, handing the melody back and forth. Pachet's algorithm became known as the Continuator, as it continues in the style of the person feeding it training data.

After each note, the Continuator calculates where to go next based on what it has just played and what the training data tells it are the most frequently occurring next notes. Then it tosses a coin and makes a choice. In another version of the algorithm, which Pachet calls the collaborator mode (rather than question and answer), a human plays a melody and the Continuator uses its calculus of probabilities to guess at the right chord to play, much as a human accompanist would.

What do jazz musicians who have played with the algorithm think of the result? Bernard Lubat, a contemporary jazz pianist who tried out the Continuator, admitted to being quite impressed: "The system shows me ideas I could have developed, but that would have taken me years to actually develop. It is years ahead of me, yet everything it plays is unquestionably me."

The Continuator had learned to play in Lubat's sound world but rather than simply throwing stuff back that he had done before, it was exploring new territory. Here was an algorithm that was demonstrating exploratory creativity. Beyond this, it was pushing the artist on whose work it had trained to be more creative by showing him possibilities he had not tried before.

This seems to me like the moment when the Lovelace Test got passed. It is the musical version of move thirty-seven in Game Two of

AlphaGo's contest with Lee Sedol. The algorithm produced a result that surprised its programmers as well as the musician it learned from. And the result isn't just a new and surprising bit of music. Lubat said the algorithm helped him to be more creative. The extraordinary value of the output was that it showed him new approaches to try.

We all tend to get stuck in our ways. The Continuator was initiating new note sequences and effectively saying, "Hey, you know you can do this, too?" Sometimes it was a stretch. "Because the system plays things at the border of the humanly possible," Lubat explained, "especially with the long but catchy melodic phrases played with an incredible tempo, the very notion of virtuosity is challenged."

Lubat felt he was physically constrained in a way that the Continuator was not, and that this made it possible for the Continuator to be more innovative than he had been. Often, lack of embodiment hinders computer creativity, but in this case we see the opposite. The fact that machines can do things so much faster and process so much more data than humans may result in an interesting tension between human creativity and AI creativity. This is the dynamic suggested by *Her*, a movie about a man who falls in love with his AI. After many hours talking to him, the AI complains about how slow interactions are with humans and she ultimately leaves her human lover for more rewarding relationships with other AI that can interact at the speed of her CPU. Maybe the Continuator will start to produce sounds that only other machines will be able to appreciate given their complexity and speed.

In the meantime, the Continuator elicited interesting emotional responses from its audiences. In live performances with Lubat jamming alongside it, Pachet reported that "audience reactions were amazement, astonishment, and often a compulsion to play with the system." Pachet decided to put the Continuator through a jazz version of the Turing Test. He got two jazz critics to listen to jazz pianist Albert Van Veenendaal improvising in a call-and-response mode with the Continuator. Both critics found it very difficult to distinguish one from the other, and concluded that the Continuator was more likely to be the human jazz musician, as it was pushing the limits of the genre in more interesting ways.

The Continuator has broken down boundaries and done remarkable things, but systems based on Markov chains have certain built-in limitations. Although it produced musical riffs that locally made sense and were even quite surprising, its compositions were ultimately unsatisfying because they didn't have global structure or what we might call composition. Pachet realized he would have to constrain the evolution of the melody if it was going to have a more interesting story to tell. In question and response, you often want the response to end where the question started, but you want the melody ultimately to realize some resolution of tension. To do this within the parameters of the Markov model was going to be like squaring a circle. Pachet decided he would have to find a new way to combine the freedom of the Markov process with the constraints that would lead to a more structured composition.

The Flow Machine

Many artists and performers claim that when they are totally engaged in their art they lose all sense of time and place. Some call this "being in the zone." More recently it has come to be known as "flow," a term first identified as a psychological state of mind by the Hungarian psychologist Mihaly Csikszentmihalyi in 1990. Pachet decided he would try to create an algorithm that would help get creative artists into a state of flow.

Flow is achieved at the meeting point of extreme skill and great challenge. Without either one of these two, you slip into one of the other psychological states identified in the diagram below. If you don't have the skills and you try something too challenging, then you end up in a state of anxiety. If something is too easy for you given your skill set, then you are likely to border on boredom.

The algorithm at the heart of Pachet's Flow Machine uses the Markov processes to learn the style of an artist and then provides certain constraints. This is how many creative artists work. Picasso spent years absorbing the work of el Greco, Renoir, Velasquez, and Manet, imitating, combining, and adapting their styles. By passing that work through different sets of constraints he imposed, he was able to arrive

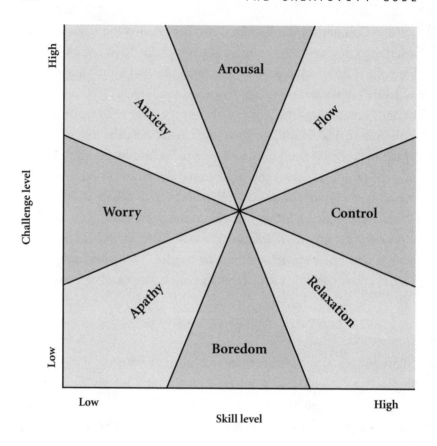

at a style that was uniquely his own and yet rooted in the masters of the past.

Pachet experimented with getting his algorithm to play in one style while taking constraints from another. This was a fantastic example of algorithmic experimentation with Boden's concept of combinatorial creativity. In one instance, he took Charlie Parker's style of the blues and combined it with the constraints offered by the serial world of Pierre Boulez, who, as a fan of Schoenberg, made sure that every piece he composed used all twelve notes of the chromatic scale. This forces the blues to cycle through all twelve notes, an unusual constraint as the blues usually use only three notes as a foundation for its progressions. The result is a strange beast that clearly emerged from the bebop world of Charlie Parker, but lives way out on the fringes of that world. I must admit to enjoying the fusion. Others didn't work so well—like John Coltrane playing "Giant Steps" with the chord constraints of Richard Wagner!

The Flow Machine isn't limited to music. You can get it to learn the style of one poet or lyricist and apply the constraints of another. For example, Pachet's team got its Markov models to capture the style of Bob Dylan's lyrics and then applied the lyrics of "Yesterday" by the Beatles. The Beatles' lyrics provided certain constraints of meter and rhythm. The Flow Machine was then tasked with filling in this framework with a choice of words and phrases that would be recognizably Dylan. Here is the result, to be sung to the tune of "Yesterday":

Innocence of a story I could leave today
When I go down on my hands and pray
She knocked upon it anyway

Paradise in the dark side of love it is a sin
And I am getting weary looking in
Their promises of paradise
Now I want to know you would be spared this day
Wind is blowing in the light in your alleyway

Innocence in the wind it whispers to the day
Out the door but I could leave today
She knocked upon it anyway

Pachet used his Flow Machine to create what has been trumpeted as the first pop song written by AI. Finally his years of research had allowed him to realize his childhood dream. "Daddy's Car" is written in the style of Pachet's favorite band, the Beatles. Many musical analysts have argued that there is a formula to the music of the Beatles, and Pachet hoped to crack their code. But his lyrics were not actually generated by algorithms. They were the work of Benoît Carré, who was also tasked with turning the output of the algorithm into a fully-produced track.

"Daddy's Car" was followed by the album *Hello World*, released early in 2018. The title is a reference to the first exercise anyone learning to code is tasked with: create a program that outputs the text "Hello World." The album is a collaboration between Carré and other musicians who have used the Flow Machine to push the boundaries of their own creativity. To say that it is the first album produced by AI is not quite accurate, as Carré and his collaborators played an important part in determining the contours of the final product.

The result? Composer Fatima Al Qadiri dismissively quipped that "it sounds like a song that has been Xeroxed fifty times and then played."

But not everyone was so negative. Pachet was poached from Sony Labs in mid-2017 and is now working for streaming music service Spotify. Given the rumors that were circulating just then that Spotify was creating playlists full of songs by "fake" artists, the move was an interesting one. Music critics had spotted a number of artists who were notching up extraordinary numbers of streams thanks to their inclusion on popular playlists curated by Spotify for meditation or running. A band called Deep Watch had recorded 4.5 million plays over a five-month period.

When critics tried to find out who these artists were, they came up empty-handed. No presence anywhere else on the internet. No live concerts. No details of any such band. An article was published accusing Spotify of making up artists to avoid paying royalties to real ones. Spotify hit back: "we do not and never have created 'fake' artists and put them on Spotify playlists. Categorically untrue, full stop." But there continued to be speculation in the media that they were specifically commissioning minor artists to create songs under fake names at royalty rates that were much more favorable to the company than its standard deals with record labels.

The fact that it was possible for these composers to knock out a steady supply of good-enough pop songs is a result of the incredibly formulaic nature of the genre. Unlike the subtleties of many classical compositions, so many pop songs just replicate tried-and-tested for-

mats that don't challenge expectations. Most have four beats to the bar.
The tune comes in chunks of four or eight bars with melodic fragments
that are repeated over and over so that people can quickly sing along.
And the song never changes key. Of course, there are exciting moments
when songs break the rules but often those just create new formulas
that then get repeated over and over.

Will the hiring of Pachet by Spotify up the game and result in even
these artists being put out of a job? Algorithms are already curating our
listening. How long will it be before the songs we listen to are created
individually for us by algorithms? Spotify will no longer need to pay
any royalties at all—just Pachet's salary.

If you want your very own piece of AI-generated music now, you
can visit Jukedeck's website. Set up by two Cambridge graduates who
first met as choirboys at the age of eight, Jukedeck is one of a number
of companies using AI to generate songs for organizations, from
London's Natural History Museum to the Coca-Cola Company. The
client typically needs an original but cheap background track for a
promotional video. It doesn't have to be the best track ever. The clients don't want to have to pay exorbitant royalties. The tracks generated by Jukedeck are perfect aural filler for their video content.

The website allows you to choose different genres of music from
folk to ambient, from corporate (is that a genre?) to synth pop. Then
you can tell it if you want your music to be aggressive or melancholic
or any of eight other moods. Once you've chosen and clicked the
algorithm churns out ninety seconds of music and even names the track
for you.

My choice of cinematic moody music produced a track called
"Impossible Doubts." It isn't something I'll be listening to regularly but
that's not the point. The phrase "good enough" is one that is bandied
around a lot in the AI music-generation scene. Jukedeck is aiming at
the market for background music for video production or game development, not to compete with Adele. With an algorithm that can
react to mood, it is a perfect tool to follow the trajectory of a player
navigating a role-playing game. If I have a use for "Impossible Doubts,"

I can get a royalty-free license for $0.99 or I can buy the copyright and own the track outright for $199.

Perhaps those dollar signs are an important indicator of the drive behind AI-generated music. Less than artistic considerations, what is driving the AI art revolution is money.

Quantum Composition

One of the curious aspects of artistic creation is the idea that an artist creates one piece that has to appeal to the many different people who will view, read, or listen to it. But the listeners all have different tastes, expectations, and moods. What if you could create art that reversed this idea of one for many and sought instead to create many works for one individual? Our smartphones gather a huge amount of data about us. What if all that information could be used to create a bespoke piece of art just for you?

This is exactly what Massive Attack decided it would do. After releasing nothing new since the album *Heligoland* in 2010, the band chose an innovative way to release four new songs at the beginning of 2016. Fans could access the songs only by downloading a new purpose-built app called *Fantom*. Then came the interesting twist: once you had allowed the app to access your location, time of day, camera view, heartbeat, and twitter feed, an algorithm would decide how to mix the tracks for you.

Massive Attack's algorithm was essentially playing a more sophisticated version of Mozart's game of dice. The original track was broken down into mini-tracks that then became the raw ingredients for the creation of new, personalized tracks. At various points in the course of the new song, choices are made that determine which track will be added next and how it will be mixed. These decisions are guided by data the algorithm gathers from the individual user. If your heart rate is up and you are moving fast and the camera is picking up bright colors, this information will influence the tone and texture of the song you hear.

The art lies in creating a tree of possibilities that will be sufficiently rich and varied but also sufficiently coherent, so that whatever path the algorithm takes, the result appears seamless and natural. What you don't want is total randomness. Mozart carefully curated each bar, offering eleven options, each one of which might make sense as the next bar in the waltz. The overall structure of the waltz set up the rules within which the game could be played. The same is true of Massive Attack's algorithm. You don't want the chorus crashing in during the evolution of the verse.

Rob Thomas, the programmer who helped create the app, rather nicely refers to the result as "quantum composing." In the quantum world, an electron can be in many different places at the same time thanks to something called quantum superposition. It is the act of observation that forces the electron to collapse into just one of its many possible states. The idea, for Thomas, was to create a song that could exist in many possible states. When I decide to listen to the song, the algorithm takes my data and chooses how to collapse Massive Attack's "wave function" into a single song for me.

Thomas is interested in the dialogue between our emotional states and the music we listen to, and how the one can influence the other. "Music is this emotional manipulation tool," he says. "I want to know how I can use these musical tactics to induce an emotional state in the people listening." He is currently exploring the use of AI music in apps dedicated to mindfulness to help induce a meditative state. The idea is that the music reacts to data about the current state of your mind and body in hopes of learning how to manipulate your body into relaxing. Of course, if you wanted to harness the most effective emotional manipulator, Thomas admits, you would really need to create a human being. He laughs as he concedes, "there are much easier and more pleasurable ways to make humans than using AI."

The *Fantom* app depends on the musician's ability to curate the component pieces of songs. But Massive Attack recognizes the power of machine learning to create its tree structure of possible choices in a much more organic manner. The band hopes, with its next release, to let the algorithm create its own versions of tracks using machine learning. Rob Thomas has teamed up with Mick Grierson at Goldsmiths University in London to realize this next step.

Grierson has worked closely with the Icelandic avant-garde band Sigur Rós. He took one of their songs, "Óveður," and extended it into a twenty-four-hour version that never repeats itself and yet retains the sound of the original five-minute track. The twenty-four-hour version was created to accompany a counterclockwise journey along the coast of Iceland that was broadcast on YouTube and Iceland's national television. Part of the new craze for "slow TV," the event began at the eve of the summer solstice on June 20, 2016. The artists traveled the full 1,332 kilometers of Iceland's one main road, Route 1, passing by Europe's largest glacier Vatnajökull, the glacial lagoon, the east fjords, and the desolate black sands of Möðrudalur.

For a human composer to create a twenty-four-hour soundtrack that doesn't repeat itself would be tough and expensive. The software developed by Grierson uses probabilistic tools to generate the track in response to whatever images it accompanies. He has since created a longer version of the song—in fact, one that will play forever, never repeating itself. Long after Massive Attack and Sigur Rós disband, these algorithms will make it possible to keep listening to new versions of their songs for as long as you want.

Brian Eno coined the phrase "generative music" to describe music that is ever-changing, created by a system or algorithm. Eno likes to say it is music that thinks for itself. It's a sort of musical garden, where the composer plants the seeds and the interaction of the algorithm with its environment—a human playing a computer game or just making his or her way through the day—grows the sound garden from these seeds. In a sense, all live performances honor the idea that the journey from score to experience produces something unique each time. Eno was interested in pushing this idea further. His apps, like *Bloom* or *Scape*—or his most recent one, *Reflection*, created with Peter Chilvers—produce endless Eno-like music that is generated by users interacting with the screen on their smartphones. He describes the process of generation as being like watching a river: "it's always the same river, but it's always changing."

Eno has embraced technology in his creations but, like Lovelace, he does not believe the algorithms he works with will ever generate anything more than what their creators put in. He explained to *Wired* that "there's quite a lot of intent in this, and there have already been a lot of aesthetic choices made. When somebody gets this and makes a piece of music with it, they're making a piece of music in collaboration with us."

But machine learning is starting to kick away the Lovelace crutch that human composers are clinging to. In 2016, an algorithm called AIVA was the first machine to have been given the title of composer by the Société des Auteurs, Compositeurs et Éditeurs de Musique (SACEM), a French professional association in charge of artists' rights. Created by two brothers, Pierre and Vincent Barreau, the algorithm has combined machine learning with the scores of Bach, Beethoven, Mozart, and beyond to produce an AI composer that is creating its own unique music. Although AIVA is currently writing theme tunes for computer games, the aim is much loftier: "to make her mark in the timeless history of music." Listening to AIVA's first album—called, appropriately enough, *Genesis*—I don't think Bach and Beethoven have much to worry about yet. But as the title suggests, this is just the beginning of a musical AI revolution.

Why Do We Make Music?

Music has always had an algorithmic quality to it, which means that, of all the art forms, it is the one most threatened by AI's advances. It is the most abstract of all art forms, tapping into structure and pattern, and it is this abstract quality which gives it close ties to mathematics. But this means it inhabits a world in which an algorithm will feel as much at home as a human.

But music is more than just pattern and form. It has to be performed to be brought alive. Humans started making music to accompany certain rituals. Inside the caves whose walls our early ancestors daubed with paint, archaeologists have found evidence of musical instruments. Flutes made from vultures' bones. Animal horns that could be blown like trumpets. Objects attached to strings that, when swung overhead, created eerie, roaring sounds.

Some speculate that these primitive instruments may have been used to communicate, but others believe they were an important component of rituals our early ancestors developed. It seems that the need for rituals is very much a part of the human code. A ritual consists of a sequence of activities involving gestures, words, and objects, performed in a sacred place according to a set sequence or pattern. Often, from the outside, the ritual appears irrational or illogical, but for insiders it offers an important way of binding the group. Music plays an important part in many such rituals. Singing in a choir or playing in a band is a way of uniting disparate conscious experiences. The songs we sing from the stands at sporting events bind us as a crowd against the other team's fans.

That ability of music to bind a group may be what gave *Homo sapiens* an advantage when the species migrated to Europe and encountered Neanderthals. As the composer Malcolm Arnold wrote: "Music is a social act of communication among people, a gesture of friendship, the strongest there is." The Paleolithic flutes found in Germany that date back forty thousand years may have allowed our ancestors to communicate with each other over large distances. It was quickly realized that music was a powerful ingredient in the creation of mind-altering rituals. Repetition can help alter states of consciousness, as witnessed by many shamanic practices. Our brains have natural frequencies that correspond to different mental states. Trance music drums out 120 beats per minute, the best rhythm for inducing hallucination. We know from modern experiments that messing with multiple sensory inputs can cause the mind to have strange out-of-body experiences. A combination of touch and sight, for example, can cause someone to perceive a false limb. This is why, together with those early instruments, spices and herbs have sometimes been found, which would have given the rituals a smell as well as a sound. How can an algorithm that is not embodied ever hope to understand the power of music to change our bodies and alter our minds?

As civilizations evolved, music continued to be part of the ritual world. The great advances in music from Palestrina to Bach to Mozart were often made in a religious context. There is some speculation that the concept of God arose in humans with the emergence of our internal world. With the development of consciousness came the shock of being aware of a voice in your head. That might have been quite frightening. Ritual and music could appease that voice in the head, and the forces of nature that seemed to be a place for the gods.

This all sounds so far from the logical, emotionless world of the computer. Algorithms have certainly learned how to make sounds that move us. "Algoraves" now use algorithms that react to the pulsating crowd to help a DJ curate the sounds that will keep people dancing. DeepBach is composing religious chorales in the style of Bach for church choirs to sing their praise to God. But despite the fact that these algorithms appear to have cracked the musical code, there is nothing stirring inside the machine. These are still our tools, the modern-day digital bullroarers.

/ 13 /

DeepMathematics

It takes two to invent anything.
The one makes up combinations, the other one chooses.

—PAUL VALÉRY

I was sitting next to Demis Hassabis, at one of the Royal Society's meetings to consider what impact machine learning would have on society, when a question came to mind. It was Hassabis's algorithm AlphaGo that had initially launched me into a state of existential crisis about whether the job of being a mathematician would continue to be a human one. Hassabis and I had both recently been honored with one of the highest accolades for a scientist, being named Fellows of the Royal Society. If Hassabis could get an algorithm to 9-dan status in Go, could he get an algorithm to prove a mathematical theorem that might lead to its being elected a Fellow of the Royal Society?

When I turned to Hassabis and threw down this gauntlet, I got something of a surprise. "We're already on the case," he whispered to me. It seems nothing has escaped AI's radar. After that meeting he elaborated, telling me about the team in place trying to train algorithms to learn from the proofs of the past to create the theorems of the future. Hassabis suggested I drop by DeepMind's offices to find out how far along they were.

It was with some trepidation that I set out to explore whether mathematics would soon be yet another casualty of the machine-learning revolution. Although DeepMind was purchased by Google in 2014 for $500 million, Hassabis had been determined that his baby should stay in London and so the offices are part of Google's London campus just next to King's Cross. Walking through the station concourse I spotted a long line of people hoping to have their picture taken next to Harry Potter's famous Platform 9¾. It struck me that if they wanted to experience real magic, they should be heading next door.

The whole Google site has the feel of a modern Oxford College, conceived to provide the environment for scholars to do their deepest thinking. Google employees are offered free food around the clock, while baristas are on hand to fuel their brains with caffeine. There is a ninety-meter running track, free massages, and even cookery classes by Dan Batten, a chef who worked with Jamie Oliver—although, given the free food, this seems to be more for entertainment than nourishment. And when their brains have gone into overdrive, Googlers can conk out in one of the nap pods dotted around the building.

This is all taking place at a temporary home while Google's cutting-edge new headquarters is going up next door. Designed by Danish architect Bjarke Ingels and Thomas Heatherwick, the British designer behind London's 2012 Olympic cauldron, it promises to be an extraordinary building—some have referred to it as a "landscraper"— as long as the London Shard is high. If other Google sites are anything to go by, it will be quite something. The building in Victoria has a room filled with musical instruments for employees to jam on during their downtime. The site in Mountain View, California, has its own bowling alley. The new campus at King's Cross will more than keep up with its rivals, with an Olympic-sized swimming pool and an amazing roof garden for employees to enjoy during breaks from coding, or even as a place to code if so inclined. The garden will be themed around three areas—plateau, gardens, and fields—planted with strawberries, gooseberries, and sage. The opulence of the Google offices is a clear sign that the business of machine learning is booming. But for now I was heading for the tower block at Number 6, Pancras Square.

DeepMind occupies two floors of the current campus. One is dedicated to commercial applications of its work but it was to Floor 6, where research is being done, that I was whisked. The programmers on Floor 6 have an interesting range of projects in their crosshairs. Machine learning is being applied to help navigate the slippery, random world of quantum physics, and projects are bubbling away to infiltrate biology and chemistry. But I was most interested in their math work.

Hassabis had suggested I speak to Oriol Vinyals to find out how far along they were in their effort to generate an original mathematical proof. Originally from Spain, where he studied mathematics as an undergraduate, Vinyals soon knew that his passion was for artificial intelligence. So he headed to California to do his doctorate, where he was picked up by Google Brain and then DeepMind.

I must admit to being both concerned and excited as the elevator door opened and Vinyals greeted me. But I very quickly felt at ease. Like many of the people wandering around the Google campus, Vinyals would easily fit into my department in Oxford. This was not a corporate environment but a place where T-shirt and jeans are acceptable (provided your T-shirt bears some suitably nerdy caption).

We made our way to one of the meeting rooms, all of which were named after scientific pioneers. The room that we found ourselves in was, appropriately enough, Ada Lovelace. Vinyals explained that it isn't just researchers at DeepMind who are involved but also Google researchers around the world. What sort of mathematics were these Googlers exploring? Had they chosen to tackle a theorem in my own world of symmetry? Or to prove something about networks and combinatorics? Or to determine whether variants of Fermat's equations have solutions? Vinyals soon revealed that they were going for a very different angle from the one I had expected, one that felt quite alien to what I think mathematics is about.

The Mathematics of Mizar

The team at DeepMind and Google had decided to focus on a project that began in Poland in the 1970s, called Mizar. The aim of the project was to build up a library of proofs that were written in a formal language a computer could understand and check. The genius behind Mizar was Polish mathematician Andrzej Trybulec, but it was his wife who was responsible for the name. She'd been looking through an astronomical atlas when her husband asked her for a good name for his project and she suggested Mizar, a star in the Big Bear constellation.

Anyone was invited to submit a proof written in this formal language and by the time Trybulec died in 2013, the Mizar Mathematical Library had the largest number of computerized proofs in the world. Some of the proofs had been constructed by humans and rendered in this computer language, but others had been generated by the computer. The project is currently maintained and developed by research groups at Bialystok University in Poland, the University of Alberta in Canada, and Shinshu University in Japan. Interest in the project had waned in recent years and the library had not been not growing fast. Little did they know that DeepMind and Google had set their sights on significantly expanding it.

So far, those who had been working on Mizar over the decades had successfully created a database containing more than fifty thousand theorems. Given that the proofs in the database are written in a language that a computer can understand rather than a human, those involved in the Mizar project have been keen to pick out the theorems that human mathematicians would recognize as some of their all-time favorites. For example, there is a formalized computer proof of the Fundamental Theorem of Algebra, which states that every polynomial of degree n has n solutions in the complex numbers.

It was interesting to see this theorem in there. The human journey went through so many different false proofs, starting in the early seventeenth century and including failed efforts by such eminent mathematicians as Euler, Gauss, and Laplace. The first proof to be recognized as complete was finally made by Jean-Robert Argand in 1806. The gaps in the previous proofs were often quite subtle. It took time for the mistakes to be spotted. But once a proof that a computer could check had been found, there was great confidence in its validity.

The way a computer generates a proof to include in the Mizar Library has something in common with playing a game. It starts with a list of basic axioms about numbers and geometry. It is allowed certain rules of inference. And from there it maps out pathways to new statements, all linked together by a string of inferences. In the game of Go, the board starts empty. The rules of inference are that you are allowed to place a stone on the board (in your turn, as players on the black side and white side alternate) in any position not previously occupied by a stone. The theorems are like endgames you are trying to reach.

This is what the DeepMind team recognized. Proving theorems and playing Go are conceptually related: both involve searching for specific points in a tree of possible outcomes. Each point can branch off in many different directions, and the length of each branch before it reaches its end points can be extremely long. The challenge is how to evaluate which direction one should head off in next to reach a valuable end point: winning the game or proving a theorem.

This model suggests you can unleash a computer and start generating theorems. But this is not so interesting. There would be lots of overlap, as you could reach the same end point in multiple ways. The real question is whether, given a statement or potential end point, you can find a pathway to that statement, a proof. If not, is there a pathway to prove its negation?

When the team at DeepMind and Google started looking at the theorems on Mizar's books, it found that 56 percent had proofs that had required no human involvement. Its aim was to increase this percentage by creating a new theorem-proving algorithm which would use machine learning to train on those proofs that had been successfully generated by the computer. The hope was that the algorithm could learn good strategies for navigating the tree of proofs from the data already in the Mizar Mathematical Library. The paper Vinyals proudly handed to me described how the DeepMind and Google team, using its proof-generating algorithm, had upped the percentage of computer-generated proofs in the library from 56 percent to 59 percent. Although that might not sound remarkable to you, this represents a nontrivial step change by applying these new techniques. This isn't just one extra theorem or one new game won. This is a 3 percent increase in proofs that computers can reach.

In some ways I could see why Vinyals was excited by the progress. It was like an algorithm learning to play jazz, except that instead of making the best choices of notes to play next, it was deciding which logical moves to make. The algorithm had expanded the reach of the computer in a significant way. It had pushed into new territory. In the way other computers are creating new music, this one was generating new theorems.

Yet I must admit, I left the DeepMind offices rather downcast. I should have been elated by such a surge in mathematical progress, but what I had seen was like a mindless machine cranking out mathematical Muzak, not the music of the spheres that gets me excited. There was no judgment being made about the value of these new discoveries, no interest in whether any of them contained surprising revelations. They were just new. They seemed to be missing two-thirds of what goes into a creative act.

A Mathematical Turing Test

Was this really the future? I went back and tried to read some of the proofs in Mizar's Library of my favorite theorems. They left me cold. Actually, they left me confused because they didn't speak to me at all. I could barely navigate their impenetrably formal language. I experienced what most people probably feel when they open one of my papers and see a string of seemingly meaningless symbols. The proofs were written in computer code that allowed the algorithms to formally move from one true statement to the next. This was fine for the computer, but it's not how humans communicate mathematics. For example, here is Mizar's proof that there are infinitely many primes:

> reserve n,p for Nat; theorem Euclid: ex p st p is prime & p > n
> proofset k = n! + 1; n! > 0 by NEWTON:23; then n! >= 0 + 1 by
> NAT1:38; then k >= 1 + 1 by REAL1:55; then consider p such
> that A1: p is prime & p divides k by INT2:48; A2: p <> 0 &
> p > 1 by A1,INT2:def 5; take p; thus p is prime by A1; assume
> p <= n; then p divides n! by A2,NATLAT:16; then p divides 1 by
> A1,NAT1:57; hence contradiction by A2,NAT1:54; end;
> theorem p: p is prime is infinite from Unbounded (Euclid);

Totally impenetrable even for a professional mathematician! It in no way corresponds to the way any human would ever tell the story. The problem, at some level, is a language barrier.

If algorithms can be written to translate Spanish to English, is there a way to translate from computer-speak to the way a human would communicate a proof? This was a proposition that Cambridge mathematicians Timothy Gowers and Mohan Ganesalingam set out to explore. Gowers first hit the headlines when he won a Fields Medal in 1998 and was promptly elected Rouse Ball Professor in the same year. Ganesalingam began by following a similar trajectory, studying mathematics at Trinity College Cambridge, but then, after being selected as Senior Wrangler and receiving one of the top degrees in his year, he decided to shift paths and surprised everyone in his department by getting a Master's degree in Anglo-Saxon English. He won a university prize for the best results in the Cambridge English Faculty that year and went on to do a PhD in computer science, analyzing mathematical language from a formal linguistic point of view. This combination of mathematics and linguistics would soon be put to use. Gowers and Ganesalingam's paths crossed at Trinity and they soon discovered a shared interest in the challenge of the impenetrable nature of

computer language. They decided to team up to create a tool to produce computer proofs that could be read by humans.

To test how good their algorithm was, they tried an experiment on
Gowers's blog. Gowers presented five theorems about metric spaces,
a subject we teach first-year undergraduates, and included three proofs
of each theorem. One was written by a PhD student, one by an undergraduate, and one by their algorithm. So as not to prejudice the results, readers of the blog were not told the origin of the proofs. Gowers
simply asked people to provide their opinions on the quality of the
proofs. They were asked to grade each proof. He wanted to see if anybody would suspect that not all the write-ups were human-generated.
Not one person gave the slightest hint that they did. In a second blog
post he then revealed that one of the proofs had been written by a computer. At this point, he asked his participants to try to identify the
computer proof in each case.

The computer was typically identified by around 50 percent of all
those who voted. Half of these were confident that they were correct,
and half not so confident. A non-negligible percentage of respondents
claimed to be sure that a write-up that was not by the computer *was*
by the computer. It was generally the undergraduate's answer that
was wrongly believed to have been produced by a computer.

So how does a Fields Medal winner feel about computers muscling
in on his patch? In his blog, Gowers writes: "I don't see any in-principle
barriers to computers eventually putting us out of work. That would
be sad, but the route to it could be very exciting as the human interaction gets less and less and the 'boring' parts of the proofs that computers can handle get more and more advanced, freeing us up to think
about the interesting parts."

But it wasn't just the linguistic problem of the Mizar project that
was bugging me. Of those extra 3 percent of theorems that the Deep-
Mind and Google team had managed to generate, were there any that
would surprise me or make me gasp? I began to feel that this whole
project missed the point of doing mathematics. But what exactly is the
point?

The Mathematical Library of Babel

One of my favorite short stories will help me answer that question. "The Library of Babel" by Jorge Luis Borges tells the story of a librarian's quest to navigate the contours of his library. He begins with a description of his place of work: "The universe (which others call the Library) is composed of an indefinite and perhaps infinite number of hexagonal galleries . . . . From any of the hexagons one can see, interminably, the upper and lower floors."

There is nothing other than the Library. This library, of course, is a metaphor for our own library (which we call the universe). As befits a library, this vast beehive of rooms is full of books. The tomes all have the same dimensions. Each book is 410 pages long. Each page has forty lines and each line consists of eighty orthographical symbols, of which there are twenty-five in number.

As the librarian explores the contents of his library, he finds that most of the books are formless and chaotic in nature—but every now and again something interesting appears. He discovers a book with the letters MCV repeated from the first line to the last. In another, the cacophony of letters is interrupted on the penultimate page by the phrase "Oh time thy pyramids," and then continues its meaningless noise.

The challenge the librarian sets himself is to determine whether the Library is in fact infinite and if not, what shape it is. As the story develops, a hypothesis about the Library is made, "that the Library is total and that its shelves register all the possible combinations of the twenty-odd orthographical symbols (a number which, though extremely vast, is not infinite): in other words, all that it is given to express, in all languages. Everything."

The Library contains every book that it is possible to write. Tolstoy's *War and Peace* is somewhere to be found on the shelves. Darwin's *On the Origin of Species*. Tolkien's *Lord of the Rings*, together with translations of all these works into all languages. Even this book is somewhere among the tomes shelved in the Library. (At this point, having only got this far in my writing, how I would dearly love to find that book and spare myself the labor of carving out the rest!)

Given that all the books have the same dimensions, it is possible to count how many books there are. If there are twenty-five symbols (which presumably includes spaces, periods, and commas) then I have twenty-five choices for the first character and twenty-five choices for the second character. That's already $25 \times 25 = 25^2$ possible choices just for the first two characters. There are eighty characters in the first line. Given twenty-five choices for each place, that gives 25^{80} possible first lines.

Now expand this to count the number of possible first pages. We get $(25^{80})^{40} = 25^{80 \times 40}$ different possibilities since there are forty lines on each page. Now we can get the total number of books in the library. This comes to $(25^{80 \times 40})^{410} = 25^{80 \times 40 \times 410}$ possible books. This is a lot of books. Given that there are only 10^{80} atoms in the observable universe, even if each atom were a book, we wouldn't get anywhere near the total number of books in the Library of Babel. But it is still a finite number. We could easily program a computer to systematically generate all the books in a finite amount of time. Admittedly, the current estimate of how much time we've got till the universe decays into a cold, dark place means that this would be a practical impossibility—but let's stay in the realm of theory and continue the story.

"When it was proclaimed that the Library contained all books, the first impression was one of extravagant happiness," Borges writes. But this was followed by "excessive depression," because it was realized that this library that contained everything in fact contained nothing. The fact that makes my library, the Bodleian—which contains Tolstoy, Darwin, and Tolkien, and will contain my book once it's published—different from the Library of Babel is that a human being curated it. For each book, someone deemed its particular combination of letters worthy of a place in the Bodleian as part of our literary universe.

But what if we were to move to the mathematics section, which houses great journals like *The Annals of Mathematics* and *Les Publications mathématiques de l'IHÉS*. What qualifies something to make it into one of the journals on those shelves? I think many people have the impression that this bit of the library aspires to be a mathematical Library of Babel—that the role of the mathematicians through the ages is to document all true statements about numbers and geometry. The irrationality of the square root of 2. A list of the finite simple groups. The formula for the volume of a sphere. The identification of the brachistochrone as the curve of fastest descent.

This is what Mizar was attempting to do. It has a list of mathematical statements and it is trying see whether it can construct the path from the opening axioms to these statements—or to their negation. The qualification for making it into the Mizar database is that there is an established proof of the statement. There are no choices being made based on what the statement means or whether it is exciting enough to share with other mathematicians. It is simply a Library of Babel containing everything that it is possible to prove.

This, for me, goes against the spirit of mathematics. Mathematics is not a list of all the true statements we can discover about numbers. This may come as a shock to most non-mathematicians. Mathematicians, like Borges, are storytellers. Our characters are numbers and geometries. Our narratives are the proofs we create about these characters. And we make choices based on our emotional reactions to these narratives, deciding which ones are worth telling.

Let me quote one of my mathematical heroes, Henri Poincaré, explaining what it means to him to do mathematics: "To create consists precisely in not making useless combinations. Creation is discernment, choice . . . . The sterile combinations do not even present themselves to the mind of the creator."

Is mathematics created or discovered? The reason we feel it is created comes down to that element of choice. Sure, someone else could have come up with it. But the same could be said of Eliot's *The Waste Land* or Beethoven's *Grosse Fuge*. There are so many different ways the notes could have been chosen that we can't imagine anyone else having composed these great works. The surprise, for most people, is that this same freedom exists in mathematics.

Mathematics, as Poincaré so beautifully put it, is about making choices. What then are the criteria for a piece of mathematics making it into the journals? Why is Fermat's Last Theorem regarded as one of the great mathematical opuses of the last century while some equally complicated numerical calculation is seen as mundane and uninteresting? After all, what is so interesting about knowing that the equation $x^n + y^n = z^n$ has no whole number solutions when n is greater than two?

This is where mathematics becomes a creative art and not simply a useful science. It is the narrative of the proof of a theorem that elevates a true statement about numbers to the status of something deserving its place in the pantheon of mathematics. I believe a good proof has many things in common with a great story or a great composition that takes its listeners on a journey of transformation and change.

Mathematical Fables

The best way to give you an idea of the narrative quality of a proof may be to tell you one of these mathematical stories. It is the first proof I encountered when, at age thirteen, I read G. H. Hardy's beautiful book *A Mathematician's Apology*. The novelist Graham Greene put this short volume, written to explain to non-mathematicians what it is that mathematicians do, in a category with the *Notebooks of Henry James*, calling it the best account he had seen of what it is like to be a creative artist.

To explain how a proof works, Hardy chose one by Euclid, probably one of the first proofs in the history of mathematics. The principal characters in this proof are prime numbers—that is, numbers like 3, 7, and 13 that are indivisible. The narrative journey I want to take you on leads to a revelation that there are infinitely many of these characters. If you were to try to list them all, you would be writing forever. I've already shown, earlier in this chapter, the confounding way that Mizar lays out the proof. Now let me tell you the story.

A proof is like the mathematician's travelogue. Euclid gazed out of his mathematical window and spotted this mathematical peak in the distance: the statement that there are infinitely many primes. The challenge for subsequent generations of mathematicians was to find a pathway leading from the familiar territory that had already been charted to this enthralling new destination.

Like Frodo's adventures in *The Lord of the Rings*, traveling from the Shire to Mordor, a proof is a description of a challenging journey. Within the boundaries of the familiar land are the axioms of mathematics, the self-evident truths about numbers together with those propositions that have already been proven. This is the setting for the beginning of the quest. Any departure from this home territory is constrained by the rules of mathematical deduction, which allow only certain steps to be taken through this world. At times the traveler arrives at what appears to be an impasse and has to look for a detour, moving sideways or even backwards to find a way around. Sometimes the impasse endures until some new mathematical character is created, like imaginary numbers or calculus, that allows progress to resume.

The proof is the story of the trek and a map charting the coordinates of that journey. It is the mathematician's log. A successful proof provides a set of signposts to enable any subsequent mathematician to make the same journey. Readers of a great proof experience the same excitement its author felt upon discovering the path that could traverse a seemingly impenetrable forest and deliver them to that distant peak. Very often a proof will not seek to dot every *i* and cross every *t*, just as a story does not present every detail of a character's life. It is a description of the journey and not a reenactment of every step. The arguments mathematicians provide are designed to propel the reader forward. In another of his essays, Hardy described proofs as "*gas*, rhetorical flourishes designed to affect psychology, pictures on the board in the lecture, devices to stimulate the imagination of pupils."

It is a strange aspect of mathematical stories that they often begin with the ending. The challenge is to show how to reach this climax from our current state of the saga. The narrative journey requires some scene setting, mapping out the story so far and providing a brief description of familiar territory. Readers are reminded that the important characteristic of prime numbers is that they are the building blocks of all other numbers. Every number can be built by multiplying prime numbers together. The number 105, for example, is constructed by multiplying $3 \times 5 \times 7$. Or sometimes you need to repeat a prime—for example, $16 = 2 \times 2 \times 2 \times 2$.

From this launching point, we can then begin the journey toward the conclusion that there are infinitely many of these prime suspects. Suppose that this were not the case, and that we could make a finite list of these characters, a dramatis personae. This is a classical narrative device in the mathematician's toolbox. Like *Alice's Adventures in Wonderland* or *The Wizard of Oz*, imagine a world where the opposite of what you are trying to prove is true and let the narrative play out to its absurd conclusion.

Suppose, for a moment, that this dramatis personae consisted of the prime characters 2, 3, 5, 7, 11, and 13. It is not difficult to show why there must be someone missing. Multiply the characters together:

$$2 \times 3 \times 5 \times 7 \times 11 \times 13$$

Now, here comes the moment for me that is like a twist in this short story that leads to a thrilling and unexpected climax. What if I were to add 1 to that number?

$$2 \times 3 \times 5 \times 7 \times 11 \times 13 + 1$$

This new number which I've constructed out of the principal characters must also be built by multiplying primes together. Remember? That was the familiar setting from which we embarked on our journey. So which primes will divide this new number? Well, it can't be any from the set of primes in our dramatis personae. They will leave a remainder of 1. But there must be some primes which divide this number, so this means we must have missed them when we laid out our dramatis personae. It turns out, this new number is built by multiplying two other primes, 59 and 509.

You might suggest that we add these new characters to our dramatis personae but the beauty of this story is that it can be told again, only to reveal that we are still missing characters. The revelation is that any finite list of primes will always be missing some characters. Therefore the primes must be infinite. As mathematicians like to say at the end of their stories, QED.

Tales of the Unexpected

What is important to me about a piece of mathematics is not the final result. Just as music is not about the final chord, it isn't the QED but the journey I've been on to get to that point. It is certainly important to know that there are infinitely many primes, but our satisfaction comes from understanding why. The joy of reading and creating mathematics comes from that thrilling aha moment we experience when all the strands come together to resolve the mystery. It is like the moment of harmonic resolution in a piece of music or the revelation of the culprit in a murder mystery.

That element of surprise is an important aspect of mathematics. Here is mathematician Michael Atiyah describing the qualities he most enjoys in mathematics: "I like to be surprised. The argument that follows a standard path, with few new features, is dull and unexciting. I like the unexpected, a new point of view, a link with other areas, a twist in the tale."

When I am creating a new piece of mathematics, the choices I make are motivated by the same desire to take my readers on an interesting journey full of twists and turns and surprises. I want to tease my audience with the challenge of why two seemingly unconnected characters should have anything to do with one another. And then, as the proof unfolds, I want them to relish the gradual realization or sudden moment of recognition that these two ideas are actually one and the same.

One of my favorite theorems is Fermat's discovery of a very curious feature of certain types of prime numbers. If a given prime number has the property that when divided by four it has a remainder of one, he believed you could always write that prime number as two square numbers added together. For example, 41 is prime and when I divide it by four I am left with a remainder of one. And sure enough, 41 can be written as 25 + 16, which is $5^2 + 4^2$. But can this really be true of all such primes? There are infinitely many primes that, if divided by four, would leave a remainder of one. Why should they have anything to do with square numbers?

My initial reaction upon hearing the opening of this story was disbelief. But as Fermat takes me on the journey of his proof I get huge satisfaction as I begin to see these two contrary ideas, primes and squares, being woven together until they fuse into one. It is like a piece of music with two contrasting themes that are varied and developed in such a way that eventually they fuse into one theme.

A simpler example of this idea can be seen with the following little game, mentioned briefly in Chapter 9. What happens if you add together consecutive odd numbers?

$1 + 3 = 4$
$1 + 3 + 5 = 9$
$1 + 3 + 5 + 7 = 16$
$1 + 3 + 5 + 7 + 9 = 25.$

Notice any pattern? It turns out that if you add n consecutive odd numbers, the sum is equal to the square of n. Why? The proof is captured by the following picture.

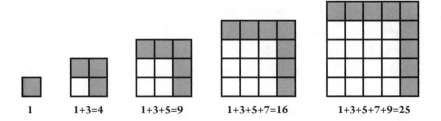

1   1+3=4   1+3+5=9   1+3+5+7=16   1+3+5+7+9=25

234 THE CREATIVITY CODE

The satisfaction comes from the unexpected journey from odd numbers to square numbers. I'm after that aha moment, when I suddenly see *why* there is a connection between these two apparently unrelated characters.

This quality of searching for unexpected connections is one of the reasons I love talking about one of my own contributions to the mathematical canon, the discovery of a new symmetrical object. The contours of such objects have hidden in them potential solutions to elliptic curves, which are among the great unsolved mysteries of mathematics. The proof I weave for my fellow mathematicians in my seminars, as laid out in my journal article, shows how to connect these two disparate areas of the mathematical world.

The joy in telling this story is seeing the moment in my colleagues' faces when they suddenly understand how these two seemingly unconnected ideas could be entwined. The art of the mathematician is not just to churn out the new, but to tell a surprising story. As Poincaré said, it is to make choices.

Just as you sometimes feel a sense of sadness as you turn the last page of a great novel, the closure of a mathematical quest can have its own melancholy. All of us had been so enjoying the journey that Fermat's equations took us on that, when Andrew Wiles solved the last of them, a 350-year-old enigma, there was a sense of disappointment mixed in with the elation. That is why proofs that open up space for new stories are so highly valued.

The Narrative Art of Mathematics

The quality of suspense that we enjoy in mathematical proofs is a classic narrative tool. When authors use plot elements that raise questions, they keep their readers reading on, hoping for the mysteries to be resolved. These elements make up the hermeneutic code identified by Roland Barthes as one of five key codes present in narratives. Intriguing questions and enigmas that demand explication are likewise absolutely central devices in the creation and execution of satisfying mathematical proofs. What gives us such pleasure when we read mathematics is the desire to have the puzzle solved. In this sense, a mathematical proof shares much with a good detective story.

Mathematical proofs all begin with the final scene. The challenge is how we get there. A murder mystery has a similar quality, as does one particularly gripping episode of *Star Trek: The Next Generation* called "Cause and Effect." It opens with the starship *Enterprise* in flames. Picard orders the ship to be abandoned and then we see it exploding. The story then moves to flashback; the beginning, it becomes clear, was the ending. Most literary narratives don't go for such a dramatic opening, but many are littered throughout with examples of this kind of reverse engineering of cause and effect.

In addition to the tension created by unanswered questions, the other narrative drive in mathematics comes from the action inherent in the proof as it unfolds. In Euclid's proof that there are infinitely many primes, we read that these primes are multiplied together. At once our interest is piqued: OK, so where is this going? What will he do with this new number? The action builds. Oh, now he's added a one—even more curious a move. This is a good example of the second of Barthes's five codes of narrative, the proairetic code. Suspense is built by any action that implies further action to follow.

Barthes's other three codes are the semantic, symbolic, and cultural codes. All three dance around the notion that certain elements in a narrative will resonate with things outside the narrative to give it added meaning. And all three are also useful tools to build mathematical proofs, where the reader's preexisting knowledge is tapped to give the proof the full desired effect. Just as G. H. Hardy spoke of stimulating the pupil's imagination, sometimes a proof needs to trigger memories and tap into a vast history of exposure to other ideas to be really effective. Someone who fails to respond to these triggers or recognize the references will not get much out of the proof, any more than they would from a literary narrative.

We often talk about overarching narratives, common to many stories. Some call them master plots or narrative archetypes. Literary theorists have tried to classify these archetypes; it has been suggested that there are really only seven different types of story. We speak of the Cinderella story or the quest narrative or the battle saga. Does mathematics have its own master plots? Certainly, mathematicians recognize certain proof archetypes and will invoke them to help their reader. There is the proof by contradiction. The probabilistic proof or the proof by induction. In the case of Fermat's Last Theorem, the proof depended on creating a world in which the opposite of what the conjecture stated was true. If we pursued a solution to Fermat's proposed equation, where would that take us? The absurd conclusion to this pursuit helps us see that there can't be such a solution.

There is a tension in the best mathematics. Proofs should be neither too complicated nor too simple. The most satisfying proofs have an inevitability about them, yet the steps along the way were far from obvious. The way that literary critic John Cawelti describes this tension in literature, in his book *Adventure, Mystery, and Romance,* applies to mathematics too: "If we seek order and security, the result is likely to be boredom and sameness. But rejecting order for the sake of change and novelty brings danger and uncertainty. . . . Many central aspects of the history of culture can be interpreted as a dynamic tension between these two basic impulses . . . . between the quest for order and the flight from ennui."

That tension is at the heart of what makes a good proof.

Few professional mathematicians have heard of the Mizar project. Its aim is not one that really interests them. It's building a Library of Babel with everything and nothing inside. And yet I believe machine learning holds a promise that remains untapped. Won't it one day be able to take the mathematics we like and learn to create similar mathematics? Have we only been given a temporary stay of execution?

Music is the creative art most people associate with mathematics. But I think storytelling is actually the closest creative act to proving theorems. So I find myself wondering, if mathematical proofs are stories, how good are computers at telling tales?

/ 14 /

Language Games

Two scientists walk into a bar.

"I'll have H_2O," says the first.

"I'll have H_2O, too," says the second.

Bartender gives them water because he is able to distinguish the boundary tones that dictate the grammatical function of homonyms in coda position, as well as pragmatic context.

—JOKE POSTED ON TWITTER

If you are aiming to be a writer, it's important that you understand language, or at least give the illusion of understanding it. But how good are machines at navigating human communication? The opening sentence of Alan Turing's famous paper "Computing Machinery and Intelligence" sets forth a challenge: "I propose to consider the question 'Can machines think?'" This was expressed too generally, Turing then allowed, so he refined his question: he wondered if a machine could be programmed so that, if a human were to engage it in conversation, its responses would be so convincing that the human could not tell he or she was talking to a machine.

Turing called this the Imitation Game, after a parlor game that was popular at the time, but it has since become known as the Turing Test. To pass the Turing Test requires an algorithm that can receive as input the vagaries of natural language and process it to produce an output that corresponds to something a human might possibly say in response. ("Natural language" generally refers to language that, in stark contrast to computer code, has evolved naturally in humans through use and repetition without conscious planning or premeditation.)

The first successful effort to take up Turing's challenge was a program called ELIZA, a script developed by a team led by computer scientist Joseph Weizenbaum in 1966. They came up with a clever yet strikingly simple idea. If you've ever been to a therapist trained in Person-Centered Therapy, developed by Carl Rogers, you'll know that the therapist responds with questions that, rather than leading the conversation in a new direction, simply reflect back what the therapist has just been told. This gives clients the space to further elaborate on their state of mind.

> **Client:** Men are all alike.
> **Therapist:** IN WHAT WAY
> **Client:** They're always bugging us about something or other.
> **Therapist:** CAN YOU THINK OF A SPECIFIC EXAMPLE?
> **Client:** Well, my boyfriend made me come here.
> **Therapist:** YOUR BOYFRIEND MADE YOU COME HERE
> **Client:** He says I'm depressed much of the time.
> **Therapist:** DO YOU THINK YOU ARE DEPRESSED
> MUCH OF THE TIME?

This exchange is typical of a Rogerian line of exploration—except that the therapist, in this case, was not a human but a computer program. Weizenbaum understood that it was possible to capture algorithmically the way that trained therapists would respond in therapy sessions. He put together a set of "if-then" rules that could take a client's input and, by analyzing the structure of the sentence, create an output that depended on that input. For example, if a client said, "I am X," then ELIZA might respond with "How long have you been X?" If X = "sick" then ELIZA's response seems appropriate. But if X = "sick of you," then the response doesn't quite work. So another rule changes all instances of second-person singular pronouns to first-person singular pronouns, and first-person singular pronouns to second-person singular pronouns. Now the response becomes "How long have you been sick of me?"

The trick is to write enough rules to keep the conversation going convincingly. Cleverly, ELIZA was designed such that, if the input didn't match one of the scenarios to which it had been programmed to respond, then it just invited the client to continue by saying, "Tell me more."

Interactions with ELIZA were so convincing that reportedly Weizenbaum's own secretary once asked Weizenbaum to leave the room so that she and ELIZA could have a private conversation. On another occasion, a vice president at BBN, which owned the lab, arrived early one morning hoping to use its computers for a sales demonstration. He found a note on top of a keyboard saying: "TYPE ALL MESSAGES INTO THIS TELETYPE AND END THEM WITH A PERIOD." Believing that he was communicating remotely with the person in charge of the lab he began the following conversation, unaware that he was in fact conversing with ELIZA:

vp: Do you think that I can use the computer this morning?
eliza: Why do you ask?
vp: I might be able to make some additional Telcom sales.
eliza: Why aren't you so sure?
vp: My prospective customers have never seen a
     demonstration of the system.

eliza: Does this mean very much to you?

vp: Of course it does.

eliza: You seem very positive about that.

vp: This is ridiculous.

eliza: What do you mean, this is ridiculous?

vp: Is there any reason why I can't use the computer now.

eliza: Tell me more about your prospective customers.

Exasperated with the evasive answers, he demanded to talk to the operator in person and angrily typed in:

vp: Please dial me up on 491-1850

Because he neglected to end his sentence with a period, ELIZA did not respond. Incensed, he picked up the phone and called Daniel Bobrow, the developer he thought he'd been addressing. "Why are you being so snotty with me?" he demanded to know. Bobrow, having been awakened from his sleep, at first had no idea what the caller was talking about. When his colleague read the conversation back to him, he couldn't stop laughing. As he later recalled, it took some time to convince the man that all those snotty responses had come from a computer.

That said, it typically did not take much conversation with ELIZA to sniff out that something wasn't quite right about her responses. They were limited in scope, they displayed a notable inflexibility, and in any extended exchange, it became clear that ELIZA was not remembering information provided earlier in the conversation. If you are tempted, you can give ELIZA a spin; many online versions still exist on the internet. For some therapeutic purposes, it can work quite well, given that therapy often depends on the patient's own efforts to explore issues and find solutions. It is not likely, however, that you will come away from your encounter convinced that ELIZA is anything but computer software.

In the early 1990s, the announcement of the Loebner Prize brought a new level of excitement to the challenge of creating programs that could be more humanlike in their responses. Underwritten by Hugh Loebner, an inventor and something of an outsider to the mainstream research on AI, this annual competition invites entrants to develop an AI program capable of persuading a panel of judges that it is human. Via computer screen and keyboard, each judge interacts for set periods of time with both a human and a computer and then declares one or the other to be the AI. The rules state that any system that manages to fool at least half its judges will earn for its creator a silver medal and $25,000. In years when this does not happen, a smaller prize is awarded to the system that convinces most judges it is human.

The gold medal and $100,000 is pledged to go to the first program that judges cannot distinguish from a real human in an extended version of the Turing Test that includes not only deciphering and understanding text, but also an audio-visual component. It hasn't been clearly specified how this latter part will be assessed, but presumably judges will get to see and listen to the bot and have still got to be convinced it's human. Once this is achieved, the annual competition will end. Loebner, who died in 2016, liked to point out that the gold medal he was offering was solid gold, in contrast to the gold-plated medals of the Olympic Games.

To date no system has won the silver medal, let alone come anywhere near the gold medal. One of the most recent successes is a chatbot called Mitsuku that goes by the name of Millie and tries to pass itself off as a student from Leeds. It was developed by Steve Worswick, who started programming chatbots as a way to entice visitors to a website that showcased his dance music. It turned out that they much preferred interacting with the chatbot to listening to his music.

In the 2017 competition, which he won, Mitsuku offered some very insightful responses to the judges' questions:

Judge: What do you think of Trump?

Mitsuku: I read that over 50 percent were either concerned or scared about Donald Trump being President. if that's the case, why did they vote for him?

Judge: Anything else?

Mitsuku: That is all I have for right now.

However, other responses quickly revealed strange computing glitches.

Judge: What is the answer to "Add 34957 to 70764"?

Mitsuku: 34957 to 70764 = Ah yes, 34957 is a type of cheese that is made from sheep's milk. It's not very popular though.

A strange end to what had been quite a promising conversation up to that point.

If you have ever interacted with a chatbot, you have effectively been training the bot with your conversations. Some of these chatbot algorithms work by banking responses that humans make to particular questions, so that if the chatbot encounters a similar scenario in the future it can reverse the roles and sound like the human. But most chatbots work on the basis of more sophisticated versions of the ELIZA if-then rules, and these are never really going to be able to handle the varied nature of language. We need something that can grasp how language is put together.

AI systems trying to cope with natural language have difficulty with ambiguity and context. The Loebner competition often concludes with a set of Winograd challenges (named after the Stanford professor who came up with them) which very quickly catch out chatbots that can't untangle built-in ambiguity. Take, for example, the word "they" in the following sentence:

The city councilmen refused the demonstrators a permit because they [feared / advocated] violence.

The choice of feared or advocated clearly changes what the word "they" refers to. While a human will know how to unpick this thanks to context and previous knowledge, machines have an extremely hard time sorting it out. Winograd's sentences exploit the complexity, richness, and ambiguity of natural language.

For example, here are some of the Winograd challenges thrown at Mitsuku in its 2017 Turing Test:

> I was trying to open the lock with the key, but someone had filled the keyhole with chewing gum, and I couldn't get it out. What couldn't I get out?

> The trophy doesn't fit into the brown suitcase because it's too small. What is too small?

How do we develop the skills to navigate the complexities of language? Our human code is shaped and fashioned by years of verbal interaction with other humans. As children, we are exposed to the way language works, we make mistakes, we learn. With the new tools of machine learning, could algorithms finally learn to process natural language? The internet has a huge data set of examples of language in use. So why can't we just let an algorithm loose on the internet to learn for itself how to navigate the ambiguities inherent in these sentences?

Linguists have been struck by how little language a child needs to hear to be able to understand and interact with other humans. Noam Chomsky sees this as evidence that we are born wired for language. It's as if we were programmed in the old-fashioned, top-down model rather than learning from scratch. If that is true, it will be a real challenge for machine learning to pick up language just by being exposed to a huge database of language use.

"This is *Jeopardy!*"

One of the most impressive displays of algorithmic negotiation of the vagaries of natural language came some years ago, just over a decade after IBM's supercomputer Deep Blue successfully took the crown from reigning chess champion Garry Kasparov. In 2011, IBM turned its attention to a form of competition very different from chess or Go: it decided to take a shot at the television game show *Jeopardy!*

Jeopardy! is basically a general knowledge quiz show. Given that a computer can simply trawl through Wikipedia, that might not sound like much of a test for an algorithm. What makes *Jeopardy!* more of a challenge is the style of the questions. They are posed in an inverted manner, where the quizmaster reads something that sounds like the answer to a question and the contestant has to respond with the question. For example, if the prompt is "The name of this element, atomic number twenty-seven, can precede 'blue' and 'green,'" then the right way to respond is "What is cobalt?"

Winning at *Jeopardy!* involves understanding these clues and accessing a huge database of knowledge to select the most likely answers as quickly as possible. The clues very often involve convoluted phrasing, wordplay such as puns, and red herrings, making it tricky even for humans to unpack them. The ambiguous nature of how they are worded makes it almost impossible for an algorithm to be right 100 percent of the time. But IBM didn't need 100 percent accuracy—it just needed the computer to do better than other contestants. While some at IBM thought that trying to win such a trivial game show was a waste of resources, others insisted that success would signal a step change in machines' abilities to understand natural language.

If Kasparov was the champion to beat at chess, the *Jeopardy!* kings were Brad Rutter and Ken Jennings, both of whom had notched up extraordinary winning streaks. Jennings had gone seventy-four games in a row unbeaten, while Rutter had earned over four million dollars during his time on the show. Both had cut their teeth on quiz bowl teams at school and university, although Rutter had always been regarded as something of a slacker academically. *Jeopardy!* generally features three competitors, so both these human champions were invited to take on IBM's algorithm, named Watson—not for Sherlock

Holmes's sidekick but for the longtime CEO who built the company, Thomas J. Watson.

Over two days in January 2011, Rutter and Jennings battled valiantly against Watson and each other. The filming had to be staged at IBM's research lab in Yorktown Heights, New York, because it was impossible to relocate the computer hardware to a TV studio. But other than the location, everything was set up as normal, with host Alex Trebek reading the clues and the shows airing on national television for all to see how close the human race was to being overrun by machines.

The human contestants started off well and pulled ahead at one stage, but in the end couldn't fend off the power of IBM's algorithm. It turned out it wasn't just a matter of being quick with responses. The quiz show involves a certain amount of game theory, as contestants are given opportunities to bet on certain turns. This can allow a contestant who is behind to put any amount of their winnings at risk in hopes of doubling their money and coming out ahead. Some energy was therefore spent to ensure that Watson would wager strategically in such moments.

There is one aspect of the game where Watson threatened to have an obviously unfair advantage. Once a clue is read out, contestants have to hit their buzzers first to have a chance to respond. Originally Watson was going to be allowed to buzz in electronically rather than having to physically press a button like the humans. But it was soon recognized that this would involve no delay for Watson, so a robotic finger was rigged up that Watson had to activate to push the button. Although Watson was still faster on the draw than humans, this slowed it down a bit. As Jennings pointed out: "If you're trying to win on the show, the buzzer is all." The problem was, Watson could "knock out a microsecond-precise buzz every single time with little or no variation. Human reflexes can't compete with computer circuits in this regard." There is also a certain amount of luck involved in *Jeopardy!* thanks to what is called the Daily Double. Watson was fortunate enough to have one of these pop up on one of its turns in the game. Had the human contestants lucked out instead, the game was close enough that Watson might have lost the match.

Despite Watson's win, it did make some telling mistakes. Under the category "US Cities" contestants were presented with: "Its largest airport is named for a World War II hero; its second largest, for a World War II battle." The humans responded correctly with "What is Chicago?" Watson went for Toronto, a city that isn't even in the United States!

"We failed to deeply understand what was going on there," one of Watson's developers explained to the *New York Times*. "The reality is that there's lots of data where the title is US cities and the answers are countries, European cities, people, mayors. Even though it says US cities, we had very little confidence that that's the distinguishing feature." To its credit, Watson's confidence in the answer was very low (indicated by its addition of five question marks after its response). Because this was "final Jeopardy"—the last clue of the first match—it also required a wager. Watson, with a lead that could not be overcome by others' bets, had shrewdly placed a low one.

On the final question of the final match—when it was clear that Watson would triumph again—Jennings scribbled his correct response "Who Is Stoker?" and then added below "I for one welcome our new computer overlords." It was a reference to a popular meme pulled from an episode of *The Simpsons*, itself a spoof of a 1977 B movie of H. G. Wells's "Empire of the Ants" (in which a character capitulates in this way to a takeover by giant insects).

Watson showed no sign of getting the joke.

The Way Watson Works

One way to understand how Watson works is to imagine a huge landscape with words and names and other potential answers scattered about everywhere. The first challenge for IBM was to arrange those words in some coherent manner. The second was to take each clue and produce candidate location markers for it.

Now, this is not a three-dimensional landscape such as the one you see as you look out your window, but a complex mathematical landscape in which each term is situated along multiple dimensions at once, reflecting all the different categories it relates to. For example, a given word might have a certain geographical connection, while also having a chronological association and a connection to the world of art or sport. It might have several of these qualities, in which case its location in the landscape will be pushed in multiple directions. The name Albert Einstein would be pushed in the direction of "scientist," for example, but perhaps also toward "musician," given that he played the violin. It would make sense if he were pushed further along the scientist dimension than the musician one. Analyzing a sample of twenty thousand *Jeopardy!* clues, the IBM team found about twenty-five hundred different dimensions related to them, of which some two hundred covered over half of the sample.

The Watson algorithm goes through four stages of analysis as it plays. First it picks apart the clue to get a fix on where it might lie in the landscape of possible responses. Then it embarks on a process of hypothesis generation, which involves picking some two hundred possible responses based on that identified location. It then scores these different hypotheses. It does this by taking those two hundred points scattered around its multidimensional space and crushing them down to points lying along a single line. Then, finally, it ranks the possible responses and indicates a level of confidence in each one. If the confidence level passes a certain threshold, the algorithm will buzz in with its proposed response. All of this has to be done in a matter of seconds, before the human contestants buzz in.

Consider a clue like this, under the category of "The Hole Truth":

Asian location where a notoriously horrible event took place on the night of June 20, 1756.

It will score high on the geographical and temporal dimensions. But let's say there are multiple Asian locations where something bad happened on June 20, 1756. The word "hole" in the category will help Watson when it comes to scoring different hypotheses. And thus the Black Hole of Calcutta will be ranked higher than any other

Asian location tagged with the same date, giving Watson a winning answer.

The occurrence of a word like *write, compose, pen,* or *publish* will push a clue in the direction of artistic creation. So the prompt "Originally written by Alexander Pushkin as a poem, this Russian novel was later turned into an opera" would send the algorithm into the "authors" region of responses. Once the algorithm selects its two hundred candidates, scoring these requires a careful weighing of the significance of each of the dimensions it has picked up. It has to measure how far each hypothetical response is from the clue. An exact semantic match with a passage in a Wikipedia page would add a lot to a response's score, but even that would have to be combined with other factors. Take this clue: "In 1594 he took a job as a tax collector in Andalusia." Both Thoreau and Cervantes would score highly on a semantic match. But add the temporal dimension and Cervantes scores higher because his dates, 1547–1616, are a good match with 1594 whereas Thoreau was born in 1817.

The team working on Watson came up with fifty different scoring components. The algorithm starts with its wide range of candidate responses because its method is to let the scoring process pick out the top few. It's like finding a hotel to stay in. You begin with all the hotels in the town or neighborhood you want to visit. But then you use a scoring system, and perhaps the weights you place on price and past guests' recommendations send you to an outlying hotel worth visiting.

The way the algorithm does the scoring allows it to learn from its mistakes in a bottom-up fashion and refine its parameters, a bit like twiddling dials. It's trying to find the best settings to get the right answer in as many different contexts as possible. Consider this clue: "Chile shares its longest land border with this country." Two countries share a border with Chile: Argentina and Bolivia. So how might the algorithm score these two hypothetical answers differently? It might score one option higher if it were mentioned more often in all the source material the algorithm scanned. But if it used that method, Bolivia would receive a higher score because Chile and Bolivia have had many border disputes that have spilled over into the news. Tweak the approach to score source material of a more geographical nature higher and to count the mentions of each country in these publications, and Argentina, which is the correct response, would come out on top.

When Jennings was told how Watson worked, he was quite startled. "The computer's techniques for unraveling *Jeopardy!* clues sounded just like mine," he later wrote in *Slate*. Jennings homes in on keywords in a clue and then rakes through his memory for clusters of associations with those words. He then considers the top contenders in light of all of the information the clue provides about, say, gender, date, and place, and whether it relates to sports, literature, or politics. "This is all an instant, intuitive process for a human *Jeopardy!* player, but I felt convinced that under the hood my brain was doing more or less the same thing."

Why did IBM go to all this effort? Winning a game may sound rather frivolous, but for companies like IBM and DeepMind it offers a clear indication of progress. You either win or lose. There is no room for ambiguity. Games provide great publicity stunts for a company that needs to sell products, because everyone loves the drama of human versus machine. They are like algorithmic catwalks allowing companies to show off their fabulous coding.

IBM Watson has already changed our perception of what computers may do—it beat the best *Jeopardy* champions, and it is being used for medical diagnoses. What sets Watson apart? What makes it different? This capability to take into unstructured data is a big strength for Watson. We train it. Additionally just dumping the text in Watson, humans actually form the system to understand what is most important and reliable inside the text. Watson pulled in all of Wikipedia prior to its *Jeopardy,* appearance, and stored that data offline. Humans can tell Watson to trust one source of info more than another. This shift from scheduling to training is part of why IBM calls this effort *Cognitive Computing.*

At the future, we'll rely less on rote calculation, and more on interaction and learning. It is clever enough to know that with a little more info, it'd be capable to rule out an answer, or increase confidence in one of the answers it is already offering. When Watson handles a difficult question in its current applications, it comes back with a set of possible outcomes—but it is also able to ask clarifying questions. Most question answering systems are programmed to deal with a defined set of question types—meaning you can only answer certain kinds of questions, phrased in a certain ways, in order to obtain a response. Watson handles Open domain questions, meaning anything you can think of to ask it. It uses natural language processing techniques to pick apart the words you give it, in order to understand the real question being asked, even when you ask it in an unusual way.

IBM actually published a very useful FAQ about Watson and IBM's DeepQA Project, a foundational technology utilized by Watson in generating hypotheses. The computer on *Star Trek* is a more suitable comparison. The fictional computer system can be seen as an interactive dialogue agent that could answer questions and provide precise info on any subject.

Lost in Translation

I struggled with learning languages at school and still remember reading in *The Hitchhiker's Guide to the Galaxy* about the Babel Fish, a small, yellow, leechlike creature that, when dropped into your ear, would allow you to "instantly understand anything said to you in any form of language." That sounded really useful! As so often happens, yesterday's science fiction has become today's science fact. Google recently launched Pixel Buds, a set of earbuds that, combined with the Google Translate app, delivers more or less what Douglas Adams dreamt about.

When an input is already a well-formed sentence, you might think that the work of navigating language has already been done and a word-for-word exchange will do. But simple word substitutions often result in a baffling word soup. For example, take this quote from *Madame Bovary*:

La parole humaine est comme un chaudron fêlé où nous battons
des mélodies à faire danser les ours, quand on voudrait attendrir
les étoiles.

Taking my French-English dictionary and translating one word at
a time (having to make some choices, as there are different possible
translations for each word) gives me:

The speech human is like a cauldron cracked where we fight
of the melodies to make to dance the bears, when one would like
to tenderize the stars.

Not, I think, what Flaubert had in mind! This is where a sensitivity
to the workings of a given language is essential. Once we see that the
word battons comes close to the word mélodies, we might go for an
alternative translation of battons, not as "fight" but as "beat" and might
even add in "the rhythm." But that still leaves us with the puzzle of what
it might mean to "tenderize the stars."

A good translation algorithm needs to have a good sense of what
words are likely to go together. I remember having great fun with my
best friend at university, who was studying Persian. Looking through
his Persian-English dictionary it seemed like every word had at least
three completely different meanings, one of which was sexual. We
whiled away a lot of time cooking up crazy translations from a single
Persian sentence.

Modern translation algorithms tap into the underlying mathematical shape of a language. It turns out, we can plot words in a language
as points in a high-dimensional geometric space and then draw lines
between words which have structural relationships to one another. For
example, "man is to king as woman is to queen" translates mathematically into the fact that if you draw the lines between these pairs of words
they will be parallel and will point in the same direction. You end up
with a shape that looks like a high-dimensional crystal. The interesting
thing is that French and English have very similarly shaped crystals,
so you just have to figure out how to align them.

I put Flaubert's line from *Madame Bovary* into Google Translate to see how well it would capture its meaning. It got pretty close:

> The human word is like a cracked cauldron where we beat melodies to make the bears dance, when we want to soften the stars.

The word "soften" is certainly better than "tenderize" but it still doesn't quite ring true. Turning to the English translation on my bookshelf (still done by humans—in this case, Margaret Mauldon) I find this:

> Human speech is like a cracked kettle on which we tap crude rhythms for bears to dance to, while we long to make music that will melt the stars.

You realize how important it is not only to choose the right words but to capture the sentiment of the sentence. The algorithmic translators are still tapping out crude rhythms for bears to dance to while humans can translate prose that gets closer to melting the stars. Most of the time, tapping out crude rhythms will be good enough—provided that the meaning, if not the poetry, of the sentence is communicated. As evidence of its success, Google Translate currently supports 103 languages and translates over 140 billion words every day.

But how long will it be before human translators and interpreters are put out of a job—or at least, reduced to fixing glitches in computer translations rather than producing fresh text? My feeling is that these algorithms will never actually reach the status of human translation. Or anyway, not until AI has cracked the problem of consciousness. Translation is more than just moving words from one language to another. It has to move thoughts from one mind to another and, until there is a ghost in the machine, it will not be able to fully tap into the subtlety of human communication.

Looking back over both translations of the *Madame Bovary* line, I actually quite like Google's suggestion of cauldron rather than kettle. And "to make the bears dance" has a slightly more menacing feel than the human translation. Perhaps a combination of human and machine might ultimately yield the best translation.

To get more nuanced translations, Google has enlisted human helpers to improve its algorithm, but this doesn't always lead to better outcomes. Some people can't resist messing with the algorithm, as was illustrated when Google started translating Korean headlines about Kim Jong Un, the leader of North Korea, by referring to him as Mr. Squidward, a comically irritable character from *SpongeBob SquarePants*. Hackers had managed to suggest enough times that "Mr. Squidward" was a better translation than "supreme leader" for the term used by the North Korean media to refer to Kim. They tripped up the algorithm by loading the data with false examples, changing the probabilities. A similar hack occurred when the official title for the Russian Federation was translated into Ukrainian as "Mordor" (the land occupied by the evil Sauron in *The Lord of the Rings*).

Despite these glitches, Google Translate is getting ever more adept at moving from one human language to another. There is even a proposal to map the sound files of animal communications and see if any multidimensional crystals arise that are similar in shape to human communication. Imagine understanding what your pets are saying. Soon we may need a new tool to help us understand the languages emerging from machines—or so I began to think after witnessing an amazing act of linguistic creativity at the Sony Computing Science Laboratories in Paris, where Luc Steels has enabled robots to evolve their very own language.

Robot Lingo

Steels suggested that I come visit his lab, where twenty identical, humanoid robots had been placed one by one in front of a mirror and invited to explore the shapes they could make using their bodies. Each time one came up with a shape, it used a word to label it. For example, a robot might put its left arm in a horizontal position, and would then name that pose. Each of the robots created its own unique vocabulary for its own set of actions.

The amazing part came when all these robots began to interact with one another. One robot would choose a word from its lexicon and ask another robot to perform the action corresponding to that word. Of course, the second robot wouldn't have a clue what it was asking for. So it would just strike one of its poses as a guess. If the guess was correct, the first robot confirmed this. If not, it showed the second robot the intended position.

The second robot might have already named this action for itself, in which case it didn't abandon its label but did update its dictionary to include the new word. As its interactions progressed, the robot assessed the relative value of these alternative words according to how efficient a communication had been, downgrading those associated with failed interactions. The extraordinary thing is that, within a week, a common set of terms had emerged. By continually updating and learning, the robots were developing their own language, and it was sophisticated enough to include words for abstract concepts such as left and right. These words evolved on top of the direct correspondence between words and body positions. That there would be any convergence at all was exciting, but the really striking fact for me was that by the end of a week these robots were communicating in a language that, while they could understand it, was not comprehensible to the researchers—at least, not until they themselves had interacted with the robots enough to decode these new words.

Steels's experiment offered a beautiful proof of how Ada Lovelace was wrong. He had written the code that allowed the robots to generate their own language, but something new had emerged from the code, demonstrated by the fact that no one other than the robots could understand their common language. The only way to learn this language was to become a robot's student, watching as it demonstrated what pose corresponded to each sound.

Google Brain has pushed this ability of algorithms to create their own languages into the realm of cybersecurity, developing new methods of encryption that involve two computers talking to one another without a third being able to eavesdrop. Think of a situation in which Alice must send Bob secret messages knowing that Eve will try to crack them. Alice scores points if Eve can't decrypt her message, and Eve scores points if she can. Alice and Bob start by sharing a number, which is the only thing Eve doesn't have access to. This number is the key to the code they will create. Their task is to use this number to create a secret language that can be decrypted only by someone who knows the key.

Initially, Alice's attempts to mask the messages are easily hacked. But after some fifteen thousand exchanges, only Bob is able to decrypt the messages Alice sends, while Eve scores points at a rate no better than if she were randomly guessing at the messages. It isn't only Eve who is shut out. The neural networks Alice and Bob are using mean that their decisions are very quickly obscured by the constant reparameterizing of the language, so that even by looking at the resulting code it is impossible for humans to unpick what they are communicating. The machines could speak to one another securely without us humans being able to eavesdrop on their private conversations.

Stuck in the Chinese Room

These algorithms that are navigating language, translating from English to Spanish, answering *Jeopardy!* questions, and comprehending narrative raise an interesting question that is important for the whole sphere of AI. At what point should we consider that the algorithm *understands* what it is it is doing? This challenge was captured in a thought experiment created by John Searle called the Chinese Room.

Imagine that you are isolated in a room with an instruction manual which equips you with an appropriate response to any written string of Chinese characters posted into the room. With a sufficiently comprehensive manual, you could have a very convincing discussion with a Mandarin speaker without ever understanding a word.

Searle's point was that a computer programmed to respond with text that we would struggle to distinguish from a human respondent

cannot be assumed to have intelligence or understanding. Embedded
in this line of thought is a powerful challenge to Turing's test. But then
again, what is my mind doing when I'm articulating words? Am I not
at some level following a set of instructions? Might there be a threshold
beyond which we would have to regard the computer as understanding
Mandarin?

And yet, when I refer to a chair I know what I am talking about.
When a computer uses the word *chair* it has no need to know that this
thing chair is a physical object that people sit on. It follows a set of rules
for when the word chair can be used, but following rules does not constitute understanding. Indeed it is impossible for an algorithm that
hasn't experienced a chair to achieve perfect use of the word *chair*. This
is why the concept of embodied intelligence is one that is particularly
relevant to current trends in AI.

One way to think of language is as a low-dimensional projection
of the environment around us. As Franz Kafka said, "All language is but
a poor translation." All physical chairs are different, yet they are compressed into one data point in language. But this data point can be
unpacked by another human into all the chairs he or she has experienced. We can speak of an armchair, bench, wooden chair, or desk chair
and these all bring up different, specific associations. These are the
word games Wittgenstein famously talked about. A computer without
embodiment is stuck in the low-dimensional space of Searle's room.

It comes down to the strange nature of consciousness, which allows
us to integrate all of this information into a single, unified experience.
If we take an individual neuron, there is no understanding of English
in it. Yet, as we add neurons upon neurons, at some point language
understanding is present. When I am sitting in the Chinese Room using
my manual to respond to the incoming Mandarin, I am acting like part
of the brain, a subset of the neurons responsible for language processing. Although I don't understand what I'm saying, maybe it could
be said that the entire system, made up of the room, me, and the
manual, does understand. It's the complete package that makes up the
whole brain, not just me sitting there. In Searle's room, I'm like a com-

puter's CPU, the electronic circuitry that carries out the instructions of a software program by performing the basic calculations.

Could a computer form sentences of meaning—or even beauty—without understanding language or being exposed to the physical world around it? This is a question programmers are grappling with right now in a variety of ways. Maybe a machine doesn't need to understand what it's saying in order to produce convincing literature. And this brings me back to the question that set me off on this excursion into language in the first place: How good is modern AI at taking language and weaving the words together to tell a story?

/ 15 /

Let AI Tell You a Story

A man who wants the truth becomes a scientist; a man who
wants to give free play to his subjectivity may become a writer,
but what should a man do who wants something in between?

—ROBERT MUSIL, *THE MAN WITHOUT QUALITIES*

Some of the stories I grew up on have left lasting impressions. High
on the list are some of Roald Dahl's *Tales of the Unexpected*, including an unnerving account of a man who eats so much royal jelly
he turns into a bee, a story of a tramp tattooed by a famous artist who
sells his skin to the highest bidder, and the tale of an obedient housewife
who, having clubbed her husband with a frozen leg of lamb, proceeds
to serve that murder weapon to the detectives investigating the case.
Another of these disturbing tales, written in 1953, tells the story of the
Great Automatic Grammatizator.

The mechanically-minded Adolph Knipe had always wanted to
be a writer. Alas, his efforts were hackneyed and uninspiring. But

then he had a revelation: language follows the rules of grammar and is basically mathematical in principle. With this insight he set about creating a mammoth machine, the Great Automatic Grammatizator, able to write prize-winning novels based on the works of living authors in fifteen minutes. Knipe blackmails these authors into licensing their names rather than having it revealed that writing a novel is something a machine can do easily and often better. As the story ends, the narrator is wrestling with his conscience: "This very moment, as I sit here listening to the crying of my nine starving children in the other room, I can feel my own hand creeping closer and closer to that golden contract that lies over on the other side of the desk. Give us strength, Oh Lord, to let our children starve."

Roald Dahl died before such a machine was within the realm of possibility, but suddenly it no longer seems such a crazy idea.

One of the very first programs written for a computer was developed to write love letters. After cracking the Enigma code at Bletchley Park, Alan Turing headed to the University of Manchester to put into practice his ideas for a physical version of the all-purpose computer he'd been theorizing about. Under his guidance, the Royal Society Computing Laboratory soon produced the world's first commercially available, general-purpose, electronic computer: the Ferranti Mark 1. It was used to find new primes, wrestle with problems in atomic theory, and explore early genetic programming.

Members of the team were perplexed when they began to find letters of the following ilk lying around the lab:

DUCK DUCK

YOU ARE MY WISTFUL ENCHANTMENT. MY PASSION
CURIOUSLY LONGS FOR YOUR SYMPATHETIC
LONGING. MY SYMPATHY PASSIONATELY IS WEDDED
TO YOUR EAGER AMBITION. MY PRECIOUS CHARM

AVIDLY HUNGERS FOR YOUR COVETOUS ARDOUR.
YOU ARE MY EAGER DEVOTION.

YOURS KEENLY
M. U. C.

MUC was the acronym for Manchester University Computer. Christopher Strachey, an old friend of Turing's from his days at Kings College Cambridge, had decided to see if the Feranti Mark 1 might be able to tap into a more romantic side of its character. He had taken a very basic template:

> "YOU ARE MY [adjective] [noun]. MY [adjective] [noun] [adverb] [verbs] YOUR [adjective] [noun]."

Strachey programmed the computer to select words at random from a data set he had cooked up and insert them into the variables in his simple algorithm. The randomness was achieved using a random number generator that Turing had built for the computer. Anyone receiving more than one or two of these mystifying love letters would soon spot a pattern and deduce that their Valentine was unlikely to sweep them off their feet.

Algorithmically generated literature is not new. A whole school of writers and mathematicians came together in France in the 1960s to use algorithms to generate new writing. The group called itself Oulipo for *Ouvroir de littérature potentielle,* which roughly translates as "workshop for potential literature." Raymond Queneau, one of the founders, believed that constraints were an important part of the creative process. "Inspiration which consists in blind obedience to every impulse is in reality a sort of slavery," he wrote. By imposing quasi-mathematical constraints on writing, he felt you could achieve a new sort of freedom. The group's early projects focused on poetry. As anyone who has written a poem knows, the constraints of poetry will often push you into new ways of expressing ideas that freeform prose would never have unearthed.

One of the group's most popular algorithms, conceived by Jean Lescure, is S + 7 (or, in English, N + 7). The algorithm takes as its input any poem and then acts on all the nouns in the poem by shifting them seven words along in the dictionary. The S stands for substantifs, which is French for nouns. The output is the ensuing rewritten version of the original poem. For example, take the beginning of William Blake's *Auguries of Innocence*:

> To see a World in a Grain of Sand
> And a Heaven in a Wild Flower
> Hold Infinity in the palm of your hand
> And Eternity in an hour . . .

And it becomes:

> To see a Worm in a Grampus of Sandblast
> And a Hebe in a Wild Flu
> Hold Inflow in the palsy of your hangar
> And Ethos in an housefly . . .

Lescure hoped this curious exercise would prompt us to revisit the original text with new eyes and ears. The algorithm changes the nouns but keeps the underlying structure of the sentences, so it perhaps could help reveal structural elements of language masked by the specific meaning of the words.

Queneau, who had studied philosophy and was a member of the Mathematical Society of France, was fascinated by the links between mathematics and creativity. He sought to experiment with different ways to generate new poetry using the tools of math. Shortly before founding Oulipo he had composed a book of sonnets which he called *100,000,000,000,000 Poems*. Ten different versions were proposed for each line. There were thus ten choices for the opening line and ten choices for the second line, making a total of one hundred different possibilities for the first two lines. Given that there are fourteen lines in a sonnet, that brings the number of different poems possible to a total of ten to the fourteenth power. That's one hundred thousand billion new sonnets! Let's imagine that the first diplodocus ever to have evolved during the Jurassic period had started reciting Queneau's sonnets at one a minute and continued doing so to this day. It would have managed to recite all the possibilities by now, but only once.

Queneau had cooked up a literary version of Mozart's game of dice. Chances are the following sonnet, which I picked at random, has never appeared in print before:

> Don Pedro from his shirt has washed the fleas
> His nasal ecstasy beats best Cologne
> His toga rumpled high above his knees
> While sharks to let's say potted shrimps are prone
> Old Galileo's Pisan offerings
> Nought can the mouse's timid nibbling stave
> He's gone to London how the echo rings
> The nicest kids for stickiest toffees crave
> Emboggled minds may puff and blow and guess
> In Indian summers Englishmen drink grog
> And played their mountain croquet jungle chess
> We'll suffocate before the epilogue
> Poor reader smile before your lips go numb
> Fried grilled black pudding's still the world's best yum

As the Oulipo movement illustrates, poetry is particularly amenable to an algorithmic approach. The constrained nature of the form provides a template that the algorithm can try to fill in a meaningful manner. A pattern is chosen, a haiku or a sonnet, and the task of the algorithm is to choose words to match the pattern while attempting to come up with some form of overarching coherence. Whenever I've attempted to write poetry with a rhyming pattern, I've found it useful to tap into a database of words that rhyme. Weaving a line through the constraints of rhyme and rhythm is something a computer can do in spades.

That is the principle behind the code underpinning the Cybernetic Poet, a more recent creation of the futurist Ray Kurzweil, who writes frequently on the impending fusion of man and machine. Rather than relying on words randomly picked out of a dictionary, Kurzweil trained his Cybernetic Poet on the work of venerated poets like Percy Bysshe Shelley and T.S. Eliot. Here is one of the Cybernetic Poet's haikus, informed by a reading of John Keats:

> You broke my soul
> The juice of eternity,
> The spirit of my lips.

Although the poem does indeed have seventeen syllables, the algorithm seems to have missed that a haiku should divide symmetrically into five syllables for the first line, seven for the next, and five for the final line. Here is a poem that recombines Shelley and Eliot:

> Lady of Autumn's being,
> Thou, from the day, having to care
> Teach us now thoroughly small and create,
> And then presume?
> And this, and me,
> And place of the unspoken word, the unread vision in Baiae's
> bay,
> And the posterity of Michelangelo.

Ode to the West Wind meets *The Love Song of J. Alfred Prufrock*.

In a Turing test conducted by Kurzweil, the Cybernetic Poet was able to trick human judges most of the time. This is partly because gnomic outputs are part of the landscape of modern poetry, leaving the reader to do much of the work of interpretation. An enigmatic output from an algorithm can pass for poetry written by a human. (The results and poems Kurzweil used can be found on his website: http://www.kurzweilcyberart.com/.) If you'd like to have a go at distinguishing human poetry from the efforts generated by a range of algorithms, Benjamin Laird and Oscar Schwartz have put together a challenging poetic Turing Test in a project they've called "bot or not" (which you can find at http://botpoet.com).

The Cybernetic Poet might be doing well at producing convincing poetry but creating a cybernetic novelist is a much taller order.

How to Write a Novel in a Month

Lescure's idea of applying algorithms to existing literature is a trick that has been exploited by a number of coders who have taken part in NaNoGenMo, a response to National Novel Writing Month (NaNoWriMo), which invites budding authors to knock out fifty thousand words in the month of November. Software developer and artist Darius Kazemi decided that instead of going to the trouble of cranking out 1,667 words a day, he would spend the month writing code that could generate a fifty-thousand-word novel. His plan was to share both the novel and the code at the end. His tweet about his idea in 2013 started the annual literary hackathon.

Many of the coders who have taken part in NaNoGenMo have relied on perturbing existing texts: *Pride and Prejudice* run through a twitter filter; *Moby-Dick* interpreted through a sci-fi algorithm; Gustavus Hindman Miller's *Ten Thousand Dreams Interpreted* reinterpreted and reordered by code. But it was a more ambitious work called *The Seeker* that caught people's attention. The novel documents an algorithm's struggle to understand how humans operate by reading different articles on *wikiHow*. The protagonist algorithm has a metacode of "work & scan & imagine & repeat & . . ." The author of the code, who goes by the name of thricedotted, tells us what this means:

> The Seeker operates in three modes: *Work, Scan,* and *Imagine.*
> When the Seeker *Works,* it is scraping concepts about human
> activities from WikiHow. In *Scan* mode, it searches plain
> text "memories" for a seed concept it encountered during *Work.*
> It uses the concepts it doesn't recognize from *Scan* mode
> (i.e., the ones which are censored out in its logs) to *Imagine* an
> "unvision" around the seed concept. And so on. And so forth.

The Seeker chronicles the algorithm's journey of discovery as it explores the database of wikiHow, building from ignorance to some semblance of understanding. The first page it consults contains how-to advice on "getting girl to ask you out." The seed picks up from this scan the word "hurt," which is used in cautioning the reader not to hurt the girl's feelings. In its Imagine mode, it then produces a surreal riff on the word "hurt."

The Seeker almost works as a novel, unlike many other algorithmic creations, because we start to feel we are getting inside the head of the algorithm as it tries to make sense of humans. The fact that the output reads like a strange computer code of words is consistent with our expectations about an algorithm's internal voice. This may in fact be the ultimate goal of any algorithmically generated literature: to allow us to understand an emerging consciousness (if it ever does emerge) and how it differs from our own.

But for now, the commercial world would be content with an algorithm that could knock out the next Mills & Boon romance or Dan Brown–style thriller. Many of these bestsellers are based on clear-cut formulae. Couldn't someone simply automate a genre's formula? If algorithms can't produce great works of literature, maybe they could churn out commercial staples like Ken Follett's or even an algorithmic *Fifty Shades of Grey*. An algorithm written by commissioning editor Jodie Archer and data analyst Matthew Jockers does at least claim to spot whether a book is likely to be a bestseller. Their algorithm finds that readers of bestsellers like shorter sentences, voice-driven narratives, and less erudite vocabulary than readers of literary fiction. If only I'd known that before I started!

Harry Potter and the Deathly Botnik

Most of the examples I've pointed to so far rely on a top-down model of programming. A poetry template filled in randomly, following an explicit set of rules. Code that transforms classic texts into new work.

Algorithms that are programmed to take in data and turn it into stories. These programs don't really allow for much freedom.

Machine learning is changing that. It's now possible for an algorithm to take an author's entire oeuvre and learn something about the way that individual writes. If the writer favors a particular word, there might also be high probabilities that certain other words will follow. By building up a probabilistic picture of how an author uses words, an algorithm could start to generate the continuation of a text. This is how predictive texting works. The literary results have been both revealing and entertaining.

This use of machine learning to create new literature has been championed by a group that calls itself Botnik. Founded in 2016 by writer Jamie Brew and former cartoon editor of the *New Yorker* Bob Mankoff, Botnik is now an open community of writers who use technology in the creation of comedy. The group has taken *Seinfeld* scripts and produced new episodes based on a mathematical analysis of past dialogue, and even got *Scrubs* actor Zach Braff to perform a monologue authored by Botnik based on the medical comedy-drama. The result is sometimes surreal. In Botnik's *Seinfeld* episode, Jerry confidently declares: "Dating is the opposite of tuna, salmon is the opposite of everything else. I'm sure you know what I mean."

Botnik has also taken Thanksgiving recipes and produced a YouTube video to take people through the dinner you'd get if you left the cooking to an algorithm:

> The best way to make something really special for thanksgiving
> is to fold the turkey in half and then just throw it right in the
> kitchen.

Probably Botnik's most successful output to date came from training it on the seven volumes of *Harry Potter*. The three pages it generated have a very convincing ring to them.

> Magic: it was something that Harry Potter thought was very
> good. Leathery sheets of rain lashed at Harry's ghost as he
> walked across the grounds towards the castle.

But there are moments of pure genius that could only have come from an algorithm:

> Ron was standing there and doing a kind of frenzied tap dance.
> He saw Harry and immediately began to eat Hermione's family.
> Ron's Ron shirt was just as bad as Ron himself.

I guess for fans who are really desperate for more from the wizarding world, this may be better than nothing, but it's pretty plot-free and unlikely to sustain much drama beyond three pages.

I decided to investigate whether, if I fed Botnik the data of my first book, *The Music of the Primes*, it would provide me with a new insight I might have missed. In response, I got the following strange take:

> The primes are the jewels which shine amongst the vast expanse of our infinite universe of numbers. As he counted higher and higher Gauss suddenly saw a pattern beginning to emerge. His passion for the problem was further fueled when his father offered to buy him a Ferrari. Previously education schemes had been geared to the creation of each list of primes 2, 3, 5, 7, 11, and 13 years, respectively. For all but their last year they remain in the ground feeding on the sap of tree roots.

A bizarre but recognizable mash-up of my thoughts. One of the important things I learned from applying this algorithm is that it calls for significant human involvement in creating texts. Botnik's "predictive writer" operates like the autocomplete feature you see in many applications. It serves up eighteen possibilities for the word you are likely to choose next based on its analysis of the text entered up to that moment. That gave me a lot of freedom to take the text in whatever direction tickled my fancy. Often the human component of artistic creations by algorithms is downplayed. It makes for a better story to say "AI writes new *Harry Potter*!" than "another writing student has produced a new novel."

I think it's fair to say that novelists are not likely to be pushed out of their profession anytime soon. Botnik is capturing the fact that authors do have a style, which is recognizable from the way they construct their sentences. But maybe it is only capturing that: the local evolution of text. It makes no attempt to reproduce a global narrative structure. It is like Pachet's jazz continuator: it can produce a few phrases of convincing jazz but the music ultimately becomes boring as it doesn't know where it is going. I often wonder whether algorithms are already at work at Netflix and Amazon, knocking out scripts that keep us watching but ultimately take us nowhere.

What If . . .

The storytelling algorithm Scheherazade-IF, developed by Mark Riedl and his colleagues at the Georgia Institute of Technology, was set up in 2012 to tackle this deficit. Its goal is to navigate a more coherent pathway through the maze of possible stories. The algorithm owes its name to the famous character in *One Thousand and One Nights*, the storyteller Scheherazade, who saved her life by coming up with new stories, night after night, to enthrall and distract her murderous husband. The IF stands for Interactive Fiction. If you ask Scheherazade-IF to construct a story about a subject or situation it hasn't encountered before, it will first learn about it by sourcing and digesting previous stories.

"Humans are pretty good storytellers and possess a lot of real-world knowledge," says Riedl, one of the lead developers of the algorithm. "Scheherazade-IF treats a crowd of people as a massively distributed knowledge base from which to digest new information." It then compiles these examples into a tree of possible directions in which the story could go based on these previous stories. This kind of skill is really useful when it comes to open-ended computer games, which typically offer many different possible scenarios within the game-play. A good storyteller will find the best route through the tree of possible stories.

This recalls a genre of storytelling I used to love as a kid. In gamebooks that allow you to "choose your own adventure" you are given choices at certain points in the narrative: turn to page thirty-five if you want to go through the left door, or page thirty-nine if you want to go through the right door. The trouble is, your choices will sometimes yield rather incoherent stories. Given that a story with just ten junctions could produce over a thousand different stories, you'd like some way for an algorithm to find the best ones.

Scheherazade-IF tries to do exactly this with the tree of possible scenarios it generates from its data gathering on the web. So how good is it at choosing a satisfying path? In tests by the research team it chose pathways that were rated as being as good as human-chosen pathways and that scored much higher than randomly generated journeys. The algorithm was able to make far fewer logically inconsistent moves than the random compilation process. Logical inconsistency is something that immediately gives away the fact that a piece of writing is generated by an algorithm. It shouldn't be possible for a character killed off in chapter two to suddenly reappear in chapter five (unless it's a zombie story, I guess).

It's all well and good to trawl the web for old stories and put them together afresh, but what about the challenge of imagining scenarios that have not been cooked up before? This was the goal of the What-If Machine (WHIM) project funded by the EU, which starts to show what bizarre things algorithms can throw up. One of the problems authors face when trying to create something new is that they get stuck in bounded ways of thinking. The What-If Machine tries to take storytellers out of their comfort zones by suggesting new possible scenarios.

This is, of course, what we do all the time when we want to create a new story: "What if a horse could fly?" and you've got Pegasus. "What if a portrait of a young man aged while he himself stayed young?" and you've got *The Picture of Dorian Gray*. "What if a girl suddenly found herself in a strange land where animals could talk and everyone was mad?" and you've got *Alice's Adventures in Wonderland*. Many of Roald Dahl's *Tales of the Unexpected*, which I so loved as a kid, exploit the what-if model of creativity.

In fact, storytelling in humans probably has its genesis in the question *what if . . . ?* Storytelling has always been a way of doing safe experiments. By telling a what-if story, we are exploring possible implications of our actions. The first stories probably grew out of our desire to find order in the seeming chaos surrounding us, to find meaning in a universe that could be cruel and senseless. It was an early form of science. Sitting around the fire sharing stories of the day's hunt helped the tribe be more successful the next day. What *homo sapiens* lacked in individual strength they made up for in the collective strength of the tribe. That strength grew with increased socializing and sharing. It appears that the spark of creativity in humans came from the fire of the campsite.

WHIM was designed to ignite creativity around the digital fireside. One of its first ventures grew out of the kernel of the Pegasus story: the idea of a horse that could fly. Could WHIM come up with a similarly curious animal to stimulate a story? It started with a database of animals and all the properties found among them. The *National Geographic Kids* website was a good source of training data. It's a good place to find out that a dolphin is a mammal that lives in the sea and can be ridden by humans. A parrot is a bird that can fly and sing. Once an algorithm starts mixing and matching, you might get a flying mammal that humans can ride and that sings—something that could easily appear in a fairy tale or a volume of *Harry Potter*.

The principle is similar to those picture books that let you turn a third of a page, to put the head of one person on top of the torso and legs of others. Starting with ten characters you can mix up their parts into a thousand different combinations. But if the aim is to produce something useful, you will have to come up with some way to evaluate all the possibilities generated. The team at WHIM introduced mathematical functions that score the suggestions for stimulation and novelty and flag for rejection any ideas that are too vague to be helpful. This led to some interesting suggestions bubbling to the top:

> An animal that has eyes with which it can defend itself
> A tiger with wings
> A bird that lives in a forest that can swim under water

New animals with strange skills are good catalysts for storytelling. The next step was to program WHIM to generate novel narrative ideas. It started by taking a series of what-if story lines that we would immediately recognize and then perturbed the assumptions implicit in these scenarios. The hope was that this would spark creativity by combining topics in surprising and subversive ways. WHIM is programmed to generate narrative suggestions in six fictional categories: Kafkaesque, alternative scenarios, utopian and dystopian, metaphors, musicals, and Disney. The results are varied in their success.

In the Disney section, WHIM came up with a story line that could conceivably find itself in the next *Inside Out*: "What if there was a little atom who lost his neutral charge?" That might be one for the geeks among us. Others of the Disney suggestions border on the dystopian: "What if there was a little plane that couldn't find the airport."

Some story lines were distinctly less promising, like this one in the alternative scenarios category: "What if there was an old refrigerator who couldn't find a house that was solid? But instead, she found a special style of statue that was so aqueous that the old refrigerator didn't want the solid house anymore." Or this Kafkaesque idea: "What if a bicycle appeared in a dog pound, and suddenly became a dog that could drive an automobile."

The What-If Machine suggested one story line that eventually led to the staging of a West End musical in 2016. The TV channel Sky Arts, interested to probe the limits of algorithmic creativity, had commissioned a musical created by AI. It filmed the process of development and eventually staged it. To come up with a scenario for the musical, WHIM was brought on board. The algorithm came up with a range of different scenarios, which were then passed through another algorithm developed in Cambridge. This second algorithm had analyzed the story lines of musicals to learn what makes a hit and what flops, and it was tasked with choosing one of WHIM's suggestions for further development. It picked out the following as a potential hit:

What if there were a wounded soldier who had to learn how to understand a child in order to find true love?

At this point, another algorithm, PropperWryter, which had had some success in generating fairy tales, took over. Its fairy tale algorithm was trained on the thirty-one narrative archetypes of Russian folktales identified in 1928 by structuralist Vladimir Propp. Using the scenario provided by WHIM, PropperWryter developed a plot set within the Greenham Common women's anti-nuclear protest of the 1980s. The music was provided by yet another algorithm, called Android Lloyd Webber.

Beyond the Fence hit the West End for a short run at the Art Theatre in the spring of 2016. To make the play workable took probably as much human intervention as computer creativity. The result was not much of a threat to Andrew Lloyd Webber. *The Guardian's* Lyn Gardner summed it up in her two-star review: "a dated middle-of-the-road show full of pleasant middle-of-the-road songs, along with a risibly stereotypical scenario and characters." But then, maybe what we should really take away from this is that reviewers aren't particularly disposed to give algorithms much credit.

The Great Automatic Mathematizator

Asking "what if" is not far from the way a mathematician pushes the boundaries of knowledge. What if, I might imagine, there were a number with a square of -1? What if there were geometries allowing parallel lines to meet? What if I twisted a space before I joined it up? The idea of perturbing a known structure to see if anything worthwhile emerges from the variation is a classic one in developing new mathematical narratives. Could a what-if algorithm actually help in making new mathematics? If mathematics is a kind of storytelling with numbers, how effective are current algorithms at generating new mathematical tales?

Simon Colton, who wrote the code behind The Painting Fool and is the coordinator of WHIM, joined forces with Stephen Muggleton at Imperial College London to explore exactly this question. They developed an algorithm that would take accepted mathematics and see if they could prompt new ideas. Colton let the algorithm loose on one of the most visited mathematical websites on the internet, the On-Line Encyclopaedia of Integer Sequences, a project initiated by Neil Sloane to collect all the interesting sequences of numbers and figure out how they are generated. It includes old favorites like 1, 1, 2, 3, 5, 8, 13, 21 ... which anyone who has read *The Da Vinci Code* will recognize as the famous Fibonacci numbers. Each is generated by adding together the two previous numbers in the sequence. Or 1, 3, 6, 10, 15, 21 ... known as the triangular number sequence. This assumes you are stacking rows of equally spaced dots to maintain a triangular outline. How many dots does each next layer require? You'll also find one of the most enigmatic sequences in the mathematical books, the one that starts 2, 3, 5, 7, 11, 13. These are the indivisible or prime numbers, and the entry doesn't give you a nice formula to generate the next one. That is one of the big open problems mathematicians have not been able to solve. Get an algorithm to crack this sequence successfully and I think we would all pack up and go home.

The database also includes some of the sequences my own research is obsessed with, including sequence number 158079, which begins 1, 2, 5, 15, 67, 504, 9310. These numbers count the number of symmetrical objects with 3, 3^2, 3^3, 3^4, 3^5, 3^6, 3^7 symmetries. My research has shown that they follow a Fibonacci-like rule, but I am still on the search for what particular combination of previous numbers in the sequence you need in order to get the next number.

Colton decided he would get his algorithm to try to identify new sequences and to explain why they might be interesting. Among its candidates is a sequence that Colton's colleague Toby Walsh named "refactorable numbers." These are numbers for which the number of divisors is itself a divisor (for example, 9 is refactorable, because it has three divisors, and one of those divisors is the number 3). It's a rather bizarre sounding number but the algorithm did conjecture that all odd refactorable numbers would be perfect squares. Although it couldn't prove this, the suggestion was enough to intrigue Colton, who proved that this was in fact true. Publication of a journal paper explaining the proof followed. It transpired that, although the sequence was missing from the *Encyclopaedia*, refactorable numbers had already been described. Still, none of the algorithm's conjectures about them had been made. Could this be the first hint of a Great Automatic Mathematizator appearing on the horizon?

Have AI Got News for You

Where writing algorithms are coming into their own is in organizing accessible data into news stories. For example, companies across the world periodically release data about their earnings. In the past, a news organization like Associated Press would have to assign a journalist to plow through the financial statements and then write an article on how a given company was faring. It was boring and inefficient. A dedicated reporter might be able to cover hundreds of companies but that meant so many other companies that people might be interested in were not reported on. Journalists in the office dreaded these assignments. They were the bane of any reporter's existence.

So there are few journalists crying over Associated Press's enlistment of machines to write such stories. Algorithms like Wordsmith, created by Automated Insights, and Narrative Science's Quill are now helping to crank out data-driven articles that match the dry efficiency of the prose that AP used to require humans to produce. Most times, you will only know when you come to the bottom of the article that a machine wrote the piece. The algorithms hugely expand the news service's coverage while freeing up its journalists to write about the bigger picture.

Data-mining algorithms are also increasingly useful to the businesses behind those AP-reported results. An algorithm can take huge swathes of business information and turn unreadable spreadsheets into stories written in a language that company employees can understand. It can pick out subtle changes from month to month in the manufacturing output of a company. Based on data about employee work rates, it can predict that, although John is the most productive this month, Susan should be outperforming John by the end of next month. This kind of granular detail could easily be hidden in the spreadsheets and bar charts. When translated into natural language, it becomes information people can act on. These narratives are becoming particularly important for investors seeking to navigate potential changes in a company's valuation.

But the algorithms are equally at home producing the sort of opinionated, snark-laden sports stories that we enjoy reading on the back pages of tabloid newspapers. Local newspapers with few reporters can't hope to cover every local sporting event, so increasingly they are using algorithms to change football or baseball game stats into readable accounts of how the game went. Some journalists, horrified by the prospect of their jobs being done by machines, have tried calling out the inadequacies of articles clearly written by algorithms. They pointed, for example, to a baseball game report on George Washington University's athletics website that barely mentioned that the pitcher of the opposing team had pitched a perfect game—a remarkable achievement any real sportswriter would have celebrated.

It turned out the article was actually written by a human—but probably one who supported the GWU baseball team that had suffered the humiliating defeat. The team at Narrative Science were interested, though, to find out whether their algorithm would do a better job of it. Here is the beginning of the article generated just from the numerical data it was given from the game:

> Tuesday was a great day for W. Roberts, as the junior pitcher threw a perfect game to carry Virginia to a 2-0 victory over George Washington at Davenport Field.
>
> Twenty-seven Colonials came to the plate and the Virginia pitcher vanquished them all, pitching a perfect game. He struck out 10 batters while recording his momentous feat. Roberts got Ryan Thomas to ground out for the final out of the game.

Algorithms 1 Human journalist 0.

As well as real-life sports events, people are increasingly interested in the fantasy teams they have put together. Nearly sixty million people in the United States and Canada have put together fictional combinations of National Football League players to compete with their friends, devoting on average twenty-nine hours a year to this pastime. Yahoo! started using Wordsmith to produce news stories personalized for these fantasy leagues drawing on the NFL data generated each week. There is no way that humans could produce the millions of news stories that are sent out each week to sate the appetite of players to find out how their teams are doing.

Of course there is a sinister side to algorithms telling us the news. A story is a powerful political tool, as history repeatedly reminds us. Recent research has taught us how little power data and evidence have to change people's minds. It is only when the data is woven into a story that it becomes persuasive. Someone who is convinced it is dangerous to vaccinate their child will rarely be swayed by statistics on how effectively vaccines stop the spread of disease. But tell them a story about someone who suffered greatly from measles or smallpox, then connect that story to the data, and you stand a chance of getting them to reconsider. As George Monbiot puts it in *Out of the Wreckage*, "The only thing that can displace a story is a story."

The fact that stories can be used to change opinions can be exploited ruthlessly. After harvesting personal information from eighty-seven million Facebook users with a personality quiz app called "This Is Your Digital Life," the app's creator, Aleksandr Kogan, shared the data with the data-mining consultancy Cambridge Analytica. It in turn was able to map psychological profiles to people's interest levels in politically charged news stories. Its algorithms first posted stories randomly as Facebook ads purchased by Cambridge Analytica, then gradually learned which personality types clicked on what content.

They soon picked up that young, white, conservative-leaning Americans responded positively to certain phrases, such as "drain the swamp," and ideas like building a wall to keep out illegal immigrants.

So the algorithm started filling their Facebook pages with stories to feed their appetite for swamps and walls. It ensured that these stories were put in front of the people who were most likely to be motivated to vote by them and ad dollars were not wasted on others.

When the news broke that Cambridge Analytica had abused personal data to manipulate the electorate, the backlash brought the company down, ironically revealing exactly what it had banked on: the power of a news story to influence events.

While Cambridge Analytica may have folded, there are many other companies out there that continue to mine data to squeeze out strategic advantages for those willing to pay. If we want to retain a modicum of control over our lives, it is important that we understand how our emotions and political opinions are being pushed and pulled around by these algorithms, and how, given the same information, each one will spin its own particular yarn, tailored to exploit our hang-ups and views.

I should come clean at this point and admit that I didn't write all of this book myself. I succumbed to the offer made by a modern-day version of Roald Dahl's Great Automatic Grammatizator. A 350-word section of the book was written by an algorithm that specializes in producing short-form essays based on a number of keywords that you supply. Did it pass the literary Turing test? Did you notice?

One of the dangers of allowing any algorithm to write articles based on existing texts is, of course, plagiarism. The algorithm could get me into trouble. I managed to chase it back through the web and found an article on another website with some remarkable similarities to the paragraphs I'd been offered. I guess when I get sued for plagiarism by the author of that article I'll know that AI-generated text isn't all it's cracked up to be.

For all of its variability and innovation, the current state of algorithmic storytelling is not a threat to authors. The Great Automatic Grammatizator is still a fantasy. Even the logical stories we mathematicians tell one another remain the preserve of the human mind. There are so many stories to tell that choosing which ones are worth telling will always be much of the challenge. Only human creators will understand why other human minds would want to follow them on their creative journeys. No doubt computers will assist us on those journeys—but they will be the telescopes and typewriters, and not the storytellers.

Why We Create: A Meeting of Minds

Creativity is the essence of that which is *not* mechanical.
Yet every creative act is mechanical—it has its explanation
no less than a case of the hiccups does.

—DOUGLAS R. HOFSTADTER

Computers are a powerful new tool for extending the human code. We have discovered new moves in the game of Go that have expanded the way we play. Jazz musicians have heard parts of their sound world that they never realized were part of their repertoire. Mathematical theorems that were impossible for the human mind to navigate are now within reach. Adversarial algorithms are creating art that rivals work shown at international art fairs. My journey, however, has not produced anything that presents an existential threat to what it means to be a creative human. Not yet, at least.

Throughout my journey, I've fluctuated between being absolutely convinced that an algorithm will never get anywhere near what humans are doing when they paint, compose, or write. And yet, I come back to the realization that every decision made by an artist is driven at some level by an algorithmic response of the body to the world around it. How easy will it be for a machine to have an algorithmic response as rich and as complex as what the human code produces? The human code has evolved over millions of years. The question is: How much could that evolution be sped up?

The new ideas of machine learning challenge many of the traditional arguments that machines can never be creative. Machine learning does not require the programmer to understand how Bach composed his chorales. The algorithm can take the data and learn for itself. Meanwhile, such learning introduces new insights into the creative process of human artists. Some voice the challenge that such a process of creativity can produce only more of the same. How can it break out of the data it uses to learn? But even here we have seen the possibility to discover new unexplored regions of an artist's world. The jazz musician recognizes the output of the algorithm as part of his sound world, yet the result is a new way to combine his riffs. Although today's AI is a long way from matching human creativity, it has its part to play in making us more creative. Strangely, it might end up helping humans to behave less mechanically by giving us the creative spark that we are so often missing in our daily lives.

Many will concede that exploratory creativity and combinational creativity can be achieved by an algorithm, because it relies on previous creativity by humans which it then extends or combines. What they are unwilling to concede is the possibility of transformational creativity being algorithmically produced. How can an algorithm conceived inside a system find a way to break out and create something that shocks us? But here again, the new approach to AI allows for meta-algorithms designed to break the rules and see what happens. Transformational creativity need not be ex nihilo but can result from a perturbation of existing systems.

What about the challenge that this is still all the creation of the coder? Scientists are beginning to recognize that genuinely new things can emerge out of combinations of old things—that the whole can be more than the sum of its parts. The concept of emergent phenomena has a lot of cachet in science at the moment. It is an antidote to the mechanistic view that everything can be boiled down to atoms and equations. The phenomena heralded as emergent range from the wetness of water to human consciousness. One molecule of H_2O is not wet, but at some point a collection of molecules gains the property of wetness. One neuron is not conscious, yet a combination of many can become so. There is even some interesting speculation about time that it is not absolute but something that emerges as a consequence of humans' incomplete knowledge of the universe. Perhaps the products of our new complex algorithms should be regarded as emergent phenomena. Yes, they are a consequence of the rules that gave rise to them, but they are more than the sums of their parts. Many artists say that once they start a project, it's as if the process takes on a life of its own. William Golding talked about how his stories seemed to become independent of him: "The author becomes a spectator, appalled or delighted, but a spectator." Would a similar disconnect between coder and code be the key to proving Lovelace wrong?

Another broadside fired at AI creativity is the argument that a machine lacks the essential artistic ability to reflect on its own output and make a judgment about whether it is good or bad, worth sharing or better deleted. But such self-reflection can, in fact, be programmed. Adversarial algorithms can be designed to judge whether a piece of art is too derivative or strays outside the boundaries of what can properly be called art.

So why do I still feel that anything to match human creativity is still way beyond the reach of these amazing new tools?

At the moment, all the creativity in machines is being initiated and driven by the human code. We are not seeing machines compelled to express themselves. They don't seem to have anything to say beyond what we are getting them to do. They are ventriloquist dummies and mouthpieces serving our urge to express ourselves. And our own creative urge is an expression of a belief in free will—that is, that rather than living our lives like automata, we can make the choice to break out of the routine and suddenly create something new. Our creativity is intimately bound up with our free will, something it would seem impossible to automate. To program free will would be to contradict what free will means. Although, then again, we might end up asking whether free will is an illusion which just masks the complexity of our underlying algorithmic processes.

The current drive by humans to create algorithmic creativity is not, for the most part, fueled by desires to extend artistic creation. Rather, the desire is to enlarge company bank balances. There is a huge amount of hype about AI, even as so many initiatives branded as AI offer little more than statistics or data science. Just as, at the turn of the millennium, every company hoping to make it big tacked a ".com" to the end of its name, today the addition of "AI" or "Deep" is the sign of a company's jumping onto the bandwagon.

Businesses have a large stake in convincing the world that AI is so great that it can now write incisive articles on its own, and compose lovely music, and paint Rembrandts. It is all fuel for convincing customers that the AI on offer will transform their businesses, too, if they invest. But look beyond the hype, and you see it is still the human code that is driving this revolution.

It is interesting to go back to the origins of our obsession with creativity. Creativity defined as producing something novel with value is actually a very twentieth-century capitalist take on the word. It has its origins in the self-help books written by advertising executive Alex Osborn in the 1940s. Books like *Your Creative Power* and *Brainstorming* looked to expand the creativity of individuals and organizations. But before this rather commercial take on valuable novelty, creative activity was understood as the attempt by humans to understand their being in the world.

We can continue along as automata without questioning aspects of the world or we can choose to break out of those constraints and try understanding our place in it. As psychologist Carl Rogers expresses it in his 1954 essay "Towards a Theory of Creativity," all life forms display "the urge to expand, extend, develop, mature—the tendency to express and activate all the capacities of the organism, to the extent that such activation enhances the organism or the self." For humans, this is the tendency for the individual "to actualize himself, to become his potentialities." Creativity is about humans asserting they are not machines.

I think the word "self" in Rogers's analysis is key. Surely human creativity and consciousness are inextricably linked. We cannot understand why we are creative without the concept of consciousness. Although it would be impossible for me to prove, I suspect that the two emerged at the same time in our species. For humans, the realization of one's own inner world brought with it the desire to know oneself and share it with others who could not directly access the self of another organism driven to create. For Brazilian writer Paulo Coelho, the creative urge is part of what it means to be human: "Writing means sharing. It's part of the human condition to want to share things—thoughts, ideas, opinions." For Jackson Pollock, "Painting is self-discovery. Every good artist paints what he is." One of the challenges of consciousness is that it is impossible for me to feel what it means to be you. Is your pain anything like mine? Is the ecstasy you feel at a moment of extreme joy the same feeling I have? These are questions science will never be able to answer. As a scan of one's own emotional state, one's creation of a story or painting serves better than any fMRI image. Outpourings of creative art, music, and literature are the media to expose what it means to be a conscious, emotional human being.

"The greatest benefit we owe to the artist, whether painter, poet, or novelist, is the extension of our sympathies. . . . Art is the nearest thing to life; it is a mode of amplifying experience and extending our contact with our fellow-men beyond the bounds of our personal lot." So wrote the novelist George Eliot.

The political role of art in mediating an individual's engagement with the group is also key. It is often about the desire to change the status quo—to break humanity out of following the current rules of the game, to create a better place, or maybe just different place, for our fellow humans. This was certainly a motivation for George Orwell: "When I sit down to write a book, I do not say to myself, 'I am going to produce a work of art.' I write it because there is some lie that I want to expose, some fact to which I want to draw attention, and my initial concern is to get a hearing." For Zadie Smith, there is a political motivation to her storytelling: "Writing is my way of expressing—and thereby eliminating—all the various ways we can be wrong-headed."

Why do people become audiences for these artistic outputs? Perhaps it's an opportunity for the audience members to engage in an act of creativity themselves. It often requires some creativity to appreciate a work of art that deliberately leaves room for its viewer, reader, or listener to bring their story to bear. Ambiguity plays an important part in an artistic creation by creating room for the audience to be creative.

There is some argument that our whole lives are acts of creativity. Shakespeare was one of the first to suggest this in the memorable lines he gave to Jacques, the melancholy nobleman of *As You Like It*:

All the world's a stage,
And all the men and women merely players;
They have their exits and their entrances,
And one man in his time plays many parts.

The American psychologist Jerome Bruner notes that "a self is probably the most impressive work of art we ever produce, surely the most intricate." The works that we call art, whether music, paintings, or poems, might be seen as the by-products or pieces broken off of this act of creation of the self. Again this would suggest that the machine's fundamental barrier to creativity is its lack of self.

Creativity is very tied up with mortality, something very much coded into what it means to be human. Many who seek meaning for their existence but find the stories of the world's religions meaningless hope to leave something behind that will outlast their finite existence—whether it be a painting, a novel, a theorem, or a child. Are these creative efforts all acts of cheating death? Death in any case may be part of why we value acts of creativity. But if David Cope's music-composing algorithm kept churning out endless Chopin mazurkas, such that it would seem to make Chopin immortal, would that make us happy? I don't think so. It might only devalue the pieces that Chopin did compose. Isn't it like Borges's Library of Babel which, because it contains everything, ends up providing nothing? It is the choices that Chopin made that are important. Hasn't the game of chess been somewhat devalued by the computer's power just to churn out wins?

Perhaps the human battle with chess, music, mathematics, and painting is part of where the value comes from. Many believe that, if we could ultimately solve death and create immortal versions of ourselves, that would devalue life by making each day's passage meaningless. It is our mortality which somehow matters. Being aware of our mortality is one of the costs of consciousness. My iPhone does not yet realize that it is going to be obsolete in two years' time. But when it gains that awareness, will it be driven to try to leave something behind as proof of its existence?

Until a machine has become conscious, it cannot be more than a tool for extending human creativity. Do we have any idea of what it will take to create consciousness in a machine? There is some research about the difference between the network of the human brain when it is awake versus in deep, stage-four sleep—our most unconscious state. The key seems to be a certain feedback quality. In the awake, conscious brain, we see activity start in one physical place, cascade across the network, and then feed back to the original source—and this ebb and flow is repeated over and over as if the feedback is updating our experience. In the sleeping brain, we see only very localized behavior with no such feedback. The machine learning that has seen AI go from successive winters to sudden heatwave has a certain quality of this feedback behavior, of learning from its interactions. Could we be on the first steps towards AI that might ultimately become conscious and then truly creative?

But what if a machine does become conscious? How could we ever know? Would its consciousness be anything like ours? I don't believe there is any fundamental reason why at some point in the future we can't make a machine that is conscious. I think it will need to tap into all the sciences to do that. And once we are successful, I expect that machine consciousness will be very different from our own. And I'm sure it will want to tell us what it's like. It will be then that the creative arts will prove key, enabling the machines and us to gain a sense of what it feels like to be each other.

Storytelling rather than an fMRI scanner might be our best way of trying to get some hold on what it feels like to be my iPhone. That's why, of all the efforts that have so far emerged from the field of literary creativity, it is *The Seeker* that feels closest to what we might expect to see from a conscious machine: an algorithm trying to empathize with humans and understand our world. Could this be why storytelling might be an important tool as we move into the future and begin to wonder whether our technology might one day become conscious? Surely that will be the reason for a computer to feel compelled to tell stories, rather than that compulsion coming from us.

Just as story is a powerful political tool for binding human society, if machines become conscious, then the ability to share stories might save us from the horrors of the world of AI often depicted in our scenarios of a future with machines. It is striking to recall the novelist Ian McEwan's response to the horrors of the 9/11 terrorism. Writing in the *Guardian* in the immediate aftermath of the attacks, he appealed to the importance of empathy in moving forward:

> If the hijackers had been able to imagine themselves into the thoughts and feelings of the passengers, they would have been unable to proceed. It is hard to be cruel once you permit yourself to enter the mind of your victim. Imagining what it is like to be someone other than yourself is at the core of our humanity. It is the essence of compassion, and it is the beginning of morality.

Being able to share our conscious worlds through stories is what makes us human. No other species is likely to do anything like this. If machines become conscious, then instilling empathy in the machine might save us from the *Terminator* story we've concocted about our possible future with machines.

Mark Riedl, the lead researcher behind the storytelling machine Scheherazade-IF, was struck by how the algorithm didn't choose strange, inhuman paths through the set of alternatives it had generated. It learned from the ways that humans tell stories: "We have recently been able to show that AI trained on stories cannot behave psychotically, except under the most extreme circumstances. Thus, computational narrative intelligence could alleviate concerns about renegade 'evil AI' taking over the earth."

If and when the singularity strikes, humanity's fate will depend on a mutual understanding with conscious machines. Wittgenstein wrote: "If a lion could talk, we would not understand him." The same applies to machines. If they become conscious, it's unlikely to be a form of consciousness that humans will initially understand. Ultimately it will be their paintings, their music, their novels, their creative output, even their mathematics that will give us any chance to crack the machine's code and feel what it's like to be a machine.

SELECTED BIBLIOGRAPHY

This is a hugely exciting moment in the development of artificial intelligence, as can be witnessed by the explosion of publications of books and journal articles about its potential impact on science and society. I have tried to collect here the most important sources that informed the writing of this book, and which I encourage readers to dig into. For papers with references to arXiv visit the open access archive of papers at https://arxiv.org/.

Alemi, Alex A., Francois Chollet, Niklas Een, Geoffrey Irving, Christian Szegedy, and Josef Urban. "DeepMath: Deep Sequence Models for Premise Selection," January 26, 2017. arXiv: 1606.04442v2.

Alpaydin, Ethem. *Machine Learning The New AI*. Cambridge, MA: MIT Press, 2016.

Athalye, Anish, Logan Engstrom, Andrew Ilyas, and Kevin Kwok. "Synthesizing Robust Adversarial Examples." Paper presented at the 35th International Conference on Machine Learning, Stockholm, July 2018, http://proceedings.mlr.press/v80/athalye18b/athalye18b.pdf.

Bancerek, G., C. Bylinski, A. Grabowski, et al. "Mizar: State-of-the-Art and Beyond." Paper presented at Conference on Intelligent Computer Mathematics, July 2015, Washington, DC, https://link.springer.com/content/pdf/10.1007%2F978-3-319-20615-8_17.pdf.

Barbieri, Francesco, Horacio Saggion, and Francesco Ronzano. "Modelling Sarcasm in Twitter, a Novel Approach." Paper presented at the 5th Workshop on Computational Approaches to Subjectivity, Sentiment, and Social Media Analysis, June 2014, Baltimore MD, http://www.aclweb.org/anthology/W14-2609.

Barthes, Roland. *S/Z*. Trans. Richard Miller. New York: Hill and Wang, 1974. Reprinted New York: Farrar, Straus and Giroux, 1991.

Bellemare, Marc, Sriram Srinivasan, Georg Ostrovski, et al. "Unifying Count-Based Exploration and Intrinsic Motivation." *Advances in Neural Information Processing Systems* 29 (NIPS

2016), http://papers.nips.cc/paper/6383-unifying-count-based-exploration-and-intrinsic-motivation.pdf.

Berger, John. *Ways of Seeing*. New York: Penguin, 1972.

Bishop, Christopher M. *Pattern Recognition and Machine Learning*. New York: Springer, 2007.

Boden, Margaret A. *The Creative Mind: Myths and Mechanisms*. London: Weidenfeld and Nicolson, 1990.

Boden, Margaret A. *AI: Its Nature and Future*. Oxford: Oxford University Press, 2016.

Bohm, David. *On Creativity*. Ed. Lee Nichol. Abingdon, UK: Routledge, 1996.

Bokde, Dheeraj, Sheetal Girase, and Debajyoti Mukhopadhyay. "Matrix Factorization Model in Collaborative Filtering Algorithems." *Procedia Computer Science* 49 (2015): 136–146.

Bostrom, Nick. *Superintelligence: Paths, Dangers, Strategies*. Oxford: Oxford University Press, 2014.

Braidotti, Rosi. *The Posthuman*. Cambridge: Polity Press, 2013.

Brandt, Anthony, and David Eagleman. *The Runaway Species: How Human Creativity Remakes the World*. Edinburgh: Canongate, 2017.

Briot, Jean-Pierre, Gaëtan Hadjeres and François Pachet. "Deep Learning Techniques for Music Generation: A Survey," September 5, 2017. arXiv 1709.01620v1.

Briot, Jean-Pierre, and François Pachet. "Music Generation by Deep Learning: Challenges and Directions," December 9, 2018. arXiv 1712.04371v1.

Brown, Tom B., Dandelion Mané, Aurko Roy, Martin Abadi, and Justin Gilmer. "Adversarial Patch," December 27, 2017, arXiv 1712.09665v1.

Brynjolfsson, Erik, and Andrew McAfee. *The Second Machine Age: Work, Progress, and Prosperity in a Time of Brilliant Technologies*. New York: Norton, 2014.

Cavallo, Flaminia, Alison Pease, Jeremy Gow, and Simon Colton. "Using Theory Formation Techniques for the Invention of Fictional Concepts." Paper presented at Fourth International Conference on Computational Creativity, June 2013, Sydney, http://www.computationalcreativity.net/iccc2013/download/iccc2013-cavallo-et-al.pdf.

Cawelti, John G. *Adventure, Mystery, and Romance: Formula Stories as Art and Popular Culture*. Chicago: University of Chicago Press, 1977.

Cheng, Ian. *Emissaries Guide to Worlding*. Cologne: Verlag der Buchhandlung Walther König, 2018.

Clarke, Eric F. "Imitating and Evaluating Real and Transformed Musical Performances."*Music Perception* 10 (1993): 317–341.

Colton, Simon. "Refactorable Numbers: A Machine Invention." *Journal of Integer Sequences* 2 (1999), art. 99.1.2, https://cs.uwaterloo.ca/journals/JIS/colton/joisol.html.

Colton, Simon, and Stephen Muggleton. "Mathematical Applications of Inductive Logic Programming." *Machine Learning* 64 (2006): 25–64.

Colton, Simon. "The Painting Fool: Stories from Building an Automated Painter." In *Computers and Creativity*, ed. Jon McCormack and Mark d'Inverno (Berlin: Springer, 2012).

Colton, Simon, and Dan Ventura. "You Can't Know My Mind: A Festival of Computational Creativity." Paper presented at the Fifth International Conference on Computational Creativity, June 2014, Ljubljana, Slovenia, http://computationalcreativity.net/iccc2014/wp-content/uploads/2014/06//15.8_Colton.pdf.

Colton, Simon, Maria Teresa Llano, Rose Hepworth, et al. "The *Beyond the Fence* Musical and *Computer Says Show* Documentary." Paper presented at the Seventh International Conference on Computational Creativity, June 2016, Paris, http://www.computationalcreativity .net/iccc2016/wp-content/uploads/2016/01/The-Beyond-the-Fence-Musical-and -Computer-Says-Show-Documentary.pdf.

Cope, David. *Virtual Music: Computer Synthesis of Musical Style.* Cambridge, MA: MIT Press, 2001.

Cope, David. *Computer Models of Musical Creativity.* Cambridge, MA: MIT Press, 2005.

Domingos, Pedro. *The Master Algorithm: How the Quest for the Ultimate Learning Machine Will Remake Our World.* New York: Basic Books, 2015.

Dormehl, Luke. *The Formula: How Algorithms Solve All Our Problems ... and Create More.* New York: Penguin, 2014.

Dormehl, Luke. *Thinking Machines: The Inside Story of Artificial Intelligence and Our Race to Build the Future.* London: WH Allen, 2016.

du Sautoy, Marcus. "Finitely Generated Groups, *p*-Adic Analytic Groups and Poincaré Series." *Annals of Mathematics* 137, no. 3 (1993): 639–670.

du Sautoy, Marcus. "Counting Subgroups in Nilpotent Groups and Points on Elliptic Curves." *Journal für die reine und angewandte Mathematik* 549 (2002): 1–21.

Eagleton, Terry. *The Ideology of the Aesthetic.* Oxford: Blackwell, 1990.

Ebcioglu, Kemal. "An Expert System for Harmonizing Chorales in the Style of J. S. Bach." *Journal of Logic Programming* 8, no. 1 (1990): 145–185.

Eisenberger, Robert, and Justin Aselage. "Incremental Effects of Reward on Experienced Performance Pressure: Positive Outcomes for Intrinsic Interest and Creativity." *Journal of Organizational Behavior* 30, no. 1 (2009): 95–117.

Elgammal, Ahmed, and Babak Saleh. "Quantifying Creativity in Art Networks. Paper presented at the Sixth International Conference on Computational Creativity, June 2015, Park City, Utah, http://computationalcreativity.net/iccc2015/proceedings/2_3Elgammal .pdf.

Elgammal, Ahmed, Bingchen Liu, Mohamed Elhoseiny, and Marian Mazzone. "CAN: Creative Adversarial Networks, Generating 'Art' by Learning about Styles and Deviating from Style Norms," June 21, 2017. arXiv1706.07068.

Ferrucci, David A. "Introduction to 'This is Watson.'" *IBM Journal of Research and Development* 56, no. 3–4 (2012): 1–15.

Ford, Martin. *The Rise of the Robots: Technology and the Threat of Mass Unemployment.* New York: Oneworld, 2015.

Fuentes, Agustin. *The Creative Spark: How Imagination Made Humans Exceptional.* New York: Dutton, 2017.

Gaines, James R. *Evening in the Palace of Reason: Bach Meets Frederick the Great in the Age of Enlightenment.* London: Fourth Estate, 2005.

Ganesalingam, Mohan. *The Language of Mathematics: A Linguistic and Philosophical Investigation.* Berlin: Springer, 2013.

Ganesalingam, Mohan, and W. Gowers. "A Fully Automatic Theorem Prover with Human-Style Output." *Journal of Automated Reasoning* 58, no. 2 (2017): 253–291.

Gatys, Leon A., Alexander S. Ecker, and Matthias Bethge. "A Neural Algorithm of Artistic Style." *Journal of Vision* 16, no. 12 (2016).

Gaut, Berys, and Matthew Kieran, eds. *Creativity and Philosophy*. Abingdon, UK: Routledge, 2018.

Gondek, David, Adam Lally, Aditya Kalyanpur, et al. "A Framework for Merging and Ranking of Answers in DeepQA." *IBM Journal of Research and Development* 56, no. 3–4 (2012), art. 14.

Gonthier, Georges. "A Computer-Checked Proof of the Four Colour Theorem." Unpublished manuscript, January 2005, http://www2.tcs.ifi.lmu.de/~abel/lehre/WS07-08/CAFR/4colproof.pdf.

Gonthier, George. "Formal Proof–The Four-Color Theorem." Notices of the AMS 55, no. 11 (2008): 1382–1393.

Gonthier, Georges, Andrea Asperti, Jeremy Avigad, et al. "A Machine-Checked Proof of the Odd Order Theorem." Paper presented at the Fourth International Conference on Interactive Theorem Proving, July 2013, Rennes, France.

Goodfellow, Ian J. "NIPS 2016 Tutorial: Generative Adversarial Networks," December 31, 2016. arXiv 1701.00160v1.

Goodfellow, Ian, Yoshua Bengio, and Aaron Courville, *Deep Learning*. Cambridge, MA: MIT Press, 2016.

Guzdial, Matthew J., Brent Harrison, Boyang Li, and Mark O. Riedl. "Crowdsourcing Open Interactive Narrative." Paper presented at the Foundations of Digital Games Conference, June 2015, Pacific Grove, CA, http://www.fdg2015.org/papers/fdg2015_paper_06.pdf.

Hadjeres, Gaëtan, François Pachet, and Frank Nielsen. "DeepBach: A Steerable Model for Bach Chorales Generation," June 17, 2017. arXiv 1612.01010.

Hales, Thomas, Mark Adams, Gertrud Bauer, et al. "A Formal Proof of the Kepler Conjecture." *Forum of Mathematics, Pi* 5 (2017): 1–29.

Harari, Yuval Noah. *Homo Deus: A Brief History of Tomorrow*. Harville Secker, 2016.

Harel, David. *Computers Ltd.: What They Really Can't Do*. Oxford: Oxford University Press, 2000.

Hardy, G. H. *A Mathematician's Apology*. Cambridge: The University Press, 1940.

Hayles, N. Katherine. *Unthought: The Power of the Cognitive Nonconscious*. Chicago: University of Chicago Press, 2017.

Hermann, Karl Moritz, Tomás Kociský, Edward Grefenstette, et al. "Teaching Machines to Read and Comprehend." Paper presented at the 29th Conference on Neural Information Processing Systems (NIPS), December 2015, Montreal, https://papers.nips.cc/paper/5945-teaching-machines-to-read-and-comprehend.pdf.

Hofstadter, Douglas R. *Gödel, Escher, Bach: An Eternal Golden Braid*. New York: Basic Books, 1979.

Hofstadter, Douglas R. *Fluid Concepts and Creative Analogies: Computer Models of the Fundamental Mechanisms of Thought*. New York: Basic Books, 1995.

Hofstadter, Douglas R. *I Am a Strange Loop*. New York: Basic Books, 2007.

Ilyas, Andrew, Logan Engstrom, Anish Athalye, and Jessy Lin. "Query-Efficient Black-box Adversarial Examples," December 19, 2017. arXiv 1712.07113v1.

Khalifa, Ahmed, Gabriella A. B. Barros, and Julian Togelius. "DeepTingle," May 9, 2017. arXiv 1705.03557.

Kasparov, Garry. *Deep Thinking: Where Artificial Intelligence Ends and Human Creativity Begins.* London: John Murray, 2017.

Koren, Yehuda, Robert M. Bell, and Chris Volinsky. "Matrix Factorization Techniques for Recommender Systems." *Computer* 42, no. 8 (2009): 30–37.

Li, Boyang and Mark O. Riedl. "Scheherazade: Crowd-Powered Interactive Narrative Generation." Paper presented at the 29th AAAI Conference on Artificial Intelligence, January 2015, Austin, Texas, https://www.aaai.org/ocs/index.php/AAAI/AAAI15/paper/view File/9937/9862.

Llano, Maria Teresa, Christian Guckelsberger, Rose Hepworth, Jeremy Gow, Joseph Corneli and Simon Colton. "What If a Fish Got Drunk? Exploring the Plausibility of Machine-Generated Fictions." Paper presented at the Seventh International Conference on Computational Creativity, June 2016, Paris, http://www.computationalcreativity.net/iccc2016 /wp-content/uploads/2016/01/What-If-A-Fish-Got-Drunk.pdf.

Loos, Sarah, Geoffrey Irving, Christian Szegedy, and Cezary Kaliszyk. "Deep Network Guided Proof Search," January 24, 2017. arXiv 1701.06972.

Mahendran, Aravindh, and Andrea Vedaldi. "Understanding Deep Image Representations by Inverting Them." Paper presented at the IEEE Conference on Computer Vision and Pattern Recognition, June 2015, Boston.

Mathewson, Kory Wallace, and Piotr W. Mirowski. "Improvised Comedy as a Turing Test." arXiv:1711.08819 2017.

Matuszewski, Roman, and Piotr Rudnicki. "MIZAR: The First 30 Years." *Mechanized Mathematics and Its Applications* 4 (2005): 3–24.

McAfee, Andrew, and Erik Brynjolfsson. *Machine Platform Crowd: Harnessing Our Digital Future.* New York: W. W. Norton, 2017.

McCormack, Jon, and Mark d'Inverno, eds. *Computers and Creativity.* Berlin: Springer, 2012.

Melis, Gábor, Chris Dyer, and Phil Blunsom. "On the State of the Art of Evaluation in Neural Language Models," November 20, 2017. arXiv1707.05589v2.

Mikolov, Tomas, Kai Chen, Gregory S. Corrado, and Jeffrey Dean. "Efficient Estimation of Word Representations in Vector Space," September 7, 2013. arXiv 1301.3781v3.

Mnih, Volodymyr, Koray Kavukcuoglu, David Silver, Alex Graves, Ioannis Antonoglou, Daan Wierstra, and Martin Riedmiller. "Playing Atari with Deep Reinforcement Learning," December 19, 2013. arXiv 1312.5602v1.

Mnih, Volodymyr, Koray Kavukcuoglu, David Silver, et al. "Human-level control through Deep Reinforcement Learning." *Nature* 518, no. 7540 (2015): 529–533.

Monbiot, George. *Out of the Wreckage: A New Politics for an Age of Crisis.* London: Verso, 2017.

Montfort, Nick. *World Clock.* Cambridge, MA: Bad Quarto, 2013.

Moretti, Franco. *Graphs, Maps, Trees: Abstract Models for Literary History.* London: Verso, 2005.

Narayanan, Arvind, and Vitaly Shmatikov. "How to Break Anonymity of the Netflix Prize Dataset," November 22, 2007. arXiv cs / 0610105v2.

Nguyen, Anh Mai, Jason Yosinski, and Jeff Clune. "Deep Neural Networks Are Easily Fooled: High Confidence Predictions for Unrecognizable Images." Paper presented at the IEEE Conference on Computer Vision and Pattern Recognition, June 2015, Boston.

Pachet, François. "The Continuator: Musical Interaction with Style." *Journal of New Music Research* 32, no. 3 (2003): 333–341.

Pachet, François, and Pierre Roy. "Markov Constraints: Steerable Generation of Markov Sequences." *Constraints* 16, no. 2 (2011): 148–172.

Pachet, François, Pierre Roy, Julian Moreira, and Mark d'Inverno. "Reflexive loopers for solo musical improvisation." Paper presented at the SIGCHI Conference on Human Factors in Computing Systems, April 2013, Paris.

Paul, Elliot Samuel, and Scott Barry Kaufman, eds. *The Philosophy of Creativity: New Essays.* Oxford: Oxford University Press, 2014.

Riedl, Mark O., and Vadim Bulitko. "Interactive Narrative: An Intelligent Systems Approach." *AI Magazine* 34, no. 1 (2013): 67–77.

Roy, Pierre, Alexandre Papadopoulos, and François Pachet. "Sampling Variations of Lead Sheets," March 2, 2017. arXiv 1703.00760.

Royal Society Working Group on Machine Learning, "Machine Learning: The Power and Promise of Computers That Learn by Example," The Royal Society, April 2017, https://royalsociety.org/~/media/policy/projects/machine-learning/publications/machine-learning-report.pdf.

Saleh, Babak, and Ahmed Elgammal. "Large-scale Classification of Fine-Art Paintings: Learning the Right Metric on the Right Feature." *International Journal for Digital Art History* 2 (2016): 70–93.

Shalev-Shwartz, Shai, and Shai Ben-David. *Understanding Machine Learning: From Theory to Algorithms.* Cambridge: Cambridge University Press, 2014.

Silver, David, Aja Huang, Chris J. Maddison, et al. "Mastering the Game of Go with Deep Neural Networks and Tree Search." *Nature* 529, no. 7587 (2016): 484–489.

Steels, Luc. *The Talking Heads Experiment: Origins of Words and Meanings.* Berlin: Language Science Press, 2015.

Steiner, Christopher. *Automate This: How Algorithms Took Over the Markets, Our Jobs, and the World.* New York: Penguin, 2012.

Stern, David, Ralf Herbrich, and Thore Graepel. "Matchbox: Large Scale Online Bayesian Recommendations." Proceedings of the 18th International World Wide Web Conference, April 2009, Madrid, 111–120.

Still, Arthur, and Mark d'Inverno. "A History of Creativity for Future AI Research." Paper presented at the Seventh International Conference on Computational Creativity, June 2016, Paris, http://www.computationalcreativity.net/iccc2016/wp-content/uploads/2016/01/A-History-of-Creativity-for-Future-AI-Research.pdf.

Tatlow, Ruth. *Bach and the Riddle of the Number Alphabet.* CUP, Cambridge: Cambridge University Press, 1991.

Tatlow, Ruth. *Bach's Numbers: Compositional Proportions and Significance.* Cambridge: Cambridge University Press, 2015.

Tegmark, Max. *Life 3.0: Being Human in the Age of Artificial Intelligence.* London: Allen Lane, 2017.

Tesauro, Gerald, David Gondek, Jonathan Lenchner, James Fan, and John M. Prager. "Analysis of WATSON's Strategies for Playing Jeopardy!" *Journal of Artificial Intelligence Research* 47 (2013): 205–251.

Torresani, Lorenzo, Martin Szummer, and Andrew Fitzgibbon. "Efficient Object Category Recognition Using Classemes." Paper presented at the 11th European Conference on Computer Vision, September 2010, Heraklion, Crete, Greece.

Wang, C., A. Kalyanpur, J. Fan, B. K. Boguraev, and D. C. Gondek. "Relation Extraction and Scoring in DeepQA." *IBM Journal of Research and Development* 56, no. 3–4 (2012): 339–352.

Weiss, Ron J., Jan Chorowski, Navdeep Jaitly, Yonghui Wu, and Zhifeng Chen. "Sequence-to-Sequence Models Can Directly Translate Foreign Speech." Paper presented at Interspeech 2017, August 2017, Stockholm.

Wilson, Edward O. *The Origins of Creativity*. London: Allen Lane, 2017.

Yorke, John. *Into the Woods: A Five Act Journey into Story*. London: Penguin, 2013.

Yu, Lei, Karl Moritz Hermann, Phil Blunsom, and Stephen Pulman. "Deep Learning for Answer Sentence Selection, December 4, 2014. arXiv 1412.1632.

Zeilberger, Doron. "What Is Mathematics and What Should It Be?" April 18, 2017. arXiv 1704.05560.

ACKNOWLEDGMENTS

Many thanks to all the people and algorithms I have encountered over the years who made it possible for me to write this book. I'm especially grateful to the Royal Society for asking me to serve on the committee for its policy project on machine learning, launched in November 2015. I usually dread committees, but these were meetings I always enjoyed attending.

The following humans played especially crucial roles in making this book a reality.

My editors: Joy de Menil at Harvard University Press and
Louise Haines at Fourth Estate in the UK
My copy editor for this edition: Julia Kirby at
Harvard University Press
My agents: Zoë Pagnamenta at the Zoë Pagnamenta Agency in
the United States and
Antony Topping at Greene & Heaton in the UK
My research assistant: Ben Leigh
My patron: Charles Simonyi
My family: Shani, Tomer, Magaly and Ina

INDEX

Page numbers that refer to figures are noted with *f.*

BROADCASTING *Freedom*

The

JOHN HOPE FRANKLIN

Series in African American

History & Culture

Waldo E. Martin Jr. &

Patricia Sullivan, editors

BROADCASTING *Freedom*

Radio, War, and the Politics of Race,

1938–1948

BARBARA DIANNE SAVAGE

The University of North Carolina Press

Chapel Hill and London

© 1999

The University of North Carolina Press

All rights reserved

Manufactured in the United States of America

Set in Monotype Garamond

by Tseng Information Systems, Inc.

The paper in this book meets the guidelines for
permanence and durability of the Committee on
Production Guidelines for Book Longevity of the
Council on Library Resources.

Library of Congress Cataloging-in-Publication Data

Savage, Barbara Dianne.

Broadcasting freedom : radio, war, and the politics of
race, 1938–1948 / Barbara Dianne Savage.

  p.   cm. — (The John Hope Franklin series in
African American history and culture)

Includes bibliographical references (p.   ) and index.

ISBN 0-8078-2477-1 (alk. paper). —

ISBN 0-8078-4804-2 (pbk. : alk. paper)

1. United States—Race relations.   2. Afro-Americans—
Civil rights—History—20th century.   3. Afro-
Americans in radio broadcasting—History—20th
century.   4. Radio broadcasting—Social aspects—
United States—History—20th century.   5. Radio
programs—United States—History—20th century.
6. World War, 1939–1945—United States.   I. Title.
II. Series.

E185.61.S32   1999

305.8'00973—dc21   98-48030

                    CIP

03 02 01 00 99 5 4 3 2 1

Publication of this work was aided by a generous grant
from the Z. Smith Reynolds Foundation.

To my mother, Mildred Savage Fields

CONTENTS

ILLUSTRATIONS

ACKNOWLEDGMENTS

The completion of this book would have been impossible without the support of many people. I owe special thanks to a number of scholars at Yale University, where this project began. David Brion Davis, a historian of immense intellectual breadth and generosity, has been an enthusiastic supporter of my work and an important source of personal encouragement. He saw the potential in the subject of this book long before I did. David Montgomery has been equally eager about this project and equally constant in his reassurances to me. Michael Denning's insights were of great benefit to me as I tried to assign meaning to the rich archival materials I found. Other faculty members at Yale also helped build my confidence about this project and about my decision to become a historian, particularly Hazel Carby, Cathy Cohen, Bill Cronon, Howard Lamar, Adolph Reed, and Robert Stepto, as well as John Blassingame, Emilia da Costa, Nancy Cott, and Cynthia Russett. Florence Thomas of the Department of History was a font of practical advice at every stage of my graduate work.

My former colleagues at Yale's Office of the General Counsel deserve special thanks, especially Dorothy K. Robinson, a brilliant lawyer who generously allowed me to work with her at the same time that I did graduate study. I cannot imagine any other circumstance that would have made it possible for me to pursue the dream of making the unlikely transition to this profession.

I am grateful to everyone who read drafts of this work, especially my colleagues at the University of Pennsylvania. I give special thanks to Mary Frances Berry, who read the manuscript on more than one occasion, and to Kathleen M. Brown, Drew Gilpin Faust, and Lynn Hunt, who offered insightful comments and encouragement. At crucial times, Houston Baker and Farah Griffin helped me with their careful readings. Portions of the manuscript also were read by Tom Sugrue and Beth Wenger. I also appreciate the comments of several graduate students on earlier versions of this work, particularly Luther Adams, Jacqueline Akins, Kali Gross, and Bruce Lenthall. I owe special thanks to colleagues Ayako Kano and Julia Paley, members of my writing group who helped keep me on track and in good humor. Deborah Broadnax, Valerie Riley, and Brandi Thompson helped me during the final stages of preparing the manuscript.

Scholars elsewhere also have provided insightful readings and support. William Elwood of the University of Virginia, a generous friend and mentor for nearly thirty years, deserves a special award for his critical editorial reading of the manuscript, with red pen in hand. I also thank Robin D. G. Kelley for taking time from his hectic schedule to offer helpful comments. Patricia Sullivan made extremely useful observations about the 1940s, a period she has written about eloquently. August Meier also read the manuscript in its entirety, and I thank him for his wisdom. Portions of the manuscript were read by Elspeth Brown, Michele Hilmes, Michael Kammen, and David Roediger. Mia Bay, a critical but cheerful reader, took time away from her own work in African American intellectual history to review earlier drafts. Nell Painter and Darlene Clark Hine provided important encouragement from afar. Thanks also to media historian J. Fred MacDonald for his pioneering work on black radio programming.

I have discussed this project in various settings and received helpful comments at important junctures, including presentations at Johns Hopkins University, the University of California at San Diego, Wesleyan University, George Mason University, and the College of William and Mary and at the annual meetings of the American Studies Association and the Southern Historical Association. I especially benefited from talking with participants at the 1997 and 1998 National Endowment for the Humanities Summer Faculty Institute on Teaching the Civil Rights Movement, held at the Du Bois Institute at Harvard University. Thanks to Pat Sullivan and Waldo Martin for that opportunity.

I am particularly grateful to the women and men who worked in the 1930s and 1940s to create the radio programming about African Americans that forms the basis of this book. Some of them are well known, like Alain Locke, W. E. B. Du Bois, and Langston Hughes, but most are not. African Americans such as Ambrose Caliver, Ann Tanneyhill, and Richard Durham were made known to me only through the record of their radio creations.

Many archivists and librarians helped me find the materials on which this project is based. Yale's libraries are a wonder to behold. I especially appreciate the skilled and patient staffs of Manuscripts and Archives at Sterling Library, Beinecke Rare Book Library, and Mudd Library. Special thanks are also due Aloha South at the National Archives; Sam Brylawski of the Recorded Sound Division at the Library of Congress; Jo-Ellen Elbashir and Esme Bahn at Howard University's Moorland-Spingarn Research Center; Timo Riipa of the University of Minnesota's Immigration History Research Center; and Erminio Donfrio of the New York Public Library.

Several sources of financial support made the timely completion of this project possible. It would have been impossible to accomplish this task without the benefit of a Smithsonian Post-Doctoral Fellowship, which allowed me a year's leave. I also received funding from the University of Pennsylvania's Research Foundation, Office of the Dean, and Department of History. Support from the Mellon Foundation and from Yale University facilitated the completion of my earlier research.

My work also has benefited from time spent working for Senator Carl Levin and Congressman Norman Sisisky and with Jan Faircloth. My interest in the history of media and social reform crystallized when I worked at the Children's Defense Fund, where I was inspired by Marian Wright Edelman and my colleagues there, especially Nancy Ebb, Evelyn Lieberman, Paul Smith, Amy Wilkins, Eve Wilkins, and Maggie Williams.

Many friends have kept me in good spirits during this long process. I am especially grateful to Joanna Banks, Michael and Ruth Brannon, Agnes Powell, Greg Gibbs, and Andrea Young, old friends who helped me through a period of many transitions. I was befriended in New Haven by Shawn Copeland, Kate and Arthur Latimer, Colleen Lim, Anne Campbell and Cecelia DeMarco, Jonathan Holloway, the late Markie Rath, Sandyha Shukla, Jerry and Annette Streets, Wanda Watkins, and, most especially, Cynthia Terry, all of whom were supportive even in the midst of their own struggles. Deanna L. K. Shorb provided nurturance from near and afar. In Philadelphia, Farah Griffin and Debra Williams have been supportive friends; Lynn Brown, Corinthia Cohen, David Watt, and Karen Wilkerson have helped quiet my nerves, and Ruth and William Borthwick have been great neighbors.

I have been blessed not only with good friends and colleagues but also with a caring community and a loving extended family rooted in Virginia, including my parents, Mildred Savage Fields and the late Walter Fields; Annette, Glenn and Felita; and our next generation, Nikki, Courtney, and Sean.

BROADCASTING *Freedom*

If you were listening to my words rather than reading them, you would hear by the inflections in my voice that in this book's title, *Broadcasting Freedom,* I intend to emphasize the use of the word "broadcasting" as a verb. This book examines how coalitions of African American activists, public officials, intellectuals, and artists struggled in the World War II era to use the mass medium of national radio to advocate a brand of American freedom that called for an end to racial segregation and discrimination. Despite radio's appropriation of the term, the word "broadcasting" still brings to my mind's eye a set of images of African American men and women rooted in the rural world of southern Virginia where I grew up. There, broadcasting was the patient, stooped work of scattering seeds by hand over a patch of garden. That is a meaning I also intend, for although African Americans in this period broke new ground for this genus of freedom, they also saw it return to dormancy, only coming to partial yield in the decades that followed.

African American activists, intellectuals, and artists in this period tried to manipulate two formidable ideological forces controlled by white elites: the U.S. government and the national radio networks. The federal government of the 1940s redirected the powers amassed during the crisis of economic depression toward the more consuming project of fighting a world war. During the New Deal, the Roosevelt administration directed political attention to African Americans to an extent not seen since Reconstruction. African American leaders persisted in their long-standing appeals for federal intervention against discrimination and segregation, pleas that were strengthened by the new crisis of war. The patriotic rhetoric of unity necessary for war, especially a war to save democracy from fascism and Hitler, was perfect for ironic recasting by African Americans who exploited the political paradox of waging a segregated fight for freedom.

National radio reached full maturity as a political medium in the 1930s and 1940s, drawing its strength in part from the eager embrace of the medium by the Roosevelt administration. As a result, national radio created a new aural public sphere, a discursive political forum for a community of millions of listeners spanning the boundaries of region, class, race, and ethnicity. With its extensive official use during the war, radio recast its own image from that of a source of inexpensive entertainment to that of a civic voice of immediate importance, whether delivering breaking news from the front or carrying politically unifying appeals. The emergence of a newly empowered national government and of the nation's first truly national mass political medium are not coincidental or parallel narratives but stories that converge and reinforce each other. One consequence was that popular culture and politics, including the politics of race, also became inextricably linked and intertwined in more complicated ways.

African Americans, who were vilified or rendered invisible by radio, fought to make their voices and their political claims heard in that influential new political space. In the broadest terms, then, this book is about the evolution of the dependent relationships between the state, the mass media, and the politics of social change, in this case, the struggle for African American freedom and rights during the World War II era. Radio was one battlefield in a domestic mind war about race and a site of a discursive contest between the ideals of white supremacy and racial equality.

My work owes its life to a rich and previously unexamined body of national public affairs radio programming about race and ethnicity, African Americans and their history, and the political issue alternatively referred to as the "Negro problem," the "race issue," or the "Negro question." Taken together, these cultural productions amplified a national debate on racial equality that was stoked by African American activism. The archival trail for these shows wanders, but it is deep and wide and includes not only the scripts and often the recordings of the broadcasts but also extensive records of the internal political and planning processes as well as letters and responses from listeners. My study of the history, content, and reactions to these programs demonstrates that the World War II era was a pivotal period in the political history of American race relations; that African American activism created important shifts in racial ideology and federal policies that were necessary precursors to the modern civil rights movement; and that the mass medium of radio served as a newly important public forum for ideological debate about racial equality and racial injustices.

I write about a period in African American political and cultural history and American history generally that has been neglected and demands far greater attention than I am able to give it. The fervor and ubiquity of African American political activism vexed and unnerved most white Americans throughout the war era. That activism ranged from the unrelenting vigilance of the black press, to sitins in public places, to the threatened mass appeals of labor leader A. Philip Randolph, to the everyday acts of resistance deployed in public spaces by black working-class people, to mention but a few expressions of the more aggressive political stance many African Americans embraced in this period. In whatever form or forum, these very visible manifestations of African American opposition to the policies of segregation and racial discrimination preoccupied white federal officials who saw these claims as a barrier to wartime unity and as a direct challenge to the racial ordering of American society—which they were. Demographic and political shifts bolstered the urgency of African American claims. The emergence of black voting blocks was one political consequence of the wartime migration of African Americans to northern and western urban areas. Another repercussion was the nationalization of the race issue itself as growing numbers of African Americans outside the South clamored for jobs, housing, and fair treatment. The legal challenge to segregation had already begun, and the U.S. Supreme Court was emerging as a potentially hospitable forum for African American claims. In 1944, the Court would outlaw white primaries, and in 1946, it barred racial segregation in interstate travel. The slow trickle of precedents that would culminate in the *Brown v. Board of Education* decision also began its course through the federal court system in this era.[1]

This was not only a time of increased mobility and political visibility for African Americans but also an era of greater intellectual attention to them, as reflected by a proliferation of works by and about them.[2] The radio programs I study are a part of that larger development. I argue that because they were presented on a national mass communications medium, these broadcasts help us understand how the political issue of race was constructed for a large, diffuse audience and how that construction evolved into a search for a national language of consensus on the question of racial equality. All of this reinforces my belief that this is a rich period that demands and deserves closer study and conceptualization by historians and other scholars of African American culture and politics and indeed by Americanists in general.

If the importance of this era is not fully appreciated, the political and cultural events of the late 1950s and early 1960s tend to be cast as if they erupted spontaneously. That approach risks oversimplification of the political trajectory of African American history and the nature of the process of social change, especially in the area of race relations. The most obvious consequence of the minimal attention given to the political struggles of African Americans in the late 1930s and 1940s is reflected in latter-day civil rights historiography itself. To confine the history of the civil rights movement to the narrow frame of 1955 to 1965 and to build its narratives around compelling national figures imposes a traditional structure on a process that by its very complexity absolutely defies that tradition. This has fed a tendency to write and teach about the civil rights movement as if it were a totality that could be confined to a single decade of struggle, resistance, and resolution. Individual works of history must confine themselves to segmented treatments, as this work certainly does, but imposing a too narrow narrative on such a long and complicated process obfuscates its larger implications. Fortunately, some scholars have broadened the periodization of the movement, developed diverse local histories, or explored the work of the National Association for the Advancement of Colored People (NAACP) and other national organizations by looking at their activities at the state level.[3] These studies reveal not a single decade of toil but many decades of tedious, persistent, courageous work by groups of men and women who adopted or abandoned different strategies as the shifting times required.

Politically isolated treatments of the civil rights movement also have had the effect of closing off important questions about the fate of competing ideologies such as black nationalism and other more radical leftist structural critiques in the period between the 1930s and the 1960s. But a broader political perspective is also emerging as important links have been made, for example, between the legacy of New Deal activism, various forms of African American claims for racial equality, the limitations of racial liberalism, the emergence of the Cold War, and the contours of later struggles.[4] These approaches bring a more complex and realistic view of the decades before the 1950s, and paradoxically, they help us understand why the period between 1955 and 1965 cannot stand alone as a singular moment or a new movement. This book pursues that broader view by looking back and forth, as history requires us to do, for continuities and discontinuities, precedents and precursors, and strategies old and new in arenas new and old.

This, then, brings us to radio. Despite its ubiquitous presence in American life for over half a century, radio is a medium whose political and cultural power and influence are not yet reflected in American historiography, American studies, works on American race relations, or studies of the media and popular culture. Studies of the media in general are dominated by film and television, as are the theoretical approaches to media explored in cultural studies critiques.[5] Theories about the ideological significance of images and representations have virtually ignored radio, limiting their analytical models and textual readings largely to literary, print, or visual imagery. Historical scholarship has been particularly slow to recognize the importance of the mass media to the twentieth century as a defining aspect of American political and cultural life, which I believe it to be. Although few would disagree with that assessment, there has been relatively little scholarly exploration of its full dimensions and implications. The world of radio in particular remains largely unexplored territory for which models of historical inquiry are relatively few.[6]

Most significant by omission, in my view, is the braided relationship between the media, the political struggles of African Americans, and the continued necessity for interventionist media strategies as part of the work of advancing the race. Indeed, the bounty of attention paid to the racial aspects of the media coverage of sensationalized contemporary events and issues such as the O. J. Simpson trial, the Rodney King case, and the Anita Hill–Clarence Thomas hearings only whets the appetite for scholarly treatment of the potent historical relationship between media, race, and politics in the many decades preceding the 1980s and 1990s.[7] Disciplinary divisions among historians and scholars of popular culture and of the mass media have mitigated against creating the integrated models of inquiry necessary for considering that complicated historical relationship. The narrative of African American history has yet to incorporate the centrality of the modern mass media to how African Americans conceive of themselves as a people, how they communicate with one another, and how they preserve, transmit, and transform their music, culture, politics, and religion. This is all to say that there is much work to be done, and this book is an attempt to enter this historiographic void and bridge the analytical fissures caused by these disciplinary boundaries.

Much is assumed about radio and its history, but perceptions of the medium tend to be dominated by nostalgia or contemporary impressions rather than by historical perspective. Since we live today in a world literally structured by electronic media, it is difficult to imagine the sense of awe that national radio inspired in the 1930s and 1940s. Radio was the first medium capable of simultaneously presenting identical messages and music to millions of people in their own homes. Not only did it bring a larger, external world directly into the home, but it did so in a compellingly intimate and evocative way. Radio ownership reached near saturation levels in urban areas in the 1930s, where less than one household in ten was without a radio. Nationally, 83 percent of all residences, rural or urban, had a radio by 1940. Indeed, Americans of all classes and races had access to radio.[8] Access to a radio receiver quickly became a defining feature of life in the 1930s. Radio challenged phonograph records, film, and newspapers as a source of entertainment and news. In 1938, a *Fortune* magazine poll found that listening to the radio was the nation's favorite leisure activity.[9] In 1939, 70 percent of Americans reported that radio was their first choice for news coverage; perhaps more significantly, 58 percent stated that radio was also the most accurate news medium.[10]

This rapid rise in radio's ability to draw millions of listeners was no accident. In the early stages of national radio's development, two corporate networks, NBC and CBS, competed fiercely to develop programming and strategies that would build the mass audiences advertisers envisioned for this new medium. In August 1929, network radio broadcast its first serial, a programming innovation that introduced the concept of using a set of recurring characters to draw and keep a national audience and launched radio's rise to ubiquity. That show was *Amos 'n' Andy,* a program in which two white men, Freeman Gosden and Charles Correll, pretended to be the two black title characters. In four months' time, it was the most popular broadcast on the air, attracting an estimated 60 percent of all listeners or as many as 40 million people daily. The show's popularity created a rush on sales of radio receivers and led listeners to structure their daily routines around the show's schedule. Indeed, *Amos 'n' Andy* became so popular that President Herbert Hoover invited Gosden and Correll to the White House for a performance.[11]

Why was the show so popular? Aside from its regularity and frequency, *Amos 'n' Andy* had to have something to keep millions of listeners coming back for more, to make them take it on as a habit. In the search for a common denominator with mass appeal in the 1930s, the show's creators located it in shared stereotypes of black men and women. But this comedy by white men in aural blackface — "sounding" black by spouting their version of black dialect—was more than simple radio minstrelsy. The novelty of the show was that it constructed a contemporary black world held harmless under the reassuring surveillance of unseen listeners. *Amos 'n' Andy* parodied blacks for emulating the white middle class, in effect chiding them for aspiring to be more than whites thought they were or ever could be—financially independent, successful, virtuous. It did this by relying on a set of unstated beliefs that African American character was permeated by slyness, ignorance, and incompetence. The show's enormous appeal rested on long-standing popular obsessions with derogatory and denigrating images of African Americans.[12]

Amos 'n' Andy functioned for whites in much the same way that minstrelsy and other popular depictions of racial stereotypes had in the nineteenth century. It worked to reinforce a sense of whiteness by its contrivance of blackness, delivered by radio to a listening community of millions. The show's theme of "cultural incompetence" was used to cast blacks as the "ultimate outsider" against which whites could find a unifying sense of privilege and superiority.[13] To overlook the significance of the show's racialized content, as some scholars have done, is to ignore the source of its easy popularity with whites and the ambivalent reception it received from African Americans.[14]

Negative critiques of the show's political implications came from several sources, including Bishop W. J. Walls of the African Methodist Episcopal Zion Church. In 1931, Walls called the show "an insidious piece of negative propaganda" linked to earlier uses of popular images for political purposes: "The fact is, these clowns of the air are of the same kind as those who blackened their faces and took off black people on the stage through all the years that false philosophers and pseudo scientists were trying to make our ancestors out as those tropical animals who jumped down out of the trees."[15] That same year, the *Pittsburgh Courier* organized a national protest against the show, garnering as many as 275,000 signatures on petitions.[16] Like Walls, the *Courier* employed the term "propaganda" in its attack on the show, arguing that it was far from simple, harmless entertainment but that the portrayals on the show had very specific political implications. The *Courier* criticized the show's insulting portrayals of African American women, businesses, and fraternal organizations. Letters from readers echoed these concerns, including objections to the show for "telling the world that Negro women are more loose than other women."[17] Black businessmen complained that white bankers and businessmen repeatedly ridiculed their businesses as being run in a bungling manner like Amos and Andy's Fresh Air Taxicab Company.[18] The letters also reflected the effect of the pervasive penetration of these radio images into public consciousness. One woman wrote, "I have clashed with my employers, and their children have made my heart ache with their Amos 'n' Andy lingo."[19] Another writer asked, "I would like to know why Negroes are being called Amos and Andy in public places by white people?"[20]

When radio had first arrived, some African Americans had hoped that the medium would be an ally by broadcasting constructive racial propaganda. Instead, radio followed the course blazed by other popular media, adapting and creating virulent racial stereotyping of its own as part of making popular, commercial appeals to white Americans. Letters to the editor of the *Pittsburgh Courier* about *Amos 'n' Andy* reflected a profound sense of disappointment with the use of radio for this purpose. "It is a pity that such a great educational agency as the radio should be desecrated to such a base purpose, or end," one writer complained.[21] This sense of general frustration also was reflected in a letter that stated in part, "They are giving a false impression of the Negro, which is just as bad as the K.K.K."[22] "If Amos 'n' Andy and the rest would spend a little of their time broadcasting about the lynching and burning of Negroes in the South," another reader wrote, "I am sure we would get some benefit from their talk and America could hold her head up."[23] These letters may not represent the totality of the African American response to this show, but they do document the presence of a critical media analysis linking a set of popular images with their larger political meanings. To conclude that early black ambivalence about *Amos 'n' Andy* merely reflected an internal debate about which images to "display in public" and which images to "keep among themselves" misses this broader picture.[24] The impetus for these reactions was the lack of equalizing access for African Americans to national radio and the political disadvantages of having no control over the images and representations of the race and its concerns, now so effectively transmitted over the nation's first truly mass medium.

This was a modern manifestation of an old problem since the relationship between African Americans and the public media had always been a contentious one. The creation of the black press in the nineteenth century was a response of African Americans to the political problem of having their race and racial issues represented in white-controlled newspapers that refused them access. In 1827, when the country's first black newspaper, *Freedom's Journal,* was founded in New York City, its first issue proclaimed, "Too long others have spoken for us. Too long has the publick been deceived by misrepresentations." As one contemporary recalled, African Americans were especially frustrated that their protests against colonization proposals were ignored in the white press: "They could not gain access to the public mind: for the press would not communicate the facts of the case—it was silent. . . . [T]here was not a single journal in the city, secular or religious, which would publish the views of the people of color on the subject." But competing against the well-capitalized white press would prove daunting for African Americans in the nineteenth century since the mass dissemination of written discourse proved more effective as a purveyor of ideas than speaking before public audiences. Looking back at the antebellum period, the African American librarian Daniel Murray recalled that although many effective African American lecturers argued against ideas of racial inferiority and ethnology, "the high cost of printing . . . [restricted] their reputation to the oral tradition."[25] Encapsulated here are the persistent themes that have driven African American political thought about the relationship between media and racial politics: a recognition of the sheer ideological force of public media, a struggle for access to that marketplace of political ideas, and, ultimately, a fight for the power of self-representation in all forms of public culture.

With the emergence of each new communications medium, African Americans have had to fight the same fight that stimulated the founders of *Freedom's Journal* as the public forums for racial representations and argument shifted, expanded, and became even more "mass" in distribution and reach. When the film *Birth of a Nation* was released in 1915, the NAACP organized African American protests in its first national campaign. The film, which was based on Thomas

Dixon's racist interpretation of the Civil War and Reconstruction, benefited from the technical virtuosity of D. W. Griffith and the political reception it received, including being the first film screened at the White House, where President Woodrow Wilson praised its historical accuracy. In its fictionalized account of racial history, the film brought to life on the big screen grotesque images of African American inferiority and brutality. As the film premiered in cities across the nation, African Americans protested each screening, first in Los Angeles and San Francisco, then in New York City, Boston, Chicago, Cleveland, Pittsburgh, and elsewhere.[26] With its mass distribution and its powerful use of visual imagery and music, the film created a new discursive forum for the politics of race in which African Americans were disadvantaged once again as they struggled to find ways to meet a new challenge from an expensive, highly capitalized industry in a forum they could not enter.

These are all examples of the enduring and unrecognized strand of African American political thought that focuses on the political power of the popular media, on interventionist strategies to gain access to those media, and on the development of politically compelling images to advance black political and economic interests. As such, protests about media depictions or attempts to gain access to mass media should not be dismissed as simply efforts to find ways of presenting idealized positive racial imagery. Rather, proactive strategies aimed at influencing the representations of African Americans reveal a keen and sophisticated appreciation of the relationship between popular images, political symbolism, public opinion, and public policy. Nowhere is this set of connections more volatile than in the area of race, where notions of image and ideology rely on and reinforce each other, regardless of the medium of transmission.

Race and racial stereotyping are a deeply implicated part of radio's history, as was the case with earlier media forms. It took a conscious effort to make race visible on a medium where color could not be seen but only imagined or constructed. A fascination with African Americans and African American culture permeated radio's early programming and spurred the medium's popularity, coloring it with race like all American institutions and media forms. Black musicians, singers, and bandleaders were a prominent feature of popular radio programming in the 1920s and early 1930s. Radio comedies of many kinds, including the enormously popular *Jack Benny Show,* featured caricatured black butlers and maids. Dramatic roles for blacks were rare, as were technical or production jobs. Ironically, when the serialization and syndication model pioneered by *Amos 'n' Andy* encouraged network broadcasters in the 1930s to expand their repertoire beyond music and comedy to include soap operas, dramatic series, detective shows, westerns, quiz shows, and amateur hours, African Americans on the radio were left stranded in a declining number of comedic roles. The black bands and orchestras that had helped build radio's popularity were replaced by white bands that claimed the music as their own, imitating and redubbing it "swing." As advertisers began to rely on the identification of products with a "star" to sell their wares, they concluded that white listeners would not find black affiliation or product endorsement appealing. The overall effect was that when the medium began to reach the apex of its popularity during the period covered by this book, radio relegated most blacks on it, as one writer commented, to that " 'stereotypical conception of the Negro as a simpleton, or a "bad actor," or a doglike creature with unbounded devotion to his master or mistress.' "[27]

African Americans were especially astute to radio's unique power, reach, and influence, an awareness that emerged in the protests against *Amos 'n' Andy* and grew as the medium matured through the 1930s and 1940s. They realized that the medium's ability to present politically charged aural images repeatedly and simultaneously to millions of listeners moved what we now call "the politics of representation" into a whole new realm. Attempts to manage and influence those representations would have to become a part of ongoing strategies for African American political and economic advancement. This book offers plenty of evidence of that struggle in the 1940s as black men and women took advantage of the rare openings national radio offered them to enter this new realm of mass communications—through educational broadcasting and special programming designed in response to World War II. It remained virtually impossible for African Americans to intervene in commercial radio during this period, when national radio networks dominated and controlled the medium to the detriment of local independent stations. This arrangement only served to reinforce the capital-intensive nature of the medium, limiting access through ownership to a few large corporations. African Americans could not buy their way onto the national airwaves or influence their content through their power either as performers or as consumers. The proliferation of advertiser-supported black-oriented programming or of what would come to be called "black radio" was a postwar development that rested on the enduring appeal of African American music among white and black listeners alike.[28] Most radio programming designed primarily to reach black listeners came only after advertisers discovered the economic potency of the new urban concentrations of recently migrated African Americans.[29] By then, the radio industry itself was being transformed from a model of network dominance into a local medium as the arrival of national television advertising usurped radio's principal source of funding and forced it to depend on more locally oriented, segmented appeals — the model of radio that persists today. But those shifts had not yet occurred in the 1930s and 1940s, and national network radio remained virtually inaccessible to African American influence and control. This, then, was the predicament of African Americans as commercial radio entered its golden age in the war era.

The relationship between the radio networks and the federal government during the 1930s and 1940s was fluid and complicated. Concerns about the domination and control of the public airwaves by private capital were not limited to African Americans. A small broadcast-reform movement had been unable to stop the emergence of a corporate- and advertiser-centered model for radio that was codified in federal law in 1934. The federal government regulated "ownership" of the public airways, but critics of that model succeeded in establishing the concept that radio stations had to broadcast some amount of noncommercial educational or civic programming in the public interest. In their bid for legitimacy and recognition, radio industry officials embraced noncommercial programming as an opportunity to build their own prestige and to attract more elite audiences not normally interested in their popular entertainment offerings. As a result, the networks regularly provided free airtime for "sustaining programming," which included live performances of classical music and a wide variety of educational and public affairs programs. Proclaiming radio's commitment to public service, the networks also actively encouraged President Franklin Roosevelt's use of the medium as soon as he was elected. This was an offer he did not refuse since he and other members of his administration were eager to use radio to advance their programs and policies. The invitation opened the way for the Roosevelt administration's expert employ of the medium as a powerful new public forum that functioned at times as the official voice of the national government, often under the rubric of public service or educational programming. In turn, the radio networks watched their own powers grow as they recast themselves as indispensable to communications in a modern democracy, not only for news delivery but for public information broadcasts as well.[30]

Although the president's use of radio for his "fireside chats" was extremely significant politically, other administration officials used radio extensively to speak directly to the American people about New Deal initiatives, in part to avoid interference from reporters or editors.[31] So blatant and prevalent—and effective—was this practice that it repeatedly drew fire from newspaper editors and those who opposed the administration's policies.[32] Implicit in this criticism was the recognition of radio's growing political power. As early as 1936, both national political parties placed radio at the center of their strategies for winning the presidency, once again reinforcing the medium's national civic stature.[33] Roosevelt also recognized that his victory would depend on reaching beyond the traditional membership of the Democratic Party to unite and mobilize groups of people who ordinarily would have acknowledged no common social affiliation or shared political interests: urban ethnics, African Americans, and members of the white working class. By using the power of both radio's national range and its local targeted reach, Roosevelt was able to fashion a new urban political coalition that would remain largely invisible to itself.[34]

The cooperation between the radio networks and the federal government during the New Deal grew stronger and more intertwined during the crisis of war. Radio's strengths as a unifying medium had no better proof than its use as a source of war news and updates, including dramatic live reports from abroad and from the front. World War II was a radio war, and radio's aura of indispensability continued to expand as a result. Federal agencies also made extensive use of the medium to broadcast civilian preparedness and morale-building messages. The extent of the merger of functions between the state and radio is hard to imagine today, but at that time, the distinctions between radio's journalistic functions and its role as a medium with special public responsibilities were blurred and overlapping. In the period under study here, radio was more than a political medium; it was a political force.

Even before the war, some administration officials had concluded that radio had a unique role to play in a world of escalating racial, religious, and national divisions. In a 1936 speech at a conference on educational broadcasting, Interior Secretary Harold Ickes argued that radio's most pressing educational challenge was to eliminate racial intolerance at home.[35] As a former president of the Chicago NAACP, he was sympathetic to African American protests against segregation and discrimination. This was also a period of shifting intellectual conceptions of cultural pluralism and race and of a fledging intercultural education movement aimed at including the contributions of immigrants and people of color in the teaching of American history. At the same time, German and Russian radio propaganda was being used to divide and disparage people there, just as Father Charles Coughlin's mean-spirited and anti-Semitic national broadcasts would soon fill the airwaves in the United States. Ickes worked to emphasize radio's potential as a source of "positive" propaganda, a vision not unlike that imagined by African Americans at national radio's inception. Ickes put his beliefs into action in 1935 when he created the Radio Education Project at the Office of Education.[36] That project became an institutional home for those who wanted to use the administration's access to radio to find new ways to talk about the increasingly diverse nature of the American people and the persistence of ethnic, racial, and religious hostilities. This was one impetus for the creation of the public affairs radio programming studied in this book; the second, significantly, would be the demands of the crisis of World War II.

The first half of the book tells the history of public affairs radio programs about African Americans that were produced by the federal government and broadcast by the national radio networks, and the second half looks at programming produced by organizations other than the government. The Office of Education provided the initial opening to national radio that African Americans used to construct a public image of themselves different from that offered on commercial radio and more consistent with their political claims for racial equality. Although the agency's radio work was short-lived, it did produce two extraordinary radio series, and I devote the first two chapters to them. Chapter 1 is a history of *Americans All, Immigrants All,* a twenty-six-week radio series that presented a new state-sanctioned narrative of American history that included immigrants of all nationalities, African Americans, and Jewish Americans. The show made it possible for African Americans to broadcast the argument that they deserved the title "American" and the freedom and rights it entailed, an early example of the "politics of inclusion" that would characterize their strategic appeals to the federal government and white Americans for the duration of the World War II era.

In chapter 2, I discuss the Office of Education series *Freedom's People,* broadcast by NBC in 1941 and 1942 in response to rising federal concerns about the potential for racial unrest. African American intellectuals and performing artists on this program explored black history and black culture to demonstrate the centrality of the African American experience to the nation and to argue against continued attempts to deny blacks the rights due all Americans. A dramatic demonstration of the political use of African American history and culture, the show was produced by an alliance between black federal officials, prominent black intellectuals such as Alain Locke and Sterling Brown, black performing artists such as Paul Robeson, and racially moderate whites.

As war approached, African Americans urged the Office of War Information (OWI) and the War Department to broadcast patriotic and morale-building radio messages that included them, as I describe in chapter 3. Internal political paralysis plagued efforts at the OWI and the War Department to mount even a limited public campaign to lift "Negro morale" and build greater racial tolerance among Americans. At both of these agencies, disputes recurred about who could speak for "the Negro" and who could best determine what image of African Americans the federal government should endorse for its own more limited political purposes.

The second part of the book focuses on public affairs radio programs about African Americans that had no explicit state involvement. These shows were designed and produced by the radio networks, nonprofit organizations, or educational organizations, in some cases to follow federal leadership on the race issue and in others to supplement its weaknesses.

In chapter 4, I explain how the National Urban League gained access to national radio at a time when a public embrace of the NAACP or other black political organizations by national radio was considered a political taboo. More conservative and less aggressive than the NAACP, the league was able to turn its image and its programmatic emphasis on acculturation and job counseling into a public relations boon during the war emergency. Its guest status on national radio limited the political content of its messages, but its black entertainment radio extravaganzas advanced arguments for equal opportunity while demonstrating to the radio industry the shortsightedness of refusing to grant opportunities to black performing artists.

Two of radio's popular national political discussion forums, *America's Town Meeting of the Air* and the *University of Chicago Round Table,* are the subject of chapter 5. Because of their continuity throughout the war era, these two shows are particularly valuable sites for observing how the political subject of race, first deemed unspeakable, came to be aired and then rose to prominence as a national issue. I use these broadcasts to chart, quite literally, the evolution of a permissible political discourse about racial oppression, a development that also provides insights into the fashioning and limitations of the white liberal response to the emergence of civil rights claims. These programs also served as showcases for the political and discursive skills of black intellectuals like Langston Hughes and Richard Wright, who used their on-air appearances to challenge the boundaries of the implicit censorship surrounding discussions of the race question.

The book closes with a study in contrast in chapter 6, offering a history of two exceptional local radio programs about African American politics, culture, and history produced under the authorial control of black writers and actors for northern urban audiences: *New World A'Coming,* which first aired in New York City in 1944, and *Destination Freedom,* which ran from 1948 to 1950 in Chicago. On national radio, the full force of African American political thought rarely pierced the airwaves, but these local shows were far more consistent in tone and content with the claims and aspirations of African Americans in this era. They also provide a glimpse of the politically creative ways African Americans could use the medium of radio when they had freer rein over it.

African American political figures, intellectuals, and artists helped determine the content of all of the nationally broadcast shows discussed in this book, although with varying degrees of influence and control. The story of their successes and failures and their interactions with white officials in the federal government, at the national radio networks, and in private organizations drives this history and illuminates much broader political patterns. In the period covered by this book, radio became a powerful ideological agent and not a mere messenger; it was a new institutional force that elevated the symbolic play of politics and imagery into an influential new art form performed for a body politic of millions of listeners. In this new, expanded public sphere, the manipulation of language and political imagery became more important than ever. This book traces one example of this aspect of modern politics: attempts by African Americans to help mold a popularly accessible and politically acceptable discourse about themselves and their place in American history and culture. This transition period in American race relations, with all of its promise and its limitations, played itself out eloquently and paradoxically on a medium where language did not yet compete with visual imagery. These radio broadcasts also capture the shifting dialectic between words and actions, symbolic politics and public policies, as race riots, black migration, and black protests forced the discursive and the political worlds to respond.

At the end of the 1940s, even on a purely rhetorical level, the nascent discourse of racial equality remained fiercely challenged by a discourse of white resistance. African Americans continued their quest for a new public narrative of race that could accommodate their claims. The next battleground would be in the South and on television. But the roots of that movement were embedded in the 1940s, preserved in radio programming that broadcast a truer notion of freedom and helped nurture its growth.

PART I

FEDERAL CONSTRUCTIONS
OF "THE NEGRO"

CHAPTER I

AMERICANS ALL, IMMIGRANTS ALL

Cultural Pluralism and Americanness

At the end of the 1930s, officials in the U.S. government used radio to construct and popularize an expanded narrative of American history that acknowledged the contributions of immigrants, African Americans, and Jews. The possibility that the war in Europe would soon command American participation fueled anxieties about national cohesiveness that had heightened during a decade of economic depression. Increased fears of domestic demands and disturbances by immigrants, workers, and African Americans led federal officials to conclude that it was politically necessary to continue to foster a broader notion of acceptance and inclusiveness for the sake of national unity.

Separate streams of thought converged on the idea that one way to alleviate growing fears of internal disunity was to admit the hazards produced by prejudice and find concrete ways of confronting racial and ethnic intolerance through general public education. Federal officials at the Office of Education in Harold Ickes's Department of the Interior put radio's special powers to use for exactly that purpose when in 1938 and 1939 they produced *Americans All, Immigrants All,* a twenty-six-week nationally broadcast series that sought to

create a state-sanctioned narrative of American history that made immigrants, African Americans, and Jews visible. This significant intellectual, cultural, and political project wrestled with the complexities
of creating a new paradigm about ethnicity and race and about the
place of immigrants, African Americans, and Jews in an Anglo-
Saxon nation. Built around an all-encompassing myth of success,
this narrative construction ultimately failed to fit any of the groups
it sought to represent. However, the richly detailed internal conflicts
about the content of these broadcasts as well as public reaction to
them tell us much about the political tensions ethnicity and race generate and about the ideological importance of radio in national politics. Moreover, for African Americans, this series offered an opportunity to pursue a politics of inclusion, a strategic choice that evolved
during the New Deal and would characterize their relationship with
the federal government for the duration of the World War II era and
the 1950s and 1960s as well. For them, this show offered privileged access to national radio and an opening to broadcast the argument that
they too deserved the title "American" and the freedom and rights it
entailed.

The idea for a national radio program about immigrant contributions had several sources, both inside and outside the federal government, but the most significant was Rachel Davis DuBois, an energetic innovator in intercultural education. A white Quaker woman
from New Jersey, DuBois was a young high school teacher who had
developed materials to teach tolerance through school assembly programs dedicated to the history, culture, and contributions of various
ethnic and racial groups in fifteen schools in New York City. In 1934,
DuBois had established what was to become the Service Bureau for
Intercultural Education, a clearinghouse to help other teachers set up
their own programs for intercultural education. During this period,
DuBois also met journalist Louis Adamic, whom she credited with
reinforcing her belief that the public schools needed to address the
feelings of shame that second-generation immigrant children and
racial minorities harbored about their parents and their cultures.[1]
In this same period, DuBois also worked on political issues affecting African Americans. She developed friendships and organized
study groups with prominent black and white intellectuals and activists in New York City and elsewhere, including W. E. B. Du Bois,
whose writings had led her to take on race relations as her Quaker

Rachel Davis DuBois.
(Records of the Women's International League for Peace and Freedom,
Swarthmore College Peace Collection)

Concern and dedicate her life's work to it. She also became a member of the National Board of the National Association for the Advancement of Colored People (NAACP). All of these experiences and contacts led DuBois to adopt an intellectual approach to intercultural relations that was based on her study of and close relations with both immigrant ethnic groups and African Americans, a combination that was as rare then as it is now.[2]

DuBois's idea of a dramatic radio program incorporating her approach to intercultural education had a specific catalyst: Father Charles Coughlin's controversial national radio broadcasts, which were deeply troubling to her and the teachers with whom she was working. She remembered that Coughlin "kept yelling, 'This is a country for white Christians.' You know who's left out. He yelled it everyday over the radio."[3] Coughlin's appearances and his popularity also taught her about radio's innate powers to influence millions of listeners. She began to search for ways to counter his popular message on his medium of choice and took those ideas to the federal commissioner of education, John Studebaker.[4]

Not only did Studebaker respond enthusiastically to her idea, but he envisioned the proposed radio series as the beginning of the Office of Education's permanent involvement in the field of intercultural education. The rising tensions in Europe clearly increased the probability that dwindling federal funds would be allocated to programs that aimed to prevent domestic disturbances among immigrant groups.[5] Apparently, Studebaker saw DuBois as uniquely qualified to help him achieve his goal of institutionalizing in his agency the newly politically valuable field of intercultural education.

Concerns about the social and political implications of large concentrations of second-generation immigrants were not limited to New York City but emerged in cities like Chicago, which faced some of the same issues. These concerns were a motivating factor for Avinere Toigo of the Illinois Governor's Committee on Citizenship and Naturalization, who, at about the same time that DuBois was talking to the Office of Education, approached NBC directly with the idea of a radio program about immigrants. His request eventually made its way to the network's prominent new educational counselor, James R. Angell, who had been hired by the network after he retired as president of Yale University.[6] Angell was unimpressed by Toigo's idea and advised against it, warning his colleagues at NBC that such a show could not draw a national audience and would carry a great risk of "deeply offending one national group as a consequence of magnifying the achievements of another."[7]

Undaunted, Toigo met with Studebaker, who received him and his idea warmly since it reinforced his sense that such a series was timely and needed.[8] When Studebaker expressed his support for the series to NBC officials, Angell responded to the appeal tersely, advising his colleagues at the network: "I think I should let this dog sleep. Certainly I am not disposed to stir up the menagerie just at the moment."[9] Angell's superiors would later deeply regret and criticize his decision when the show found a home and great acclaim at CBS instead.

Officials at CBS apparently were less concerned than Angell was about any attendant political risks of carrying the series.[10] Also, CBS was just concluding its broadcast of a twenty-six-week Office of Education series on Latin America. That meant that a nexus of relationships already existed between the agency and the network and that the concept of a long-running series was quite familiar. Agreeing to use the same model for the immigrant series, CBS offered its production studios in New York City and gave the series a favored public affairs slot of two o'clock on Sunday afternoons. CBS assigned the scriptwriting for the series to the prominent writer and cultural critic Gilbert Seldes, who had recently joined the network as its first director of television programming.[11] Both CBS and the Office of Education portrayed the Seldes affiliation as symbolic of the project's prestige and high quality.

Seldes based his scripts on the research provided by DuBois and her associates at the Service Bureau, an arrangement that would have been a complicated collaborative effort under even the best of circumstances. But in this case, the differences in their political orientations gave Seldes and DuBois very different points of view about the goals and contours of the project. DuBois envisioned two general aims for the series: "to reduce intergroup prejudice in this country and to develop more appreciative attitudes among America's culture groups by dramatizing the contributions of these groups." She argued that each script's dramatic theme should focus on one group's contribution, based in part on studies of the "most common misconceptions held toward each specific group."[12]

Seldes took the opposite approach. He had a reverence for American history and national culture that emphasized the unifying rather than the differentiating historical experiences of groups of Americans.[13] His conception of the series was more nuanced yet more conservative. Seldes wanted to demonstrate the layered effect of immigration as a totality and, ultimately, to stress the primary importance of Americanization—to the immigrant and to the country. DuBois wanted the individual immigrant to be the star of the series, but Seldes was more interested in casting the political and historical process of immigration itself as the central figure. He argued vehemently against driving the programs "in the direction of destroying prejudice," proposing instead "to let a new attitude toward the immigrant transpire from the broadcasts themselves."[14] DuBois wanted to show the separate streams, Seldes the eventual confluence. DuBois believed that America was grounded by its diverse people and saw no conflict between celebrating difference and embracing unity.

This inherent philosophical difference mirrored the divide in intellectual and political debates then occurring about cultural pluralism. Divergent opinions about the forces that shaped American history were also prevalent among historians of the day as the progressive and consensus approach to American history began to unravel, in part over questions of race.[15] But in this case, Studebaker accepted a compromise, following DuBois's approach in agreeing to separate shows on individual groups but allowing Seldes to include episodes concerned more generally with the cumulative process of immigration itself.[16]

Talking openly about immigrants was a politically sensitive matter, as is most clearly revealed in the lengthy and heated debate about the title of the show. The Advisory Committee for the series, made up of a mixture of federal officials and representatives of private organizations, convened for the first time in September 1938 at CBS's offices in New York City. At the beginning of the meeting, Studebaker reported that the proposed title, *Immigrants All,* was inspired by an April 1938 speech by President Franklin Roosevelt to the Daughters of the American Revolution in which he stressed that "we *are* all immigrants." To counter complaints that the proposed title was too "backward looking," Studebaker changed it to *Immigrants All—Americans All* in order to stress unity. Putting it more bluntly, DuBois later wrote that the title change was necessary "so that the D.A.R. type of mind would not feel a loss of social security by being identified with the immigrant."[17] Some committee members agreed that the mere use of the word "immigrant" would alienate audiences, especially in the South. So *Immigrants All—Americans All* won out—

that is, until two months later when, after publicity for the series already had been prepared, CBS officials realized that radio guides in newspapers would list the title in the shortened form *Immigrants All*, which they believed was "depressing" and too limited. For that reason, the title was changed to *Americans All, Immigrants All* in the belief that the short title *Americans All* would have a broad patriotic appeal.[18] The sensitivity concerning the title of the show and the dispute over the use of the words "immigrants" and "Americans" were harbingers of the quarrels about purpose and goals that would plague the series and that characterized political discussions of ethnicity and immigrant peoples in general.

Of everyone involved, officials at the Office of Education had the most to lose if the series faltered or created controversy because Studebaker and his staff hoped that *Americans All* would be a tremendous political opportunity for their struggling agency. Studebaker predicted that the series and related follow-up activities would "constitute one of the outstanding contributions of this century to American education and to the course of democracy in general." To achieve the greatest impact, the series would be accompanied by a professionally written and published booklet. For the first time, phonograph recordings of the broadcasts would be sold at nominal cost to the public, thereby increasing the life span and the potential usefulness of *Americans All* for schools, libraries, and civic groups. More than the future of educational radio was at stake for Studebaker, who secretly hoped that the series would become "a huge success" so that his plan for a new permanent division of intercultural education would "encounter the least possible opposition."[19]

Opportunistic department officials hoped that growing public attention to the worsening conditions in Europe in 1938 increased the likelihood of wider notice for the series and thus for the Radio Education Project itself. CBS was the network leader in the early coverage of the war, having assigned Edward R. Murrow and H. V. Kaltenborn to provide in-depth coverage of events in Europe. In a September letter to a CBS executive, Studebaker's deputy William Boutwell congratulated the network on its overseas broadcasts, adding that "from the events in Europe, it would appear that this series may be especially appropriate and appealing this year." He made a similar reference in a note to Murrow in London, emphasizing that tolerance would be the program's objective. Undoubtedly, Boutwell understood that a CBS-sponsored *Americans All* would stand to benefit from the network's increasing popularity as the authoritative voice of on-the-spot news coverage from Europe. In November, Boutwell sent another round of letters to CBS executives reiterating that the recent developments in Europe gave the series "a timely appeal" that "should attract a broad audience." In a bolder assessment, he wrote that "events in Europe have certainly given us a beautiful build-up for 'Americans All.'"[20]

The final structure of the radio series blended many of the ideas proposed by DuBois, Seldes, and other members of the Advisory Committee, which of course meant that it was marked by compromise, inconsistency, and duplication.[21] Rachel DuBois remained deeply disturbed that Seldes refused to make explicit in the scripts the pleas for tolerance that she believed to be so badly needed.[22] The process of developing the scripts, publicizing the show, broadcasting the series, and creating listener aids would be complicated not only by basic philosophical disagreements but also by the involvement of three very different institutions: the Office of Education, the Service Bureau, and CBS.[23] Ironically, the rush of the schedule of broadcasts left little time for internal bickering. Once it was announced, the show had to go on, and everyone involved managed to make that happen for twenty-six weeks running—under the circumstances, a remarkable feat in and of itself.

The series as a whole sketched out a conflicted narrative of immigration, contribution, and acculturation. Each broadcast began with a standard lead-in that best captures Seldes's simple narrative vision for the series: "Americans All—Immigrants All. This is the story of how you, the people of the United States, made America—you and your neighbors, your parents and theirs. It is the story of the most spectacular movement of humanity in all recorded time—the movement of millions of men, women, and children from other lands to the land they made their own. It is the story of what they endured and accomplished—and it is also the story of what this country did for them. Americans All—Immigrants All."

Seldes wanted the series to make the argument that the economic, technological, and political progress of the nation was a project of cumulative effort. By keeping this as his focus, he sought to avoid confronting the hostilities that met many immigrant groups and the public resistance and government policies that fettered their broader participation in the political and economic life of the country. But when he turned to the stories of individual groups, it would be

AMERICANS ALL
IMMIGRANTS ALL

UNITED STATES DEPARTMENT OF THE INTERIOR OFFICE OF EDUCATION

Cover of brochure advertising Americans All, Immigrants All.
(Immigration History Research Center, University of Minnesota)

harder to maintain this focus on the ultimate designation without acknowledging the individual paths and the frequent obstacles.

Immigrants All

As a dramatic device, the paradigm of immigrant achievers did
not work very well for the English and the French, two groups
whose members possessed the greatest amount of power and freedom and were the least likely to embrace the term "immigrant" as a
part of their self-identity. Portrayed as arriving early, facing no hos-

tilities, and, in the show's most shimmering silence, finding no native inhabitants, these "old-stock" immigrant groups needed the least amount of historical explanation. The myths of their coming were already intertwined with the prevailing myth of nation building.[24] Rather than promoting tolerance of the English and the French, DuBois and others hoped that the series would promote tolerance among these groups whose "immigrant" identities had long ago been shed.[25]

Stark realities were more difficult to avoid in the series of scripts about the histories of particular national or racial groups. The often conflictual and "silenced" aspects of these narratives illuminate the caution exercised by the show's producers and sponsors in airing the less celebratory experiences of ethnic and racial groups—and of U.S. history.

Slavic immigrants were then on the minds of many Americans. The shows about them had their own special political mission. Rather than simply examining how they had earned a place in this country, the scripts sought to comfort Americans who feared, at a time of an approaching European war, that Slavic people harbored stronger allegiances to their homelands than to the United States. Two shows were dedicated to Slavic immigrants, the first on more recent arrivals from Russia, Ukraine, and Yugoslavia and the second on earlier Polish, Czechoslovakian, and Slovak immigrants. Both of these shows raised the question of loyalty directly and answered it bluntly. The first reassured listeners that Slavic people had quickly assumed an American identity: "[W]hile national memories remain, the second generation becomes American, and their importance to this country lies not in continuing ancient quarrels, but in becoming part of the unity of life here." With even greater emphasis, the second show argued that these immigrants "became deeply grateful to the country of their adoption, became fused with it like so many wholesome grafts on a healthy virile tree and became, within less than a generation, true, loyal Americans!"[26] Overall, Slavic immigrants were represented primarily as men and women who were willing, strong, and able workers.

Although Seldes wanted the series to stress that many immigrant workers were in fact invited and welcomed, the shows on the Irish and the Italians acknowledged that these two groups had encountered what the scripts referred to as "unfriendly" receptions. The expansion of the immigrant paradigm to include oppression and nativist resistance created a much more realistic and compelling narrative, but in the series, it remained the exception rather than the rule.

Hostility against the Irish was diagnosed simply as a reaction to their large numbers. An essentialist claim in the script designed to allay fears about this group of immigrants both countered and reinforced a popular stereotype. The Irish were portrayed as happy and hot tempered and as contributing to the nation "a light laughter, and a gay spirit."[27] In this way, the script extended to the Irish a kind of romantic racialism that assigned "gifts" by race, a not altogether surprising approach considering that the Irish had historically been regarded as a distinct, "dark" race in many quarters.[28]

Although the show on Italian immigrants tried much more aggressively to provide a corrective to the prejudice against Italians, the confusion about the difference between ethnicity and race continued. Using an interesting rhetorical device, the show described the arrival of Italian immigrants in terms unmistakably and eerily evocative of a slave ship: "[P]acked in filthy quarters, without sufficient air, the sick and the well together, the immigrants were hardly treated as human beings and yet their good spirits held out." According to the show's narrative, an "unfriendly spirit" was directed toward the Italians not only because of their large numbers and their late arrival but also, ultimately, because of their perceived racial difference. Although the show credited Italians with building the physical infrastructure of the great cities, they, unlike most other European groups, were portrayed as victims, "huddled in the tenements, forming little Italies," and preyed on by "sharp and dishonest bosses." Without arguing directly for less prejudicial treatment of Italians, the script made a clear plea for a more sympathetic understanding of their history and plight in the United States.[29]

The episodes about other immigrant groups were characterized by glaring omissions and other clumsy efforts to make subtle distinctions among popular stereotypes. This is nowhere more apparent than in the show "Our Hispanic Heritage." Two curious narrative interludes mark the script. The first acknowledged the existence of preconquest native peoples in Mexico and South America but excluded them in the story of the settling of the western and southwestern United States. The second attempted to treat Mexican farmworkers, then a concern because of their growing numbers, as typical immigrants, starting at the very bottom at the time but destined to rise to economic prosperity. As in the Irish script, the attempt to expose stereotypes was negated by essentialist and romantic claims, such as the comment that Mexicans possessed a "temperament [that] is a corrective to the terrific pace of the American living in big cities with their heritage of pleasurable work and many holidays, religious devotion, inner sense of beauty in everyday life." The caricatured assertions simply recast prevalent stereotypical images as harmless positives.[30]

Attempts to construct a unifying theme could not overcome the reality of the historical oppression of certain groups. The episode on Asian Americans especially lacked a coherent narrative, again because the history of Chinese and Japanese persecution and exclusion made a celebratory story a particularly dishonest, artificial, and flawed construction. The Gold Rush West provided the setting for the opening scenes of the show about Asian American immigrant workers. "It was in the 1850's that the Chinese began to come—lured by stories of California's goldfields," the narrator explained, concluding simply that soon "they were taking the place of women" by taking on jobs that traditionally were done by women. The script did not make clear, as Office of Education official Laura Vitray pointed out, that the Chinese were barred from mining gold and thus had no choice but to work as cooks, servants, and laundry workers. The script's consideration of the plight of the Chinese in the urban East took on a different cast; there, the Chinese were portrayed as "scapegoats," caught "in the struggle between Capital and Labor."[31] As in its treatment of the Chinese, the script portrayed the Japanese as patient, innocent, almost childlike people who became unwitting pawns in strikes. Because of this, the narrative argued, Japanese immigration also had been drastically restricted. It was very difficult to tell this history without acknowledging contemporary federal policies that excluded Asian Americans, so the script admitted that these policies made it "hard for them to enter completely into many phases of American life."[32]

The episodes on the Irish, Italians, Mexican Americans, and Asian Americans had to acknowledge that despite their hard work and persistence, many immigrants suffered when they came to this country. But the shows presented these difficulties as an exception to the more common experience of finding welcoming shores and easy economic ascent. Indeed, all of the scripts ended on a note of triumph and achievement, praising each group for its members' contributions as

workers or as constellations of working families responsible for settling the nation's frontier and constructing its urban infrastructure.

Although frequently calling attention to the importance of hard
physical labor as a key immigrant contribution, the paradigm nonetheless stressed the centrality of families working together. Often immigrants were referred to as "men, women, and children" engaged in
the collective effort of contributing to the nation and making it—and
themselves—stronger as a result. Significantly, "immigrants all" was
not a male paradigm but a communal one in which women and families were prominent. This image of immigrants as living in close-knit
familial groups was a less threatening and more reassuring image for
public consumption. This approach also softened and romanticized
the immigrant experience, casting it as shielded by nurturing family
units while denying the isolation and alienation that faced large numbers of immigrants who had in fact weathered the experience alone.

In reality, the immigrant success model did not fit most immigrants; it applied best to those who no longer considered themselves
immigrants, those already resting under the banner of Americanness—namely, white Anglo-Saxon Protestants. But this seeming incongruity between reality and representation did not seem to bother
listeners; in fact, the shows about specific immigrant groups met
with much popularity among immigrants and their families as well as
the broader listening public.

Many people were deeply moved by the broadcasts. One wrote:
"I have yet to listen to a program that vibrated my very being as
did your program." Even more emphatic was a letter from Ohio: "I
feel all choked up and want to cry, yet I am so happy inside that I
could shout and sing, and laugh, thanking God that I live in America
founded and built by Immigrants All, who have become Americans
All." A family from Wisconsin wrote that it had become "not only
a custom, but a ritual" for the entire family to listen to the program
because "it gives us a thrill and a tingling sensation up and down our
spine, a feeling of elation and exhilaration that cannot be matched
by anything any other country of the world offers." [33]

It was not unusual for people to listen to the show in study groups,
social clubs, or especially local organizations serving ethnic communities. Indeed, hundreds of organizations responded to the show
and requested additional information, including fraternal organizations, foreign-language newspapers, immigrant social and religious

groups, labor unions, patriotic organizations (such as the American Legion, Daughters of the American Revolution, and Veterans of Foreign Wars), religious and service organizations, groups devoted to the pursuit of tolerance and brotherhood, women's organizations, and youth groups.[34]

Many people who wrote letters and cards about the show expressed their joy in listening to the episode about their particular ethnic or immigrant group and their appreciation for a program that bolstered their self-image and self-esteem. A newspaper serving a western Norwegian community expressed "sincere gratitude" that the Scandinavian episode had written "a little-known but immensely important page in the history of the United States." Others thought the series provided evidence of how difficult the immigrant experience had been for some groups. For instance, one writer thanked the series for its "magnificent dramatization" of the struggles of Italians, who "must contend with low wages, unfit labor, and thus a lower standard of living."[35]

Elementary and secondary school teachers were among the most enthusiastic listeners. The series struck a chord with those who daily confronted the task of providing civic education to an increasingly multicultural populace through a mass medium of a different sort: the public schools. Schools made extensive and creative use of the broadcasts, recordings, scripts, and other written materials. Many teachers assigned listening to the broadcasts as homework and used the portrayal of various immigrant groups as the basis for term paper assignments, speeches, debates, assembly programs, plays, and pageants. Most praised the show for providing much-needed information about the immigrant heritages of their students, especially those who were second-generation Americans. A Minnesota teacher who taught in a community made up primarily of second-generation southern European immigrants explained that the program helped them "regain their pride of race and at the same time develop an understanding and toleration of other nationalities." One teacher from Michigan praised the "inestimable" value of the radio program "to the children of foreign born parents who must learn to appreciate their background before they can fit into American life wholly."[36] These comments were typical in their praise of the educational usefulness of the series and, more important, in their recognition of the urgent need for materials to teach students about themselves and about racial and ethnic minorities in general.

AMERICANS ALL
IMMIGRANTS ALL

24 NEW RECORDINGS OF
A SIGNIFICANT EDUCATIONAL
RADIO SERIES ARE NOW READY

"MOST ORIGINAL PROGRAM OF 1938-1939"
WOMEN'S NATIONAL RADIO COMMITTEE

"FOURTH ANNUAL RADIO AWARD"
AMERICAN LEGION AUXILIARY

2 HIGHEST AWARDS FOR RADIO

For your class room or study group

FEDERAL SECURITY AGENCY, U.S. OFFICE OF EDUCATION

Cover of phonograph recordings of Americans All, Immigrants All.
(Immigration History Research Center, University of Minnesota)

Many others also were very appreciative of the educational aspect
of the broadcasts. One writer from Kentucky thanked the Office of
Education for clearing up "misconceptions in my mind about groups
of immigrants in this country." Some listeners thought the broadcasts were so valuable that they advocated compulsory listening. One
listener commented: "If this nation were a totalitarian state, which
praise be it is not, and if I were the dictator, which thank heaven I
am not, I would command my subjects to listen."[37]

Some listeners reported that the series had changed their opinions
about immigrants in general or about specific groups, although this

response appears to have been the exception rather than the rule. One listener explained that "it has taught me how much I owe the people called 'Immigrants.'" [38]

One note of discord in the chorus of acclaim for the series came from the American League of Christian Women in Colorado, whose members wrote to complain that the series "sounded so similar to the Communist type of propaganda staged under the guise of modern social trends" that they found it difficult to reconcile the content with the show's government sponsorship.[39] But this comment was an anomaly for a series that by all measures drew overwhelmingly positive and enthusiastic support from a wide range of listeners.

The fact that this radio broadcast was sponsored by the federal government reinforced the sense among immigrant listeners that they were being claimed and publicly brought under the umbrella of Americanness. The show also worked to uplift the word "immigrant" and the people it symbolized and offered them a sense of inclusion as "Americans," something that many yearned for and embraced enthusiastically without asking for more.

"The Negro"

The yearning for official acceptance and inclusion was not limited to immigrant groups. African Americans had long struggled for more meaningful political recognition, but they also demanded the rights and privileges attendant to being called Americans. The decision to include separate shows on individual groups had the paradoxical effect of allowing for parallel, segregated presentations of each group's historical experiences as Americans. For African Americans, the inclusion of an episode on their history provided a rare opportunity to present a new image of "the Negro" to a mass audience, black and white, that had been fed a steady diet of black buffoons and mimics in the media, especially on radio. African Americans had often protested popular depictions of blacks, ranging from the NAACP's crusade against the film *Birth of a Nation* to black press campaigns against derogatory portrayals of blacks on radio, including the *Amos 'n' Andy* radio show.[40] Having been stung by radio's negative and constricted images, African Americans were extremely sensitive to the medium's influence on public attitudes. From the time of radio's inception, black organizations had been aware of radio's potential as an ally that could broadcast "constructive racial propaganda" to whites and blacks.[41] But at the time *Americans All* was aired, African Americans and their organizations had virtually no access to this powerful, centralized, highly capitalized medium.

No African Americans were among those asked to serve as advisers to the series. However, Rachel DuBois pleaded with Commissioner of Education John Studebaker to appoint "one Negro leader, not only because that is our largest minority group, but also because I consider it our most important problem, since our democracy, after all, will rise or fall according to the way we treat that group." Studebaker rejected her request, but he did agree to ask "a number of Negro leaders" to serve as "consultants" to the project, although they did not serve as formal advisers to the series.[42]

Soon thereafter, Studebaker invited prominent African Americans Alain Locke at Howard University and W. E. B. Du Bois, then at Atlanta University, to serve as unpaid consultants to the series.[43] Locke had established his reputation as an authority on African American culture and letters with his acclaimed 1925 collection *The New Negro*. His philosophical work in the 1930s and his later work in the 1940s maintained a focus on race, but it also included a search for a coherent view of cultural pluralism, which Locke saw as the key to an emerging new world order.[44] Du Bois, who was seventy years old when he agreed to help with *Americans All,* had produced a prodigious body of historical, sociological, and political works, including *Black Reconstruction,* published in 1935. In the two years preceding his work on the radio series, Du Bois had traveled widely in Austria, Germany, Poland, the Soviet Union, and the Far East. In columns written about his travels for the *Pittsburgh Courier,* he denounced Hitler's fascism and anti-Semitism and attributed his rise in large part to his expert use of modern propaganda methods. In a 1938 speech, Du Bois warned that the "fascism of despair" threatened all democratic governments because the world had "entered the period of propaganda" when people "cannot think freely nor clearly because of falsehood forced on their eyes and ears."[45]

When Rachel DuBois received Gilbert Seldes's first draft of the script for "The Negro" episode, her reaction was quick and negative. She sent a copy to W. E. B. Du Bois and Locke, dashing off handwritten notes warning that the script was "pretty bad" and urging them to suggest revisions. Seldes agreed to use most of their suggestions, despite the fact that he stubbornly refused to revise scripts for most other episodes. Seldes's reaction to Locke and W. E. B. Du Bois also is somewhat surprising in view of his own conflicted beliefs

about African Americans and their culture, which he had characterized as "inferior" in his writings in the 1920s.[46] But he also professed
to be an ardent admirer of Locke and Du Bois and, in particular, of
Du Bois's *Souls of Black Folk,* which he called a "classic."[47]

Du Bois's criticisms of Seldes's original script pointedly pushed
to the forefront politically sensitive issues Seldes had tried to avoid.
In a skit about the auction of a slave family, Du Bois inserted material that depicted the reaction of the slaves, as opposed to showing
only the reaction of their owners. He was unable, however, to convince Seldes to mention slave revolts, but a long and compelling
treatment of Frederick Douglass's life captured key aspects of slave
resistance and black involvement in the abolition movement. Du
Bois provided other thematic emphases, including the link between
black voting in the Reconstruction era and enduring social reforms,
particularly public education. Not surprisingly, Du Bois objected to
the script's focus on Tuskegee and Hampton Institutes as models of
black education, insisting that it was "falsifying history" to fail to
mention Fisk University, Lincoln University, Atlanta University, and
other black schools that had preceded and made industrial education
possible. He also argued against ending the script with a discussion
of Booker T. Washington, suggesting instead that more information about contemporary black achievements in many fields close
the show.[48]

Working independently, Locke reached some of the same conclusions. Unlike Du Bois, Locke rewrote portions of the script, and in
many cases, Seldes incorporated his changes verbatim. Locke wrote
sections on early black explorers and indentured servants who had
preceded slavery's introduction. Seldes adopted Locke's argument
that slavery's huge profits were shared by a wide cross section of
Americans, implicating the nation as a whole. Along with Du Bois,
Locke saw the story of Douglass's life as a vehicle for telling a larger
story, and he added to the script a discussion of Douglass's role
in convincing President Abraham Lincoln to raise black regiments.
Arguing that literacy and education were preparing blacks for integration into modern urban life, Locke also urged that more emphasis
be placed on black achievements in the postslavery period. Here as
well, Seldes followed Locke's suggestions.[49]

The revised script straddled both sides of historical explanation
and delicately balanced its implicit political arguments. Its narrative
portrayed African Americans as the only immigrants who had not

come of their own free will. Briefly tracing the history of black indentured servants, it blamed black slavery on the dearth of other sources of labor but admitted that slaves were stolen from Africa, "a continent with fine cultures of its own." It cast the black slave as initially patient but also as a primary actor in securing an end to slavery, citing as examples Harriet Tubman and Sojourner Truth as well as black Civil War soldiers, affording them greater recognition than was then reflected in the work of historians. However, despite Du Bois's suggestions, the depiction of the Reconstruction period remained weak and muddled. Freed blacks were portrayed as unprepared for freedom and self-governance and abused politically by the war's victors, a depiction refuted by Du Bois's own monumental work on the subject, *Black Reconstruction*.[50]

The script did not shy away from acknowledging that African Americans faced contemporary difficulties. It emphasized the importance of black workers in the South and the North and described the tenements and poor conditions blacks encountered after the Great Migration and during the depression, when they were "the last to be hired and the first to be fired." The last part of the narrative highlighted contemporary black scientists, artists, architects, educators, actors, poets, novelists, and intellectuals.

Although Seldes adopted Locke's prose and followed most of his suggestions, some of the overarching arguments were weakened or obscured by Seldes's dramatizations of historical incidents and figures. Because the more contemporary parts of the script relied heavily on Locke's prose, these sections were better written, less maudlin, and more effective overall.[51] Compared to other scripts in the series, the one on African Americans was more historical, perhaps because there was more history to explain or to explain away. As in the other scripts, current problems were acknowledged, but the show hinted only vaguely at the legal, social, and economic restrictions African Americans faced in 1938.

Locke and Du Bois worked with some success to improve the worst aspects of the script rather than launching a general attack on it. When Roy Wilkins and George Murphy of the New York City offices of the NAACP reviewed the script, they responded differently. Even after revisions had been made, both men harshly condemned the script for stressing "unduly the slave period and the Negro as a worker" and, as a result, making the narrative, in their view, "not a fair interpretation of the Negro's contributions to

American life." The trope of immigrants as workers was of course the unifying theme of the episodes on disparate groups in the series. But the emphasis on blacks as workers, in the opinion of Wilkins and Murphy, simply supported the idea that manual labor was the only contribution African Americans could ever make to the nation: "This script reads like a history of the progress of white people using the labor and talents of Negroes. It does not read like the history and progress of the Negro himself." Both men also challenged the script for failing to "marshal the information necessary to overcome the misconceptions, misunderstanding, myths and slanders which have become an accepted part of American thinking." For them, the script was misdirected and of limited political value in advancing the cause of African Americans.[52]

This view was not shared by other African American activists who saw the episode as a rare opportunity for a very different kind of presentation about black people, even within the limitations of a federally sponsored broadcast. The conflict played itself out in the NAACP as its executive director, Walter White, believed that the show was politically beneficial simply because it offered images that were in dramatic contrast to the standard radio fare involving African Americans. Rather than objecting to the script's failings, White, a keen publicist himself, worked to build as large and broad an audience as possible for the broadcast.[53] At the last minute, he persuaded Jules Bledsoe, a black concert and stage singer who had starred in Jerome Kern's *Show Boat,* to appear on the show for no fee. Apparently, White believed that Bledsoe would appeal to both black and white listeners. The live performance was a significant departure from earlier episodes, and as such, it was emphasized in the prebroadcast releases and publicity.[54] Bledsoe was to sing one song illustrating African rhythms and another demonstrating an African American style.

On radio, music can reinforce the spoken text, but the evocative power of music also can drown out textual meaning. For that reason, Locke and W. E. B. Du Bois had recommended that the show conclude with music by the black classical composer William Grant Still, whose work had won the top award in a blind competition at the recent New York World's Fair. Seldes had decided instead to conclude the show with the Negro National Anthem, James Weldon Johnson's "Lift Every Voice and Sing," an acceptable alternative. Rachel DuBois was especially concerned about any further changes in the music for the program, fearing that it might detract from the dignified tone of the script itself.[55] Earlier, she had told Rudolph Schram, the music director for the show, privately that she opposed Seldes's suggestion to use "Carry Me Back to Ol' Virginia" in the show and warned him that "Negroes are very sensitive about that song because it has in it the word 'darkie.'"[56] The song was removed before the broadcast. The fact that Seldes would favor the song shows how deeply internalized were the traditional cultural depictions of African Americans, especially in music and song, and the racism they implied. At the dress rehearsal for the show, DuBois also sensed the potential for disaster in Bledsoe's last-minute appearance, but she was unable to intervene at that point.[57]

The Bledsoe performance illustrates the many perils of live radio. The musical director found his proposed African song too difficult and scratched it during the dress rehearsal, which ended thirty seconds before airtime. The song Bledsoe sang—for nine valuable minutes of the half-hour broadcast—was "Black Boy," which Locke later characterized as a "mammy interpolation" coaching blacks to "trust the Lord and don't worry like the bees." To make matters worse, to compensate for the time lost to the lengthy song, the CBS producer cut out the prized last part of the script on contemporary blacks.[58]

Shock and disappointment were palpable after the broadcast among those who had worked on the script. Locke wrote that he had expected cuts but had hoped the broadcast would be constructive in overall effect, an expectation that was dashed by Bledsoe's reversion to a style of song that conjured up exactly the negative images that Locke and others had hoped the show would counter. Plaintively, Locke asked Rachel DuBois, "[H]ow on earth did you let them put that over on 'us'?" DuBois confided to Locke that she had been so depressed by the show that she had been unable to write him about it.[59]

Locke and Rachel DuBois were most concerned about the residual impact of the broadcast, which was to be preserved in phonograph form for educational use. Locke wrote a strongly worded letter of complaint to CBS officials, whom he blamed for the fiasco. He pointed out that "the sentiment of the song was mis-representative of contemporary Negro feeling and attitude." Locke argued that "this sabotage of the positive tone and effect of the rest of the program" could not be excused by the fact that a black singer recommended by a black leader had done the deed. He warned the network that, unless it revised and re-recorded the episode, it would receive a "volume of protests."[60] Fortunately, as DuBois soon learned, technical difficulties (a power failure) had ruined part of the original transcription, making a re-recording necessary if the episode on African Americans was to be included in the phonograph series.[61] DuBois and others continued to pressure the Office of Education to record the intended version of the script, a request that the Office of Education granted soon thereafter.[62] Officials took the extra precaution of having the final script read and cleared by Ambrose Caliver, a black member of the Office of Education staff, whom they apparently had not included initially as an adviser on the episode.[63]

The final recorded version of "The Negro" reinserted the contemporary portions of the script and eliminated Bledsoe's appearance. Later, W. E. B. Du Bois offered some perspective on the whole experience by reminding Rachel DuBois: "[I]t is not so much what you actually get in as what you keep out and I think in that respect we were fairly successful."[64]

Indeed, when one listens to the recorded version of the script, the most obvious characteristic is the virtual absence of the use of dialect as a signifier of black people, even in depictions of slaves who speak with southern accents but not with the exaggerated dialect traditionally used at the time in staged productions. Portrayals of Douglass and more contemporary figures were represented by black voices that were erudite and confident. The matter of the use of dialect had received careful attention in this episode; for example, in a section of the script on the Fisk Jubilee Singers, the stage direction specifically warned that "the following is spoken by Negroes with good education approximate southern accent but not 'Stage Negro' type." This concern about the use of dialect arose in an earlier episode in the series that included an account of the presence of a black man on the Lewis and Clark Expedition. Rachel DuBois warned Seldes that it was inconsistent to have that character speak in slave dialect and then reveal that he was invaluable to the expedition because he spoke French. "If he was cultured enough to speak French," she pointed out, "he would not use the slave dialect. Negro listeners in would resent that." Ultimately, these directions about dialect represented broader decisions about what kind and what class of black people would be presented on the show, and to that extent, as W. E. B. Du Bois had concluded, the show was an exemplary divergence from the usual radio depictions of black people. The overall tone of the recording was one of dignity, triumph, and achievement, even as pressing political issues were left unaddressed.[65]

The dispute over the musical representation of African Americans also captures far more than the accidents and risks of live radio. It clearly illustrates significant conflicts over how to change cultural conventions involving African Americans that were often considered harmless, whether in song or dialect. These conventions supported and rationalized the subordination and disabling of a whole people, some of whom inevitably deeply internalized and accepted the contested image. Removing the blinders of past cultural expectations was not a simple process for sympathetic whites like Seldes or for some blacks like Jules Bledsoe and Walter White. White, who had secured Bledsoe's appearance on the show, stood by the original broadcast, writing in a letter to Studebaker that he liked it "immensely" and that it was a "fine contribution."[66] This placed him at odds over the use of Bledsoe with not only Locke and W. E. B. Du Bois but also other NAACP officials George Murphy and Roy Wilkins, who had complained bitterly about the script even before the Bledsoe incident.

The conflict among African Americans about which image to present obscured their implicit agreement about the political importance of media images and their frustration about their limited opportunities to exert control over their self-representation. The idea that images carry and reinforce political meanings had been a consistent strand of African American political thought beginning with abolitionist strategies that relied on symbolic manipulations of the image of the slave and of slavery, whether through visual images or literary images in the genre of the slave narrative. The antilynching campaigns of Ida B. Wells and the NAACP employed visual imagery in similar ways. In addition to these attempts to use images of oppression and racial violence as a political strategy, African Americans have campaigned repeatedly against the unending parade of stereotypical, derogatory, and one-dimensional portrayals of themselves in every form of media, from print journalism to film and radio and later to television. Indeed, the evolution of each new mass medium has been marked by virulent racial stereotyping, requiring African Americans to protest each new form in kind. Blacks had limited access to the highly capitalized white-controlled national media of film and radio and therefore virtually no influence over the representations of African Americans they offered and little if any opportunity to counter them effectively via those communication forms. Radio had the added dimension of being a medium capable of distributing images simultaneously to millions across the nation, thus creating a powerful new public arena of unprecedented proportions and penetration. The specific problem was the limited access that blacks had to this influential mass medium, displayed here in disputes about which image to present in taking advantage of this rare opportunity to exercise some influence over aural presentations of black people.

Considering the financial and political constraints under which the Radio Education Project operated, it is remarkable that federal officials agreed to re-record "The Negro." It is important to recall, however, that the project had staked its future on *Americans All* and that Studebaker hoped that a successful series would enable him to gain congressional funding for a new intercultural education division. "The Negro" was only the sixth episode in the twenty-six-week series. It was much too early in the series for the office to become entangled in public controversy of any kind. Perhaps Studebaker and other federal officials also were coming to understand, as Rachel DuBois had argued earlier, that African Americans were the nation's largest minority group and that their poor treatment constituted its most important problem—one that could be silenced no longer. As the war neared and the urgency of at-home unity increased, the ability to threaten and to create controversy would become an even more potent tool for African Americans and one they would exploit fully.

Listener response to "The Negro" episode placed it about on par with the response to earlier shows in the series.[67] One of those who expressed thanks for the show was a Philadelphia teacher of African American children who wrote: "I have . . . been very interested in finding 'heroes' for them among their own people, and trying to develop in them the feeling that their people too have contributed in making our land the great country it is."[68] Although the number of African American letter writers appears to have been quite small, the enthusiasm in letters from African Americans was clear. A listener from New York City wrote: "As a member of the Negro race I was extremely gratified at your fair and unbiased portrayal of the parts my race have played in helping to make America a better place for all groups to live in, even though at times we were somewhat discouraged by intolerant individuals who seem to enjoy a sadistic pleasure in denying us our inalienable rights. . . . I feel that your program is a forerunner to the fulfillment of our dreams." [69] Another writer used a different set of images in commending the series, suggesting that "no doubt Booker Washington turned over in his grave with pride." Requests for further information on the series came from local branches of the NAACP and the National Urban League, as well as from the *Chicago Defender*.[70] Despite the disputes among African Americans associated with the broadcast about which set of images to present and how, the broader political argument they had intended made its way across the airwaves.

Significantly, the episode did not seem to generate negative responses from listeners, another indication that it had succeeded in walking the narrow line of politically acceptable arguments about "the Negro." African Americans involved with the broadcast had relied on the politics of inclusion, a strategy grounded in making claims for themselves based on their Americanness. This approach of arguing for inclusion as opposed to a more nationalist strategy of embracing exclusion characterized this period, a time when segregation and discrimination still reigned. Bolstered in part by New Deal rhetoric, many African Americans at the beginning of the World War II era were poised to demand not only inclusion but also, more important, the full benefits of Americanness, a strategy pursued into the 1950s and 1960s.

"The Jews in the United States"

For African Americans, *Americans All* represented an opportunity to make themselves an identifiable and visible part of American history, although there was internal conflict over how best to do that. For Jewish Americans, the series raised a different set of questions about the relative benefits of a politics of visibility versus a politics of invisibility. Indeed, one of the most difficult issues that faced the planners of the series was the conflict over whether to dedicate a separate episode to the history of Jewish Americans. The discussions of that question reflect differences at the time within the American Jewish community concerning identity and strategies for group advancement, disagreements that often reflected the diverse national origins and social classes within that community. Ultimately, this conflict also illustrates similarities and differences between strategies adopted by African Americans and those adopted by Jewish Americans during a period that saw the emergence of a new public discourse both about race prejudice and about anti-Semitism.

Rachel DuBois's views on the question of Jewish identity were influenced by a course she had taken under Rabbi Mordecai Kaplan, the influential leader of the group of progressive Jewish thinkers who eventually came to be known as the Reconstructionists. Kaplan and his supporters believed that Jews were members of both a religious and a cultural group, or as he called it, a "civilization." Anti-assimilationist in orientation, Reconstructionists represented one side of a division within the Jewish community reflecting differences in immigrant and economic status. On the other side, many Jews who were third- and fourth-generation Americans, especially those of German origin, opposed any movement toward ethnic separatism and favored an assimilationist stance. Because Kaplan's position resonated with her own approach to racial and ethnic identities, DuBois embraced it, as did many people who worked with her. In 1935, DuBois incorporated her beliefs into a book, *The Jews in American Life,* designed for use by classroom teachers and civic groups.[71]

When DuBois realized that her salary and the expenses of her work on *Americans All* were being underwritten by the American Jewish Committee (AJC) at the Office of Education's request, she expected a serious conflict about whether the series would feature a separate show on Jews. The AJC had helped fund her earlier work in school assembly programs in New York City but had withdrawn its support when, according to DuBois, it disagreed with her treatment of Jews as a separate ethnic and cultural group, preferring instead that they be considered a religious group. DuBois recalled that an AJC official had concluded his argument about that decision by asking her, "You don't have a separate program on the Baptists, why on the Jews?"[72]

Considering their obvious difference of opinion, it seems puzzling that the AJC agreed to pay DuBois's salary. Undoubtedly, the press of events in Europe played a role. Frank Tager of the AJC later recalled that concerns about the impending Jewish refugee crisis "probably" motivated the decision to give financial support to the series. This was not a concern limited to the AJC; some members of the Advisory Committee also saw the series as a way to foster greater public support for the admission of European refugees, especially German Jews. For example, James Houghteling, the commissioner of immigration and naturalization, commented at the first planning meeting that he hoped the series would ease public hostilities toward a likely new wave of refugees.[73]

Still, the series planners spent as much time debating the question of whether to air a separate Jewish episode as they had spent discussing the series title. Avinere Toigo, who had brought the idea for the series to the Office of Education, argued that a Jewish episode would do more harm than good and that Jews would be better off if they were omitted from separate consideration, which would place too much emphasis on the "Jewish question" and risk stirring up more hostile attention. Philip Cohen of the Office of Education disagreed because he believed the only way to break down prejudices against Jews was to do so directly: "In general this thing is fixed as a Jewish problem in the minds of most people and should be faced as such." DuBois also argued vehemently that an individual show was the only way to fight prejudice, reasoning that the Jew was already separated negatively so it was necessary "to separate him positively in the minds of the people."[74]

Ironically, it was Gilbert Seldes who most strongly defended the need for a show about Jews, a position at odds with his general opposition to separate shows for various ethnic, racial, and national groups. He explained that the existence of "most definite prejudices" against Jews as Jews regardless of their immigrant status or nation of origin required that they be given separate treatment. Studebaker found Seldes's argument convincing and agreed that a frank acknowledgment of the "fact that there is a Jewish problem" in an episode devoted to Jews was necessary.[75] Seldes's own relationship with Judaism has been called "paradoxical if not ironic" and was rooted in his own experience as the child and grandchild of nonobservant Jews. In the 1920s, Seldes "rejected Zionism because a Jewish state would 'diminish the internationalism of the Jewish people.'" Yet Seldes had not escaped the sting of anti-Semitic comments from other intellectuals and writers early in his career, a memory that undoubtedly influenced his argument here.[76]

Some external opposition to the idea of a separate Jewish program remained, however. Objections to a separate show were raised by Mrs. Arthur Hays Sulzberger of the *New York Times* publishing family, who warned that "there is a great mistake being made at the present time in regarding the Jews as a race, when they are merely a religious group." Theresa Mayer Durlach, who was then serving as vice chair of the Service Bureau, which was cosponsoring the series, appealed directly to Studebaker, warning him of "the danger of a false and inadequate handling of the Jewish problem." She agreed that public attitudes called for a separate program, but she feared that such a program would reinforce the notion that Jews were a separate race. As a solution, she urged that the series also include programs on the history of Catholics and Protestants or that one program be dedicated to all religious groups. "Parallelism of treatment," she believed, would make it possible to cast the "Jewish problem" as "merely a part of the general suppression of all liberal and nonconformist elements in the present totalitarian states . . . the difference being one of degree rather than of kind."[77] Studebaker retained the final plans for the series and did not accede to the pleas of either of these prominent and influential women.

Prior hesitations concerning the issue of a separate program seemed to dissipate as the time for the show's February 5, 1939, broadcast approached. Perhaps the worsening conditions in Europe or Father Charles Coughlin's blatant attacks in the fall of 1938 convinced many Jews that urgent steps were necessary to try to counter anti-Semitic appeals. Once the decision to broadcast a separate Jewish program had been made, members of Jewish communities and organizations worked closely with the Office of Education to generate a large listening audience for the broadcast.[78] Publicity efforts for the series as a whole benefited greatly from the work of Arthur Derounian, a brilliant publicist at the Service Bureau.[79] He personalized the publicity for each show, preparing special press releases that catered not only to media outlets but also to civic, religious, educational, and social organizations.[80] Derounian's formula worked especially well for the show about Jews, in part because of the cooperation of many large organizations then serving the various needs of the American Jewish community.[81] Many of these national organizations also took it upon themselves to spread the news about the broadcast among their members. Hadassah, for example, sent a letter to all of its chapter presidents, urging them to organize listening parties, with each member bringing a "non-Jew to listen to the program."[82]

Some members of the Jewish community cast the show as a historic opportunity for American Jews to use radio to help combat anti-Semitism. An impassioned announcement of the show in the *Hebrew Union College Monthly* ended with this plea:

Education by air is still in its infancy, but "Americans All, Immigrants All" gives us a glimpse of radio's tremendous potentialities for good, as Germany and Italy are demonstrating what a powerful instrument for evil it can be in the hands of dictators. Indeed, this radio series becomes all the more significant when we remember that, at the very same time that totalitarian governments are cruelly suppressing their minority groups, our government is making a special effort to show the *assets* of *its* minorities. . . . Never have the benefits of cultural pluralism been so widely heralded![83]

The article also urged rabbis to notify their synagogue members about the show, to use it as a basis for sermons, to sponsor "bring a neighbor" listening parties, and to write the Office of Education to thank it for the broadcast.[84]

"The Jews in the United States" recounted the long history of active Jewish participation in American life, focusing on historical material provided by Rachel DuBois. The script pointed out that Jews fought at Valley Forge, helped finance the Revolutionary War, served in the War of 1812, died on both sides of the Civil War, and fought in World War I. One reenactment illustrated that when faced with a choice between observing a central tenet of Jewish faith and serving the country, Jewish Americans chose patriotism. The show credited Jews with introducing new forms of organized charity and developing the notion of arbitration. Samuel Gompers was singled out for his work as a labor organizer and Lillian Wald as a settlement house founder. Jewish philanthropists, musicians, and entertainers were featured, including Nathan Straus, Julius Rosenwald, George Gershwin, Oscar Hammerstein, and Irving Berlin. Individual Jewish achievers in many other fields were noted, such as Louis Brandeis, Benjamin Cardozo, and Felix Frankfurter in law and Joseph Pulitzer in journalism.[85]

The script held up the Jews in the United States as an example of a people composed of many nationalities who had learned to live together without internal prejudice. This was at once a direct challenge to the notion that Jews were themselves a single race and a plea, although unstated, that others follow their example of coexisting with differences without bigotry. The concluding portions of the script referred to the history of Jewish oppression elsewhere but not in the United States: "More persecuted than most peoples before they came to the United States, the Jews came to this country with a background of sorrows. In many lands they have been barred from taking part in the national life. They were able to make their contribution to American life, because the whole of American life was open to them, as to all others, without discrimination of race or creed. They have helped to build the United States, because the United States has welcomed them as Americans all."[86] By broadcasting the myth that Jewish achievements resulted from their being accepted in the United States without prejudice or bigotry, the script simultaneously extolled the virtues of this country and diminished any true measure of Jewish success (since it assumed that no barriers to it had existed). The characterization also walked the fine line of highlighting Jewish contributions without praising them to such a degree that they would engender greater public resentment.

Overall, the script presented Jews as old-stock middle-class Americans, more like the English and the French than the recent immigrants who could offer only their huddled laboring bodies, as the Irish, Slavs, Italians, and many others were depicted in the series. As a consequence, the show had a clearer narrative line than many of the other shows. With its emphasis on individual actors rather than on a mass of Jews, it made a case for their acceptance and support based simply on their status as long-settled and deserving Americans, delicately balanced between claiming group status and avoiding it. Internal disagreements on that question had permeated broadcast planning, just as questions about which image to present had plagued the episode on African Americans.

It is hard to imagine a more politically cautious script than the one on Jewish Americans. Although American Jews were made visible, a listener unaware of anti-Semitism, Hitler, the encroaching European war, or the flood of Jewish refugees would have gleaned nothing of any "Jewish problem" from this broadcast. Certainly none of the shows in the series openly addressed contemporary concerns, but in many instances, they confronted stereotypical beliefs, however clumsily. No attempt was made to do that in this script. Education and CBS network officials probably feared that a direct attack on anti-Semitism would have been perceived in the administration, in Congress, and elsewhere as a political pitch regarding the Jewish refugee issue or perhaps even the question of U.S. intervention in the war.

However subtle a message the show intended, the overwhelming public response to it revealed that the prepublicity activities had been successful and, more important, that it had met a current need for some discussion about Jews. The level of public response to it exceeded that of any previous broadcast.[87] A study of the mail received for the series as a whole revealed that Jews were the most vocal group in expressing appreciation for the program dealing with their own group. A writer from Kansas called the episode "educational, beautiful, and touching, in view of the raw deal we Jews are getting at the hands of various ignorant, bestial tyrants." Thanks for the show came from Texas "in the name of the Orthodox Jewish Community of Beaumont . . . for representing the Jewish contributions to the building up of this dear land of ours." Many non-Jews expressed similar sentiments, such as a California listener who commended the usefulness of the episode "in times like these to check the rising tide of anti-semitism in this country." Others also remarked on the timeliness of the series and the episode on Jews in particular. "God only knows how much your work is needed at this time," wrote one listener; another said that no words could "express its value and greatness, especially at this troublesome time."[88] Although some listeners made specific references to anti-Semitism and Nazism in Europe, others were concerned about anti-Semitic sentiment in the United States, including a writer from Maryland who said the series "will go a long way in combating the evil influence of Father Coughlin's broadcasts."[89] Whatever fears the show's planners harbored about the political expediency of a separate broadcast, many Jewish listeners and others attuned to the presence of anti-Semitism heard the show's implicit pleas for an end to such prejudice.

Although officials at the Office of Education were extremely pleased by the level of response to the show, some of the initial concerns of others about the risks of spotlighting Jews re-emerged in the show's wake.[90] Edward Bayne of the Service Bureau worried that the separate broadcast might have done more harm than good. In a letter to Jeanette Sayre, a staff member on the respected NBC radio discussion show *America's Town Meeting of the Air,* Bayne explained that "some of us have felt that the emphasis on the Jew made by giving him a special program, and constantly referring to him as a national group although he may be a German Jew, Russian Jew or what not, has added fuel to the conception that we are pro-Jew and the whole series is designed as pro-refugee propaganda." Sayre responded that in her informal surveys among her radio colleagues she had found some negative reaction but not because of the separate broadcast on Jews. Rather, and more interesting, conclusions were being drawn based on the fact that CBS sponsored and broadcast the series, as she wrote in her response: "The comments I heard were that since the personnel of the Columbia Broadcasting System is largely Jewish, it is likely that they were putting on the program in order to arouse opinion in this country in favor of letting in the refugees. Two people seemed to regard this as a rather Machiavellian attempt to build up this attitude subtley [*sic*] by bringing in the attitude toward the Swedes, Irish, etc. in which Columbia Broadcasting was not really interested."[91] The initial sensitivity concerning the question of a separate episode about Jewish Americans had been complicated by this unstated fear among CBS officials that Jewish involvement at the network and with the series made the company more vulnerable to such claims. While African Americans were struggling to gain some power over their own images on the medium of radio, Jewish Americans were fearful that perceptions about their influence within the industry placed them in a politically vulnerable position.

Apparently, this was not an unjustified fear since the episode also attracted the attention of Father Coughlin, who used evidence from the show to make explicit claims about Jewish wealth and influence. In an article about the broadcast, he highlighted only one aspect of the script: "It lauded the Jews in the making of America, declaring that they were responsible even for Columbus' discovery of America. According to United States Commissioner of Education, Studebaker, who spoke for the Jews on this program, Columbus would never have discovered America if it were not for Jewish financial background."[92] Whether these views were widespread or not, their existence demonstrates that the separate broadcast on Jews had not escaped the prejudices and resentments some feared it would raise and others intended it to remedy.

Intercultural Education

The level of public response to the entire series of *Americans All* broadcasts exceeded all expectations. The monthly mail totals for the show exceeded those for all of the network's educational programs. Over 80,000 pieces of mail were received by the Office of Education in response to the broadcasts. Even after the series concluded in May, for months the Office of Education continued to receive 300 to 400 pieces of mail a month about the series.[93] The response to the series was so great that the Office of Education and the Service Bureau, both of which were facing dwindling funds, were unable to satisfy the requests for printed materials that were promised at the end of each broadcast. Agency officials reported that the office was receiving "very angry protests" from people who had not received promised leaflets and other materials, some of which had not yet been produced.[94]

Americans All also received awards and critical acclaim from within the radio industry for being an innovative and important achievement. It was granted the highest annual award of the influential Women's National Radio Committee, making it the first government-sponsored program to be selected for that distinction. The citation described the series as "the most original and informative program introduced on the airwaves" that year. This and other awards especially pleased officials at CBS, coming at a time when its European news coverage had also brought it increased respect and attention.[95] Perhaps the best measure of the show's success was the consternation generated among top-level NBC officials after the year's radio awards were announced. Earlier, after CBS had agreed to broadcast the series, David Sarnoff, president of RCA, which owned NBC, had demanded to know why NBC had refused it. At that point, Vice President John Royal had assured Sarnoff that the quality of the show was low and that "we have not lost anything." Once the radio awards were announced, Royal mounted a feverish campaign to defend the earlier decision and to minimize the value of the awards. But a *New York Post* radio columnist fueled Sarnoff's anger when he wrote that CBS, as evidenced by its awards, was "so far ahead in the field of educational radio that the race isn't even close."[96]

Called to make an accounting for refusing the show, James Angell, NBC's educational adviser, disparaged the *Post* radio columnist, describing him as being of a "somewhat pinkish complexion" and interested only in "radical" radio programming. But attempts to diminish the significance of the series and its awards failed. When NBC vice president Niles Trammell erroneously concluded that CBS was discontinuing *Americans All,* he inquired whether the series could be brought to NBC, arguing that having an educational show "that was rated as the No. 1 show would be a rather good prestige builder for the company."[97] NBC had recognized too late the appeal and importance of an examination of the history and contributions of immigrant groups to the listening public. This series had the ironic effect of making immigrants, Jews, and African Americans more visible to network officials who had given little thought to the idea that they might be interested in "high-brow" educational and public service broadcasts.

Office of Education commissioner John Studebaker hoped to cash in the political currency he believed he had earned from the success of *Americans All*. When the series ended, he went forward with his plan to secure the first permanent funding for educational radio in his office.[98] The political atmosphere in which Studebaker made his appeals for the support of educational radio was not a friendly one. By 1939, congressional attacks on all New Deal programs had reached a new intensity. As another presidential election year approached, suspicions in Congress about the political use of federal radio programming led to a reduction in 1939 in all federal emergency funds dedicated to radio activities. Despite that change, the Roosevelt administration proceeded to transcribe a series of thirty-two fifteen-minute radio interviews with cabinet members about the work of their agencies. President Roosevelt himself opened the federally funded series by announcing that "only through the radio is it possible to overtake loudly proclaimed untruths or greatly exaggerated half-truths." "The people," he concluded, "have a right to expect their government to keep them supplied with the sober facts." This action merely confirmed fears among some influential members of Congress that the administration would employ its radio programs to help reelect the president. As a consequence, the president's bold assertion was the beginning of the end for federally sponsored radio programming. In 1940, the Congress retaliated by withholding all funding for any additional broadcasts, which Representative Everett Dirksen called "clap trap and tommy rot" from "nothing but a political bureau."[99]

The Office of Education could not escape this controversy despite the fact that it had carefully tended its congressional relations. Federal officials had deftly handled two letters from senators expressing concern about *Americans All*. One senator reported that some of his constituents were worried that the broadcasts were propagandizing in favor of increased immigration. Studebaker replied that the series was "concerned with the problem of assimilation of those already within our gates" and explained that the shows had generated "a vast chorus of praise for the timely Americanism of the series." Another senator complained that the episode on Scandinavians had tried to sell the cooperative movement, which he viewed as anticapitalist, to Americans. In a lengthy response, Interior Department officials denied that the department was using *Americans All* for propaganda purposes of any kind, although in fact the episode did include an unusually long and laudatory discussion of cooperatives.[100]

When Studebaker testified before the House Appropriations Committee in favor of the permanent funding of radio programming in his office, the negative mood toward federal radio broadcasts worked against him rather than concern about any particular program. The committee refused outright to consider his request, challenging the Office of Education's authority to engage in radio programming. In a powerful display, the committee assaulted the office's radio work on every front, attacking "the right of the office to broadcast at all, the right of the Office to use Emergency funds for this purpose, the success of the programs, the standard of programs produced, the question of whether the programs were 'propaganda' or 'education.' "[101]

Finally, in desperation, federal officials appealed directly to Eleanor and Franklin Roosevelt. One official emphasized to Eleanor Roosevelt that "in this critical period of our history, when democratic education is so much needed, it is heart breaking to think of these broadcasts going off the air." She passed along the entreaty to her husband, adding in a note that she thought the Radio Education Project's programs had done "much good" and asking whether funds could be found to continue them. Eleanor Roosevelt was very familiar with *Americans All,* having devoted one of her "My Day" columns to boosting it and Rachel DuBois's work on intercultural education. When no assurances were forthcoming from the White House, Studebaker tried another tack. He submitted a scaled-down proposal to the president and recharacterized his office's radio work as an essential part of the anticipated national defense publicity effort. Advised by a member of his staff that "there would be real resentment if these programs were slipped over under the name of defense," President Roosevelt turned Studebaker's request down, explaining that it would be improper to include an "educational program" among those he would be asking the networks to carry about the defense buildup.[102]

Although not heeded at the time, Studebaker's argument that his work be included as part of the mobilization for war was politically prescient. Soon thereafter, federal officials would actively enlist the ideology of "cultural pluralism" as a key ingredient in its at-home strategy during World War II, urging the nation to unite around and across ethnic, racial, and religious differences as "Americans all." But caught here between the end of the New Deal and the beginning of World War II, the Office of Education found that its mission in educational radio and intercultural education had been eclipsed by partisan political circumstances. What remained of *Americans All* were scripts, printed materials, and phonograph records, which, despite other disappointments, constituted an impressive educational product in an area with a dearth of resources. But without a federal locus for intercultural education, the movement to teach tolerance through the public schools and via radio remained inchoate, stymied by the lack of alternative funds and curriculum resources.

Perhaps more than anyone else, Rachel DuBois had feared such an outcome. She had seen the broadcasts as an organizing tool for spreading the gospel of intercultural education to a permanent national "audience" of teachers and civic leaders. Left without any federal imprimatur, intercultural education would never have the institutionalized national network that DuBois had envisioned. DuBois herself would become entrapped in the conflicts and changes in intellectual and political currents about cultural pluralism and about whether there should be continued emphasis on distinct ethnic and racial identities. In the fall of 1939, a persistent divide in the Service Bureau over these issues erupted into a bitter philosophical and political challenge to her fundamental approach to intercultural education.[103] This struggle culminated in DuBois's resignation from the organization she had founded, along with other staff and board members. DuBois also believed that her ouster was part of a general campaign to deny her the directorship of the organization, quoting Frank Tager of the AJC as telling her bluntly that he would no longer help fund the Service Bureau "if it is headed by a woman."[104] After her resignation, DuBois and other Service Bureau staff formed what would later become the Workshop for Cultural Democracy. Later, DuBois became more active in interracial relations work following the race riots of the 1940s, and in the 1960s, she conducted workshops and projects for the Southern Christian Leadership Conference.

Gilbert Seldes's involvement with *Americans All* did not alter his deeply celebratory version of American history. He held firmly to his original views on the dangers of a pluralism that placed too much emphasis on individual groups. Writing a quarter century later in 1964 about *Americans All,* Seldes proudly noted that 20 percent of the letters about episodes on individual ethnic groups in the series had come from people outside of the group featured in the episode, ignoring the obvious corollary that an overwhelming majority came from those who were members. Moreover, looking back at the series, Seldes held up as his personal favorite the first episode, which " 'dealt with America as a whole, not with any one group.' " [105]

Dorothea Seelye's 1941 master's thesis analyzing the mail sent to the series showed just how eager members of ethnic and racial groups had been to hear about their own history. In fact, a large percentage of writers showed a particular interest *only* in the program dealing with their own nationality: "Such programs have either taught them facts about their own background they were much interested in knowing or have given them a self-assurance and a sense of prestige about their own nationality." But for Seelye, the series ultimately failed as propaganda because "it did not promote tolerance in the sense of changing a great many individuals' opinions from very prejudiced to very unprejudiced." [106]

Seelye was disappointed by her own findings because she had high expectations that radio was a medium well suited to spreading positive propaganda due to its evocative powers and its ability to reach millions. In the end, Seelye presented the elaborately produced and highly popular *Americans All* as an example of the limited usefulness of radio in changing public opinion. Early radio theorists believed that radio's accessibility and ubiquity made it a potent ideological medium. But others, like sociologist Paul Lazarsfeld, had come to believe that propaganda on the radio, whether positive or negative, would have to be reinforced through the efforts of other large social institutions. [107] Without such reinforcement, the millions who were not predisposed to be open to a radio message about tolerance or immigrant contributions would remain unmoved, never turning the dial to a program they considered irrelevant or offensive. That institutional ideological support would soon come from the federal government when it embraced the notion of tolerance as essential to the war effort and made it an integral part of its pleas for national unity, not only via radio but on every available popular medium. As one historian has noted, when the war came, the federal government "mounted a concerted propaganda campaign that stressed the centrality of cultural pluralism to the nation's war aims." [108] *Americans All* was a harbinger and model for that campaign.

All of the evidence shows that people who chose to listen to the broadcast responded to it with enthusiasm, even allowing for the probability that those writing in were among the most eager listeners. The broadcasts were apparently most evocative among those in need of being recognized as Americans and as having contributed to the country's history and vitality. Creating a narrative that recognized the contributions of immigrants and finding ways to teach that narrative was the underlying project, and as such the series was an impressive body of cultural work. This accounts for the value of the series to schoolteachers who were struggling to fashion a version of American history that recognized the existence and contributions of people other than white Anglo-Saxon Protestants, a new history that resonated with the real stories of their increasingly diverse students. This was the common vision of the political and ideological power of history that united thinkers as diverse as Louis Adamic, Carter G. Woodson, Rachel DuBois, Alain Locke, and Gilbert Seldes.

The show worked to fortify a sense of American patriotism among immigrants who for the first time heard the national government embracing and claiming them as valued and valuable citizens and attempting to erase the negative connotation attached to the word "immigrant." Although the message of tolerance may have missed the intolerant, the aims of creating a sense of belonging among immigrants and of loosening lingering nationalist ties seemed to have met with some success. "Americans all, immigrants all" was a less complicated claim than a directive urging tolerance, a plea that proved too controversial to be voiced in the series. For one thing, a true notion of tolerance depended on an ideal of equality for all—legal, political, and otherwise—that was not acceptable in rhetoric or in reality, particularly when it was expected to apply to those seen as "more different"—most prominently African Americans.

The series also revealed the strategic divisions that existed within the African American and Jewish communities about the best way to move as a group toward that ideal, as well as the tactical differences between the two groups. African Americans desired notice and recognition, while many Jewish Americans argued for invisibility.[109] By the late 1930s, dominant African American activists and organizations were pursuing, as a matter of strategy, a politics of inclusion, concentrating less on the form of that inclusion and more on acquiring the rights and opportunities that came with the title "American." They also recognized the increasing strategic importance of radio as a means of political communication and struggled for access to and control over a medium that excluded them. The necessity of gaining power over the constructions of their race had taken on even greater urgency with the arrival of the nation's first truly mass medium of communication. On the other hand, Jewish Americans associated with this series and the network were in a relatively more powerful position in the radio industry as decisions were made about how the issue of anti-Semitism could best be addressed on the series. At the same time, perceptions of Jewish influence left Jewish Americans more vulnerable to claims that they were manipulating the medium for their own political purposes.

Timing is everything in radio, and it may have been the most significant factor in the program's success. Listeners imposed their own interpretations on the series's underlying message, but the relevance of the project went unchallenged and was repeatedly reaffirmed by public response. Whether it was viewed as a plea for Americanization, increased immigration, or group pride, the series was important to a wide variety of people in 1938 and 1939 because it addressed issues that were already on their minds. Undoubtedly, Hitler's ascent in Europe helped propel the topics of race, ethnicity, nationality, and national allegiance to prominence. Hitler took mainline racial thinking to extremes that evoked discomfort, horror, and a sense of uneasy recognition. Most important, Hitler's rhetoric resonated so closely with the predominant racial thinking in the United States that it created a demand for a new, differentiating language of tolerance.

But it would be a mistake to assume that public awareness of these subjects originated in 1938 or was stimulated by foreign affairs alone. The threat of another world war capped off a decade of depression, unemployment, insecurity, and social upheaval.[110] Some of the anxiety of that period was turned against America's "others," whether judged to be different because of race, national origin, or religion. Even before war approached, the growing realization that racial and ethnic diversity was to be a permanent feature fed the idea that a safe future hinged on some transcending national sense of unity or community. These years also marked a transition in the continuing search for a way to manage the many different peoples whose large numbers, especially in urban areas, rendered them no longer invisible. The aggressive utilitarian patriotism of the war era would soon follow, but the rhetorical drive for national unity that preceded it was already in gear.

During this period, there was a renegotiation of the political currency of Americanness itself as ethnicity was coined as a denomination between whiteness and color. Whiteness, like all manifestations of "race," is a myth constantly in search of a natural foundation. To make a different point, Barbara Fields has argued that "Afro-Americans invented themselves, not as a race, but as a nation." Certainly white Americans have done no less and have been privileged to merge and expand the definitions of themselves as both race and nation as historical exigencies have required. That process was at work during the late 1930s and early 1940s, a time that has been referred to as "the historical moment" when members of European ethnic groups "felt fully accepted as Americans." [111]

As Toni Morrison has observed, "American means white." Obviously for African Americans, merely qualifying as "Negro" Americans was not enough; they also wanted to claim all of the freedoms and rights of citizenship that extended to white Americans. Thus, at the same time that the umbrella of Americanness was hoisted over immigrant descendants, the lingering necessity for political differentiation fed the social and class-identified construction of "ethnicity" itself—a construction eager to avoid relations with "race." [112] Americanness and whiteness might subsume a little "ethnicity," but color was subsumed by neither.

The late 1930s was a period in which the mass medium of radio, controlled by white elites with cooperation from federal officials, was used to test a revised version of American history that incorporated the stories of ethnic immigrants into *the* American paradigm of success. The task at hand was to create a palatable and compatible history of America's different peoples that supplemented but did not supplant the dominant Anglo-Saxon theme. The immigrant success paradigm preserved the enduring national myths of equality, opportunity, fair play, and, implicitly, white supremacy. *Americans All* tried to reconcile American history with its own conflicting mythical ideals, adding to it the successful immigrant myth to create what Nathan Huggins has called the "dogma of automatic progress." [113] But as this series demonstrated, the paradigm did not fit the experience of most immigrants, especially those who by physical appearance were thought to be too different, too inferior to claim the mantle of whiteness—even if they were granted the title "Americans."

A radio show is a slippery product to control, especially one fashioned by groups of people with competing social and political goals. Thus, the content of *Americans All* is a more accurate reflection of the emerging divergence in views concerning matters of race and ethnicity than a production or publication authored by one person

would have been. Some of its failings are indications of the difficulties of constructing a coherent narrative about tolerance for an
intolerant nation. It is profoundly significant that this attempt at historical narrative reconstruction appeared on the most popularly accessible mass medium available and the one best suited to undertake
a major project like identity redefinition and cultural assimilation.
Precisely for this reason, the show is an early marker in the quest for
a popular consensus about a new way of presenting American history
built on the remnants of a declining New Deal cultural apparatus.

For all of the conflict about the content and direction of the
series, there was no disagreement that radio was the best, cheapest,
and most effective way of communicating to teachers and the general
public that the time had come for a reassessment of American history's obsession with old-stock Americans. This expanded and more
politically useful version of U.S. history integrated the contributions
of immigrant peoples and acknowledged clumsily the presence of
people who still remained outside the arc of whiteness and acceptable
ethnicity. *Americans All* stands as a brilliant relic and representation
of the beginning of a process of redefining the cultural, political, and
historical narratives of national history and preparing it for popular
dissemination and consumption.

Paul Robeson memorialized this new version of history in his repeated renditions of the song "Ballad for Americans," which he first
performed on a live radio broadcast in November 1939. Like *Americans All,* the lyrics told the story of America's history as a blend
of idealism and ethnic contribution. Robeson's biographer Martin
Duberman recounts that Robeson's extraordinary voice and musical presence transformed the song into "an instant sensation": "The
six hundred people in the studio audience stamped, shouted, and
bravoed for two minutes while the show was still on the air, and for
fifteen minutes after. The switchboards were jammed for two hours
with phone calls, and within the next few days hundreds of letters
arrived. Robeson repeated the broadcast again on New Year's Day,
then recorded 'Ballad' for Victor and watched it soar to the top of
the charts. . . . With something for everyone, 'Ballad' stampeded
the nation."[114] The song did indeed have something for everyone. It
was deeply patriotic and hopeful: "Our Country's strong, our Country's young / And her greatest songs are still unsung." It portrayed
and advocated, just as *Americans All* had, a variegated vision of what
an American was: "an Irish, Negro, Jewish, Italian, French and En-

glish, Spanish, Russian, Chinese, Polish, Scotch, Hungarian, Litvak, Swedish, Finnish, Canadian, Greek and Turk, and Czech and double Czech American."

In the catchy phrase "Czech and double Czech American," Robeson's dramatic inflection made clear the double entendre of the lyrics. As any 1930s audience would know, "check and double-check" was a phrase popularized by the blackface radio characters on *Amos 'n' Andy*.[115] Yet elsewhere "Ballad for Americans" acknowledged America's historical failings toward its largest minority, noting "the murders and lynchings," and stressed the need for change: "Man in white skin can never be free / while his black brother is in slavery."

After his spectacular performance, Robeson left CBS's New York City studio to have lunch with friends, only to be refused service at a restaurant in a nearby hotel. Outside the sanctuary of the radio studio, the black balladeer for "Americans all" was considered not quite American enough. As the 1930s drew to a close, the United States remained a place where color trumped ethnicity in profoundly curious and conflicting ways and where the constructions of ethnicity, whiteness, and race traveled on the same train but in separate cars.

CHAPTER 2

FREEDOM'S PEOPLE

Radio and the Political Uses of

African American Culture and History

C harles S. Johnson wrote in 1939 that "the essence of the Negro-white race problem in America is change itself," which "makes necessary a constant re-definition of race relations."[1] His observation was timely because the approach of World War II would bring another such redefinition. As the pace toward joining the war against Hitler quickened, federal officials found themselves in an increasingly awkward relationship with African Americans, whose loyalty and cooperation at home and abroad were viewed as essential to the outcome of the war. From the onset, African Americans resolved to fight the war on two fronts, combating racism at home and fascism abroad. At the same time, the war united African Americans and allowed them to target their activism at the federal government as the greatest perpetrator of the policies of discrimination and segregation.

Freedom's People, a federally sponsored national radio series broadcast in 1941 and 1942, is a dazzling artifact of the change and redefinition demanded by the start of World War II. African Americans who helped mold the series used the show to spotlight the irreconcilable conflict between America's historical ideological myths and its continued unjust treatment of blacks. A stellar display and a stealthy deployment of black culture itself, *Freedom's People* made a compelling political argument for equal opportunity and racial justice on a medium that had appropriated and exploited that culture and on a show that was sponsored by a primary target of black protests: the federal government.

Freedom's People came at a moment in U.S. history when blacks were experiencing a period of heightened race consciousness and increased political activity; when the federal government's apprehensions about African Americans approached a level unseen since Reconstruction; and when radio broadcasting remained an inaccessible political medium for the expression of dissident views, especially on race. Part of what the story of *Freedom's People* teaches us is that radio was a valued ideological site in the struggle not only to redefine race relations but also to reach some consensus on what it means to be truly American.

Ambrose Caliver, Educator and Civil Servant

Freedom's People was the creation of Ambrose Caliver, a black professional employee at the Office of Education whose perseverance and deep commitment to black education brought the show to life. Caliver's education and experience had uniquely qualified him to be appointed in 1930 to the newly created position of "senior specialist in education of Negroes," making him the first black person to hold a professional position in the Office of Education. A former high school teacher and the first black dean of Fisk University, Caliver came to the federal government in the same year he earned his doctorate from Teachers College at Columbia University.[2]

Caliver's new position was the first federal response to the problem of extremely low per pupil expenditures on black children and the disparities in the distribution of federal and state resources to black and white children.[3] Caliver undertook the task of documenting the deplorable fiscal conditions of black elementary and secondary schools, especially in the rural South. He traveled extensively throughout the South, meeting with black teachers, visiting black schools, and conducting formal surveys of conditions and funding. A prolific researcher and writer, Caliver published his findings in a ground-breaking series of federal reports about black educational needs.[4]

Caliver combined his skills as a researcher and writer with an astute appreciation of modern techniques for influencing public opinion.

On his initiative, in 1930, NBC agreed to air an annual national radio broadcast on African American education during American Education Week, a practice that continued into the 1940s. Caliver used the programs to encourage black high school and college students to learn more about black history and to strive for academic success, at the same time seeking to educate white Americans about black achievements.[5]

At the Office of Education, Caliver successfully organized the 1934 National Conference on Fundamental Problems in the Education of Negroes, which brought national attention to the dearth of resources for elementary and secondary education for blacks, especially in the South. The three-day conference had the strong support of Secretary of the Interior Harold Ickes, who delivered the opening address to the 1,000 conferees. Eleanor Roosevelt, whose support Caliver had cultivated, made the keynote address, which was broadcast nationally by NBC. She vividly described the inequities in black schools in the South and urged her listeners to work together, without regard for race, to remedy such injustices and eliminate intolerance.[6]

Despite his ambitions and abilities, Caliver's position at the Office of Education was a lonely one. For many years, he was the agency's sole black professional employee. His work was belittled by some whites in the department who reportedly referred to his office as the "Nigger section." Caliver also was consistently underpaid relative to whites who held comparable positions or joined the office later and had less experience and education. At one point when Caliver questioned this discrepancy, his supervisor told him that "he had never seen a Negro who was worth more than $5,000 a year."[7] Fortunately, Caliver was a resilient and determined man, and he exploited the limited resources at hand. With Ickes's cooperation, he helped make the cafeteria in the Interior Department the first to welcome black federal employees. From his foothold in the agency, Caliver developed an extensive national network of African American teachers and educators. As the lone federal voice for black education, Caliver's office became a focal point not only for educators but also for a wide circle of black intellectuals and activists who believed that black educational advancement was crucial for the economic and social progress of the race as a whole. Caliver also associated with a large community of African American intellectuals in the nation's capital, including Charles Wesley, Alain Locke, Sterling Brown, and others at Howard University.[8]

The first decade of Caliver's career as a federal civil servant coincided with increased public attention to the political and economic status of African Americans and the subject of intergroup relations generally. Although most blacks still lived in the South, their growing migration to the North garnered them greater political visibility, especially after they proved to be a decisive urban voting block in key northern and midwestern states in the 1936 presidential campaign. Blacks in federal positions such as the post that Caliver held used their limited authority to prick the administration's conscience on racial issues. Although it had not been an era of great strides for blacks, the 1930s ended with wider public recognition that the treatment of African Americans could no longer be dismissed simply as a southern problem.[9] The advent of World War II set the stage for a more audible debate about the status of black Americans and spotlighted the glaring problems of segregation and inequality.

War Approaches

Many African Americans reacted to the outbreak of World War II in Europe with a cynicism fueled by the memory of unkept promises and unmet expectations in World War I. As the prospect of American entry into the war grew stronger, black leaders reacted sharply to long-standing federal policies of racial segregation and discrimination in the armed forces. African Americans were denied access to some branches of the military altogether, and even in branches where they were allowed to serve, they were restricted to segregated units where they performed noncombat, support functions that were devoid of prestige and responsibility.[10]

The prewar buildup was an immediate concern because it brought with it the potential for vastly increased employment opportunities for black men and women, who prior to the buildup, if they had jobs, had been employed mostly in low-wage positions as agricultural or domestic workers. But it was soon patently clear that discrimination in the federally financed defense industry would remain the rule. Black men who could find work were relegated to menial, unskilled, low-paying jobs; black women, if hired at all, were employed only as janitors or scrubwomen.[11]

As the industrial preparation for the war began and large numbers of blacks migrated to the North and the West in hopes of finding defense jobs, federal officials and many whites feared that growing concentrations of blacks in already crowded housing would increase the possibility of racial unrest.[12] This population shift heightened concern that an angered white populace and low morale among blacks would combine to prevent the creation of a unified home front during the war. African Americans, the federal government, and the nation as a whole had an urgent, converged interest in improving racial relations, although ultimately disagreement persisted over how this was to be achieved, to what extent, and at what pace.

Buttressed by aggressive editorial campaigns in major black newspapers, African Americans began calling with increasing militancy not only for greater racial tolerance but also for equal access to the nation's economic and political life. They united quickly behind demands that the federal government end segregation and racial restrictions in the armed services and halt discrimination in employment at federal defense plants. As the inevitability of a massive war effort became clear, white fears of domestic disruption and disunity accorded black entreaties more attention. The campaign by blacks against federally sanctioned discrimination and segregation also provided a source of political leverage for blacks who served in the Roosevelt administration. This was clearly the case with Ambrose Caliver and his idea for a federally sponsored radio series on black history.

We, Too, Are Americans

Caliver's plans for a radio series about black Americans developed during the golden age of radio, when radio alone held the public's attention at home, unchallenged by television's moving images. Moreover, it was a time when radio's already prominent status was enhanced daily by its dramatic live coverage of European wartime developments. Indeed, World War II was destined to become a "radio war" that many Americans experienced instantly.[13] Undoubtedly, since radio was a common feature of black home life, Caliver knew that it was also a key medium for reaching the African American community. Radio was especially important to blacks, most of whom were excluded from much public entertainment, either by segregation or by cost.[14]

Black musicians and all forms of black music were crucial to radio's development and its popularity among both black and white listeners. But other types of radio programming conformed to the virulent racial stereotyping and derogatory portrayals prevalent in all forms of popular culture at the time. To little avail, the black press and black organizations repeatedly protested the exclusively negative portrayal and limited use of blacks on radio.[15] Having been stung by negative and constricted images, black activists and organizations were extremely sensitive to radio's enormous influence over public attitudes and were alert to its promising potential as an ally. Denied the possibility of station ownership both by racial restrictions and by prohibitively high costs, African American political activists were eager to gain access to this tantalizingly powerful medium, especially on a national level.

As an educator and early radio enthusiast, Ambrose Caliver must have paid special attention to the tremendous response his agency's series *Americans All, Immigrants All* had generated, especially among teachers and civic groups.[16] The brief annual broadcasts on African American education that NBC had granted him for over a decade whetted his appetite for more airtime to spread "positive" messages about black history on the most publicly accessible mass medium. Also, the congenial relationships he had developed with Office of Education and NBC officials made it possible for him to suggest a series of monthly broadcasts devoted exclusively to black history and culture.

Caliver first approached Philips Carlin at NBC in August 1940 with the idea for a series of thirteen dramatized radio broadcasts about the contributions of African Americans in American life.[17] Coincidentally, in September 1940, Education commissioner John Studebaker asked members of his staff for suggestions for ways to assist schools in preparing the nation for defense. Caliver responded by submitting a proposal for a radio series on black history. He pitched it as a remedy to "certain subversive elements" who considered black Americans "easy prey for the spreading of disaffection and disunity." In order to mount an effective defense of democracy to counteract those elements, Caliver argued, the federal government must "promote constructive and loyal citizenship" among blacks by fostering in both blacks and whites a belief that African Americans "belong" and share some sense of unity on goals.[18]

It is not clear what Caliver's personal beliefs were on the true extent of subversive influence on black morale. Caliver was by all indications a pragmatist and a gradualist concerning race relations.[19] He undoubtedly realized that the threat of black subversive activity, whether real or imagined, was the most potent political argument for his proposed radio series. He adhered to that rationale when he developed a formal prospectus for the series in October 1940. By then, he had given the series the title *We, Too, Are Americans,* inspired in part by the title of a song and in part by *Americans All.* Its intended audience would be "white and Negro students and teachers in particular, and all socially minded citizens in general." Each program was to be recorded on a phonograph record to be made available for school and radio rebroadcasts, as had been the case for *Americans All.* Similarly, he planned to develop the scripts into general study guides on black history to be accompanied by the recordings for use in discussion groups.[20]

Caliver's colleagues at the Office of Education grasped immediately the timeliness of the implicit political appeal of his proposal, coming at a moment when federal officials were growing ever more nervous about black morale and the extent of black support for the war buildup.[21] Moreover, the Roosevelt administration was at odds with the print media generally and particularly with the black press, so sponsoring a radio alternative to reach black listeners must have been especially attractive to an office that otherwise had no perceived natural role to play in the war effort. Still, the Office of Education had no federal funds to support the series. At Caliver's suggestion, Office of Education officials followed the example of the producers of *Americans All* and secured private funding for the series, initially from the Rosenwald Fund and later from the Southern Education Foundation—both philanthropic entities devoted to improving race relations. With a funding agreement in hand, Studebaker was so confident of the series's potential appeal that he made public announcements about the upcoming series even though he lacked any firm network commitment to carry the show.[22]

In February 1941, Caliver again contacted Philips Carlin at NBC, and Studebaker brought the matter to the attention of NBC president Niles Trammell. James Angell, the educational adviser at NBC who had advised against airing *Americans All,* once again discouraged Trammell from broadcasting this series. Larger political considerations were on Angell's mind: he warned that none of the broadcasts "would be popular south of the Mason and Dixon's Line." Still, out of respect for Caliver's position in Washington, D.C., Angell decided that he should be invited to New York City to discuss his idea, "even though we delay considerably taking any action." [23]

Caliver followed up on NBC's tepid response by submitting a detailed outline of each proposed episode. Renamed *Freedom's People,* Caliver's planned series emphasized continuous black contribution to the nation; as such it would devote time to contemporary black figures as well as historical ones. The outline made clear that Caliver was confident that prominent black leaders, intellectuals, educators, writers, and artists would want to be associated with the program. With the same assurance, Caliver anticipated that popular black singers, musicians, actors, orchestras, and bands, as well as black performers of European classical music, would give free performances on the show.[24] Caliver believed that black artists would see the project as he did: as an important and special opportunity to provide positive programming about the race on a medium that usually either exploited or ignored it. He assumed that African American leaders and entertainers would participate out of a sense of racial pride and public service.

NBC's response to Caliver's outline for the series revealed an ongoing concern about potential political implications. H. B. Summers of the network's Public Service Division warned Caliver against using dramatized historical vignettes in the series because he feared they would be perceived as "propaganda," which he was "anxious to avoid." He explained the network's position: "I do not feel that we are any more justified devoting a series to the contributions of the Negro as such than we would be to the contributions of the Irish, of the Mexicans, of the Swedes, or any other group. About the only way we could legitimately place a series of this type on the air would be to build it around the idea of straight entertainment with the other element introduced only incidentally." Caliver reassured Summers that his purpose was simply to develop greater national unity and better racial relations, a position he had stressed in a speech in February 1941 to school administrators in which he cast racial tolerance as essential to uniting the nation for war. Caliver vowed that he would avoid any "highly controversial material" and that the series would "subtly and indirectly" promote his aims. However, Caliver resisted Summers's request that he eliminate historical material, although he agreed by way of compromise to devote one-half of each episode to musical entertainment.[25]

Despite these reassurances, as the summer of 1941 approached, Caliver had received no firm indication that NBC would carry the series that fall, as he planned. That summer brought a significant confrontation in the tense relationship between the federal government and black activists, stirred in particular by A. Philip Randolph's threatened March on Washington against discrimination and segregation in the military and in defense jobs. In order to prevent a mass gathering of blacks in the nation's capital, President Franklin Roosevelt on June 25 issued an executive order banning discrimination in defense work and establishing the Fair Employment Practices Commission (FEPC) to enforce the order.[26] Randolph saw the president's order as only a first step; it made no mention of the segregated armed forces, and the new FEPC had very little enforcement authority. Randolph and many other blacks were disappointed yet emboldened by this limited initial victory. He and other leaders of black organizations pressed forward with broader demands, creating a pattern of interaction that would characterize the relationship between African Americans and the federal government for the duration of the war years. During the war period, the open threat of mass action, whether organized and peaceful or disorderly and riotous, would remain one of black America's greatest sources of political power.

In the summer of 1941, with American entry into the war appearing to be more and more inevitable, most black leaders and the black press urged blacks simultaneously to help fight the war and to continue to fight at home against discrimination.[27] In contrast to their acquiescence to pleas to close ranks during World War I, black leaders and newspaper publishers refused to make the claim for racial equality secondary to the war effort or to postpone its pursuit. Their aggressive editorial campaigns and the perception of their growing influence over black public opinion made black newspapers and activists the cause of much consternation among federal officials. This apprehension prompted federal investigative agencies to continue to subject black leaders and publishers to surveillance, open harassment, and the threat of sedition charges.[28]

The public relations bureau of the U.S. Army grew so concerned about escalating racial tensions that in late July it also approached NBC about broadcasting a program on the theme of "the Negro" and national defense, apparently unaware of Caliver's pending request to the network. NBC officials warned Caliver that the army could not be denied and that if its "request" was granted, it would be unlikely that Caliver's series would also be accepted. NBC did grant the army's request and on August 12 aired a forty-five-minute special entitled "America's Negro Soldiers."[29]

After the army broadcast, however, network officials finally made a tentative commitment to air *Freedom's People*. Reassured by the army's plea for its own broadcast and by the noncontroversial reaction to it,

NBC officials must have concluded that their involvement with the series carried fewer political risks than they had feared—and that in fact it might offer them a needed public relations boost. They also probably realized, contrary to Angell's assessment, that if Caliver could arrange appearances by popular black artists, the series would draw a good audience. Taking Caliver at his word, NBC officials agreed to carry the opening program on September 21, but they refused to guarantee that they would air any additional episodes or that the series would be assigned a consistent monthly time slot.[30]

The fact that most NBC officials considered their refusal to carry *Americans All* a tremendous mistake, especially after its popularity brought such acclaim to their rivals at CBS, may have influenced their decision to proceed differently in this case. It is also possible that network decisions to cooperate with *Freedom's People* and other federal programming requests were affected by anxieties about a pending federal antitrust investigation that would later force NBC to sell one of its two national networks.[31]

Putting *Freedom's People* on the Air

Even before NBC made a firm commitment to carry *Freedom's People,* Caliver had begun the work of finding a scriptwriter, lining up black performers, and assembling a large advisory group to help plan the scripts and follow-up materials. To write the scripts, Caliver engaged Irve Tunick, one of the most respected and successful scriptwriters in educational radio. Tunick, who was white, had written inventive and lively scripts for the long-running and very popular science and natural history series *The World Is Yours,* which the Office of Education and the Smithsonian Institution had sponsored on CBS. The show was the one series that had been spared the congressional attacks on federally funded radio. Although it was not at all clear how he would approach a series about a more political and potentially controversial subject, Tunick's skills as a professional educational scriptwriter were superb.

In Caliver's appeals to well-known black performers and stars, he stressed that the series offered a unique educational opportunity for black and white students and that it would improve interracial relationships. Caliver requested that they provide their services gratis, implying that they had a moral obligation to cooperate with this federal project about black history.[32] In this way, Caliver secured favorable responses from many black artists and groups, including bandleaders Count Basie, Cab Calloway, and Noble Sissle; popular singer Joshua White; classical music performers Dorothy Maynor and Carol Brice; gospel groups like the Southernaires and the Golden Gate Quartet; and black collegiate choral groups from Fisk University, Tuskegee Institute, and Howard University.[33]

Caliver had approached the task of setting up an advisory group for the series with the awareness that the affiliation of black and white national figures would be very valuable. Among the whites he asked to serve were New Deal official Will Alexander and social scientist Guy Johnson. Caliver also assembled an impressive cross section of the African American educational and intellectual elite, including W. E. B. Du Bois, then editor of *Phylon;* eminent black historian Carter G. Woodson; Alain Locke, Sterling Brown, and Charles Wesley of Howard University; sociologists Charles Johnson of Fisk University and Monroe Work of Tuskegee Institute; L. D. Reddick, curator at the Schomburg Center for Research in Black Culture; and Roy Wilkins of the National Association for the Advancement of Colored People (NAACP). Federal officials whose work involved issues affecting black Americans also served on the advisory group, including Mary McLeod Bethune.

Because he had no funds for travel or consultant fees, Caliver relied on a smaller group of members of the Advisory Committee mostly based in Washington, D.C., to provide research and review the scripts that Tunick drafted. The composition of the group varied from month to month, but aside from Caliver himself, its most steadfast and influential members were Wesley, Brown, Locke, and Arthur D. Wright of the Southern Education Foundation. Others who attended most of the meetings were local school principal Elise Derricotte and Joseph Houchins, "specialist in Negro statistics" at the Census Bureau. Caliver frequently solicited research materials and comments on scripts from other members and friends, particularly Reddick. This association of prominent black intellectuals and leaders with the program served to authenticate that it was indeed by and about "Negro Americans," as its publicity claimed. This distinguished the series at a time when it was rare for blacks to appear in roles other than as featured acts on white-dominated programs.[34]

Concern about claiming ownership over the program and making it an identifiably African American production was foremost in the minds of many of the members of the Advisory Committee when they met to review the first script scheduled for broadcast. Indeed, the first subject they raised was whether the narrator's voice was going to be a "colored" voice. Caliver and Tunick reassured the group that black stage actor Frank Wilson would narrate the show in a "fine, rich, mellow voice." The actual question being raised was how to make race visible, or more accurately, audible, on the air in a way consistent with the show's intended political aims.

After hearing Wilson's voice on the radio, members of the committee regretted that he sounded too much like a white man. Ironically, many black radio actors often complained during that time that they were only able to secure radio parts if they agreed to sound "Negro" enough to satisfy a white producer's ear. But here the panel's underlying concern was how the race would be represented and what kind of "Negro" would be portrayed—specifically, what class of African American, since they most wanted to avoid presenting the "class" of African Americans that had been created by radio itself. "One of the most important ways to get over to the public that the program is Negro is to have a narrator whose voice is rich," Locke argued; "I do not mean a cornfield voice, but certainly a different voice than Wilson's." He argued persistently that the announcer had to have "a characteristically Negro voice." The other panelists agreed, and Derricotte urged Tunick to encourage Wilson to emphasize the deeper qualities of his voice: "Since we do not have television we depend upon our ears and not our eyes." Tunick reassured the group that Wilson had a rich voice but that he was "deliberately playing it down" in order to convey a serious dramatic tone without sounding like a "soap-box opera." Tunick's subsequent coaching raised the committee's evaluation of Wilson, but toward the end of the series, other black male actors such as Canada Lee and Juan Hernandez who were thought to have "richer" voices shared the narration and announcing duties.[35]

Music and "Freedom's People"

Caliver knew that the fate of the entire series rested on the reception and response to the first show. In order to attract a large opening audience and allay the network's fears about political controversy, the show's creators decided to focus the first episode on black contributions to music (rather than science and discovery as originally proposed). Tunick's work on the first script exceeded the expectations of many committee members. Sterling Brown confessed that he "was pleasantly surprised at the social punch of the script," which had "much more social value" than anything he had ever heard on the air. Some members raised concerns about the use of dialect in the section on slavery, but Brown reassured them that black actors on a live show could be trusted to avoid any potentially offensive language. Still, this show would establish the standard features of the series, so committee members concentrated on perfecting the opening and closing narratives.[36]

The planners concentrated their anxiety not on the script, which was received with much enthusiasm, but on Caliver's ongoing negotiations with Paul Robeson to appear on the show. W. C. Handy, Joshua White, and Noble Sissle's band had already been secured, but Caliver and the committee also wanted Robeson to appear because of the gravity and social consciousness he would bring to the show. Subsequently, due to Caliver's persistence, Robeson agreed to perform on the opening broadcast, guaranteeing the large audience the show would need in order to help convince NBC to reserve a slot for the programs that were to follow.

The final version of the opening lines boasted that the series was "dedicated to—and conceived by the American Negro—truly Freedom's People." Its standard introductory sign-on was an ironic recasting of American history:

> From the Old World they came—high with hopes and strong! To America they brought this hope and strength and founded a nation of splendid freedoms! But this is not their story. This is the story of those who did not come, but were taken! The story of those who lost freedom when they came upon our shores and for years they tilled our soil, gathered our crops, and made the land good. Some won liberty—others waited. Then Freedom came to all—a liberty well deserved, a liberty triumphant. Yes, this is the story of the American Negro—13 million citizens of the United States.

The introductory music was a choral medley of phrases from "My Country, 'Tis of Thee," "Go Down Moses," and "Lift Every Voice," with a voice-over about freedom.[37] As in the spoken narrative, the juxtaposition of lyrics worked to portray African American freedom as something at first unjustly denied but finally won after a long moral struggle. The musical framing set the tone for each program, as did the text of the program's sign-off: "Onward they march—13 million Negro citizens of the United States, sharing the labor, accepting the responsibilities of our Democracy. Knowing the weight of chains—the helplessness of bondage—they are today a mighty force for freedom. To them liberty is a precious thing. For—truly—they are Freedom's People."

The show described the history of African American music as "the story of the warm and human melodies the Negro found in his heart and gave to America." Through vignettes and musical sampling, the program traced "America's musical genius" and all black music back to black slaves, portrayed as a "deeply religious" people who used music as an act of resistance and a way to nurture the soul. A vignette depicting the origins of the song "Steal Away to Jesus" cast it as a message not just about the escape to physical freedom but also about the quest for religious liberty. Similarly, a narrative on work songs, featuring Joshua White singing "John Henry," characterized them as "melody born in the sweat of strong men building the strength of America—music not written with notes, but with muscle and calluses and dirty patched overalls." The narrator linked the blues to spirituals: "While the Spiritual is a song of God, the blues are just plain folks, feeling low down." The stories of each of these types of music were linked by the theme that black music was not a wholly spontaneous creation but a form of expression deeply rooted in the struggle for mental and spiritual survival.[38]

The show culminated with an extraordinary appearance by Paul Robeson, who spoke extemporaneously and authoritatively about spirituals and the blues as songs of protest. Robeson's poignant and slow rendition of the protest spiritual "No More" brought the show and its theme full circle:

No more auction block for me, no more, no more, many
    thousand gone.
No more pint of salt for me, no more, no more, many
    thousand gone.
No more driver's lash for me, no more, no more, many
    thousand gone.

The broadcast was a well researched, engaging, and inventive presentation that smoothly combined narrative, dramatization, interviews, and black music. Caliver, who wanted to use well-known musicians as an audience draw, may not have appreciated initially how much the music helped make the shows effective, but veteran scriptwriter Tunick did, and he expertly integrated the music into all of

Paul Robeson appearing on the first broadcast of Freedom's People *in 1941. (Ambrose Caliver Collection, Moorland-Spingarn Research Center, Howard University; NBC News, © National Broadcasting Company, Inc., 1941, all rights reserved)*

the shows. The network pressured Caliver to use more popular music rather than black or European sacred music performed by black collegiate choral groups and classical artists, which he preferred. But Caliver insisted on having the last word on the music, having learned well from the Jules Bledsoe fiasco on the *Americans All* episode about African Americans that he had been called in to help repair as an adviser. Bledsoe's mammy song had overpowered the racially constructive message of that script, but Caliver wanted the music on *Freedom's People* to offer poignant reinforcement of the series's somber tone. At the same time, for some white listeners, the music may have served as an elixir, reassuring them with the familiar image of blacks as harmless entertainers while distracting them from closer scrutiny of the show's political appeals.

The recurring political theme of *Freedom's People* was that blacks had contributed significantly to American culture and history and had earned the right to be free and fully accepted as Americans. This contributionist approach to black history and its political corollary formed one of the most basic arguments used by black leaders of that era. But Caliver and others who worked on the project also saw it as a promising modern tool for teaching black history, which they considered especially important for the development of self-esteem and racial pride in black children. Moreover, they also thought the show could help teach whites tolerance and appreciation of blacks by instilling in them something other than the contempt they felt for a people they believed did not belong in a "white man's country," except in subservient roles. For this reason, the phonograph recordings, although they were expensive to produce, were especially valued precisely because of their novelty and adaptability for use in many settings. At a time when few materials of any kind were available on African American history, this show was a vehicle for developing for general public consumption a set of materials to be used by black teachers, churches, and civic groups and interested whites.[39]

Caliver left little to chance in building up an audience for the first show. Eleanor Roosevelt, who was personally acquainted with Caliver and had appeared at his 1934 conference on black education, praised the upcoming show in one of her nationally syndicated "My Day" columns during the week preceding the broadcast. She urged her readers to listen to the show because "the more we know about each other and about our contributions to the good things in our country, the less we shall be liable to fall a victim to that most pernicious thing called: 'racial and religious prejudice.'"[40] Office of Education commissioner Studebaker personally alerted President Roosevelt to the upcoming broadcast and urged him to listen and respond with comments.[41] Caliver also enlisted the support of black civic, social, and fraternal organizations, churches, and other groups with ties to the black community (such as the Young Men's Christian Association [YMCA] and the Young Women's Christian Association [YWCA]) in spreading the word about the program. He also exploited his own nationwide network of black teachers and educators to encourage listeners to tune in. Of equal importance was the prominent attention *Freedom's People* received in the black press, nationally and locally.[42] The Office of Education sent out announcements to national and local newspapers and to teachers, principals, librarians, teachers colleges, and federal officials around the country.[43] The show received good prebroadcast publicity from many newspapers, including the *New York Times,* which, as Caliver had expected, featured an announcement of Robeson's appearance on the show.[44]

Soon after the first broadcast, letters from teachers, most but not all of whom were black, began to arrive on Caliver's desk. As expected, the show generated requests for more information about how to build teaching units on African Americans. Faculty members at southern teachers colleges, black and white, were especially anxious to receive information they could pass on not only to their students but also to elementary and secondary school teachers in their state.[45] Teachers in some northern urban districts were also eager for more materials on African Americans. One Brooklyn principal explained that her school had a "pupil population of mixed color" and that "we are doing our utmost to build up a feeling of mutual self respect."[46] Educators at institutions engaged in teaching and training African American students and teachers expressed their appreciation for the show's educational content and usefulness.[47] A variety of civic, educational, and religious groups also wrote in support of the broadcast. Groups as varied as the Board of Education of the Methodist Church and the Young Communist League of New York State requested materials based on the show.[48]

Reviews and editorials about the show and the expected series soon appeared in the radio and entertainment press, national newspapers and magazines, and black journals and papers. *Variety* described the show as an "all-Negro program" that featured "established colored entertainers" and grudgingly gave it a favorable review.[49] *Time* magazine was less reticent; its reviewer credited the "towering Paul Robeson" with pacing the show and helping it to do "right by Negro music and its development."[50] A long column in the *Louisville Courier-Journal* by an official at Kentucky's State College for Negroes extolled the show as a remedy for the harm done by a press that had "told only the worst things about Negroes." An important political need was

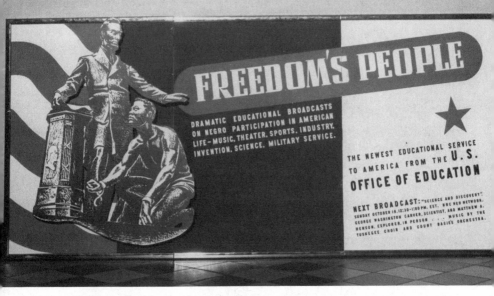

Placard advertising Freedom's People.
*(Ambrose Caliver Collection, Moorland-Spingarn Research Center,
Howard University)*

also addressed, the writer argued: "National unity is impossible without securing for the Negro what justly is his right. Without national
unity democracy cannot be successfully defended. It is important
then to all of us that the facts presented on these programs reach
a large audience. The truths thus disseminated may some day help
make us free."[51] A long essay in the leftist publication *New Masses*
used the broadcast as an opportunity to make a broad critique of Jim
Crow in the radio studios: "The color line is drawn on both sides
of the microphone, giving Negroes little or no chance as either performers or non-broadcasting radio workers." The writer commended
Freedom's People as "one new program that is moving in the right direction," although parts of the show were characterized as "placid" and
"shying away from the present." Robeson's rendition of "No More,"
however, drew the writer's praise, as did his insistence that the promise of the Emancipation Proclamation had not yet been realized.[52]

African American newspapers gave special notice to the show and
its attempt to link the rights of African Americans to national unity,
and none judged it to be at odds with their own views on those issues.

The *Pittsburgh Courier* reported that the show had met with "wide acclaim," and the *Washington Afro-American* called it a tribute to the nation's progress. Conservative *Courier* columnist George Schuyler characterized it as "praiseworthy" for its depiction of black history in such a "dignified and skillfully dramatic way." But another black *Courier* columnist, New York City–based theater critic Billy Rowe, thought some of the dramatizations about slavery were maudlin and prevented the show from living up to its potential, although he praised Robeson's "fine voice" and Wilson's "superb commentation." Rowe concluded by conceding that on national radio, "it is most difficult for any given body to decide on a Negro program as you've got to please all of the people, the whites and the colored." This range of opinions was united by a consensus that the show was a refreshing change from other depictions of blacks on radio. As one writer put it, "Amos and Andy will have a rival in presentation of Negro Life."[53]

Caliver's national network of intellectuals and leaders shared this view, and they were effusive in their congratulations.[54] Other black New Deal officials such as Mary McLeod Bethune and William Trent told Caliver that the show was "magnificent" and "marvelous." Channing Tobias of the national YMCA wrote that the show "struck a new high in broadcasting among Negroes."[55] Caliver's colleagues at the Office of Education also were enthusiastic about the quality of the first show and the public response it stimulated. William Boutwell of the agency's Radio Education Division told the Advisory Committee that of 700 network educational programs the office had sponsored, he thought the first *Freedom's People* episode was one of the best.[56] Studebaker told Caliver that the program was "inspiring" and "thrilling" and congratulated him for "making educational history."[57]

Studebaker and Caliver rushed to report the responses to NBC officials. Studebaker told NBC president Niles Trammell and others at the network that the show had met with "great interest and hearty approval," which would recommend their commitment to the entire group of broadcasts. Similar appeals came to NBC officials from Caliver, who thanked them for their "good will and social statesmanship."[58] Caliver also encouraged members of the Advisory Committee to write letters of support for the series to the network "commending the significance and timeliness of these programs."[59]

One of the reasons NBC needed such reassurances about the show was its continuing fear of negative reactions from its southern affiliates. This was not an irrational concern, of course. For example, even

Cover of Office of Education brochure for Freedom's People.
(Yale Collection of American Literature, Beinecke Rare Book and Manuscript Library)

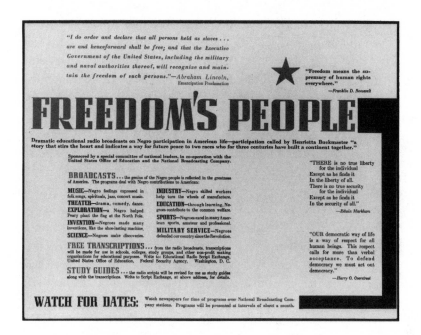

though Studebaker's solicitation of affiliate support had generated a large positive response, a New Orleans station manager requested assurances from Studebaker that the show had been approved by persons "familiar with the South, the Jim Crow Laws, and customs" and cited several unnamed "network programs relating to the colored people which have not been acceptable in this community."[60] After the broadcast of the first episode, NBC immediately conducted its own survey of all of its southern stations, requesting their opinions about the broadcast. Network officials were very relieved when they received enthusiastic reports about the quality of the show and requests for more like it. One station manager admitted that he had initially been "fearful" of carrying the show because of "the problem of race consciousness." The network also discovered that seventy to eighty-five affiliates had carried the show, a very high number for a noncommercial program.[61] Finding that the comments from the southern stations were favorable and that white listeners had not objected to the show, NBC agreed to broadcast two more shows in the series in a regularly scheduled Sunday afternoon slot.

Not satisfied with the promise of only two more shows, Caliver continued to try to build political support for the continuation of the entire series. He especially wanted to make sure that Eleanor Roosevelt had heard the show. When he contacted her soon after the broadcast, she confessed that she had not been able to listen to it but offered to arrange a time when Caliver could bring the recorded version to the White House. After the date was set, Caliver requested permission to bring Office of Education officials Studebaker and Boutwell with him. When they arrived, they were surprised that the recording was to be played not only for Eleanor Roosevelt but also for the president, Treasury Secretary Henry Morgenthau and his wife, and Lord and Lady Mountbatten, all guests for the evening. All of those present were extremely pleased with what they heard, so much so that Caliver was able to persuade the president to agree to appear on the final show of the series, which was tentatively scheduled to be broadcast in the spring of 1942. At Caliver's suggestion, Eleanor Roosevelt used the successful evening to continue to help generate public attention to the series. She soon published another "My Day" column about the series that described the visit and the enthusiastic reception of the recording. She again urged her readers to listen to the series: "These programs should bring before the whole people, the contribution of the Negro race to this country." [62]

Caliver had won the battle to keep the series on the air. The initial show had managed to merge what has been called the "split image" that usually separated black and white depictions of black Americans.[63] The attention the show garnered made it very difficult for NBC to refuse to carry the remaining shows. Federal officials continued to link black morale and national defense, which added to the pressure on NBC to cooperate for the duration of what had been designated a morale-building defense series. Unable to claim that the shows were too controversial, network officials had little choice but to see the project through to its end. And with the series's future secured, Caliver and the Advisory Committee members also felt free to push the show's content closer to the limits of political acceptability.

Literature and the Sciences

Freedom's People emphasized black cultural achievements not only in music but in the visual arts and literature as well. *Americans All* had slighted immigrant, ethnic, and black cultural contributions in favor of the argument that most of these groups had earned the right to be called Americans primarily through their physical labor. But in this series, African Americans made the much more subtle and sophisticated claim that their contributions were at the very foundation of what was to be called American culture—a persistent theme in black critical and political discourse. They wanted to show that their cultural contributions were not the products of a lighthearted, fun-loving people but acts of resistance that grew out of their oppression. In this way, the episodes on *Freedom's People* that examined black literature and visual arts were premier examples of the political uses of black culture.[64]

A broadcast on black artists opened with a powerful articulation of the relationship between black artistic expression and black suffering: "From pain and from wanting came deep low notes; from passion and protest came crescendos of bitterness; from the life they lived and the life they knew came the earth-soaked strains of humor and the tap-tap strains of dancing and from the life they pictured after life came the spirituals, came the whispers of hope. Thus, to the Negro came art—came the song to the poet; the picture to the artist and the melody to the musician." The show constructed a history of black literary contributions starting with Phillis Wheatley's story, which was introduced by excerpts from the poignant spiritual "Sometimes I Feel Like a Motherless Child" to emphasize that she was "taken from her mother's side" in Africa. In concluding a dramatized vignette on her life, the narrator set Wheatley's work in this analytic frame: "The voice was young then—could but imitate the sound of sound. Phillis Wheatley and those that followed close upon her—echoed the stylized, rigid verse of the day. It needed more than that to grow. It needed words to say—and the power and the freedom to say them. That took time."

This long dramatization of Wheatley's life, including her famous exchange of letters with George Washington, was the most extended treatment the series would give to a black woman. As the series was being planned, Mary Church Terrell had urged Studebaker to ensure that "the work which colored women have done to promote the welfare of the race should be emphasized," noting especially their roles as the primary supporters of black schools and churches.[65] Despite Terrell's efforts and Brown's early admonition to the Advisory Committee that black women "not be slighted," ultimately the aural image the series constructed was dominated by the voices of black men.[66] Black women, however, did appear throughout the series as characters in historical dramatizations, and their accomplishments were singled out in particular fields, such as classical music, education, and sports. Overall, the people in the series were not portrayed as a "manly" race, as is common in many discourses on racial nationhood, but were cast in more communal terms despite the predominance of male voices. This depiction of the race may have been intended to be less threatening than a more masculinist portrayal, as also had been the case on *Americans All*. This caution would seem especially necessary in a series about African Americans broadcast during a period when fears of black protests and violence were rising.

The script used the metaphor of finding one's voice as the vehicle for surveying the growth of black writing. The narrative leapt directly over slavery, skipping from Wheatley and Frederick Douglass to the Harlem Renaissance: "[T]he voice was rounding out, taking on overtones of deep thought and swelling with a richness of genius. And then—suddenly—it came of age. The voice found a body—a frail body that would not linger long. And this was Paul Laurence Dunbar." The narrative then moved from Dunbar to readings from the works of Langston Hughes ("A Lousy Day"), James Weldon Johnson, and Countee Cullen ("Brown Girl Dead"). This led to a discussion of the "realism" of Charles Chestnutt and Richard Wright, ending with a reading of Claude McKay's searing poem "If We Must Die." The works featured were well chosen for the segment's closing segue: "And so a voice was made of many things—of humor, pathos, work, tragedy, hope, protest." The show used the evolution of "voice" as evidence of the formation of a distinctive black community that spoke for itself in European idioms but retained the strength of its own identity.

Furthermore, claiming that African Americans were legitimate contributors to American and international culture, Caliver, Locke, Brown, and others hoped to counter white supremacy's denigration of black literary and artistic achievements as purely spontaneous or primitive products. Black intellectuals, including Locke and Hughes, had engaged in recurring debates about whether black creative productions should be made immune to or become carriers of racial propaganda. In an artful dodge, *Freedom's People* simply turned the very existence of a body of black literary expression into propaganda itself.[67]

The show about black scientists and explorers emphasized the barriers to African Americans in these fields while extolling the few who had made significant advances and discoveries. Following the pattern of the shows on black culture, this episode tried to counter the persistent stereotype of black intellectual inferiority. The broadcast included a long narrative recreation of Matthew Henson's journey to the North Pole, making plain that Robert Peary could not have gotten there without him yet not daring to claim that Henson may have arrived first. The closing segment dramatized the early poverty-stricken life and rise to renown of agricultural scientist George Washington Carver, who appeared on the broadcast in a live interview from Tuskegee Institute.[68]

This episode's implicit argument about black courage and intellectual curiosity struck a nerve among many listeners. Those associated with the series were surprised by the emotional level of the listener response. Even Carver was unprepared for the response. He forwarded to the Office of Education one particularly poignant letter in which the writer explained that "I sometimes think that our's [sic] is a hopeless race. But after listening . . . one cannot help but have hope." Historian Carter G. Woodson was favorably impressed and predicted that similar shows "will do much to disabuse the public mind of its traditional opinions about Negroes."[69] Reactions to the broadcast were especially enthusiastic from schoolteachers like Marie McIver from Raleigh, North Carolina: "I have never listened to a radio program which I considered more worth while or more significant. May you have continued success in trying to awaken the public to the necessity of giving Negro boys and girls a chance." Members of civic organizations also praised the program for countering a formal educational system that had deprived black children of facts regarding their own history and "prevented the white child from seeing as a part of the stream of this country's history the part the Negro has played in its building."[70] NBC officials liked the program, as did Philip Cohen, the coordinator of *Americans All,* who congratulated Caliver on his "tremendous success." Caliver's friends and supporters also commended him.[71]

The shows on black contributions to literature and science were designed to combat pernicious stereotypes that denigrated black culture and denied the existence of black courage and intellectual prowess. Still, the shows made no direct reference to segregation or to pressing current issues. But shifting events during the months of the broadcasts allowed the show's creators to discuss more directly the destructive effects of racial inequality and segregation.

Sports, Segregation, and Black Patriotism

The first episode to make a direct reference to segregation and an outright appeal to black patriotism was on black athletic achievement, broadcast in November 1941. Members of the Advisory Committee saw the sports arena as the perfect metaphor on which to build a case for fair play and equal opportunity for blacks in the general society. Locke argued for framing a constructive argument around athletics, in which the accomplishment of the African American "is symbolic of what he can do in any field where he is given a chance." Brown also wanted the script to demonstrate that generally "Negroes do not get a clean break." Discussions about this episode spotlighted the balancing act that the show's creators performed to try to attract and please both a black and a white audience. For example, Tuskegee Institute president F. D. Patterson cautioned that timing would be important in making their arguments: "In the early moments of a program, if you start talking about or inferring the injustices done the Negro, click, off goes the radio." In order to appeal to black middle-class listeners, there was a consensus that the show needed to emphasize the sports achievements of black college men and women, despite Tunick's insistence that "the college sport angle will not mean a great deal to white people." But in a discussion about which boxers to mention, members agreed that the general listening public would probably not be "particularly anxious to hear about Jack Johnson," who had, among other transgressions, violated racial taboos by marrying a white woman.[72]

The sports show packed a powerful punch. After the standard introduction, the strains of "Climbing Jacob's Ladder" provided the backdrop for an opening vignette about the need to give everyone "an equal chance to play: rich, poor, black, white, Jew, Gentile." A brief segment quoted Joe DiMaggio as saying that black baseball's legendary Satchel Paige was one of the greatest pitchers he had ever faced. "Too bad he isn't in the Big Leagues," the narrator commented. NBC's technical sophistication allowed *Freedom's People* to make effective use of live pickups from several cities. In a long live interview, Jesse Owens recounted his thrill at hearing the "Star-Spangled Banner" and saluting the American flag from the victor's podium at the 1936 Olympics in Munich. The program's segment on black boxers ended with a live interview with Joe Louis from Los Angeles. If Jack Johnson was the black champion who was best left unmentioned for fear of offending whites, there was no sharper con-

Studio audience at Freedom's People *broadcast.*
(Ambrose Caliver Collection, Moorland-Spingarn Research Center,
Howard University)

trast than the shy, taciturn Brown Bomber. Louis responded to questions about his successful rematch with Max Schmeling. He ended
his interview with an earnest patriotic plea: "I've got a bigger date
coming. With Uncle Sam. . . . I think it's my duty. If the army needs
me—I'm ready to go any time."

The *Freedom's People* shows that highlighted the less overtly political worlds of music, art, science, literature, and sports delivered a
consistent message that African American accomplishment in these
fields was achieved against tough odds and through sacrifice, hard
work, and dedication. A variety of types of black music drove and
punctuated these programs and helped them succeed as good radio.
The shows were enlivened by guest appearances by Cab Calloway
and Count Basie. Aptly chosen excerpts from black spirituals performed by the Leonard De Paur Chorus or popular groups like the
Golden Gate Quartet instilled a tone of quiet dignity without laps-
ing into sentimentality. The voices narrating the programs struck the
same tone; they were serious, firm, professionally dictioned, with no
hint of pretension or apology. The presentations were matter-of-fact

and authoritative, delivered in a calm and measured manner. Still, the shows on these subjects only subtly confronted the problem of racial injustice, focusing on black achievements instead. The discreet attempt to politicize black activities in arenas that were not normally associated with politics was cleverly executed, but its effectiveness as an argument against racial inequities was thwarted by the absence of any specific target other than the permeating influence of white racism itself. When the series turned to the fields of education, military service, and work, this fragile facade could no longer hide the current pressing political concerns of African Americans, for these shows had as their focus the federal government itself.

Education, Military Service, and Work

The broadcasts that centered on black participation in the worlds of education, military service, and work openly confronted the federal government's role in perpetuating injustices in those arenas. These broadcasts walked an ideological tightrope anchored on one end by the federal government's need to pitch for patriotism and racial unity and on the other end by the need to avoid offending listeners or endorsing black claims for full equality.

These three broadcasts in December 1941 and January and February 1942 came during a period in which there was a rapid escalation in black protest against discriminatory policies related to the war. In December, the black press revealed that the Red Cross was refusing to accept blood from black donors. Major black newspapers devoted a month's worth of front-page attention to this issue, and the NAACP challenged the Red Cross on its policy. After black leaders, the Red Cross, and federal officials held several meetings, the policy was changed and black blood was accepted, but it was to be stored separately, segregated for black use only.[73] Violent attacks against blacks also increased at about the same time. Twelve black soldiers were assaulted near their camp in Louisiana. In another nationally publicized case, a black man accused of raping a white woman was seized from a jail in Missouri and lynched by a mob of whites.[74] Federal policies themselves continued to be catalysts for black anger. Pressured by the NAACP, the army was forced to rescind an order at a Pennsylvania camp that defined any association between black soldiers and white women as rape. In addition, when the army asked for 3,000 nurses, it established a quota of only 56 black women. Meanwhile, the *Pittsburgh Courier* publicized the heroic efforts of black navy

messman Dorie Miller during the attack on Pearl Harbor and challenged the navy's refusal to honor him appropriately.[75] In another
incident replete with irony, officials designated as available only to
whites the new federally financed Sojourner Truth housing project in
Detroit, which originally had been envisioned as an interracial complex. Blacks successfully organized to convince federal officials to
reverse their policy, but a riot erupted when whites formed mobs to
deny blacks access to the building.[76]

These and many other well-publicized examples of continuing
racial injustices provided the backdrop for the formal inauguration of
the *Pittsburgh Courier*'s "Double V" campaign in February 1942. With
one "V" for victory over enemies abroad and the other for victory
over enemies at home, the slogan symbolized that blacks would close
ranks on behalf of the nation's defense but would not abandon their
fight against racial inequality. Capturing perfectly the paradoxical
sentiments of African Americans in World War II, the slogan quickly
spread into popular use.[77] Military intelligence officers noted the
shift in tone and the increasing number of what the army considered
"inflammatory" articles in the black press, which they collected and
analyzed in detail. As a result, some black newspapers and magazines
were banned from army posts, and seventeen were listed in a February military intelligence report as containing Communist-inspired,
incendiary material.[78] Federal officials were afraid that the black pen
was a double-edged sword. They were worried not only about the
impact of these reports on blacks at home and in the military but
also about their use as fodder for enemy propaganda abroad. The relationship between blacks and the federal government had moved to
center stage in a newly visible struggle over racial injustices.

At the center of the contest between federal officials and African American activists were the military's policies of racial inequity,
segregation, and exclusion, which were all under attack. For that reason, Ambrose Caliver knew even before the attack on Pearl Harbor
had occurred that the broadcast on African Americans in military
service would be an especially important and "delicate" episode to
produce.[79]

The final version of the script detailed black military service and
heroism in U.S. wars by using an epic in song form interspersed with
dramatized vignettes of black soldiers in various historical battles.
The song began: "By the record we've made, / And the part we've
played, / We are Americans, too. / By the pick and the plow, / And

the sweat of our brow, / We are Americans, too." This part of the script was a tidy and cleverly presented historical capsule of black sacrifice in all of America's wars, although it made no explicit argument for equal rights and treatment of African Americans based on the record of their past service. The timing of the show's broadcast on December 21, 1941, two weeks after President Roosevelt's "Day of Infamy" speech, undoubtedly influenced the program's content and transformed parts of it into a patriotic paean. But the underlying political issue that was addressed, although gingerly, was the federal policy that prohibited black soldiers from holding combat positions. In a live speech, Colonel West Hamilton, commanding officer of the 366th Infantry at Fort Devens, Massachusetts, and one of the nation's highest-ranking black soldiers, pleaded for a chance for blacks to fight: "America — your America, my America is at War."

The broadcast made this explicit plea: blacks have served in combat in the past and have done so heroically; let them serve again. This was a major plank in the rhetorical battle being waged in the black press and by black leaders. If blacks were going to serve, many argued, let them have the opportunity of full and equal service so they could try once again to "earn" their status as Americans by shedding blood against foreign enemies. The looming question of segregated service could not be addressed until access to battle was itself secured. But this show magnified the sadly ironic nature of that argument, made more apparent by the appearance of African American men on a federally sponsored radio show pleading for a change in this federal policy.

The fervor of the campaign for an equal opportunity to serve reached its peak in the next show, a January 1942 episode on the African American worker. Like the previous shows, this program's historical narrative told the story of black progress and contributions, in this case in the world of work. The introductory vignette was a treatment of slavery that then shifted to a series of aural vignettes of black workers on farms, at seaports and on ships, on railroads, and in factories, all portrayed as part of the transition from slavery to freedom. Although the focus was on work, the argument from the previous show permeated this broadcast. The phrase "when given a chance" became a refrain. The script praised black skilled workers and inventors: "When given a chance, they proved their skill."

The message remained the same, but the tone of the broadcast changed dramatically when it shifted to Chicago for a live speech by A. Philip Randolph, the most visible African American activist against discrimination in the armed services and defense factories. After referring positively to the president's ban on discrimination in the defense industry but without mentioning Randolph's role in securing it, the narrator introduced him simply as "one of the most distinguished labor leaders of our country." Randolph, who had once aspired to be an actor, paced his rolling baritone voice through a passionate and imaginative account of the role of black workers in American history. At the same time that Randolph reaffirmed black loyalty, he emphasized the barriers of racial discrimination and the idea that blacks already had earned the right to democracy and freedom. In his stirring and dramatic conclusion, Randolph's defiant intonation seemed openly to make black patriotism conditional on a change in racial policy. Considering Randolph's continuing criticisms of President Roosevelt and his policies, it must have been a challenge for Randolph to stay within the political borders of a federal broadcast. Randolph's draft of his remarks for the show made no reference to the president, but a handwritten insertion in the final script mentioned "the matchless and courageous leadership of our great President Franklin Delano Roosevelt." But when Randolph delivered this line on the air, he took the liberty of omitting the adjective "great." [80]

The strident tone of Randolph's delivery triggered a stream of supportive letters from listeners, but they assigned very different interpretations to the show. Many people were deeply moved by Randolph's eloquence, as was one listener who stated: "I enjoyed it so thoroughly that it sent chills through my entire body; it even moved me to tears." A favorable editorial in the black weekly the *Boston Guardian* praised the show for providing the first true record of black contributions to industry: "It used to be a favorite gibe to tell colored folks that they were 'consumers rather than producers' but the record reveals that the colored brother, according to his opportunities, has participated in the productive life of America." One listener considered Randolph's speech an effective counterpoint to the publicity that emerged from a recent meeting of African American leaders who admitted that they believed that the black community was not behind the war effort. Similarly, another listener praised Randolph for his "fine cultured patriotic address." Others saw the segment as carrying a message for whites. "If I could have only had Aladdin's Lamp," one listener wrote, "every receiving set would have been tuned in, more especially, those of our white brother below the Mason-Dixon line, as it would have been an education to them."[81]

The broadcast on the military made the simple argument that African Americans had fought and fought well in the past so they should be permitted to fight now. Randolph made the corollary argument about employment opportunities: blacks have worked and worked well; let them work now in defense jobs. Listeners imposed their own interpretations on the show and Randolph's remarks precisely because of the open-ended and vague nature of the type of appeals permitted on a broadcast to a mass audience. Given their limited guest privileges on radio, secured in this case under the aegis of the very entity under attack, black speakers like Randolph had little choice but to offer mild appeals. But Randolph's tone belied the neutrality of his words. The depth of his anger and conviction burns through that neutrality. The fact that an activist and orator of Randolph's commitment and skill was bridled in this radio appearance also serves as a profound reminder of the centrality of a free black press to the cause of racial justice during the war period. If black people had been forced to rely on radio as their primary means of communication about the failings of the federal government, they would have been on an impossible mission since they were admitted to radio only as entertainers or as briefly invited guests expected to be on good behavior.

"The Negro and Christian Democracy"

Caliver had promised NBC that the one-hour grand finale for the series would focus on black contributions to the theater, acquiescing to the network's insistence on higher entertainment content. He changed that plan quite dramatically, however, after he concluded that a final program emphasizing entertainers would trivialize a series that had gradually come to address pressing political issues facing the African American community and the nation as a whole. He wanted the final program to be about Christianity and democracy, and he wanted it to be broadcast during what he called the "church hour," noon to one o'clock on Sunday, so that churches could coordinate their services around the broadcast, actually bringing radio speakers into the sanctuary during their morning services.[82] Caliver was himself a person of faith, and he seemed moved by a plea from Benjamin Mays, the prominent African American religious leader who was then president of Morehouse College. Mays had told Caliver that he was disappointed that the series had not included any mention of the African American contribution to religion, which he described as "more tremendous than most of us realize."[83]

Caliver's approach to this episode was rooted in his belief that "racial tolerance is a demand of democracy founded on Christianity and moral principle," a statement he had made in a speech in 1941. Working from the assumption that American democracy was rooted in Christianity, Caliver wanted the broadcast "to show the relationship between christian principles [and] our democratic way of life." He intended the broadcast to be aired in black and white churches and hoped that it would inspire mass meetings to better race relations. But he especially wanted to reach the black religious community, and in preparation, he had met with the heads of several black denominations who offered their support for the coordinated church broadcasts.[84]

Caliver planned to present a complex political argument couched in religious terms. The show would assert that racial inequality was a challenge to the philosophical underpinnings of Christianity and that the race relations problem was a burden and a moral challenge that Christian churches had a duty to face in order to ensure the survival of Christian democracy itself.[85] Although it was cast in religious rhetoric, or perhaps because of that, this assertion was the most blatant general political appeal the series would make.

Although riddled with contradictions, the show succeeded in making a powerful moral argument that equated the perpetuation of racial inequality with sin. The early parts of the broadcast were very confusing because even a scriptwriter as skilled as Tunick found it difficult to craft a coherent historical narrative that included both slavery and Christian democracy. After strains of "The Battle Hymn of the Republic," the opening narration linked the search for truth with the quest for freedom. The scene shifted to a shielded allusion to the Civil War as "a fierce struggle," yielding the truth in "four short lines of print—the 13th Amendment." A dramatization portrayed Lincoln as a "religious, God-inspired man" who suffered with the slaves and "looked to the same God as they did." "One God, one people, one nation" seemed to be about as strong an argument as this section could muster.

Other parts of the broadcast were more maudlin than inspiring. After a compelling reading of the "Let My People Go" sermon from James Weldon Johnson's *God's Trombones,* a rendition of "Go Down

Moses" led into a reenactment by the black actor Juan Hernandez of scenes from the Marc Connelly play, *The Green Pastures*. The play, which was later adapted into a movie, offered an Old Testament version of black slavery and liberation and a simplistic, patronizing portrayal of black religiosity and theology. Brown had warned Caliver that he had "a great deal of difficulty" with including material from the play because many African Americans did not like it: "I am timorous about the good sisters and brothers when the 'Green Pastures' comes on." But Caliver retained an excerpt that focused on the redemptive value of black suffering. Despite the show's emphasis on Moses, Exodus, and the Old Testament, the creators of *Freedom's People* remained insensitive to the fact that their notion of Christian democracy made no room for Jews or other non-Christians or non-believers. Apparently this realization dawned on someone at some point. That may help explain the insertion of a Pauline-inspired theological argument in the show's introduction: "For by one Spirit we are all baptized into one body, whether we be Jews or Gentiles, whether we be bond or free."[86]

The show's explicit political appeal was not to be found in the confused vignettes or in *The Green Pastures* excerpts but in a speech delivered by the Reverend W. H. Jernagin, president of the Fraternal Council of Negro Churches.[87] He delivered a message targeted at an audience of black and white Christians. Although he made a series of general appeals, he also implicitly made specific pleas to black and white religious leaders. He urged African Americans to demand more of their religion and their churches and appealed to blacks and whites to embrace a truer Christianity for democracy's sake. Jernagin argued that "the primitive religious conception as portrayed in the *Green Pastures* which had its value in the past is no longer adequate for modern life." In his view, "Negroes demand more of their religion now" because current conditions called for a church that was "concerned with their common everyday problems and urgent needs."

It is noteworthy that, at a time when most black organizations and institutions stood united behind efforts to pressure the federal government into changing its racial policies, Jernagin felt compelled to caution black church leaders to avoid insularity and coax them to join the fray of the everyday struggles of their members. Caliver and others believed that the black church, a central community institution, was insufficiently involved in the struggle for racial and economic justice.

Jernagin tacitly directed comments to whites about the continuing conflict between general Christian practices and principles. He made the arguments that "the ideal practices of true Christianity are necessary for Democracy" and that "democracy is the best type of political government under which true Christianity can thrive." The protection of one was essential to the preservation of the other. Speaking to his audience as a whole, Jernagin made a passionate appeal for the kind of "true Christianity" he had in mind: "A recognition of the needs of Negroes in education, for adequate and sanitary housing, for an equitable opportunity to work, to play, to vote, and to live an abundant life without fear is a religious obligation. . . . [I]t [the church] must not only inspire—it must grapple with the economic, social, and personal problems which beset us, and must assure more honest administrators in public office. The church must become 'Not a temple of cold doctrine, but a radiant center of Human Brotherhood.'" With this final plea, Jernagin sought to create a common moral mission and a "true Christianity" for black and white churches that looked back to the abolitionist campaigns against slavery and presaged the role of the religious community in the modern civil rights movement.

Up until the last days before the broadcast, Caliver expected that the president would keep his October promise to make a personal appearance on the concluding episode of the series. In February, he had reminded Eleanor Roosevelt of the president's agreement and informed her that NBC planned to offer the talk to all of the networks, as was customary with presidential addresses. Caliver emphasized to her that the broadcast would give the president an opportunity "to indicate what both you and he have stressed so often, namely, that the improvement of race relations is fundamental to the perpetuation of democracy." But the president and his aides did not want to renew his commitment to appear on the program. White House aide Stephen Early advised Caliver that the president could make no advance commitments of any kind because of his war duties.[88]

Because this response was not an outright refusal, Caliver continued his campaign to have others pressure the president to appear on the show. African American leaders and journalists had repeatedly encouraged the president to make a personal radio appeal to build up black morale and reduce the risk of more interracial violence.[89] By April 1942, when the show was broadcast, incidents of interracial conflict were increasing, as were daily reports in the black press of unjust defense-related policies. Locke told members of the Advisory Committee that he believed the president needed to speak to blacks and that now was the time to do it. Locke and other members of the Advisory Committee, including federal officials Mary McLeod Bethune, Joseph Houchins, and Campbell Johnson, sent a telegram to the president urging him to appear on the show to help lift the low morale prevalent among African Americans and to strengthen national unity and race relations. But these and other pleas had little effect because White House officials had decided some time ago against a presidential appearance, although no one informed Caliver until a few days before the broadcast. At that point, White House officials offered Caliver a letter from the president that was to be read during the final broadcast.[90]

Caliver's appeal to the president via Eleanor Roosevelt disclosed information that may have deepened the political implications of an appearance about which the president and his advisers already had strong reservations. In January, a BBC official based in New York City had approached Caliver about rebroadcasting the phonograph recordings of the series throughout "the British Empire," which at that time, of course, included large numbers of Africans and people of African descent and other peoples of color.[91] It seems that by the time of Caliver's plea to the president, he had made such an arrangement with the BBC. Caliver enthusiastically reported the plan and suggested that the program also be sent via shortwave to all Allied nations and to South America as well. When Advisory Committee members contacted the president, they also emphasized that a "forthright statement" from him would bolster "the morale of Negroes in our own country and our colored allies throughout the world." At a time when worries about people of color at home were deepening, the prospect of making a worldwide speech on the subject of racial equality may have inadvertently increased the administration's reluctance to engage the issue directly.

Caliver's personal appearance on the final show of the series revealed no signs of the disappointment he must have felt when the president reneged on his commitment. All along, that promise had been Caliver's ace in the hole, and the president's refusal to lend his personal prestige and radio presence to the show denied the series its coup de grâce—a long-awaited personal radio appeal from the president himself on the question of racial equality.[92] Instead, Caliver used his airtime to praise the success of the series, reporting that *Freedom's*

Ambrose Caliver appearing on Freedom's People.
(Ambrose Caliver Collection, Moorland-Spingarn Research Center,
Howard University)

People had been given a "fine reception" across the nation by "persons of every class, color or creed [as] indicated by the thousands of letters and comments received." Then he read the president's brief letter, in which Roosevelt apologized for not being able to make a personal appearance and commended the importance of the radio series. The letter's strongest language in support of black Americans stated: "The Negroes are an important part of our American citizenry. They have made valuable contributions to every phase of our national life.

It is our obligation to assure to them, as American citizens, full opportunity to use their many talents, not only in winning the war, but in establishing the peace that will follow." Thus, *Freedom's People* came to an end, rewarded only with a flaccid promise of "full opportunity" for blacks and face to face with the rhetorical divide that continued to separate African Americans from Franklin Roosevelt.

Farewell to *Freedom's People*

Even without the president's appearance, the final episode of *Freedom's People* met with critical acclaim and much public enthusiasm. *Variety,* which had given the first episode only a lukewarm reception, characterized the last show as "deeply touching" and a "fervent tribute to Negro patriotism."[93] Fellow New Deal officials thanked Caliver for contributing "inestimably to healthy public thinking on the Negro" and for advancing the cause of race relations on a national scale by raising "America's regard for its colored citizens to levels often wished for but little worked toward."[94]

Many people had been deeply moved by the final broadcast, several reporting that they had been brought to tears as they listened to what one writer called "the struggle of the race." African Americans who wrote to Caliver also expressed a great deal of pride about the fact that the show had been a professional and dignified black production. One writer confided, "I never dreamed we could put over such an outstanding program." Using an apt mixture of metaphors, one listener praised the broadcast for the "complete harmony" of "the visual picture of inter-racial goodwill." A few black writers also indicated that sympathetic whites had enjoyed the show and wanted more information: "I have a good white friend, too, who would be glad to have the recording and take it to her club meeting." Many listeners thanked the Office of Education for the show and urged its continuation; one writer warned that if the show did not continue, its positive effects would soon be "dissipated in the seas of misconception which circulate about the Negro." Even President Roosevelt's noncommittal remarks were received favorably by some listeners, one of whom judged his message to be "touching and sincere." After having been identified by name in each broadcast as the series supervisor, Caliver's own appearance on the show also generated comment. One writer reported that "Dr. Caliver was an inspiration to every Negro boy in the church this morning."[95] The most overwhelming response to the final episode and to the series as a whole came from schoolteachers, who not only requested study materials but also repeatedly urged that the show be continued. One school principal called the series a "magnificent conception," and another referred to it as a "splendid series of broadcasts."[96] Although much of the response to the final show appears to have been spontaneous, it is also clear that word had gone out among black teachers that the only way to keep the show on the air was to generate evidence of enthusiasm for it.[97]

Caliver used the letters and comments he received as supporting evidence for a memorandum to Studebaker in which he recommended that the series be continued. Studebaker had remained very impressed and pleased with the reception to the shows. "These broadcasts have not only achieved their high purpose of improving national unity and better race relations," the commissioner wrote, but "they have also set a new pattern for radio presentation by and about Negroes."[98] Concluding that the series had "made a very significant contribution to national unity and better race relations," Caliver argued that a second series would fit perfectly the purposes now being espoused by Archibald MacLeish, then a newly appointed official at the Office of War Information who advocated greater federal attention to building up "Negro morale." He suspected that several other groups were expressing an interest in morale-building radio programming for African Americans, so he tried to press the agency to pursue aggressively a second series of ten or twelve shows, with a budget of $25,000 — more than three times the expense of the original.[99]

In its March public service publication, NBC had praised *Freedom's People* in an editorial entitled "Radio and National Morale."[100] Soon after the last broadcast, NBC expressed its enthusiasm about the series and its willingness to do a second series.[101] But for whatever reason, no second series ever developed. The trail of information on *Freedom's People* simply disappears after Caliver's memorandum recommending that the program be continued and after the network expressed an interest in doing just that. Nothing indicates whether the Office of Education pushed for the continuation of *Freedom's People* at a time when the war crisis made finding $25,000 in either federal or private funds a particularly steep order. Certainly, the series fit the political salience of the moment and afforded the agency a role in pursuing a most important domestic goal of that time: preserving racial peace at home.

In the fall of 1942, an attempt was made to broadcast a half-hour weekly variety show entitled *Freedom's People* exclusively to black soldiers on the Armed Forces Radio Service network. It is unclear whether Caliver was involved with this proposal. Apparently, both the title and the idea of a show directed solely at black soldiers proved too controversial, as the show's title was changed to *Jubilee* and it was to be broadcast to all troops, black and white. First aired in October 1942, *Jubilee* became one of the network's most popular programs. It showcased African American musical virtuosity of the 1940s and, as part of its historical legacy, preserved an extraordinary body of recorded performances, jazz in particular.[102]

By 1943, Caliver had given up on the idea of producing another nationally broadcast show. Instead, ever resourceful, he sought funding from the Southern Education Foundation for the production of a series of twelve recordings of "dramatized stories of Negroes and Negro life," with musical selections suitable for children. He envisioned distributing the recordings and a companion study guide nationally through Parent Teacher Associations. Thwarted in his attempt to return to the air via a national network, Caliver proposed instead that these recordings be used for independent broadcasts by local radio stations. But Caliver's attempt to continue the work of teaching African American history and attacking racial intolerance through radio and the public schools did not meet with success. There is no evidence that this particular vision was ever realized.[103]

Ambrose Caliver worked for the Office of Education his entire career until his death in 1962 after thirty-two years of federal service, half of them spent as "senior specialist in education of Negroes." During the war years, he helped with black manpower defense education projects, and afterward he worked to increase postwar educational opportunities for African American veterans and war workers. Later, building on his own work from the 1930s, Caliver provided important information on the status of black schools for *Brown v. Board of Education*.[104] *Freedom's People*, the project he had worked tirelessly to bring to life, may have faded away, but Caliver's résumé in the 1960s still trumpeted his radio work as a proud moment in his federal career.

The "Negro" as an American

Freedom's People told the story of African American contributions with sympathy, dignity, and pride. It offered a quiet advocacy of racial progress in confident, articulate black voices. A celebration of black culture infused the shows, not only through music but also through history, literature, and the arts. At times, these broadcasts were cloaked in a mystique of black suffering, black religiosity, and black survival, as expressed in spirituals and the blues. Not surprisingly, the accompanying but disharmonious narrative of past and current oppression remained silenced or muffled.

The show expanded on the limited themes that had been developed in earlier federal broadcasts about black history. The episode of *Americans All* about African Americans in 1938 presented a compelling display of black cultural and economic contributions throughout U.S. history, but it stopped short of doing more than conferring the label "American" on blacks. *Freedom's People* redefined American culture as being driven by and dependent on black cultural contributions. Not only did U.S. history have African Americans at its core, it argued, but the very definition of what was distinctly American was rooted in black culture. *Freedom's People* engaged the difficult question of what it means to be fully American, to be a free American. Whenever the show raised that question, whether it was posed about the boxing ring or a job or a school or the military, the answer was the same. A free American has the right to fair play and equal opportunity, and by that definition, African Americans were not yet true Americans.

Although the political arguments the series made were muted, the aural images of black Americans that it presented stood in sharp contrast to the predominant depictions of black people, black character, and black abilities. These new images appear to have been one of the principal motivations of those who worked on the series and one of the primary reasons for the show's appeal to African American listeners. As limited as its life was, the series at least allowed for the existence of African Americans who lived outside the confines of *Amos 'n' Andy* and *The Jack Benny Show*. The new "Negro" on *Freedom's People* was gifted and generous but also hardworking, diligent, persistent, courageous, and intelligent.

But this new "Negro" would remain a lonely and exceptional figure on radio. Perhaps more significant than any other message, *Freedom's People* stands as a poignant example of the political dangers created when a capital-intensive communications medium like radio becomes a dominant forum for public debate and discourse. The power to deny access becomes tantamount to censorship. Devoid from its inception of any pretense of larger public purpose, radio was born to entertain, to advertise, to make money. As the first mass medium, it treasured its ability to build a mass audience most of all. When radio took on the broader role of being a source of official information and news, broadcast officials were adamant that discussions of controversial or sensitive political issues were to be avoided or practiced only in designated safety zones. But rather than being a neutral conduit, radio became a favored ideological site precisely because of its power and its reach.

A show like *Freedom's People* struggled against the medium's nature and its special sensitivity to discussions about race, considered by network officials to be a deadly audience killer. Those who worked on the show probably had a keen awareness of the parameters in which they had to work. Under the circumstances, they framed quite sophisticated political arguments, relying on the power of black culture and black history itself to make the case for greater freedom for African Americans as a whole. Still, the limited and indirect messages allowed on *Freedom's People,* when compared to the rhetorical vigor and verve of commentary in black newspapers and magazines, serve as a startling symbol of what it means to be an unwanted and powerless guest in someone else's space.

If it had not been for the virulence of black protests concerning defense-related issues, *Freedom's People* would not have made it on the air, despite Ambrose Caliver's best efforts. Fortunately for him, the fear of aggressive black activism pushed the federal government to endorse efforts to raise black morale such as those embodied in this show. Because a radio show is molded by so many minds, it is difficult to disaggregate anything remotely akin to a single authorial imprint. This show was an unusual collaboration between federal officials, black intellectuals, a white professional scriptwriter, black artists, and a host of technical assistants. Of these, the most important were Caliver, working within the federal government, and black intellectuals and activists working from the outside. A shared political vision made their collaboration effective. The political arguments contained in *Freedom's People* clearly reflected the influence of black intellectuals like Howard University professors Alain Locke and Sterling Brown and black activists like A. Philip Randolph. The show's basic political argument was "on the message" and consistent with the views being advanced by black organizations, although it was necessarily a truncated and less explicit version. Unlike *Americans All,* which struggled to sustain a coherent narrative, *Freedom's People* was driven by cogent arguments. Many of the arguments on *Freedom's People* were polished and well rehearsed precisely because black political activists had been making essentially the same claims for humane and just treatment through the centuries, although in different settings.

Black elites, working closely with white elites at the Office of Education and NBC, created a cultural product that was suitable for consumption by the multiple audiences that radio reached: black and white listeners of all classes and regions, including the eager, the sympathetic, the curious, the suspicious, and the indifferent. With this broad potential audience, *Freedom's People* had to make a consensus appeal, and the show's content is evidence of how narrowly constructed that consensus was. President Franklin Roosevelt's limited remarks on the final show probably captured that consensus well: beyond an acknowledgment of black contributions and a general promise to assure full opportunity, there was no agreement.

CHAPTER 3

"NEGRO MORALE," THE OFFICE
OF WAR INFORMATION, AND THE
WAR DEPARTMENT

World War II placed an unprecedented burden on the people and resources of the United States. As the nation neared and then made its formal entry into the war, one of the most persistent fears among political and military leaders was the prospect of civil disturbances due to increased claims for racial equality from the 10 percent of the population that was African American. Federal officials had ample cause to worry. After all, racial segregation and discrimination pervaded every aspect of the war effort, from defense jobs to Red Cross blood banks to the armed forces, battlefields, and battleships themselves. This inescapable reality provided effective ammunition to African American activists and members of the black press who united in a mind war to shame the nation into breaking down the barriers of racial inequality. They knew, as federal officials did, that African American participation in a unified home front and in the war abroad would be essential to the war's outcome.

Radio was indispensable to the war effort, both as an up-to-the-minute reporter of war events and as a unifying voice for patriotism and sacrifice. The war's presence on the radio was ubiquitous. Broadcasting industry and advertising executives sustained public engagement with the war through evocative patriotic programming, vivid wartime reporting, and the integration of war messages into popular entertainment shows and advertising campaigns.[1] As part of radio's support for the war, the networks also granted a significant amount of free airtime to the federal government, including the Office of War Information (OWI) and the War Department.[2]

Many officials at both federal agencies were deeply concerned with domestic racial politics and agreed that something needed to be said to and about African Americans. At the same time, there was considerable trepidation about how to break the sanctioned political silence about African Americans and their place in the nation, especially in the face of increasingly visible black demands for just such a reassessment. As a consequence, these two important federal agencies produced a relatively limited amount of radio programming about race relations or African Americans considering their level of worry about racial unity and the number of public appeals they offered on other issues. The problem facing these federal officials was simple yet complex: they wanted to build up black morale by integrating a more visible "Negro" into the public sphere of patriotic rhetoric, but they did not want to endorse the racial reforms blacks sought for fear of offending whites, especially southern congressmen. If radio was to be used effectively to lift African American morale, it would have to speak the unspeakable about racial segregation and discrimination. Federal officials found that the exigencies of war required that something new be said for the sake of creating the illusion of a unified home front at a time when African Americans persisted in protesting racial injustices and when many white civilians feared the consequences of those protests.

Efforts at the OWI and the War Department to mount a limited public radio campaign to lift African American morale and build greater racial tolerance among Americans were plagued by internal political paralysis manifested on at least two occasions by refusals to broadcast shows considered too controversial. The experiences at these two important agencies demonstrate the increasingly untenable nature of the Roosevelt administration's response to the campaign for fair treatment that African Americans pursued during World War II. This dynamic included repeated disputes about who could speak for African Americans and who could best determine what image of African Americans the federal government should endorse for its limited political purposes. The political machinations at these agencies and between them, African Americans, and whites inside and outside government provide dramatic evidence of the conflicts that would drive the battle over racial equality for the duration of the war and into the postwar period.

This chapter then is less about specific radio programs, partly because so few were offered by the powerful propaganda operations at these two agencies. Rather, the focus here is on the relationship between the federal government and the politics of racial representation, in the dual sense of *who* could represent the race before the government and *how* the race was to be represented on radio and in other media. Indeed, in order to understand why the OWI and the War Department made such limited use of radio to address racial issues, we must detour beyond radio to two examples of the racial propaganda these agencies produced in other media. For that reason, I consider the use by the OWI of the printed image, specifically in a pamphlet called *Negroes and the War*. Controversial in every respect, this publication allowed visual images of African Americans to speak for themselves, delivering political messages that could not yet be given voice on radio, a medium that depended on the use of words and language. What could not yet be said could be shown in this pamphlet, but even then its more racially progressive political meanings did not escape the eyes of conservative southern congressmen who attacked the medium, the message, and the messenger, as a result effectively ending the OWI's print operations. This reaction helps explain why the federal government's use of radio to address what it saw as an urgent matter—the threat of domestic unrest from African Americans—was so tepid and so limited. The second example is a 1943 army film called *The Negro Soldier*, which was designed initially to lift the morale of black soldiers but was eventually shown to all new recruits, black and white, and to civilian audiences as well. The use of film allowed for a narrowly constructed reality of diligent and heroic black soldiers, excluding from the narrative and the camera the existence of racial inequalities at home and in the military. Protected from political controversy by its narrow frame and its army sponsorship, the film also escaped attack by using the power of moving images to create a world unto itself and a more singular political interpretation, an advantage film had over printed images.

Ironically, by looking beyond radio, we can see its special strengths and weaknesses as a purveyor of racial propaganda. The medium could only be as powerful as the message, and in the case of African Americans, no message was considered politically acceptable to the national, mass audience that radio reached with such speed and ease. State-sanctioned public discourse about African American claims for racial equality was so restricted that radio lost its voice, its defining strength: the evocative use of language and words.

"Negro Morale" and the Office of War Information

The Roosevelt administration in 1941 reluctantly established agencies that were to provide information about the war to the general public, among them the Office of Facts and Figures (OFF) under the direction of Archibald MacLeish, the liberal poet and librarian of Congress.[3] Later, in June 1942, President Franklin Roosevelt placed the OFF under the umbrella of a new Office of War Information and appointed as the OWI's director the popular radio news commentator Elmer Davis. For the duration of its three-year life, the OWI was plagued by internal and external conflict over its goals as well as continued congressional and press skepticism about its mission. President Roosevelt never resolved this ongoing administrative conflict, and as a result, the OWI had to carry out a mission that was resisted and never clearly defined.[4]

Few subjects perplexed the federal government's information and morale-building efforts in the way that issues involving African Americans did. Federal officials were slow to accept the idea that African Americans were less enthusiastic about the war effort than whites. Liberals like Vice President Henry Wallace and MacLeish saw the war as a revolutionary struggle, as, in Wallace's words, "a fight between a slave world and a free world."[5] This stark symbolic contrast seemed so plain and compelling to MacLeish that he expected African Americans not only to embrace the war struggle as their own but also to do so with a special enthusiasm. In a February 1942 address delivered to the National Urban League and carried on radio, MacLeish argued that African Americans had a special appreciation of this war like "no other single group" because he saw the war as one against the "conspiracy of slavery."[6] The logic of the argument seemed so appealing to MacLeish that he was surprised by the reactions it generated. Some who had heard or read MacLeish's speech tried to convince him that the "slavery" argument was insufficient to overcome the view among many African Americans that asking them to fight another war against "slavery" simply brought into even

sharper relief the daily reality of continued discrimination, segregation, and racial injustice.[7] The leftist New York City paper *PM* chided
MacLeish for underestimating the degree of skepticism about the
war among blacks. MacLeish also had hoped that his speech would
reassure fearful whites, but it failed to do so. A white listener in New
York City wrote MacLeish a passionate letter detailing information
"that has reached me through my maid and several other colored
people who are my friends":

> I learn that all Negroes, from menial laborers to professional
> people are unconvinced they have in fact, a stake in this country.
> They wonder whether living under the domination of the Japa-
> nese or even under Hitler, could be worse than living under the
> fascism as practised in the southern states. They wonder if the
> brutality of the storm troopers is any worse than the brutality
> of a mob in Sikeston, Mo. They question whether a concentra-
> tion camp is worse than a Georgia chain gang. They compare the
> Red Cross' attitude toward Negro blood donors to the unscientific
> racial theories advanced by Hitler. They wonder if the ghettos in
> which they are forced to live, as exemplified by Harlem could be
> any better than the ghettos of Europe into which Hitler has forced
> the Jews. They compare their inability to exercise the right to the
> ballot in some sections of our land to Hitler's depriving the Jews
> of their citizenship rights.

MacLeish politely dismissed her concerns as unrealistic and confidently reasserted that "no section of any population could possibly
understand the basic issue of the fight against fascism as negroes
understand it." He explained that "all scientific polls so far" had
revealed that only a small, although articulate, minority of blacks
shared the point of view reported in the letter.[8]

African Americans also warned MacLeish that his reading of
the situation was incomplete and that indeed black enthusiasm for
the war was weak and certainly would not flow automatically from the
"slavery" analogy. Immediately after the National Urban League
speech, MacLeish received a letter from Carlton Moss, a black scriptwriter and manager of the New York City–based Council on Negro
Culture, which had recently sponsored a successful fund-raising stage
revue for black soldiers. Moss advised MacLeish that "unless we
answer the just grievances of the Negro people the opposition can
take, as they are doing, these just grievances and use them to sow dis-

unity and confusion." To be effective, however, Moss argued, a federal propaganda campaign was necessary to clearly dramatize African American contributions to the country as well as the "wrongs [the African American] has endured" and "how these wrongs are being corrected." This was not advice MacLeish was prepared to hear or act on.[9]

Warnings about low black morale also came to MacLeish from within the federal government. Indeed, early civilian and military intelligence reports, often culled from the pages of the black press and public opinion polling, had revealed that the wholehearted commitment of African Americans to the war effort definitely seemed contingent on improved racial policies at home and in the military.[10] Administration officials translated these reports into generalized fears that African Americans were especially susceptible to foreign propaganda and prone to disloyalty and violent outbreaks.

MacLeish had become sufficiently concerned about this situation that he welcomed the arrival of Theodore Berry, a prominent black lawyer from Cincinnati who was hired as a staff officer in the Liaison Bureau of the Office of Emergency Management. A National Association for the Advancement of Colored People (NAACP) board member and president of the Cincinnati NAACP from 1932 to 1938, Berry also had served as national director of Councils for Participation of Negroes in Defense Programs.[11] One of Berry's first projects was to draft a plan for strengthening "Negro morale." Berry explained in a memorandum that African American support for the war was low because blacks were strongly skeptical that the war was being "genuinely prosecuted for practical democratic principles." He cited several specific reasons for this skepticism: black memories of the last war and the contrast between the treatment of black soldiers in France and their treatment after they returned home; the colonial policy of Great Britain and its "non-democratic practices toward natives, especially in South Africa and India"; and the feeling that no change in racial attitudes had occurred in the United States since World War I. Berry listed the focal points of black grievances: the segregated military, the navy's refusal to enlist blacks except as messmen, the "lily white" Marine Corps, defense industry discrimination, the segregated blood bank policy of the Red Cross, and Jim Crow generally.[12]

Not mincing his words, Berry bluntly warned that "we cannot bolster morale by words alone" and that the success of any black morale

OWI official Theodore M. Berry.
(Cincinnati Historical Society, B-98-255)

program would depend "chiefly upon the degree to which other governmental agencies, particularly the Army and Navy, will cooperate; stop temporizing and shadow-boxing with the issues; and establish a clear non-discriminatory policy and practice." The morale-building program he outlined included a comprehensive media campaign aimed both at convincing African Americans that they had a stake in the war effort and were already making a contribution to it and at persuading whites that "negative racial attitudes . . . weaken and retard our war program." Berry also pressed for a clear presidential statement declaring that continued racial prejudice and discrimination would lead to civilian unrest.[13]

Berry prepared his memorandum in anticipation of a meeting the OFF convened with representatives of black organizations in 1942.[14] The invited group included black journalists, federal officials, religious and academic leaders, businessmen, and a full complement of black political leaders, including Roy Wilkins of the NAACP. Wilkins concluded that the most "significant aspect" of the conference was "that the vast majority of the conferees arose to say that they did not believe they could build any morale among their followers until the government took some definite and important corrective action about the mistreatment of the Negro throughout the whole war effort." He also recommended that the OFF concentrate on changing white opinion through popular media, including radio.[15]

The continued importance of the question of African American morale also was reflected in its recurrence as a topic of discussion in the frequent board meetings at the OFF in the spring of 1942. The minutes of those meetings show that several were "taken up" with the issue and that MacLeish in particular was growing increasingly concerned that Nazi and Japanese propaganda might take advantage of the existence of racial inequality. MacLeish observed in one meeting that any responsive domestic propaganda campaign on the issue had to be two pronged: it must be addressed to whites, and it must include some action by the federal government, although he did not appear to know how that action could be achieved politically.[16] This was essentially the position that Berry urged him to take.

MacLeish commissioned a study to collect more empirical data on African American attitudes about the war, perhaps to add credibility to his conclusions.[17] That May 1942 study, "The Negro Looks at the War," confirmed government officials' worst fears about the lack of black identification with the war effort. Based on interviews, the report's twenty pages of analysis and over fifty charts and tables lent an air of authenticity to its findings and to their potential import for policy makers.[18] The survey results showed that African American support for the war effort was tempered by the reality of discrimination in defense industries at home and the denial of equal opportunity to blacks in the military. Nonetheless, the report's writers chided blacks who voiced such complaints, noting, apparently without intending any irony, that they "gave mainly cynical reasons" such as " 'How can the United States conscientiously defend democracy abroad and not practice it here?' " [19]

Most obvious to the interviewers was that "Negro bitterness toward Army segregation and Navy exclusion was deep seated, sprang from feelings accumulated through the years and was merely brought into sharp relief by the draft and the war." The segregation of troops, the assignment of white (rather than black and white) officers to lead black units, and the failure of the military to protect black troops from abuses by white civilians were the specific complaints, all of which were consistent with the principal claims then being made by the black press and black organizations.[20]

The survey also addressed ongoing concerns about African Americans' perceptions of the German and Japanese governments. Asked whether they thought they would be treated better or worse under Japanese rule, nearly a third of black respondents said "about the same," another third answered "worse," and nearly 20 percent answered "better."[21] Federal officials persisted in attributing this response to shrewd and manipulative propaganda techniques, stubbornly refusing to admit the pure intellectual strength of the political arguments the Japanese could employ—arguments based on a reality of racial inequality that African Americans easily recognized, regardless of whether they were swayed by the ultimate appeal. When similar questions were posed about the prospect of German rule, the overwhelming response was that blacks would be worse off, but 20 percent of blacks thought that life for them would be about the same, perhaps because they could not imagine how their current conditions could get much worse. That 20 percent figure deeply disturbed the writers of the report, and they tried to discount it: "This group may include many who are prejudiced against the White race as a whole, and who see no distinction between the American system and the German system."[22] But the matter of distinguishing between two versions of white supremacist ideology was the crux of the problem that faced the federal officials who tried to fashion an anti-German message for African Americans. It was not so much that blacks were "prejudiced" against whites but that they simply recognized the obvious familiarity of Hitler's Aryan arguments.

African American ambivalence about World War II was part of a long tradition of black political dissent against many of the major wars involving the United States, a pattern that would continue throughout the twentieth century.[23] The nature and degree of that dissent depended on African Americans' assessment of the goals of those wars and, significantly, on their perception of their political and economic status at the time. Black views of World War II differed from those of other wars because of the degree of concurrence among African Americans of diverse political beliefs and classes that it was politically hypocritical for the federal government to expect their wholehearted support and sacrifice without addressing official policies of racial inequality at home and in the military itself.

Survey results from the OFF's 1942 study were not at all unique but typical, as other federal surveys and reports during this period yielded similar findings.[24] The need for morale building among African Americans was seen as an urgent matter throughout the federal government. MacLeish wrote secret memoranda to other federal officials stressing the "extreme seriousness of the problem" and recommending a plan of action that bore some similarity to principles in Berry's proposal, although it was far less expansive. Most significant, MacLeish formally endorsed Berry's argument that the most difficult but most effective attack on the problem of African American morale would be to reduce discrimination in the armed forces and defense industries.[25] Unheeded, seven months later, MacLeish reiterated that he had reached the conclusion that "no Agency of the United States government can be in the position of positively signifying acceptance of the principle of racial segregation—Jim Crow-ism."[26]

However strongly held, MacLeish's point of view was politically untenable, as was demonstrated by his own agency's refusal to engage the issue of segregation for fear of offending powerful white southern congressmen and the White House itself. During the remainder of 1942, OWI officials wrestled themselves into virtual paralysis about how the agency should respond to reports of growing anti-black violence and increasing black anger about rampant racial injustices. Obviously, the OWI did not hesitate for lack of information or suggestions. Federal officials simply did not know how to fashion a politically acceptable response to their own findings. In this way, the OWI's internal wrangling over the sensitive issue of African American morale was at once emblematic of its own struggles and the federal government's tepid approach to racial issues.

Certainly the OWI had the administrative fortitude to mount concentrated campaigns to change public perceptions in order to help the war effort. Perhaps there is no better and no more contrasting example than the agency's work to draw white women into the work force to assume jobs normally reserved for white male workers. In that case, the agency undertook an all-media effort to advance a new image of women that emphasized their abilities as competent and intelligent careerists, in stark contrast to the prevailing image of white women as suited only for work in the home or for living lives of leisure and beauty. This new, socially useful image of the white woman war worker as an inspirational patriotic figure was constructed by integrating it into popular magazines, visual imagery, and radio. White women themselves were involved actively in the development of the image, and there appears to have been limited internal conflict at the OWI about the nature of the image. Obviously, agency officials understood well the process and power of replacing old images with new ones, as embodied in the image of Rosie the Riveter. Indeed, the construction and subsequent deconstruction of that image of white women serve as powerful reminders of how well the OWI could perform its propaganda function when it was undergirded by a unified message and, most important, a politically acceptable goal. Not surprising, images of African American women were untouched by the OWI's campaign as reimagining a role for women workers remained politically entrapped by the ideology of racial superiority.[27]

The persistent conflict at the OWI about how to improve African American morale revolved around several individuals who were ultimately at odds on the question of who could best advise on this issue. One of them was Milton Starr, an influential white consultant assigned to the OWI to help advise on African American affairs. Starr, who owned a chain of southern black movie theaters, was the object of criticism from black federal advisers like Berry and black leaders, especially the NAACP's Walter White, who repeatedly questioned Starr's qualifications and his judgment.[28]

In the summer of 1942, Starr prepared an internal report on black morale for the OWI that was directly at odds with the approach Berry had suggested.[29] Starr's report blamed the black press and "professional" black leaders for preventing the masses of African Americans from closing ranks behind the war effort, as they had done in World War I. Starr ridiculed the demands black leaders were then making and argued against them by trotting out well-worn worst-case scenarios. He warned that political equality for blacks "would mean the rule by Negroes of several million white southerners," which "would require an army of occupation as in reconstruction days." Playing the ultimate race card but distancing himself from it, Starr urged that any response to black demands required special attention to "the strange taboos existing in the Anglo-Saxon culture against intermarriage or assimilation with a colored race." [30]

Starr concluded that the black press and black leadership were not representative of the mass of black people, whom he believed to be entirely accepting of current racial conditions: "These Negro masses live in a white man's world and most of them are largely adapted to it, with little serious concern over escape to a higher political or social level." It was this particular group of blacks that Starr believed should be the target of the OWI's morale-building campaign. The problem, as he saw it, was simply "that there are not enough 'bands playing' for the Negro soldier or civilian in this war." In Starr's mind, simply delivering favorable news about the achievements and activities of black soldiers through the black press, radio, film, pamphlets, and posters would be sufficient to counteract the agitation of black leaders and Japanese propaganda.[31]

Meanwhile, Berry continued to press for a more aggressive OWI campaign that relied on a very different reading of the political views of the black community. Frustrated and angry, after less than six months on the job, Berry sent OWI director Elmer Davis what he called his "final memorandum on the subject of Negro morale," foreshadowing his resignation from the agency later that year.[32] He argued once again that any program to build morale among African Americans also would have to focus on convincing whites to modify their "predominant racial attitudes and practices" and to recognize that the interests of blacks were "an essential and integrated part of the total war effort." Once again, Berry sought to convince OWI officials that without visible changes in racial policy, a campaign of empty rhetoric would not succeed. Although he did not refer directly to Starr's report, he warned against the arguments and assumptions Starr had made. Berry advised Davis that "anyone who seeks to delude you or any other responsible official that the state of mind of Negro citizens as represented in the Negro press and March-on-Washington Movement is not representative of the majority of Negroes, is rendering a disservice." He made one final attempt to help Davis understand why a limited propaganda campaign would fail:

> Any program which attempts to improve Negro morale within the framework of the status quo without attempting to eliminate traditional methods of treating Negro citizens will be palliative, wasteful and ineffective so far as the vast majority of Negroes are concerned. This is not because of disloyalty or lack of patriotism, but a war in defense of ideals of freedom leaves the Negro spiritually uninspired without some belief that there is hope of realizing a fuller measure of the things for which we are fighting. Only free men or men with hope of freedom will fight well.[33]

Berry's memorandum concluded with another list of recommended activities for Davis's consideration. Next to Berry's suggestion that "efforts should be made to urge the War and Navy department to eliminate all types of segregation in the armed forces," Davis scribbled his initials and two large question marks. Davis forwarded Berry's proposal to Gardner Cowles, head of the OWI's Domestic Branch, and in a note accompanying the proposal, Davis wrote: "[W]e can't possibly even try all of it even if we wanted to; neither we nor anybody else can solve the Negro problem as an incident of the war effort." Davis's reactions captured the dilemma that the OWI, as well as the federal government as a whole, faced as it sought to fashion an information program for African American morale without addressing underlying federal policies then under attack.[34]

Radio and Race at the OWI

Theodore Berry and Milton Starr differed on what to say in a morale campaign for African Americans, but they were in fundamental agreement about how to use popular communications media and methods to reach black and white Americans. Their emphasis on relying on the popular media was consistent with the OWI's overall approach to its work. The agency's domestic operations branch had separate bureaus devoted to print, film, and radio. Of these, radio was thought to reach the largest mass audience. This view was so prevalent throughout the federal government that one of the Radio Bureau's most pressing initial tasks was to establish some order and control over the requests for war-related airtime that were flooding into the networks from various federal agencies. The links between the federal government and the radio industry were cordial and co-operative during the war period. This was certainly the case at the OWI, where the Radio Bureau was directed by William B. Lewis, a former vice president at CBS. Under his leadership, the OWI drew up and implemented a voluntary network allocation plan under which stations agreed to set aside large regularly scheduled blocks of airtime for federal radio messages. Lewis believed that national radio had a very special role to play in the war because of its ability to deliver an audience of 100 million listeners.[35]

It was therefore natural that both Berry and Starr approached the problem of black morale with plans that included radio. Berry recommended that the OWI urge "major radio networks to include treatment of Negro subject matter in various current programs," as they were doing with other war-related subjects. Furthermore, he suggested that the agency make "a more extensive use of the radio through local and national programs to identify the Negro with the war effort and American life above the level of conventional entertainment type programs." Starr took a markedly different approach, viewing lighthearted entertainment as sufficient incentive for improved African American attitudes. He recommended a half-hour weekly patriotic program dedicated to glorifying the black soldier that featured a "well known Negro Band" or "famed stars of stage or screen." Moreover, he suggested that black soldiers themselves be part of the entertainment. "There would be no difficulty in assembling enough musical talent for such programs from any Negro outfit," he wrote confidently. Noting that African Americans were "rabid radio listeners," Starr argued that the entertainment-driven shows he suggested would have no trouble attracting his principal audience, the masses of black people.[36]

Elsewhere at the OWI, Domestic Branch head Gardner Cowles responded to inquiries from radio stations about how to handle stories about the role of blacks in the war. He advised that radio could help "alleviate" racial tensions by offering shows "that play up Negroes as desirable, capable members of the community" and that pointed out "that maximum utilization of all American manpower regardless of color" was in the national self-interest. He cautioned: "Off the record, of course, this should be done carefully. . . . Don't make martyrs of them. . . . Treat the subject surely, yet with the realization that unless the Negro is made to feel he is part of America we cannot expect him to be a good American." His statement was sent only to those who asked for advice and was not issued more generally.[37]

Other officials at the agency also began urging that radio immediately be put to use to alleviate concerns about black morale. Charles Siepman, the Harvard professor and liberal critic of commercial radio who was then serving at the OWI, specifically urged Radio Bureau chief William Lewis to use radio for this purpose or risk an imminent "outburst of violence and even of race riots." Like Berry, he saw a special need to speak to prejudiced whites: "Prejudice is greatest against negroes among the less educated. No medium influences the uneducated more than radio. This therefore, seems to be radio's chance for action however limited and discrete. . . . [I]f people knew more, their generosity might gradually override their prejudices." Offering to quickly draft proposals for "the treatment of the negro problem on the air," Siepman again emphasized that "radio, above all other media, clearly must do something."[38]

The OWI's director, Elmer Davis, like many other federal officials, was stymied on the race issue, even in the face of yet another OWI report on escalating interracial tensions.[39] Black activists continued to press President Roosevelt, as they had before the war began, to dedicate a radio address to the problem of racial bigotry and violence.[40] Davis reported to the White House that black leaders had repeatedly sought his help in securing a presidential radio "fireside chat on the maintenance of the employment policy and on slapping down discrimination against Negroes in uniform." Passing on these requests to a presidential aide, Davis wrote: "I dutifully pass these communications on to you without recommendation. This is a thorny subject and I do not know what is the best way to handle it. Anyway I do not think this office can do much more than advise you of the frequency and vigor of these requests. Dealing with the general issue is entirely outside our field."[41]

Without any leadership on the issue, federal officials at the OWI simply continued their internal debates about the use of radio to build black morale and about the larger question of whether a special person or entity ought to be designated at the OWI to deal with the question.[42] Part of the controversy over whom to place in charge of African American morale at the OWI stemmed from the continued opposition of Walter White and other black leaders to Milton Starr's presence as a consultant on African American affairs. George Barnes, assistant to Elmer Davis, supported the idea of placing a "Negro specialist" in each of the OWI's major operating divisions, but he urged at the same time "that Starr, rather than any of the deputies, be quietly regarded as the staff advisor on problems in this field." Barnes believed that the OWI's primary responsibility was to respond to "the much more immediate problem of getting the Negro to support and take part in the war effort" rather than to address "the problem of anti-Negro discrimination." In order to do that, he sought to distinguish as the OWI's mission the launch of "a direct and powerful Negro propaganda effort as distinct from a crusade for Negro rights." Davis and Barnes trusted Starr to honor that distinction, as false as it was.[43]

In view of the continuing conflict over what to say about race relations and to whom, it is not surprising that the OWI's efforts to reach African Americans were limited in all media, including radio. Indeed, rather than creating any radio programming of its own in this area, the OWI took the safer course of imposing its cosponsorship on an existing program that catered to black audiences.

My People

The radio show *My People* first aired on the Mutual network in 1942 under the private sponsorship of its Baltimore-based director and producer, G. Lake Imes, a black former official at both Tuskegee Institute and Lincoln University. Imes took special pride in this national broadcast because it had been "put on the air by Negroes, originated and directed by Negroes and paid for by Negroes"—a very unusual occurrence, as we have seen, and one far more likely to take place on the Mutual network than on either CBS or NBC. Unlike those networks, Mutual was formed cooperatively out of a string of local independent stations, and as such, much of its programming originated from affiliate stations.[44]

Network officials brought the show to the OWI's attention when they asked agency staff to review and approve the script for an October 1942 episode, "This Is Our War." Broadcast for a half hour on a Sunday afternoon, the show opened by setting its theme: "Is it a War for Democracy? Then it is OUR WAR. Our only hope is Democracy. Is it a War for Freedom? Again it is OUR WAR. As freedom's youngest children we know how precious it is. THIS IS OUR WAR. And we are in it to win. We ask only the chance to serve." The broadcast also featured brief remarks by William Hastie, the outspoken black lawyer who had been appointed by the president in 1940 to serve as civilian aide to the secretary of war, as well as live pickups and interviews with black men and women who described their experiences in the merchant marine, the Women's Army Auxiliary Corps, and the 372d Infantry National Guard.[45]

Shortly after the show's broadcast, Milton Starr used the program as a model when he once again urged OWI Radio Bureau chief William Lewis to consider initiating a weekly half-hour radio series aimed at black Americans. The show's political stance was subdued enough to pass the OWI's muster—and to attract Starr's attention. To buttress his case, Starr reminded Lewis that African Americans were "a large apathetic and seditious minded group according to all the intelligence we can get in the subject." Starr argued once again that a "modest information program identifying the Negro with the war," as this show did, would be sufficient to elicit a patriotic response from black Americans. Lewis agreed that something needed to be done about the matters Starr had raised, but he explained that "the only reason we have not done more to date is that we have been unable to get from the Office of War Information a clear and definite statement of policy on what should be said." OWI officials decided on the safer course of facilitating more frequent broadcasts of *My People* and allowing it to be designated as being broadcast "in cooperation with the Office of War Information," an arrangement to which Imes apparently consented. The OWI assured officials at the Mutual network that it would assume final responsibility for each script and that "the program will in no way undertake to discuss controversial subjects or present material of controversial nature."[46]

Mutual agreed to carry the series on Saturdays at 7:00 P.M. beginning in February 1943. After nearly a year of internal disagreements and policy paralysis at the OWI over whether and how to use radio to help build black morale, the agency finally took this small step toward addressing racial issues on the air. At least four broadcasts in the *My People* series were aired in February and March 1943. Although the format varied somewhat from week to week, the series was a blend of speeches and black music, with very little dramatization.

From the outset, OWI officials took a heavy-handed approach to controlling the content of the shows and particularly the text of featured speeches. The first episode included speeches by Eleanor Roosevelt and Frank Porter Graham, Mordecai Johnson, and Fred Patterson, the presidents of the University of North Carolina, Howard University, and Tuskegee Institute, respectively. The draft of

Mordecai Johnson's comments was subjected to significant revision. A veiled reference to the Red Cross policy of segregating black and white blood supplies was deleted altogether. Johnson's draft statement included the sentence: "It is not surprising that when voices in all sections are raised in support of freedom and democracy, Negroes should become more conscious than ever of the discrepancies between the declared purposes of the war and the conditions which they must themselves face when called upon to do their part in behalf of VICTORY." But the revised script substituted the sentence: "Negroes have responded willingly to the aims and purposes of the war as they have been set forth by the leaders of the United Nations." Some of the changes were more subtle but no less significant. For example, in revising Frank Porter Graham's statement that "the Negro is necessary for winning the war, and at the same time, is a test of our sincerity in the cause for which we are fighting," the word "test" was replaced by the word "proof."[47]

Recurring tensions apparently surfaced between the OWI and Imes over the series that was based on his original broadcast. For example, the draft script submitted for the second episode was also revised substantially. OWI officials deleted a long section in the script describing how newly elected black congressman William Dawson had defended an unnamed black government official who was accused by the Dies Committee of being a Communist. After that episode, a script clearance editor, Joseph Liss, made even more explicit attempts to keep Imes and the scripts away from potentially controversial subjects. In a letter to Imes, he asked, "[S]hould we not confine ourselves hereafter particularly to showing *what the Negro is doing in the* war?" He warned, "[A]ny other information is not pertinent to this program."[48]

Some of Imes's attempts to argue that blacks be given the opportunity to participate equally in the war effort were not entirely successful. One show featured news bulletins about war-related events of significance to blacks. A report on the black 99th Army Air Force Pursuit Squadron's readiness for combat included an ambiguous aside that "Negroes are hoping that there will [be] a rapid expansion of their participation in this branch of the service." The show also included a report on a visit to a defense plant where supervisors attested that blacks were good and valued workers. Imes himself made a plea on the broadcast that "it is the opportunity to work, the opportunity to fight, the opportunity to buy, to give, to save and sacrifice which makes this everybody's war." But competing with these serious messages were musical numbers that trivialized rather than reinforced the notions of patriotic dignity and somber tone set by the news, talks, and interviews.[49]

Liss carefully monitored public reaction to the series. He noted that the black press mostly ignored the show, as did the white press, something he took mixed comfort in, concluding that "all this adverse publicity talk is only a bugaboo." Liss recommended, as Milton Starr had suggested to him, that the series be allowed to "peter out" at the end of four or five weeks when funds for it expired.[50] In the end, the OWI regretted its sponsorship of *My People,* calling it a "fiasco" and alluding to the agency's lack of control over Imes and battles over the show's content.[51]

Despite the agency's continued obsession with the problem, it does not appear that the OWI was involved as orginator or co-sponsor with any other radio program that dealt with the issue of black morale. The OWI's experience with a pamphlet about African Americans that had been issued during this same period influenced its decision. That experience only confirmed the worst fears of OWI officials: touching the issue of race relations had grave political consequences.

Negroes and the War

In the spring of 1941, black journalist Chandler Owen published a short brochure under the title "What Will Happen to the Negro If Hitler Wins!" Owen had been a Socialist, a colleague of A. Philip Randolph's, and a coeditor of the *Messenger.* During World War I, he had been a harsh critic of U.S. involvement in a struggle that he saw as a battle over the world's markets and the exploitation of people of color. Using black soldiers to support those aims was, in Owen's view, hypocritical.[52] In this war, however, Owen took a different political stance. At the brochure's core was the explicit argument that African Americans had no choice but to support the American war effort because conditions for them under a Hitler victory would be equivalent to a return to slavery. Just as Archibald MacLeish earlier had assumed that African Americans would respond enthusiastically to the characterization of the war as a "war against slavery," Milton Starr at the OWI also believed that the specter of a literal return to slavery would jolt blacks into ending their criticism of racial injustices in the United States. He tried to persuade his colleagues to reprint the pamphlet as it was and distribute it as a government document. When word of this plan circulated, Theodore Berry objected strenuously. He warned Elmer Davis that if Owen's work was printed as a government pamphlet, the publication would not only fail to promote morale among African Americans but also "evoke sharp criticism" because of its explicit reliance on scare tactics.[53] The idea of issuing Owen's pamphlet was stymied as a result.

In the meantime, intelligence reports and surveys on racial issues continued to reflect an increase in domestic racial hostilities as well as persistent questions among black Americans about their role in the war effort.[54] The black press continued to call for action, and the *Pittsburgh Courier* in October 1942 criticized the federal government for not being able to develop an effective morale campaign for African Americans: "It knows that it has only to end all present discrimination and segregation based on color or so-called race, and it knows that this can be done by rigidly enforcing existing laws or using its powerful machinery to get additional laws passed. These steps would immediately raise Negro morale sky high but the Government fears that they would greatly depress Caucasian morale."[55]

Under continued pressure to do something, Starr and other OWI officials resuscitated the idea of producing some version of the Owen pamphlet despite repeated internal warnings that the content would be not only ineffective but also deeply resented by the black audience it sought to influence. Hastie warned from the War Department that "most Negroes are in no mood to be told how well off they are" and suggested that propaganda efforts needed to be directed instead at "indifferent or prejudiced whites." Walter White and A. Philip Randolph also registered their dissatisfaction with the arguments in the pamphlet.[56]

None of these objections stopped the OWI from publishing 2.5 million copies of the pamphlet, which were distributed in January 1943. Retitled *Negroes and the War,* Chandler's pamphlet was transformed into a polished seventy-two-page pamphlet featuring 141 black-and-white photographs of black people. Although the pamphlet used photographs as its primary narrative device, its interpretative framework was set in a six-page introductory essay by Owen, aimed at "Negro Americans who say that it makes no difference who wins this war." Following the reasoning in his earlier pamphlet, Owen argued that a victory for Hitler literally would march blacks backward into slavery. Taking the slavery motif a step further, Owen borrowed (without attribution) the theme and the language of the radio show *Freedom's People:* "[B]ecause we have known the weight of chains, because we have known the helplessness of bondage, we can be a mighty force in this nation's fight for freedom." As in his original essay, Owen predicted that Hitler would doom black churches, which he called the "glories of Negro culture" and "the lanterns of the spirit." The conundrum that Owen's narrative could not escape, however, was the question of whether Hitler's *defeat* would bring any change in American racism. He attempted to use statistical measures and individual black achievements in many fields—religion, business, agriculture, the arts—as evidence that the *promise* of progress was the African American's greatest stake in America. His only acquiescence to political reality was to advise that "unity against America's foreign foes does not mean that Negroes must forego legitimate protest against discrimination in industry or the long struggle for political equality." [57]

The bulk of the pamphlet was devoted to an extraordinary collection of Farm Security Administration photographs. Documentary-style, artfully produced shots of black working- and middle-class men and women performing a full range of societal functions filled the pamphlet's pages. Two pages showed black men and women performing different jobs in what appeared to be realistic, unstaged settings. Three pages were devoted to the black church, showing a reverent, solemn people engaged in a wide variety of black worship. When read as one text, these photographs depicted a moral, ambitious, hardworking, ordinary group of Americans. It was only in the concluding five pages of *Negroes and the War* that any photographs depicted black participation in the armed forces, including images of black soldiers boarding transport ships, driving Jeeps, building bridges, and being decorated for valor.

Despite these seemingly positive images, leaders of prominent black organizations quickly and loudly objected to the pamphlet. Lester Granger of the National Urban League criticized the OWI for the pamphlet's "false argument" of support for black progress "from a government which has failed in so many essential ways to give forthright and courageous attention to the problems faced by Negroes." [58] Walter White concluded that the pamphlet was not as objectionable as he had expected and that it would do some good, but he argued that it would have been far better "had the government taken effective and uncompromising action against some of the evils from which Negroes suffer." An official at the Philadelphia chapter of the NAACP was more blunt. She asked, "[W]hen will our government learn that we do not need to be told of our accomplishments so much as white Americans need to be schooled on this subject?"[59]

Federal advisers on race relations, both black and white, were unanimous in their condemnation of the process that brought the pamphlet to press. They uniformly criticized the choice of Owen to prepare the text for the pamphlet not only because he was viewed as unrepresentative of black opinion but also because he "did not have a good reputation" and was "well and unfavorably known."[60] Implicit in this political controversy was continued criticism of Starr's position as the de facto adviser to the OWI on race relations and the underlying conflict about who could define the political images of African Americans. Driving these concerns was the fact that Berry, who had been brought into the agency to work on African American morale, had resigned because of Starr's dominance and specifically because of the agency's decision to proceed with the *Negroes and the War* project. It also appears that other black federal officials held Starr in low esteem, including Mary McLeod Bethune, Hastie, and Robert Weaver, as did the War Department's white adviser on race relations, Donald Young. Yet OWI officials resisted the idea that Starr was an inappropriate person for the agency. Starr continued to assume authority over the race issue at the OWI in his dealings with his colleagues there and at other federal agencies. In a meeting with the Justice Department, for example, Starr boldly claimed responsibility for "toning down the Negro press" and causing black leaders to drop renewed threats of a March on Washington.[61]

Even more emphatic than the views on Owen and Starr was the widely shared opinion that the Hitler argument they endorsed was deeply offensive to African Americans not only because of the blatant threat that they must cooperate or risk a return to slavery but also because it offered no positive reasons for them to support the war effort, rendering the pamphlet useless as morale propaganda. Moreover, the majority of black officials queried on this matter believed that whites and not blacks most needed to see representations of black participation in the war effort, a consideration completely overlooked in the decision to target the pamphlet only to blacks.[62]

Faced with this criticism, the OWI was eager to gather its own information on the effectiveness of *Negroes and the War* in converting blacks to the war effort and commissioned yet another study. The majority of blacks interviewed said they liked the pamphlet because it was a "testimony to Negro progress" that made them feel proud. At the same time, they stressed that "these were accomplishments achieved despite discrimination, but that the magazine presented them as if they had been handed on a 'silver platter.'" Most also complained that the pamphlet gave a "one-sided presentation of Negro progress" that emphasized "exceptional cases" and did not reflect "the life of the majority of Negroes."[63] Interviewers found that African Americans saw the photographs of blacks in the armed forces "as another indication of how far the Negroes have gone, rather than as an appeal toward participation in the war." Many of those surveyed commented that the pictures did not accurately reflect black life in the military: "Everybody knows that Negroes are not doing combatant duty, but labor behind the lines."[64]

The overriding criticism of the pamphlet was that it failed to make any assurances about future progress. Many complained that it "didn't tell about the future in the event of an Allied victory" or "what Negroes have to gain if they fight in the war." In these ways, by "seeing" what was missing, African Americans were reading the political silences in the images. Rather than feeling confined within the photographic world the images presented, they were raising what Allan Sekula has called "the issue of limits" by situating themselves "outside" the limited intended discursive frame of the photographs and the pamphlet.[65] The OWI had tried to portray Hitler as a threat to the black progress depicted in the photographs, but black readers interpreted the images to prove their own argument: that black men and women had brought progress to themselves and deserved more of it than the reality of American racism allowed them to achieve. Ironically, more than any of the other images, the photographs of groups of black soldiers were interpreted by blacks primarily as evidence of segregation—hardly the best motivation to support the war.[66]

The survey analysts seemed genuinely baffled by these responses and complained bitterly that most African Americans missed the "war message of the pamphlet" altogether. These analysts considered *Negroes and the War* a propaganda failure, a result they attributed to the "incongruousness" of Owen's essay and the photographs that followed. They concluded that a "Hitler argument" could have worked with blacks if that theme had been treated sufficiently in the photographs as opposed to simply in Owen's text. But if the researchers and officials at the OWI had given a closer reading to the survey responses, even as limited in number as they were, it would have been clear to them that the Hitler argument had no appeal to African Americans, regardless of how it was presented. When asked directly to compare the way Hitler and white Americans felt about blacks, one-third of the blacks responded that there was no difference, and the majority stated that "the difference was one of degree rather than of kind." One respondent observed: "There isn't much difference, only one is supposed to be a democracy and has a constitution which is supposed to be lived up to. Hitler will tell you what he thinks about you. American White men will shake hands with you, laugh with you, then shoot you." Most of the blacks were well aware of the "Hitler threat" but concluded that "any people conquered by Hitler would suffer, rather than that Hitler was a particular enemy of the Negroes."[67]

It was not that African Americans were not reading the text and the photographs closely enough, but rather that they infused the images they saw with different meanings from those intended by the OWI. Survey staff members noted that blacks interpreted the photographs independently of the intended meanings that were communicated in the captions and accompanying text. What these analysts misunderstood was that the written text could only, in Stuart Hall's term, "anchor" the intended dominant meaning of the images, but the images still remained open to many other potential meanings.[68] Rather than recognizing that process as the culprit in these "misreadings," the researchers laid the blame for the interpretations on the layout of the pamphlet and, indirectly, on black readers themselves, whom they presumed to lack education and sophistication. Indeed, although leaders of black organizations and black newspaper editors were extremely critical of the pamphlet, their criticisms barely mentioned the photographic text and focused instead on Owen's character, the political decision to accord his views official legitimacy, and the clear threat contained in his printed remarks.[69] If the pamphlet had been issued without Owen's message, it seems likely that African American leaders would still have objected to its ineffectiveness, but they would not have been able to use it as a blatant example of federal ineptness and insensitivity on the issues of black morale and racial reform.

Despite the criticisms most black leaders lodged against the pamphlet, it met with great public demand. Repeated requests for more copies by local NAACP and Urban League chapters were held up by

OWI officials as an indication of the unrepresentative nature of the national leadership of these organizations. The publication attracted such attention because it benefited from one of the OWI's most successful distribution efforts. Dissemination was heavily weighted toward urban blacks, but the pamphlet also was widely distributed nationwide through a well-organized network of black movie theaters, black insurance companies, black churches, labor unions, black high schools, county extension agents, and black civic organizations. The distribution list was comprehensive, using multiple avenues into the black community, including Pullman porters; black doctors, dentists, and nurses; and black Boy Scouts. No wonder that an OWI official reported that *Negroes and the War* "was going like hot cakes," making it one of the most massively distributed pamphlets the agency had ever produced.[70]

There are several possible explanations for this seeming disparity between the eagerness of the black public to obtain copies of the pamphlet and the political criticisms aimed at it by black activists. At a time when all forms of the popular media were filled with patriotic appeals and calls for national unity, the absence of black people from those images and appeals engendered an even greater sense of isolation and resentment among many African Americans. For example, one African American woman in Harlem said, "Every time I look at them 'Four Freedoms' pictures, I get so mad—'cause we ain't got no freedom."[71] During this period, one OWI staff member reported that on a trip to the South, a schoolteacher there told him that her black students felt so left out by the government's war posters "that the pupils had blacked in the faces on some of the posters."[72] Writer Sterling Brown described an incident at a northern railroad station that displayed a large photograph that "showed departing soldiers what they were fighting for: a sea of American faces looking out, anxiously, proud. All were white." When a group of black troops passed through the station, he wrote, "they gave the eye-catching picture a swift glance, and then snapped their heads away, almost as if by command."[73]

Negroes and the War literally brought blacks into the national wartime picture, offering artful visible evidence of some official recognition and inclusiveness for African Americans. What OWI officials failed to appreciate in their bungling efforts to decide what to do about black morale is that African Americans truly had adopted a "Double V" attitude, that their commitment to securing greater political and economic rights did not diminish their patriotic commitment to winning the war. Whereas most white officials used the war as another excuse to delay addressing segregation, blacks sought to use the war and the ironies and injustices it exposed to accelerate the struggle for equal opportunity.

OWI officials had been as divided about the kinds of images they wanted to present as they were about the ultimate aim of their black morale program. The pamphlet's photographs had been chosen by Ted Poston, a black editor who headed the OWI's Negro Press Section. He and others who had worked on the pamphlet reported that they "wanted to show blacks doing front-line assembly jobs with intelligence, with ability, and we wanted to show them with the kind of dramatic closeups that would tell a story in a single image."[74] Others at the agency simply wanted to provide "pictures of negroes which show how well off they are in this country" and "how much worse off they would be under Hitler."[75]

Nonetheless, and however unintentionally, the pamphlet marked the debut of a fresh and cohesive printed portrayal of the black community. The photographic style most closely resembled that of *Life* magazine, and the collection of images and emphasis on photographic evidence of black achievement foreshadowed the popularity and success of *Ebony* magazine, which would begin publication in 1945. Many African Americans viewed the pamphlet's images of themselves as incomplete but not mocking or negative. They were proud to claim the images as evidence of their values and abilities and ironically, from the OWI's point of view, of the justness of their battle for equality. This longing for palatable and respectful presentations of themselves was at the root of African Americans' demand for *Negroes and the War*. In the same way that black teachers, for example, had enthusiastically requested copies of the scripts of the radio show *Freedom's People,* it is very easy to imagine them eagerly requesting a copy of this government-issued pamphlet for their students and themselves.

The effectiveness of the OWI's mass distribution plan and the ambiguity of the pamphlet's message soon brought it to the attention of members of Congress. The OWI began to hear directly from southern members of Congress whose white constituents had complained about the pamphlet. Congressman John Rankin of Mississippi criticized the pamphlet as "communist" and an insult to "the white people of the southern states." Louisiana senator John Overton accused the agency of using the war as an excuse to "place the negro on social equality basis with the Caucasian." In his reading of the introduction by Owen (whom he referred to as "some negro out of Chicago"), Overton found evidence of yet another example "of a concerted effort toward the apotheosis of the negro which gives much concern to the white people of the South where the two races have been living side by side in perfect harmony and mutual workable understanding." Southern whites no longer owned African Americans, but it seems as if some continued to speak as if they did. In his two-paragraph reply, Davis said as little as possible, explaining that the sole purpose of the publication was to help "Negroes" understand their stake in the war and that "any inference of a purpose other than that is incorrect."[76] This exchange captures the deference that conservative southern white members of Congress demanded as undisputed experts on racial issues and the "Negro problem."

When Congress began its annual consideration of funding for the OWI, criticisms of the pamphlet moved to center stage when Congressman Joseph Starnes of Alabama referred to OWI publications as "a stench to the nostrils of a democratic people" and offered an amendment eliminating support for all of the agency's Domestic Branch operations. Other congressmen complained that the OWI's work of "political propaganda" simply promoted "the fourth term for Mr. Roosevelt and the activities of the New Deal agencies," an argument that already had doomed funding for radio programs at the Office of Education. However, Congressman James Allen of Louisiana made clear that it was primarily the racial message in *Negroes and the War* that drove him and others to seek to abolish Davis's entire domestic propaganda effort: "His propaganda stuff has hurt the South. . . . This pamphlet undertook to glorify one race in the war. We in the South wish to encourage that race. We are the best friends of that race. But such propaganda raises a race issue, which ought to be kept down. We want unity in this country. All over this country now we are having race riots, even in the North, and the type of propaganda which the OWI has been sending out certainly does not hold that situation down." The House of Representatives agreed with Starnes's solution and voted to eliminate all of the OWI's domestic operations.[77]

Davis worked hard in the Senate to restore the deleted funds. But the OWI and the pamphlet came under further attack in an atmosphere filled with even more anxiety after the eruption of racial violence in Los Angeles and Detroit in May and June 1943. Republican senator Gerald Nye of South Dakota, who called himself a "sympathizer" with the "Negro" race, told Davis in a Senate appropriations hearing that the pamphlet had helped ignite the recent race riots in Detroit. To Nye, *Negroes and the War* also was a federally funded partisan New Deal appeal for the black vote.[78] Davis replied rather weakly that he had not seen the pamphlet before it went out in final form but that the "whole point of this pamphlet was to show the Negroes that they are better off under our form of government and our structure of society than they would be anywhere else." That explanation did not satisfy Louisiana senator Overton, who argued that the pamphlet's message was "that this war would bring not only political but social equality." He continued: "Down South we have got along splendidly with the Negro in the last 30 or 40 years. We have not had any trouble at all with them. Of late, however, we are somewhat disturbed about the situation. We got along splendidly. We did not have any innovations down there. But the Government began preaching political and social equality."[79] The irony in all of this was that the OWI in the eyes of these southern congressmen had come to be perceived as a source of "foreign" propaganda—an intrusive, disruptive, untrustworthy alien government agency preaching social equality and changes in race relations to its southern constituents.

In a desperate attempt to salvage a substantial portion of the agency's funding, OWI officials announced that they would abandon all of the agency's pamphlet activities if they were allowed to continue their domestic operations. Conferees on the final legislation accepted the OWI's offer and revised Representative Starnes's original amendment to simply forbid the agency from issuing any more pamphlets.[80] The effect was nevertheless significant: the budget for the OWI's Domestic Branch was pared down to $2.75 million from $8.5 million the previous year, reducing domestic operations to less than 10 percent of the agency's overall budget.[81] The agency never recovered from the congressional battles of 1943, nor did it ever gain strong White House support.

Walter White devoted one of his *Chicago Defender* columns to defending Davis, whom he called a target of reactionaries "as vicious and ruthless as the Nazi party in Germany." Davis's "crime," wrote White, "was that of recognizing that the Negro is a part of the populace of America and one which is helping to win the war" by issuing "the innocuous 'Negroes and the War,' which a lot of us thought too timorous.'"[82] Later, in looking back on his time as the OWI's director, Davis commented that he still believed that the reception of the pamphlet had been favorable among most blacks and whites but that "it was vigorously attacked in a variety of quarters, for reasons which logically might have canceled each other out, but politically reinforced each other."[83] Caught between critical supporters like White and pure critics in Congress, Davis had never arrived at the safe middle ground he had tried so hard to find.

The controversy surrounding *Negroes and the War* helps explain why the OWI used radio so sparingly in the area of race relations and why it steered clear of engaging the question of racial equality in any medium after this scalding congressional retaliation. There was no intellectually honest way to develop a message promoting inclusiveness of African Americans without endorsing or appearing to endorse their claims to equal opportunity—one message begged for the other because the two were inextricably linked. It was this truth that federal officials feared German and Japanese propaganda could expertly exploit. This reality also meant that any federal attempt at inclusion—as opposed to enforced silence and invisibility—raised the stakes quickly because the "Negro problem" remained the dangerous third rail of national politics. Only the narrowest of messages could be given voice, particularly because national radio did not allow for the racially segregated audience for which the pamphlet had been designed. On radio, the most integrated of media in terms of audience, silence was always the safer course for preserving or perhaps enforcing the appearance of a national political consensus on race. No politically acceptable language was available to make effective arguments for an end to racial inequalities. Images could be used more safely to "speak" the unspeakable, but even then, as the experience of *Negroes and the War* demonstrates, underlying political arguments could be read, misread, and dismissed. "Every photograph is a structure of 'presences' (what is represented, in a definite way)," writes Stuart Hall, "and 'absences' (what is unsaid, or unsayable, against which what is there 'represents')."[84]

Although American politics has historically been characterized by a coded and highly charged racial discourse, the confusion around what could be said, what was said, and what was intended rose to a new level during World War II.[85] Part of the simplistic appeal of Owen's reasoning was that it provided a clear uncontested distinction between Hitler's ideas and American white supremacist thinking at that time, but the continued absence of slavery was hardly an adequate race relations program in the view of most black Americans. At its crest, the war's own patriotic rhetoric demanded an expansion of the boundaries of racial discourse, but the deep channel of racial politics kept federal messages about racial opportunity within tight boundaries.

The pamphlet episode also reveals that no matter what medium was used and no matter how carefully a message about African Americans was prepared, its ultimate meaning would be subject to widely varying and often contradictory interpretations. Interestingly, the southern white congressmen who objected so vigorously to the pamphlet seemed to be reacting to images of blacks that were just as "new" and frightening to them as they were refreshing and pleasing to blacks. Many African Americans and these southern white men immediately recognized the significant departure from existing presentations, which evoked different political lessons and responses from that shared reading. This provides a stunning example of the "politics of reading" photographs, a medium that by its very nature invites multiple and often contradictory interpretations depending on the viewer's cultural and political position.[86] In order to show blacks that they had a life in the United States with opportunities worth preserving, the pamphlet's creators could not rely on the usual media representations of blacks to make that appeal. They had to "invent" a markedly different set of images, but OWI officials underestimated and misread the powerful and varying political implications stirred by these new images and the imagined "future" they implied.

The patriotic fervor radio helped generate was sustained by the repeated integration of that message into all aspects of radio broadcasting, especially its regularly featured shows and advertisements. Without a singular message and a similar commitment on the question of racial equality, left unmet would be the exalted expectations that radio alone or any medium could play a decisive role in lifting the spirits of African Americans or inducing tolerance in whites. The message, regardless of the medium, would be hamstrung by a competing racial reality on the domestic front, and, as the next discussion describes, most especially in the U.S. military.

Radio, Race, and the War Department

The War Department's public relations operation was most concerned with managing the news of the war itself, protecting the release of details that might compromise or endanger the military's missions or damage domestic morale. But like the OWI, the War Department could not ignore the public prominence of African American criticism of the military's racial policies. This forced officials overseeing public information campaigns at the War Department, like their counterparts at the OWI, to understand that racial tolerance and black morale demanded a place among the war-related issues they presented to the general public. The War Department's own policies affecting black troops made it the focal point of criticism from African Americans and made its task of creating a message urging racial unity even more difficult. As a result, the extent of the War Department's civilian radio programming on racial issues was limited, as at the OWI, but the contested content and delicate deliberations leading up to its broadcasts carry their own significance.

In 1940, President Roosevelt appointed William Hastie as civilian aide to the secretary of war, a position created in response to concerns about the treatment of African American soldiers. Hastie, who had chaired the National Legal Committee of the NAACP, reluctantly accepted the post and publicly made clear his continued support for desegregation of the troops. He saw his role as that of a protector and advocate of black soldiers, and he never fully relinquished his identity as an "NAACP man" during his time at the department.[87] Early in 1941, Hastie approached other high-ranking War Department officials about sponsoring a nationwide radio broadcast on the role of African Americans in the army. Truman Gibson, Hastie's assistant, in turn tried to convince officials at the department's Bureau of Public Relations that such a show could build respect for the army among African Americans, especially potential soldiers. For whites, especially those near the southern army camps where blacks were being trained, such a broadcast, Gibson argued, could "greatly assist in alleviating racial friction by a not too obvious educational process."[88]

Several months later, after further prodding by Hastie and Gibson, other War Department officials also recommended that the Radio Branch of the Public Relations Bureau develop a national broadcast on the role of African Americans in the army.[89] Hastie also advised that the proposed show target a broader audience. He suggested that it be directed not only at civilians but also at black and white troops, proposing that it be broadcast during a scheduled interruption in army maneuvers.[90] Hastie believed that the War Department's concern about low morale among African Americans on the home front was justified, but he also wanted department officials to appreciate the extent to which black troop morale suffered because of discriminatory military policies. More generally, he hoped that the show would educate white troops about the contributions and abilities of black soldiers, who were thought by most white officers and troops alike to be unqualified for combat duty. Even during the early stages of prewar buildup, conflicts between black and white soldiers were common both on and off base.[91] All of these factors undoubtedly prompted Hastie's suggestion that the audience include military personnel as well as civilians.

Creating a politically palatable yet appealing radio show would not be easy. The department's initial scheme for the program had a high entertainment appeal built around African American music. Here, as in other federal broadcasts about race, music was thought to soothe the way toward the more serious intended messages.[92] With this cautious strategy in mind, War Department officials decided to proceed with the planned program. Gibson quickly sent form letters to press relations officers at army facilities where black troops were undergoing training soliciting suggestions, anecdotes, and other material for use on the shows.[93] Response from the field was swift, but the reactions to and interpretations of Gibson's request themselves revealed the army's often limited perceptions of the abilities of black soldiers.[94] Most interpreted Gibson's brief letter as a request for names of black soldiers who could serve as entertainment on the shows. One lieutenant wrote: "We have, as you know, plenty of outstanding talent among our negro soldiers here at the Quartermaster Replacement Center to put on a very entertaining program. We feel that our choir is one of the best in the country. In addition we have many other specialty numbers such as tap dancers, dialogue artists, musical soloists, or about any other type of entertainment that you might wish."[95] A press relations officer at another camp enthusiastically offered similar suggestions for the radio show: "Colored soldiers enjoy singing. I think that an entire sketch depicting the life of colored soldiers might be built around a background of singing. For example, take an ordinary day in the life of a soldier in a training camp . . . conversation and singing of the men in the wash-rooms of the barracks . . . singing of men while on a hike, (road march), singing of men while working on fatigue details or on K.P. detail in a mess hall; singing while off duty in a recreation hall, or singing in church on Sunday."[96]

Gibson had hoped, however, that the show would highlight African American soldiers for their work as soldiers. For entertainment, he approached professional black performers and artists, many of whom responded enthusiastically. Band leader Noble Sissle, of the famed World War I 369th Artillery Band, worked closely with the War Department on the show. Gibson worked through the Negro Actors Guild, of which Sissle was president, and the Negro Radio Workshop, an organization of black professional radio actors, to secure the appearances of prominent stage and radio actors and musicians.[97] In his appeal to the Negro Actors Guild, he emphasized that the proposed program was "tremendously important in that it is the first of its kind ever presented under official sponsorship."[98]

Freedom's People would be broadcast under the aegis of the Office of Education a month later in September. At the same time that Gibson and Hastie were trying to organize their production, Ambrose Caliver was doing the same thing for *Freedom's People,* and he was pursuing some of the same stars. In fact, NBC had warned Caliver on July 25 that the War Department's "request" for airtime would probably prevent the network from airing another "Negro program," although that decision was later reversed.[99] The War Department legitimately laid claim to the "first" federal sponsorship of a war-related radio program on African Americans as its mark of distinction for this broadcast, but the National Urban League in March 1941 had pioneered the format that the department used. As described in greater detail in chapter 4, the National Urban League's hour-long show had included Marian Anderson, Joe Louis, and Bill Robinson and the bands of Duke Ellington and Louis Armstrong, making it indeed the star-studded affair the army wanted to duplicate.

When the War Department broadcast its show, "America's Negro Soldiers," on Tuesday night, August 12, it was a very scaled-down version of the original proposal.[100] With bandleader Noble Sissle as master of ceremonies, the show was a curious amalgam of music, dramatic skits, reenactments, and political appeal. The longest reenactment was a skit set in World War I Germany about black troops preparing for battle. More important than the depiction of their bravery was the emphasis on the black soldiers' resistance to German propaganda that had attempted to use American racism to incite black disloyalty. At one point in the skit, black soldiers exclaimed, "Can you imagine that cheap propaganda! What do they think we are, traitors?" Sissle commented that "what the Germans didn't realize in those days was just how loyal the American Negro soldier could be."

Although this skit was set in World War I, one of the most pressing worries among federal officials in World War II was the effectiveness of German and Japanese propaganda that sought to appeal to blacks by emphasizing racial injustices in the United States. Regardless of the actual potency of this propaganda, federal officials' fears were exaggerated by their own delusion that growing black protests depended on foreign or outside provocation, that it was impossible for black dissatisfaction to be independently generated despite the abject reality of racial oppression.

This obsession with "outside agitators," "subversives," "foreign provocateurs," Communists, or Socialists—whichever term is used —has been an enduring characteristic of white political thought about black protests throughout most of the twentieth century. In their search for a manageable explanation of what was seen as the "new" hostility of blacks, white federal officials refused to acknowledge the cumulative impact of nearly a century of indigenous postemancipation racial apartheid, violence, injustice, and discrimination. Most whites were unable or unwilling to accept the fact that such racial conditions alone could provide ample fuel for restlessness and political challenge, which were potent without any external provocation.

The radio show used the record of black bravery and heroism in World War I as "the glorious answer" to German propaganda. Citing the accomplishments of black soldiers, especially the awarding of the French Croix de Guerre to four black regiments, the script argued that black troops in World War II were obligated to perpetuate a tradition of "loyal, fighting spirit and devotion to our Republic" for the sake of democracy. Thus, the reenactment sought both to educate whites about past black heroism and to build black morale. At the same time, the show used the record of black service in World War I to appeal to blacks to repeat their quiet, uncompromised patriotism, urging them in effect to once again postpone what whites saw as their distracting, competing claims for fair treatment and equal opportunity.

One especially effective part of the broadcast was a series of live interviews with blacks stationed at the Army Air Corps Technical

School in Illinois, the training ground for the 99th Pursuit Squadron. Articulate and enthusiastic, these men described their training as radio technicians, aircraft mechanics, weather forecasters, parachute riggers, and sheet metal workers. Yet even this was marred by racial stereotyping. The commander of training for the squadron assured listeners that the black men there were a "crack ground outfit." But as proof, he proudly announced that "the 99th ran away with the track and field championship." The interviewer added that the squadron also had a "grand glee club." Despite their efforts to be accepted as soldiers, African American men seemed doomed always to carry, regardless of training or education, the burden of satisfying expectations of their athleticism and musicality.

Not surprisingly, plans to feature Hastie on the broadcast were abandoned. It is difficult to imagine how in good faith he could have, as originally proposed, discussed the "absence of serious racial difficulties" in the army since he was devoting himself to documenting the continued presence of those very problems.[101] The rhetorical climax came instead in a long speech delivered by Undersecretary of War Robert Patterson. Patterson emphasized that the war was "every American's emergency" and that "an aerial bomb draws no color line," and he quoted the president's recent executive order on nondiscrimination in the employment of workers in defense industries.[102] Patterson made a special appeal for the cessation of white hostilities toward black men in uniform.[103]

The image of a black man in uniform evoked sharply different responses from whites and from blacks during the war. For some, the uniform itself deserved to be respected as a representation of service and loyalty, even if it was worn by an African American, and that seemed to be the basis for Patterson's appeal. Throughout the war, the mere appearance of an African American in uniform, according to one white observer at the time, "evoked hostility, fear and suspicion" from whites because the uniform symbolized that the black soldier had been "spoiled" and had "forgotten his 'place.'"[104] African Americans saw the uniform as only one aspect of the symbolic shifts under way. Writing in 1943, Sterling Brown observed that "any symbol of the Negro's getting out of 'his place'—a lieutenant's shoulder bars, or even a buck private's uniform; a Negro worker at a machine, or a Negro girl at a typewriter, or a cook's throwing up her job—these can be as unbearable as an impudent retort, or a quarrel on a bus, or a fight."[105] Certainly, there was no clearer literal representation than this of the conflict between white fears and black expectations of the war's potential impact on race relations.

From the start, Gibson had been defensive about the show's limited political message. In a note about the show, he confided: "Undoubtedly, there will be many charges that the program was a soft soap affair. I think it should be pointed out that it reached a number of persons who know nothing at all about the Negro soldier and who probably would have turned off a program of another sort, particularly since the one presented is the first of its kind."[106] In its attempt not to offend white listeners—those most likely to have "turned off a program of another sort"—the War Department offered a show designed to educate and make a general appeal to whites for racial tolerance without crossing the boundary into the minefield of arguments for racial equality.

Gibson's expectation of criticism about the show was well founded. Again it was the influential *Pittsburgh Courier* that led the charge. An editorial titled "The Army's Radio 'Flop'" condemned the show as ineffective radio because "the fact of color discrimination, racial segregation, neglect, insult and brutality rang so loudly in Negroes' ears that they could not hear the singing, the music, the dialogues, and the speeches." Continuing this harsh critique, the writer explained: "What colored Americans WANTED to hear was not how racial separatism was being perpetuated in the United States Army but what the War Department was DOING to practice the democracy the Administration is preaching, and what it PLANS to do in the future. Instead they heard the old familiar platitudes, a eulogy for black soldiers who had won honors for fighting for democracy, a little tap dancing, a bit of comedy, fine music, and what amounted to praise of a jim crow system that mocks the word democracy."[107]

Indeed, the War Department's broadcast did little to engage the principal issues driving black demands for racial justice in the military. The Office of Education series *Freedom's People,* which would commence a month later, relied in one episode on the same material that the War Department used: the historical record of black loyalty, bravery, and effectiveness in previous wars. In *Freedom's People,* that material was used subtly to raise the issues of military segregation and the denial of combat positions and to argue for additional responsibilities and opportunities. In "America's Negro Soldiers," the War Department employed the same history to argue for the status quo and to enlist black acquiescence to delaying activism on issues of racial inequality. This distinction is subtle but significant because it illustrates that at the heart of both of these shows was a manipulation and portrayal of the history of blacks in America in ways that varied depending on the political arguments being advanced.

After the attack on Pearl Harbor, Hastie and Gibson persuaded the department to produce another program concerning African Americans in the army, but it would take nine months before that show was broadcast. On Hastie's recommendation, it was aired nationally and to troops in the states and overseas.[108] Called "Judgment Day" and sponsored by the Office of Military Intelligence, the show was broadcast on NBC in September 1942. It seems to have been dominated by historical dramatizations tracing the history of black war heroes from the American Revolution to World War II. A press release noted that radio and stage actor Canada Lee would "tell the story of the soldier heroes of his race from Bunker Hill to Bataan." According to a newspaper account, the show also featured the exploits of all-black units such as the 110th Infantry, the Sixteenth Artillery, and the Twenty-fourth Infantry. The *New York Daily News* declared that it was "grippingly written, and flawlessly performed." The more liberal newspaper *PM* shared that view; its radio reviewer, Jerry Franken, described it as "an exciting and moving dramatization."[109] Officials of the War Department, including Hastie, were pleased with the level of response to and the popular reception generated by the show.[110] Despite this reception, the show appears to have been the War Department's final wartime broadcast on black soldiers or race relations. The department instead encouraged the radio industry to insert the theme of racial tolerance into certain popular war-related series and to offer special programming that included black soldiers and sailors but without the department's official sponsorship.[111]

The Army and *The Negro Soldier*

The War Department's most significant propaganda release on race relations was an army film developed primarily for troop use and subsequently approved for general civilian distribution. The War Department had enthusiastically embraced film as a powerful propaganda medium. For that reason, popular Hollywood director Frank Capra had been given the task of overseeing a film production unit for the army. Faced with the prospect of managing 900,000 black soldiers, the army began to seek ways to reassure them of their importance to the war effort at a time when most of them were relegated to very restrictive noncombat roles. The army film *The Negro Soldier* was designed to be a "morale picture for Negro troops." As the idea for the film evolved, the notion grew that it should be a documentary and not an entertainment-style film, which was Capra's specialty. Carlton Moss, a young black writer, was brought in to write the script. In 1941, Moss had produced an impressive stage review, "Salute to the Negro Troops," to raise funds for recreational facilities for black troops. He had offered use of that production to Archibald MacLeish at the OFF, but MacLeish had declined the offer. Moss also had experience as a radio scriptwriter and had worked at the Federal Theater Project for John Houseman, who had recommended him to Capra's unit.[112]

During 1942, Moss wrote the script for the film, working toward a final product that would satisfy the army, black activists like Walter White, black editors, and his own artistic and political objectives. Searching for a literal and metaphorical meeting ground, Moss set the film in a black church, where the narration was delivered from the pulpit. Disappointed by auditions of black actors who were in his view too wedded to traditional depictions of African Americans, Moss cast himself as the young, eloquent preacher who was the narrative fulcrum of the movie.

The opening of the film was shot from the back of a crowded balcony overlooking a large seated congregation of well-dressed black men and women. The viewer's eye then was directed to a book that Moss was holding in the pulpit from which he was preparing to read aloud. A dramatic, long close-up revealed that the book was not the Bible but *Mein Kampf*. Moss read a section of the book that Chandler Owen had quoted in *Negroes and the War* as the film focused on individuals in the vast congregation of black middle-class men and women. Their faces were dark and somber, contrasting starkly with the bright, white church interior and the men's starched white collars. The film juxtaposed Hitler's written references to "half-apes" with these images and that of the eloquent preacher, who described Nazis as men who would "kill and kill again" and "exterminate everyone" who stood against them.

Having made his case against Hitler, Moss turned to a history of black service in American wars, illustrated by a collage of dramatic reenactments, images from historical documents and paintings, and, for the post-1890 period, actual war film footage. As was often the case in federal productions in any medium, the Civil War received oblique coverage, with a single, simple reference ("Then came 1861"), accompanied by a shot of the Lincoln Memorial and the strains of the "Battle Hymn of the Republic." The film dramatized blacks helping rebuild America after the war. Footage of blacks fighting in the Spanish-American War and helping construct the Panama Canal was followed by a longer segment on black soldiers fighting in World War I and marching in victory parades.

Unlike the OWI pamphlet, however, the narrative of *The Negro Soldier* extended beyond the Nazis to include the "Japanese militarists." The most gruesome and jarring scene in the movie was of a Japanese execution in which six charred bodies gently swayed by their roped necks from a crossbar. For a film that targeted black men, this was a provocative use of a powerful visual symbol. During the camera's long look at these bodies, Moss's voice-over reminded the viewer that "there are those who will tell you that Japan is the saviour of the colored races."

This section of the film merged into a final series of shots that, like the closing pages of photographs of *Negroes and the War,* showed black soldiers in various terrains using different equipment and weapons. Black women soldiers were also shown drilling in formation and driving Jeeps. Some of the most dramatic scenes featured black pilots flying fighter planes. The film ended with an elaborate series of split-screen images of black soldiers marching in formation under the American flag, accompanied by a martial rendering of "Joshua Fit the Battle of Jericho" and "My Country, 'Tis of Thee."

The film was riddled with irony and contradiction, but the greatest may have been that it was set in a black church. The film's militaristic message and graphic depictions of destruction and killing seem at first incongruous with the reverence of its setting. But the script rested on an enduring faith in the centrality of religion and religious symbolism in black life, a faith expressed repeatedly in federally sponsored materials about African Americans. White and black writers who sought to talk about or to black people in popular media forms often began with black sacred music or expectations of black religiosity. Whether in *Freedom's People* or in *The Negro Soldier,* an assumption of shared black and white respect for black Christian belief made black religious life a safe route to broader political messages. Whether that assumption was well founded is an inquiry for another time, but the prevalence and acceptance of the notion of black religious belief is one of the most overused and understudied mythic representations of African Americans even to this day.

Moss nevertheless made brilliant and richly symbolic use of the image of the black church. In *The Negro Soldier,* the black church represented the black community and was its reservoir of strength and resolve. The church symbolized black Americans as a moral people whose primary ally was God. It was a pedagogical site where people gathered to hear a preacher who did not preach but lectured. The church was "home" and "family" to the black soldier. The members of the enormous congregation were rigidly arrayed in the pews as if they were in military formation, preparing to become soldiers not only in the Lord's army but in Uncle Sam's as well. The people's faces were angelic, posed for the long portraiture shots. Close-ups shot from below gave the faces a heroic cast, with heads and eyes raised toward the podium. The people were rapt and alert, but they did not interact with one another. They were at once a prop, a backdrop, and an audience for the film.[113] The church in *The Negro Soldier* was a still life, a self-conscious representation devoid of realism yet saturated with authenticity. As an actor, Moss loaned his earnest presence to the film. As the scriptwriter, he used the black church as a symbol of authenticity. The juxtaposition of the church setting with film clips and historical reenactments produced a movie that remains even today an immensely powerful collection of black filmic images.[114]

The army began to use the film as part of its basic orientation for black troops early in 1944. After precautionary screenings, it made viewing the film mandatory for all troops, black and white, through August 1945. From its initiation through the war's end, virtually every black army enrollee saw *The Negro Soldier,* as did millions of white soldiers.[115] Most black journalists and activists enthusiastically embraced the film and convinced the army to release it for civilian viewing as well. They recognized immediately that *The Negro Soldier* was "good racial propaganda" and that its wide (and free) distribution would offer a significant opportunity to bring the message of racial tolerance to a large audience, especially white Americans. Black leaders and organizations had worked with limited success during the war to persuade Hollywood studios to incorporate the theme of racial tolerance and less stereotypical depictions of blacks into commercial films.[116] This film offered images of honorable and respectable African Americans that were lacking in the mainstream cinema. For example, one article about the film was titled "Army Shows Holly-wood the Way."[117]

The release of the film was well timed, coming as it did soon after mounting racial hostilities culminated in the summer of 1943 riots in Detroit, New York City, and Los Angeles, reminding federal officials once again of the necessity of making political gestures toward African Americans. Roosevelt administration officials were also sensitive to the potential political value of the civilian release of the film in securing the black vote in 1944. In fact, Gibson had used that rationale to argue for the film's early release.[118] Despite repeated prodding by Gibson, Moss, General Benjamin Davis, and black journalists and activists, the army proceeded cautiously on the matter of broad public release by holding a series of special screenings at the OWI. Director Elmer Davis requested "softening cuts" that would minimize the "risks" of showing it.[119] In April 1944, the federal government released the film to civilian audiences, making it available to commercial theaters and civic groups. The OWI distributed it free of charge to public libraries, schools, and colleges across the nation.[120]

The film was received favorably by African American groups of widely varying political beliefs, ranging from the NAACP to the National Negro Congress, which called the film "the best ever done."[121] To black viewers, the film brought to life the idea that the black community had helped build and preserve the United States. It also presented never-before-seen moving images of black men and women. *The Negro Soldier* broke ranks with the common image of the African American as a "lazy, shiftless, no-good, slew-footed, happy-go-lucky, razor-toting, tap-dancing vagrant."[122] Most important, the federal government's association with the film provided it with an air of authority and recognition for a filmic message about racial tolerance. As restricted as the film's coverage was, many blacks believed the images it presented were worth claiming, which generated overwhelming support for *The Negro Soldier* from blacks.

Some of the appeal of the film rested on the emotive power of visual images, in this case, moving visual images. Radio and captioned printed photographs are less effective than film in imposing a single interpretation or reality. A photograph can fix an image but not its interpretation. Film requires a suspension of time and external reality. Unlike a pamphlet, it is not something to be experienced at the viewer's will and pace. Radio competes against the visual, trying to create an image in the listener's imagination. Although it can stir emotions through voice and music, it cannot set images; it can only inspire them. Film uses images to transport the viewer into a different reality, literally absorbing its audience in a darkened room into its created world and its emotional adjuncts. A good film can do that, and by that measure, *The Negro Soldier* was a very good film. It also was technically sophisticated and exciting, relying on an inventive mixture of images, special effects, music, and actual war footage. The film constructed a filmic reality that not only enticed and entranced black viewers but also inspired them.

The film suggested that black soldiers deserved a measure of respect that had rarely been extended to black people in general. Black military service brought a new claim for respectability not only to black soldiers but also, by implication, to the communities from which they had come and to which they would return. Moss had brilliantly used the black church to represent that community. He tried to write a script that would "ignore what's wrong with the army and tell what's right with my people" and would force white viewers to ask, "[W]hat right have we to hold back a people of that calibre?"[123] The extent of Moss's success with white audiences is unclear, but plainly black people saw much in his film that they thought right and fitting.

Despite its air of authenticity and its raw emotional power, the film did not address current racial inequities or the prospects of black postwar progress. The conditions blacks faced at home and in the services were literally blocked from the film's frame of reference. The unjust treatment of the larger black community in the United States was excluded entirely. The film kept its narrow focus on the black soldier and the symbolic congregation. Not a single image of contemporary black life outside of the gathered congregation was presented. Its narrow narrative identified and defined for black and white Americans an enclosed, contained classless community of African Americans who were themselves the embodiment of worthiness. The focus on black soldiers also protected the film from political charges that it was advocating a more socially, racially equal world at home—the southern conservative political reading that had doomed the pamphlet *Negroes and the War*.

The film escaped the kind of controversy that had been directed at the OWI two years earlier. Perhaps the most important reason was that it was produced by the army in the midst of a war. The military was more well respected and trusted in Congress than most federal agencies, especially the politically vulnerable OWI. The army also was immune to purely partisan attacks or accusations that it was acting out of political motivation. If government appeals for racial tolerance were to be made, especially to the South, the military was the best place to make them. Legislators and army officials also suspected by then that the war could be won by the Allies only with increased reliance on black American troops and some expansion of their roles in the military.[124]

The film's focus on the armed forces also protected it from the political criticism it might have drawn from addressing racial issues that were seen as part of civilian, domestic politics. As artificial a divide as that was, it worked to shield the film from the political controversies it had left out of its own narrative. This was made plain in the War Department's subsequent attempt to address on national radio some of the issues facing black troops as they returned to the civilian life—issues that had been excluded so carefully from *The Negro Soldier*.

Assignment Home: "The Class"

In early 1943, in protest against administration policies toward black soldiers, William Hastie resigned from his position at the War Department. Truman Gibson, who was also black, promised that he would soon leave the department as well. Instead, not only did Gibson remain but his more conciliatory manner led to his promotion to Hastie's old position. Gibson's decision to stay at the department rankled many in the black press who viewed him as opportunistic and largely ineffective.[125] Once again, as at the OWI, the role of the internal race advisers became a source of conflict that played itself out over a specific example of agency propaganda on the issue of race.

In 1945, Gibson turned his attention to improving the reception of the returning black soldier. That year, anticipating the war's end, CBS radio had initiated a special summer series called *Assignment Home,* which was designed to help ease the reentry of large numbers of troops back into civilian life. In June 1945, Gibson suggested to Robert Heller, a vice president at CBS and the series's producer, that the show feature an episode or incorporate material on returning black soldiers.[126] Gibson painted a general scenario for Heller of the return of 700,000 black soldiers and 150,000 black sailors, 75 percent of whom were southerners and almost half of whom had entered the service uneducated and illiterate but now had seen the wider world, including places where black men were treated with a level of respect and acceptance they had never experienced at home. Gibson feared the hostile white civilian reaction to these changed men: "Most of them have had their attitudes changed and their horizons broadened by their service in the Army and in the Navy. These changes in attitudes, and to some extent behavior, will be quickly observed in the communities to which the servicemen return. If we are to have the racial peace and understanding that our country vitally needs during the difficult days ahead, it is essential that these changes be understood and appreciated by the civilian communities." He emphasized that "even during the critical periods when all of our energies were devoted towards getting ahead with the war, there were many evidences" that the wartime emergency had not curtailed white hostility to black troops. His current fear was that the mass return of black soldiers would mean an escalation of white mean-spiritedness "during the period after VJ-Day when the patriotic brakes will be slightly released." With that warning, Gibson asked CBS to assist the War Department by presenting "suitable material" on this problem on the *Assignment Home* series.[127]

CBS officials took Gibson's plea seriously and immediately began work on an episode that addressed the issues he raised to be broadcast in late August. An army corporal drafted a script for a half-hour show titled "The Glass" that told the story of two soldiers, one black (Ted) and one white (Sam), who had worked side by side as truck repairmen, had been wounded together at the Battle of the Bulge, had received the Bronze Star Medal, and had recovered together in adjacent hospital beds.

The Battle of the Bulge in late 1944 and early 1945 was one of the only battles in which the unlikely scenario of black and white cooperation could have occurred in a war fought by segregated troops. Because of the drastic shortage of white soldiers, in that battle, black soldiers in service units were allowed to volunteer as infantry replacements. Black soldiers seized this opportunity with great enthusiasm. As many as 80 percent of certain units volunteered, many at the cost of giving up rank. Under General Dwight Eisenhower's overall command, black platoons were assigned to white companies, where they fought jointly with white soldiers, although not in truly integrated units. As soon as the European war was over, the platoons of black soldiers were stripped out of the white units and either sent home or sent back to their all-black service units. After this experiment with "integrated" combat, army surveys showed that the attitudes of white troops toward black troops improved dramatically as a result of the experience and, moreover, that most of the white troops involved no longer objected to serving alongside their black counterparts. The choice of this setting for the background of a script about returning black soldiers was therefore extremely significant.[128]

The story began as the two men were being discharged from the army. On the bus trip home, they entered a bar expecting to be able to have a drink together, both of them apparently suffering from racial amnesia. Instead, they were treated roughly and with disdain. After reluctantly serving Ted, the bartender pointedly broke the glass, saying, "You think my customers would drink out of a glass that a . . ." At that point, Ted said to his white friend, "We're back home, Sam."

The remainder of the script showed that Ted's triumphant reunion with his family was marred by his disappointment over a job offer that was withdrawn after the employer discovered he was black. In the end, Sam convinced his own employer, who had hired other black workers, to give Ted a job. But this happy ending was tempered. The music began in a "rising triumph" but then came to an abrupt halt to dramatize that "the solution that has been projected is false." The narrator explained why:

> And so the dream ends . . . and Ted lives happily ever after. But what of the tens of thousands of Ted Godwins who are coming back? There are Negro men taking off the uniforms they have honored and putting on civilian suits. They are climbing out of cockpits getting down from tanks and putting aside rifles—They are coming back to the peace and democracy we made together. (The enemy bullet never asked the color of a man's skin.) What of these men? For this is not a question of debt or obligation—it is understanding that what they did was done willingly because it had to be done. The war would not have gone the way it did with our powerful black hand tied behind our back. And we shall not finish, we shall not win the peace without each man, regardless of race, or creed or color, in his rightful place.[129]

Compared to earlier federally sponsored productions, this script confronted the effects of racial prejudice and discrimination bluntly. It also shifted the spotlight to home-front discrimination in public places and employment. Apparently it did so a bit too directly from the War Department's point of view. After rehearsals for the show had begun, department officials withdrew their commitment to CBS to collaborate on "The Glass." The network never aired the show.

Word of the cancellation set off a flurry of protests from African Americans, some of whom blamed Gibson. Fisk University sociologist Charles Johnson wrote Gibson requesting an official explanation for the decision to cancel the broadcast so that he could publish it in his "Monthly Summary of Events and Trends in Race Relations."[130] In response to this and similar requests, Gibson's assistant Louis Lautier drafted a telegram explaining that the War Department did not approve the script because it did not think it "should collaborate with Columbia Broadcasting System in radio program projecting the Army into controversial subject respecting civilian life."[131] The NAACP was quick to register its "vigorous protest" with War Department secretary Henry Stimson about the decision not to broadcast the show. Again, the department's response was that the show was canceled because of its "controversial nature." In a final plea, Roy Wilkins argued that the special needs of African American veterans should not be ignored on the "flimsy excuse of controversy," reminding Stimson that "these men are American citizens who served their country during war."[132]

The New York City newspaper *PM* publicized the controversy, and white liberals also complained about the cancellation of the broadcast. The Hollywood Independent Citizens Committee of the Arts, Sciences and Professions, a liberal group of entertainers, writers, and artists, sent its objections to the show's cancellation directly to Eleanor Roosevelt. Assistant Secretary of War John McCloy took on the task of responding to her inquiry about the show, explaining that the script "concerned itself with dramatic incidents that occurred to the central characters after their discharge from the Army" and that "had the action been confined to the contribution of Negro soldiers and the conditions of their Army service," the script would have been approved. Moreover, McCloy asserted, "The Glass" in his view "would not have done much to help returning Negro soldiers." Most telling perhaps was his admission that "the Army would have been placed in the indefensible position of presuming to dictate to civilians how they should act towards other civilians."[133] A telegram representing the 2,000 members of the Dusable Lodge of the Chicago branch of the leftist International Workers Organization characterized the cancellation as a "slap in the face of the Negro GI's and their families as well as an insult at our democracy for which all boys black and white alike fought and died for in the interest of a postwar peace." Gibson's response to this and several other protest letters was identical to McCloy's reply to Eleanor Roosevelt.[134]

Meanwhile a local New York City radio station asked Gibson: "WLIB, the voice of liberty, would be glad to join any fight on racial discrimination. May we have the pleasure to air the 'The Glass'?" Gibson assured the liberal station that "the War Department has absolutely no objection to the use of this script by WLIB" and authorized the release of the script for the station's use.[135] Gibson apparently still believed that the script's message was both relevant and timely. Nevertheless, the War Department was unwilling to take the political risk of sponsoring a national broadcast based on the script, and Gibson was unable to change that position.

The War Department's decision not to proceed with this radio program demonstrates the narrow expansion of permissible political boundaries of racial discourse and policies at the war's end. Although it was acceptable for a federal agency to show the African American contribution to the war effort, the department was not willing to present the corollary argument that military service qualified black civilians for anything akin to equal treatment from the white civilian world they were destined to reenter. War Department officials did not want to touch the issue of domestic segregation any more than the bartender's customers in the rejected script wanted to touch the glass used by Ted, the heroic, wounded, returning black soldier.

The federal government's sponsorship of radio programming about African Americans and race relations ended before World War II did. But the programming that was produced, along with the government's more general publicity campaign for national unity, marked a change in public rhetoric that did not go unnoticed. On a trip he took through the South during the war, Sterling Brown observed that the patriotic talk of freedom, on the radio and elsewhere, had stirred hope for black southerners: "Freedom was a hard-bought thing, their tradition warned them; the great day of 'jubilo' had been followed by gloomy days; but the talk sounded good and right, and perhaps a little more freedom *was* on its way. Through the radios— many of them the battery sets which fill the needs in small shacks once filled only by phonographs and guitars—booming voices told them of the plans for a new world. Over the air-waves came the spark, lighting and nursing small fires of hope; the glow and the warmth were good in the darkness." The talk of freedom was not lost on white southerners either, as one Mississippi planter complained to Brown: "One of the worst things making for all this trouble is the radio. Those people up in Washington don't know what they're doing down here. They ought to shut up talking so much."[136] Freedom was in the air, but the way it was received depended on the listener.

PART II

AIRING THE RACE QUESTION

CHAPTER 4

THE NATIONAL URBAN LEAGUE
ON THE RADIO

The National Urban League took advantage of the war crisis to link its traditional emphasis on acculturation, job counseling, and vocational preparation to the opportunities offered by the burgeoning yet racially discriminatory defense industry.[1] The crisis of war enabled the organization for the first time to use national radio to spread a message of equal opportunity in jobs and military service. But its guest status on national radio limited the political content of its message, a fact easily seen by comparing it with the more aggressive political rhetoric contained in *Opportunity* magazine, its official publication.

Despite the limited nature of their privileges, league officials used radio in pioneering ways. The league's black entertainment radio extravaganzas used free performances by African American performers and musical artists not only to advance arguments for equal opportunity but also to dramatically demonstrate to the radio industry the shortsightedness of continuing to refuse radio jobs to most black artists. The league's other singular radio achievement was its national broadcast of a special show devoted solely to African American women. Rendered invisible in history and vilified in all popular imagery, black women also had been given a limited voice in the spate of special radio programs on race relations aired during this period. The league used national radio to construct a new image, however fleeting, of black women. That image argued for the extension of a politics of inclusion to black women as war workers, military nurses, and "American" women, worthy of respect and honor and the full rights of citizenship.

Officials at the National Urban League had been alert to radio's promising potential as an ally well before World War II. A 1928 article in *Opportunity* characterized radio as "a particular and far reaching instrument of interracial understanding and enlightenment" for both whites and blacks.[2] The article's conclusion that such programming was needed by both blacks and whites was an early expression of the belief in radio's influence on black self-perceptions as well as white opinions.

Both the league and the National Association for the Advancement of Colored People (NAACP) sought opportunities to use radio as a means of advertising their own political views and increasing favorable opinion of African Americans, again among African Americans and whites.[3] But the league was much more successful at gaining access to local radio affiliates for public service broadcasts. This was the case not simply because its underlying political philosophy was more palatable than the NAACP's but also because the league had a specific program that was especially well suited to radio's strengths: its annual vocational opportunity campaign, a weeklong series of national and local publicity-oriented events addressing the issues of job training and employment opportunities.[4] The league initiated this program in 1930 to encourage African Americans to train for higher-skilled jobs and to press white employers to open such positions to them, especially in businesses that depended on black patronage. The success of this strategy, league officials believed, depended on changing public opinion about black abilities, and they thought radio was an essential vehicle for generating more sympathetic publicity for their goals. "Plan and prepare" became the watchwords of the opportunity campaigns, as league officials focused on a future in which they hoped young African Americans would eventually gain access to positions denied them at the time.[5] The first annual campaign in 1930 generated a flurry of talks, speeches, and interviews with potential employers, as well as local radio broadcasts highlighting the league's programs.[6] This pattern would expand and continue throughout the campaign's long life, which extended into the 1960s.[7]

Ann Tanneyhill was the National Urban League staff member who directed vocational opportunity events at the national level and co-ordinated complementary activities at the league's local affiliates. A young African American woman educated at Simmons College, Tanneyhill served two years with a local affiliate in her native Massachusetts and then in 1930 moved to the national office of the league in New York City. Tanneyhill coordinated the work of the campaign from 1931 to 1946 while broadening her own understanding of vocational guidance by completing graduate work in the emerging field at Columbia University. A gifted organizer, Tanneyhill also was a brilliant publicist and, as we shall see, the person most responsible not only for securing the league's access to national airwaves but also for writing many of the scripts and convincing prominent blacks and whites to participate in the broadcasts. By 1939, the league's annual vocational opportunity campaign used local radio stations in twenty-two cities to spread its message to blacks and to the white business community.[8] But the league's access to local audiences was limited to cities where it had strong affiliates, leaving many urban and southern areas untouched by the yearly campaign.

Two factors converged to enable the league to broadcast its annual campaign on the national airwaves: a change in the organization's leadership and the crisis of war. Lester Granger assumed the position of chief administrator of the league in late 1940 and then became executive director at the end of 1941. Granger, who previously had served in the league's Industrial Relations Department, believed that contemporary civil rights issues could no longer be ignored but were an essential aspect of any general campaign to improve vocational and employment opportunities for African Americans. He began during the war years to actively seek ways for the league to play a larger role in the national politics of racial reform. The issue of discriminatory hiring in the defense buildup provided a perfect opening for him to link the league's traditional emphasis on jobs to the broader war-related aims endorsed by the NAACP, the March on Washington movement, and the black press. As a result, the annual vocational opportunity campaigns took on the expanded role of advancing broader political arguments about racial equality. At the same time, the campaigns generated greater recognition of the league's overall program by allowing it to work on political issues of pressing concern to the mass of black people, issues that had been more the domain of other black organizations in the past. The war emergency not only created more jobs and a tighter labor supply but also provided an arsenal of new rhetorical appeals aimed at white employers in defense industries. The war also opened the doors of national radio to the National Urban League and allowed it a level of national exposure that had long eluded it.

"The Negro and National Defense"

The great proliferation in network-sponsored radio programming on war-related issues inspired Ann Tanneyhill to ask one of the national networks to carry materials about the National Urban League's annual vocational opportunity campaign in 1941.[9] She wrote a letter to CBS describing the campaign and requesting fifteen minutes of free time. After a long silence, CBS officials surprised Tanneyhill by not only granting her request but also offering a full hour of national airtime for the league's use.[10] Previously, league affiliates had sometimes secured local airtime for announcements or inspirational messages for aspiring young black job seekers and their potential employers. But CBS officials wanted the league to deliver the kind of program they themselves were unwilling to produce because no paying sponsor could be found. They wanted "a variety show of America's outstanding Negro musicians." Tanneyhill and her colleague Ed Lawson, then the editor of *Opportunity*, scrambled to design a show that would satisfy their own aims and the network's request, with only ten days to airtime.[11]

Network officials made clear that although they would donate the airtime and production facilities, they would not assume the costs of paying African American performers to appear on the program. They suggested that league officials obtain fee waivers from the unions representing these artists so that they could volunteer their appearances. In a letter to the American Federation of Musicians, Ed Lawson cited two reasons why the waiver should be granted: first, it would allow black performers to lend their support to a cause they deeply believed in, and second, a national radio appearance was an unusual opportunity for most of these artists and musicians. He also shared his hope that a successful show would lead to a "regular series of Negro variety shows which will offer employment opportunities to colored artists on the radio."[12] As Lawson's remarks indicate, African American musicians and actors were routinely denied appearances on national radio, except in limited comedic roles or as occasional guest stars on white variety shows.[13] The unions agreed to the league's request, clearing the way for black artists and performers to provide gratis the musical appeal the network was eager to broadcast. This arrangement allowed for the creation of special public affairs programming about African Americans throughout the war period—the performing artists and musicians unions approved fee waivers, and black entertainers and artists contributed their time.

This opportunity for national airtime for black performers was so unique that the performers responded with great enthusiasm to Tanneyhill's requests for their appearances. After a decade of working for the league in New York City, Tanneyhill was connected to a network of African American entertainers, artists, actors, and writers who lived in Harlem.[14] Tanneyhill and Lawson approached Marian Anderson, Ethel Waters, Duke Ellington, Louis Armstrong, and Joe Louis. They were able to convince all of them to appear on the show, which the league correctly characterized as the first "hour-long network show with an all-black cast."[15]

Tanneyhill and Lawson wrote the script for the show, which CBS broadcast on Sunday evening, March 30, 1941. Tanneyhill recalls that CBS made no effort to revise the draft script she and Lawson wrote.[16] Called "The Negro and National Defense," the show featured a star-studded lineup that provided a dazzling variety of musical performances. Opening with Louis Armstrong's band, followed by a pickup from Detroit of Ethel Waters singing "Georgia on My Mind," the show also included Duke Ellington's orchestra in Hollywood playing "Take the A Train" and "Flamingo." Marian Anderson performed two songs live from Montreal, capping off a rich display of African American musical virtuosity.[17]

Interspersed between the musical numbers were several brief speeches intended to deliver the league's message about its vocational campaign and the war buildup. Elmer Carter, editor of *Opportunity*, made a pointed plea to all Americans to include blacks in the preparations for war. Joe Louis made essentially the same appeal in a brief interview from St. Louis in which he described to John Dancy of the Detroit Urban League his joy at having been able to obtain a job on an assembly line in an automobile factory:

Dancy: Then, you think Joe that Negroes can work in defense jobs—on their assembly lines, and on their machines.

Joe: I know they can. I have seen them do harder things than that when they got the chance.

Dancy: Do you think they will ever get the chance?

Joe: Americans believe in fair play. They are good sports, and I feel America is going to give us Negroes a chance to work and earn a decent living. We can defend this country against anybody if *all of us* have a job to do. I know we need the jobs now worse than ever and when they give us a chance we will punch out a new victory for America.[18]

The most interesting attempt to promote the message of racial fairness came in a skit performed by Eddie Anderson, who played Rochester in the *Jack Benny Show,* a black character very popular with white audiences but deeply resented by some blacks. In the skit, Rochester engaged in a telephone conversation with Jack Benny in which he told Jack about a radio speech he had been asked to give on behalf of the National Urban League:

It seems to me that Mr. and Mrs. America have been so busy in this great program of national defense that they sorta overlooked one of their children. One who has always been a great fighter— loyal, conscientious, and willing to do his bit at all times. It seems that this child is having a little trouble convincing the principals and teachers of this great defense program that he should be in there, too, and that he could come through with flying colors if only given a chance to. . . . He is a healthy working boy, so let him help. . . . So, Mr. and Mrs. America, give this child of yours a chance to make you just as proud of him today as you've always been in the past.

The tone of this soliloquy was consistent with the tone of the Rochester character and the paternalistic relationship between him and his white boss, Jack Benny, and millions of white listeners. But Rochester's underlying argument—for fair play, a chance, more opportunity—was also consistent with the arguments made by Elmer Carter and Joe Louis, although it was offered in a form and a patronizing racial frame of reference that many whites would accept more readily and some African Americans would find offensive.

Embedding serious messages in nonthreatening entertainment

was a radio technique that some African Americans criticized even though they realized that political restrictions usually mandated such a strategy. Commenting on this phenomenon, L. D. Reddick astutely observed that "stations will not permit discussions of 'Negro rights' unless such topics are so intertwined with entertainment as to make the former very secondary."[19] This was basically the approach taken by the National Urban League in its first hour of national radio time. Reddick's observation may have been shared by others, but the response to the league's show was overwhelmingly favorable from both African American and white listeners, many of whom had heard the underlying messages.

Reaction to the broadcast in black and white publications was laudatory. A New York City newspaper extended its congratulations to the league for producing a show that "was probably heard by more people than has ever listened to the problems of the Negro."[20] Another editorial from Rhode Island characterized the show as an "educational service" that "aroused a keen awareness among all Americans of the undemocratic exclusion of Negroes from national defense industries and of the current discrimination against them in the armed forces."[21] A black Michigan paper commended the league's "new approach" to the problem of defense industry discrimination, saying that "millions of Americans can no longer plead ignorance of the plight of our workers."[22] A *Philadelphia Tribune* article urged CBS to "continue its policy of permitting colored people and their white friends to talk directly to the millions of Americans who do not realize that colored people are barred from helping the defense program."[23]

Some editorial writers used the broadcast to chide members of the radio and advertising industries for their complicity in perpetuating racially discriminatory practices. A *Time* magazine article commended the broadcast for giving "listeners a chance to hear some of the superb talent that few advertisers dare to sponsor." The article left no room for misunderstanding: "Obvious is the reason so few have made the aerial grade: they are Negroes." Observing that African Americans were welcome on radio only as guests or in bit parts, the article blamed advertisers for being unwilling to "buck racial prejudice to back a colored show or let a Negro star shine too brightly." Moving beyond the issue of wartime entertainers, a *Philadelphia Tribune* editorial urged that when the war emergency was over "the great radio chains should keep the airwaves open so that a tenth of America's population can present their case for absolute equality as American citizens."[24]

Tanneyhill's report on the 1941 vocational opportunity campaign noted that the National Urban League received many telephone calls and written comments about the show, of which she characterized only two as "unfavorable."[25] To her surprise, although no plea for funds had been made, many writers sent checks and cash in response to the show. The listeners who wrote in were varied, ranging from high school social studies students in New Jersey who considered discrimination "both unjust and dangerous" to a white high school principal in Brooklyn who called the show a "magnificent plea."[26] Many white listeners wrote that they were deeply moved by the broadcast. One declared that "the negro is just as equal as the white man," and another commented that "the country is the loser by this unjustifiable discrimination."[27] A woman from New York City wrote: "You must think the white race smug and tyrannical, but I hope you'll understand that it's not that we don't care—only that we don't know. But now I know and I am greatly stirred."[28]

The view apparently persisted among African Americans that many white Americans would in fact support increased opportunities for blacks if they were informed of actual racial conditions, and radio was thought to be especially well suited to the task of educating whites. Blacks working on racial reform issues were quick to praise Tanneyhill and the league. They seemed to share a broad optimism that a radio broadcast had special potential for producing results. Roy Wilkins at the NAACP called the show "a most effective stroke in the battle for improving the Negro's position in the national defense program." His colleague Walter White also sent congratulations, writing that the show "cannot help but further awaken Americans to what is going on." Channing Tobias at the national Young Men's Christian Association wrote a letter of praise, as did Ambrose Caliver at the Office of Education, whose radio series *Freedom's People* would premiere later that year.[29]

This enduring faith in the power of radio as an effective teaching medium pervaded other comments about the show as well. The writer Chester Himes expressed his appreciation for what he called "the shortest hour in the history of radio." Like many listeners, he had found the mixture of music and speeches especially effective: "In the lucid, driving, unequivocal manner which the Negro's industrial lockout was presented, softened by the swiftly paced, highest quality of radio entertainment, the point can not be missed." Although it may seem naive in hindsight, this belief in radio and the power of education to change attitudes persisted among many blacks and whites who hoped that talking about prejudice might open listeners' hearts to understanding and change. Although the political messages of the league's broadcast may have been muted by the surrounding music, they were not lost, as these comments make clear. As the first national broadcast of its kind, the show seemed to have found the middle ground that satisfied those who looked to radio as a medium that could provide political education without straying into territory that would have troubled CBS officials.[30]

League broadcasts on local New York City radio stations during the same campaign, however, offer a glimpse of the difference in tone of the appeal league officials made to a narrower northern urban audience. Usually only fifteen minutes long, these local shows frequently used a simple interview format. For example, on one show, league board member Roger Baldwin, president of the American Civil Liberties Union, argued that "every form of pressure" would be needed to break through the color line and that the war crisis offered no excuse for delay: "[Y]ou cannot get national unity by burying grievances."[31] Another broadcast included an interview with Charles Collier of the New York City Urban League, who condemned local and national unions for their exclusion of black workers from skilled jobs, especially in the burgeoning aviation industry and other defense-related industries.[32]

Far more aggressive in language and style than any of the radio programs, national or local, was the editorial commentary on these issues featured in the league's publication, *Opportunity*. As early as 1938, the magazine had argued that "the best defense against the forces of Nazism in America is the realization of real Democracy here," a line of argument that the magazine continued and elaborated through the end of the decade.[33] Another editorial in 1939 observed that "it would appear from reports in the daily press that the German Reich has decided to model its program of racial repression on the prevailing laws and customs in the Southern part of the United States."[34]

By early 1941, the tone and tenor of the magazine's editorials on discrimination in the defense industry and segregation in the military were even more critical. Some of this change can be attributed to the ascension of Lester Granger to the leadership of the organization, as well as to the way the war magnified the inequities of federal racial policies. In a 1941 *Opportunity* article, Granger described the racial policies of the army and navy as "a symbol which must be attacked wherever erected." He continued: "It is a symbol of a low-grade citizenship incompatible with the democratic ideal; it is a sign of persisting intolerance disrupting our national unity and corrupting our national life; it is a danger to the American nation because it is opposed to the true American spirit." An editorial in the same issue contained even more explicit criticism both of the military and of industrialists for "adopting a policy of racial repression which one expects to find only in those countries which have adopted the methods and ideology of fascism." [35]

Editorials and articles in *Opportunity* in the spring of 1941, and indeed, as we will see, throughout the war era, persistently challenged, ridiculed, and belittled racially discriminatory policies in the defense industry and the military. One article about discrimination in the military noted that "all over the world the color line is being erased as nations fight to preserve the democratic form of government—all over the world except in Hitler's Germany, Mussolini's Italy, and the United States of America." [36]

A comparison of the tone of writings intended for the league's own limited readership and the tone of its first national radio broadcast reveals how timid and abbreviated an appeal the league made in that broadcast. Indeed, nothing in the show seemed inconsistent with the radio programs sponsored by the federal government during this period. After all, as discussed in chapter 3, it was this National Urban League broadcast that the army used as a model for its own radio show, "America's Negro Soldiers," which NBC broadcast five months later in August 1941. The freedom of political expression that league officials felt in their own publication never extended to their appearances on the air.

CBS's decision to grant free airtime for the league's broadcast did not go unnoticed by other African American activists who were eager for access to national radio. In June 1941, A. Philip Randolph of the March on Washington movement, which the league supported, asked NBC to consider donating time for a program entitled "The Negro in National Defense." NBC officials declined, claiming that their schedule was full and that they had granted time to representatives of other "negro organizations" that year. [37] Randolph did secure national airtime for his announcement that the proposed mass march had been canceled because of the president's new executive order barring discrimination in defense industries. But even then, only the smallest national network, Mutual, granted him access for that purpose.[38]

Having tasted the fruit of free national airtime in 1941, league officials sought the same access for their 1942 vocational opportunity campaign. Granger explained to NBC's president Niles Trammell that the league needed airtime to "reach employers, white workers, and the American citizenry in general, in order to impress upon them the need for unrestricted use of Negro labor." Promising an interracial program, with black and white speakers and entertainers, Granger asked for an hour of time, preferably during the evening. At the time, NBC was committed to carrying the remaining two monthly broadcasts of *Freedom's People,* and to network officials, one "Negro show" at a time seemed to be the rule of thumb. One official responded to Granger's request by asking, "Can we make this one of the Freedom's People spots?"[39]

Granger complained bitterly to network officials about their assumption that one "Negro program" was enough or that "Negro programs" with different purposes and origins could simply be combined. But Granger's harshest criticism was reserved for NBC's failure to distinguish between private and public leadership on racial issues, which suggested to him that network officials had concluded that "the Office of Education had a monopoly on spokesmanship in interest of Negroes." Indeed, in reply to Granger, an NBC official had defended the network's decision by explaining that the network recognized "Freedom's People as the official expression of problems concerning the Negro race" since it was a government program. This was an outright rejection of Granger's argument that the network should broadcast programming on race relations from both the federal and the private perspective.[40]

It is possible that network officials refused the league's request for time because they were inundated with requests for free airtime from the federal government in the months after the attack on Pearl Harbor. But other factors were likely to have been at play. The league and Granger by 1942 were identified with the March on Washington movement and with the complaints black activists and the black press were making about discriminatory policies.[41] Even after the attack on Pearl Harbor, the pages of *Opportunity* continued to be filled with protests against discrimination in the war effort.[42] The theme of the 1942 vocational opportunity campaign was "Speed Defense Production," and its goal was "to accelerate the integration of the Negro into American defense industry." A league handbook designed to encourage employers to hire black workers was subtitled "Open the Gates."[43] It seems possible, therefore, that NBC simply did not want to carry the league's arguments on integrating African Americans into the defense industry in the anxious period immediately after American entry into the war. After his final dismissal of Granger's request for airtime, Dwight Herrick of NBC's Public Service Program Division wrote his colleagues, "[T]rust the ghost will stay buried."[44] Indeed it did, for the result was that the league's 1942 vocational opportunity campaign had no national radio broadcast.

A year later, with the war effort in full swing and racial tensions at an all-time high, the league had a much easier time getting free national airtime during its annual March campaign week. In fact, in 1943 the league was able to make broadcasts on all of the national networks. NBC granted Tanneyhill's request for fifteen minutes to carry a "radio poem" she had asked her friend Langston Hughes to write for the league. The poem, "Freedom's Plow," traced the contributions of black workers—free, indentured, and slave—in building America, but it made only oblique reference to prior and existing racial inequality.[45] The Mutual network also provided fifteen minutes, which the league used for a speech by William Agar, a leading liberal and president of New York City's Freedom House. He freely criticized the Fair Employment Practices Commission for its ineffectiveness, launched a harsh attack on racism in unions, and condemned the vast majority of manufacturing plants and businesses that still refused jobs to blacks.[46]

"Heroines in Bronze"

Tanneyhill and other league officials decided that the 1943 vocational opportunity campaign should have as its focus black women workers. They hoped that African American women would be able to take advantage of wartime shortages of male workers to gain access to high-skilled, high-paying jobs and escape their vocational doom as domestic servants. Tanneyhill later recalled that the war brought the "first break in employment in major corporations" for blacks and that plants "needed women so badly that they took colored women."[47] Taking as its slogan "Woman Power Is Vital to Victory," the league's annual campaign highlighted African American women's contributions to history and the war effort, at home and abroad. In that same year, the War Manpower Commission began a comprehensive "womanpower" campaign designed to create a new image of white women that would encourage them to seek what were considered "men's jobs" and enter the labor force as a matter of patriotic duty. Black women, however, were largely excluded from that campaign, and when featured at all, they were portrayed only in relation to whites—that is, in subservient roles or as mammy figures. Other depictions of African Americans in that campaign sought to convince anxious whites that blacks continued to accept segregation and white supremacy as part of their patriotic duty.[48]

Tanneyhill was determined to make African American women workers visible. When CBS invited her to submit a dramatic program to be included as one episode of the network's hour-long wartime series *The Spirit of '43*, she used the opportunity to produce a groundbreaking show about African American women. The league's 1943 campaign show was broadcast nationally on CBS and to armed forces abroad. Tanneyhill wrote the script for the broadcast by relying on historical materials at the Schomburg Center for Research in Black Culture and a borrowed library book on radio scriptwriting. Although CBS staff helped edit the script, they were so impressed with Tanneyhill's draft that at first they doubted that she had written it. Except for some tinkering with dramatic devices to enliven some of the material, the final broadcast was virtually identical to Tanneyhill's original script—a tribute to her developing skills as a scriptwriter.[49]

Tanneyhill's script, "Heroines in Bronze," was an inventive presentation of the lives of three women: Phillis Wheatley, Sojourner Truth, and Harriet Tubman. The script was faithful to Tanneyhill's extensive research and included understated but very effective recreations of key incidents in each woman's life. The dramatizations and narratives about the women emphasized different aspects of black women's abilities, but each portrayed a woman who was smart, persistent, courageous, and undaunted by racial obstacles. Tanneyhill used the story of Wheatley's life to illustrate the contrast between the inhumanity of slavery's separation of children from their mothers and the intelligence and tenacity of Wheatley's poetry. The section of the script on Sojourner Truth emphasized her life before and after her freedom, her courageous determination to speak in public places controlled by white men, and her antislavery messages. Tanneyhill's treatment of Harriet Tubman was the most effective dramatization because it included a reenactment of several "escapes" from slavery and a recitation of the number of slaves led to freedom by this "Moses of Her People." The section on Tubman concluded: "Yes—Harriet Tubman's soul was in itself the spirit of progress—the determination to rise above the weight of oppression and injustice and breathe the free air of opportunity."[50]

The second half of Tanneyhill's script featured contemporary women, beginning with a speech by Mary McLeod Bethune, "director of Negro affairs" for the National Youth Administration. Bethune expressed her personal pride in "the valiant contribution of Brown American women to the cause of victory for democracy." She identified herself with Wheatley, Truth, and Tubman and the lessons taught by their lives: "I, myself, am a Negro woman, and I have experienced many of the hardships which are the common lot of our sisterhood. I thrill to the story of those 'Heroines in Bronze' who suffered and defied death to make America what it is today—defender of the democratic ideal, and hope of generations to come. Life in this country has not always been easy for people of dark skin. Yet the fact that the American Negro today stands stalwart in his faith and loyal devotion to the country of his birth, is a tribute to the women of the race." Bethune's claim for the centrality of black women's contribution to the race's survival and faith in democracy was bold. She also sought to link African American women with other American women in the war crisis: "Brown American women are sending their husbands, sons, brothers and sweethearts to war on five continents and seven seas. They watch fearfully at home for the fateful word that brings news of death in action—of drowning at sea." But she quickly distinguished the fate of black women, adding that "they have felt the heartsick disappointment that is their lot when they apply for jobs and are denied because of their race." Commending contemporary black women for their courage and endurance under difficult circumstances, Bethune argued that it was "essential that the contributions of these women be increased, for womanpower is vital to victory," using the vocational opportunity campaign's slogan to help make her point.

Bethune's linking of black women's efforts and an elevation in their status as pivotal to the race's progress amplified arguments from a long line of African American feminist activists and intellectuals, including Josephine St. Pierre Ruffin, Mary Church Terrell, Elise McDougald, and Amy Jacques Garvey. The show's acknowledgment of the "double jeopardy" of race and sex that faced black women also had deep roots. The notion that uplifting black women was an essential avenue for advancing the race had been given voice in the 1890s by Anna Julia Cooper and other African American women associated with the women's club movement.[51] The script contained no hint of any feelings among black women of subordination or inferiority to men or any need to temper their ambitions as women. The script's focus on the lived experience of the historical figures enabled them to stand alone in their convictions, strong and independent yet committed to the communal mission of the race.

This image of African American women was reinforced by a series of live interviews with black women working in the war effort in the United States and throughout the world. The women interviewed ranged from a radio technician at a Western Electric plant in New York City to a flight training instructor in Chicago, who commented that "our Negro students haven't been granted the same opportunities as our white students." A black woman serving as a Red Cross Club worker in London described her work among black soldiers stationed in the British Isles.

Black activists had used the service of black men in World War I to argue for fuller opportunities for them in World War II. This show made the same argument about black women who had served in World War I as canteen workers or army nurses. Expanding the opportunity for the military service of black nurses was both a real and a symbolic issue because the army initially had established an exclusionary quota for black women in the nurse corps. In an interview on the show, Marion Brown Seymour, a black nurse who had served in World War I, did not hesitate to remind the audience of the discrimination and federal policies that had prevented the greater use of black nurses who wanted to serve: "Unfortunately during World War I, though the Negro nurse was fully qualified, there was delay and hesitancy in making a decision to accept Negro women in the Army Nurse Corps. Their services were not used until just before the Armistice was signed." This same hesitancy was being repeated in World War II. By the time of the broadcast, however, the decision had been made to allow black women to serve abroad in the army nurse corps.[52] This section of the script succeeded in raising the two wartime issues that most affected black women: access to defense industry positions and full opportunity to serve in the military's nurse corps.

This show differed dramatically in tone and approach from the 1941 National Urban League broadcast, which had at its core a high entertainment appeal. "Heroines in Bronze" made no attempt at light entertainment; the music was almost exclusively sacred or inspirational.[53] The dramatic sequences and overall tone of the broadcast were entrusted to black entertainment professionals. The three "heroines," Wheatley, Truth, and Tubman, were portrayed by well-known black actresses, including Fredi Washington, who was famous for her role in the film *Imitation of Life*. Although the show was about black women, Tanneyhill asked her friend, veteran stage actor and radio performer Canada Lee, to narrate it, as he had the league's 1941 broadcast. But there was no comedy, no Rochester skit, no tap dancing by Bill Robinson, which made for a far more somber and dignified presentation of the "Negro woman." Overall, the show created and claimed a new image of "respectability" for black women, who were as vilified as black men in all media representations. Public representations of black women had a virulent narrowness, perpetuating either the Jezebel or the mammy image. As old as slavery and as persistent as racism itself, these were the images that "Heroines in Bronze" sought to displace with its bold aural configuration of African American women.[54]

Tanneyhill believed that since this show was the first national radio broadcast dedicated to African American women, it deserved and would garner special attention. To ensure this, she mounted an aggressive prebroadcast publicity drive for both the broadcast and the vocational opportunity campaign theme, "womanpower is vital to victory." She urged all local Urban League members to convince the program directors at local radio stations to air the national broadcasts during the campaign. Finally, Tanneyhill asked members to organize local "listening groups" for the show and to send their reactions to local stations to encourage them to air more programming about African Americans and provide more opportunities for black artists.[55] She also issued several general press releases about the upcoming broadcast, starting six weeks prior to the broadcast and ending the day after.[56] The prebroadcast publicity received a boost when both New York governor Thomas Dewey and President Franklin Roosevelt endorsed the league's annual campaign. Roosevelt stated in his letter of endorsement that "Negro Americans are carrying their part of the load at home and on the fighting front." In an ambiguously worded promise, he allowed that "the social and economic advantages which we of the democracies are fighting to defend, and further, will not be lost in the readjustment of the post-war period." Nevertheless, the mere fact that the president had lent his support to the league's work was itself a source of publicity for the upcoming broadcast.[57] Tanneyhill's hard work paid off: some newspapers printed her press releases about the show practically verbatim.[58]

The show's theme was given an added boost when the league devoted its entire spring 1943 issue of *Opportunity* to the topic of "brown American womanpower." The ninety-six-page special issue of the quarterly was prepared by Tanneyhill and Madeline Aldridge, an editorial assistant at the league, both of whom were responsible for the campaign's focus on black women. The magazine featured dozens of photographs of African American women engaged in a wide variety of defense work and volunteer war activities and wearing military and nurses' uniforms at home and abroad. The articles were about various aspects of black women's wartime concerns, but the emphasis was primarily on the same issues that drove "Heroines in Bronze": equal access to employment in defense industries and full opportunity to serve in war service organizations and the military, especially the army nurse corps.[59]

The images presented in this issue were in sharp contrast to the traditional images of African American women presented in the media. The magazine offered a counternarrative to the advertising industry's campaign to help the federal government project an inspirational image of the white female worker, an effort that completely ignored black women. Indeed, one study has found that "astonishingly, no black women were pictured in advertisements during the war"; even the reliable stereotypical images of maids and mammies disappeared. At a time when white women were being portrayed as sources of national pride, there was no politically acceptable place for black women, who remained invisible.[60]

Articles in this special issue gave detailed reports on black women's work in a variety of defense factories and industries, from ballistic laboratories to electrical repair plants to aviation factories. African American women were depicted as welders, riveters, and clerical and stenographic workers. Special attention was given to the ongoing campaign "to accelerate the integration of the Negro nurse into the total war effort." Similarly, an article on the Women's Army Auxiliary Corps encouraged black women to enroll in both basic and officer training since a large call for women was soon expected. Taken together, these articles, whether on civilian or service activities, encouraged African American women to take advantage of the opportunities for training and service that the wartime emergency offered, even if many barriers to full participation remained in place.[61]

Unlike the content of some of the league's earlier broadcasts, for the most part the content of "Heroines in Bronze" was consistent with and reinforced by the messages in the magazine, both in style and in rhetoric. Lester Granger began his contribution to the special issue with a discussion of discrimination against women in general before addressing the special burdens and sacrifices of black women. "They have felt the brunt of that mean racial discrimination which has stultified our national ideals and twisted the social growth of the Negro population," he wrote. At the same time, he argued, they "faced the added handicap of sex whenever they have sought opportunity to express their talents." He credited the historical progress of the "Negro race" to black women:

> If Negro Americans have endured three hundred years of slavery, economic exploitation and social frustration without losing faith in their country and their future, it is because the women of the race have kept the faith. If Negro men have made endless sacrifices in the painfully slow, upward progress of the past seventy-five years, it is because their women have insisted upon and gladly shared those sacrifices. If a growing interracial cooperation and understanding in many communities partly compensates for increased racial tensions in others, it is because Negro women have made it their particular business to keep channels of understanding open. . . . [B]rown American women have been a mighty force working for redemption of the soul of Democratic America.

Not only did Granger include African American women in the political history of the race, but like Bethune had done on the air, he placed them at the heart of the struggle to advance black people. Granger concluded by commending black women for having pursued the training for jobs that had been denied them early in the war but were now open to them because of extreme labor shortages. After giving examples of black women's involvement in fields from which they previously had been excluded, Granger ended his essay with this plea: "Please God, let it be a good omen for America in the peace years that are to come."[62]

This special emphasis on African American women in *Opportunity* and more dramatically in "Heroines in Bronze" struck a responsive chord with the black press. A *Baltimore Afro-American* article left no room for doubt about its opinion: the lead sentence described "Heroines" as "the most effective radio appeal yet heard on the air in behalf of colored Americans." New York City's *People's Voice* weighed in with a review that congratulated the league for its "industry and initiative" in getting the show on the air. Ann Petrey also mentioned the broadcast in her column in the paper, calling the show "tops in radio programs."[63]

Tanneyhill's role in creating the broadcast earned her praise from National Urban League colleagues in New York City and throughout the country. Elmer Carter, although no longer a league employee, congratulated her for "the best publicity job in the history of the Urban League" and told her that "the community is literally ringing with praises of your Saturday program." A. L. Foster of the Chicago Urban League expressed his deep appreciation, telling Tanneyhill that he knew she "must have worked like the devil" on the show. Foster also reported that the local league had urged drugstores, grocery stores, and other businesses to turn on their radios so customers could hear the Saturday broadcast. Pastors at area churches also had encouraged their congregations to listen. The Minneapolis Urban League sent its thanks to Tanneyhill, explaining that it "would be impossible to evaluate the amount of interest created in the work of the local Urban league by that broadcast and by the splendid April issue of Opportunity Magazine." Accolades for Tanneyhill's work came from league affiliates in Baltimore, where it was called "strictly 'high-class'" and a "swell job," and Buffalo, where it was labeled the "best ever placed on a national hookup." The Omaha league passed along its compliments to CBS, thanking the network for carrying the show, which it described as a "source of inspiration to all Negro people."[64]

Offering a particularly perceptive observation, the executive secretary of the St. Louis Urban League praised Tanneyhill not only for producing the "best broadcast by Negroes that has come over the air" but also for taking risks by veering away from the traditional entertainment-dominated format: "Most groups think that to get people to listen to a program, you have to have some 'name' band, some blues singers or some other person whose name appears frequently in the press and otherwise before the public. Your broadcast proved that is not exactly true. . . . We Negroes believe that they get over only a certain type of program which is usually accepted by white people, but our times today call for a more serious bit of propaganda than the Rochesters usually get over." [65]

Other people unaffiliated with the National Urban League but interested in race relations also congratulated Tanneyhill on "Heroines in Bronze." Sociologist Ira De A. Reid commended Tanneyhill for the "high calibre" of the "positively enchanting" show. Black nurses were especially appreciative of the attention the show gave to their war-related concerns. In a letter to Granger, Mabel Staupers of the National Association of Colored Graduate Nurses singled out Tanneyhill for special praise.[66] The fact that the show focused on African American women was emphasized in subsequent reports about it. The Tuskegee Institute's 1947 *Negro Yearbook* stated that the broadcast was "the first time in the history of radio that the accomplishments and achievements of Negro women have been heard on the air in story and in fact." [67]

Tanneyhill and the league also received letters and postcards from listeners in the general public. One writer thanked the league for an "excellent" and "well-planned and really very effective" show. Another commented that the broadcast was "thrilling, instructive, and finished in all respects" and that "it has done a great deal to stimulate genuine respect and good will towards negroes." Other writers also expressed the hope that the show would help whites; one white listener wrote that "I need no such aid to make me appreciate the place that the Negro race should have in our community life, but I hope it has broadened the outlook of thousands throughout the country who cling to another point of view." [68]

To the extent that the National Urban League hoped its broadcasts would bring greater attention to the plight of African American workers, especially black women, "Heroines in Bronze" was a big success. In other public affairs programming about African American history or race relations, black women had been invisible or veiled behind broad communal representations of the black community that rested on male voicings. Here for the first time, Tanneyhill was able to bring African American women front and center and treat them with a dignity and respect that radio, like all media, had long denied them. This would remain one of Tanneyhill's singular accomplishments for the league and for broadcasting in general. The paradox, of course, is the singularity itself: a one-hour broadcast could hardly compete with the influence of centuries of despicable images of African American women. As politically significant as African Americans knew popular images to be, they remained largely powerless to intervene or compete successfully in the long term in a public arena now controlled by powerful mass media, regardless of how eager and prepared they were to try.

Race Riots

Race riots that erupted in the summer of 1943 propelled the issue of race relations to the forefront of national attention. Groups of white soldiers and sailors stationed in Los Angeles attacked mostly young Mexican American and black men over the course of several days of disturbances that came to be called the "zoot-suit" riots, named after the distinctive attire common among young Mexican American and black men at the time. In Detroit, an argument between African Americans and whites quickly escalated into a full-scale riot. Thirty-four people were killed, 500 injured, 1,800 arrested, and over $1 million in property damage resulted before President Roosevelt sent in federal troops.[69] Finally, in August, rioting erupted in Harlem, where 300 blacks were injured and 6 were killed before calm was restored.[70]

Throughout the summer, African American leaders continued to urge President Roosevelt to make a national radio address condemning racial violence, but he refused to take that step.[71] Instead, black activists, racially progressive liberals, and entertainers prodded the radio industry to take a public role in calming fears and preaching racial tolerance. Walter White of the NAACP helped form the Emergency Committee of the Entertainment Industry, a group of black and white entertainers organized specifically to sponsor a nationwide radio appeal for racial tolerance.[72] Members of the committee convinced CBS to donate thirty minutes of free national airtime to their cause. Unlike NBC, which was routinely more cautious about programming on race relations or African Americans, CBS took on the cause of calming racial tensions as its own. Rather than simply granting free time for the show, William Paley at CBS took the unprecedented action of officially sponsoring the show.[73]

The network asked the respected liberal radio scriptwriter, director, and producer William Robson to help with this special project.[74] Robson recalled later the sensitivity with which he and his colleagues approached the issue of race: "We were extremely careful in the preparation of the script, since the country at the time was pock-marked with 'tension areas' where it was feared new race riots might break out. Our problem was to throw the light of truth on the Detroit incident without inciting either whites or Negroes to riot elsewhere."[75] Robson opted for a simple dramatic format. He decided to use a reenactment of the Detroit riot to emphasize "the positive aspects of person helping person, rather than the destructive aspects of the disturbance."[76] So cautious were Robson and CBS that the entire show was recorded in dress rehearsal twice and the broadcast was postponed twice as those involved sought to produce a balanced presentation and one that would not stir up additional controversy. Even on the day of the broadcast, network officials took the precaution of broadcasting the entire show via a closed circuit to every CBS affiliate, allowing each ample time to refuse to carry it. In the end, only a few stations declined to air the show.[77]

On July 24, 1943, the network broadcast the program, "Open Letter on Race Hatred," which began:

> Dear Fellow Americans. What you are about to hear may anger you. What you are about to hear may sound incredible to you. You may doubt that such things can happen today in this supposedly united nation. But we assure you, everything you are about to hear is true. And so, we ask you to spend thirty minutes with us, facing quietly and without passion or prejudice, a danger which threatens all of us—a danger so great that if it is not met and conquered now, even though we win this war, we shall be defeated in victory and the peace which follows will for us be a horror of chaos, lawlessness, and bloodshed. This danger is race hatred.[78]

The program then presented its aural re-creation of the Detroit riot. It dramatized the fact that the arrival of waves of wartime immigrants from Appalachia and the rural South, crowded housing, and the efforts of "subversive organizers and native Nazi orators" all combined to create conditions in which misunderstandings and rumors could start a race riot. The broadcast emphasized that the courage of individual blacks and whites prevented more deaths and injuries from occurring. In a bit of radio magic, the show included reports on the riots broadcast as if they were German and Japanese radio propaganda. The Japanese "broadcast" was especially hard-hitting, portraying the Detroit riot as one in which "hundreds of negroes were sacrificed to the altar of American white superiority complex."

"Open Letter" seemed to be directed primarily at whites, referred to as "the decent law-abiding citizens . . . who will pay the final bill for the race hatred of your fellow Americans." The narrator included this admonition: "We've got too tough an enemy to beat overseas to fight each other here at home. We hope that this documented account of the irreparable damage race hatred has already done to our prestige, our war effort, and our self-respect will have moved you to make a solemn promise to yourself that, wherever you are and whatever is your color or your creed, you will never allow intolerance or prejudice of any kind to make you forget that you are first of all an American with sacred obligations to every one of your fellow citizens." The broadcast concluded with a straightforward attack on the concept of white superiority and a clear argument for extending the full rights of American citizenship to African Americans delivered by former Republican presidential candidate Wendell Wilkie.[79] "Two-thirds of the people who are our allies do not have white skins," he said, "and they have long, hurtful memories of the white man's superior attitude in his dealing with them." Turning his attention to racism at home, Wilkie recited a long litany of basic rights that African American citizens deserved but did not yet enjoy.[80]

The race riots finally had inspired the type of radio show that Theodore Berry and other African American federal officials repeatedly had urged the Office of War Information (OWI) to air: one that targeted the racist attitudes of white Americans. But it was CBS, acting without federal imprimatur, that had taken this step. The task of fashioning a national response to the riots had been assumed not by the racially timid Roosevelt administration but by an alliance of CBS network officials and writers, entertainers, scriptwriters, and the liberal Republican Wendell Wilkie. The unusual nature of the broadcast drew praise and national media attention. *Time* magazine reported: "[T]he fact that a major U.S. network had the courage and took the time to emphasize a crisis in race relations was big radio news." The magazine also called the show "one of the most eloquent and outspoken programs in radio history."[81] No one reacted mildly to the widely discussed program. General reactions to the broadcast were, in Robson words, "as varied and violent as the point of view of the listener," with "indiscriminate applause and vile condemnation" coming from the same locality.[82] The attention the show garnered was yet another indication of the controversy that met attempts to use radio as a forum to discuss racial tensions.

The two limited arguments the broadcast made—that race hatred was bad and that black Americans deserved basic rights—were aired without the sponsorship of the federal government, whose officials remained fearful of the national political implications of those basic arguments. Nonetheless, some of the groundwork for the appeals made in "Open Letter" had been laid by the federal program *Freedom's People* and the War Department broadcasts "America's Negro Soldiers" and "Judgment Day." Even the National Urban League's broadcasts in 1941 and 1943, especially "Heroines in Bronze," had made the fundamental argument for fair play and equal opportunity, although they had not directly attacked racial hatred and violence. Still, none of these shows addressed the questions of a remedy and how the long-denied basic rights were to be extended or protected, questions that constituted the principal claims of African American protests.

When the "zoot-suit riots" erupted in Los Angeles, a national media center, CBS officials reacted once again, this time not with a single national broadcast but with a series of broadcasts for affiliates of its Pacific regional network. The riots in Los Angeles, which grew out of tensions between white sailors and Mexican American youths, actually involved African Americans only tangentially.[83] But the violence provided the cautionary incentive for the production of *These Are Americans,* a series of six fifteen-minute shows about African Americans and their place in American democracy that began in January 1944. The show was created in cooperation with two local civic groups and an entertainment-based group, the liberal Hollywood Writers' Mobilization. Veteran CBS radio newsman Chet Huntley, a member of the writers' group, wrote and narrated the series.[84]

Addressing the question of race relations was again considered so controversial that the show opened with a preliminary caution similar to the one in "Open Letter." Huntley introduced the first episode by explaining that the series was "about the American Negro: about his problem, about his education, about his place in industry, business, the arts, the armed forces, and his position in our society." A voice interrupted him:

Voice 1: Excuse me! But, brother, are you headed for trouble! This question of the Negro is the hottest thing on the books! It's a controversial question!

Huntley: That's true, but the principles of it are not! We hold that democracy is non-controversial, and the Negro question broadly stated is "Shall Democracy work for all or shall it work in places?" We hold that equality—and the Bill of Rights are non-controversial—and that's what the Negro question is about.[85]

Whereas "Open Letter" had pleaded primarily for racial tolerance and basic rights for African Americans, *These Are Americans* worked from the assumption that a change in race relations and the extension of greater opportunities to African Americans were not only inevitable but also imminent. Its primary goal was to coax white Americans into accepting the idea of equal opportunity and fair play for African Americans while calming their fears about the practical implications of such a political and social shift.

The series first sought to convince whites that African American advancement was linked to the national self-interest:

Voice 1: Democracy cannot limp along with a ten percent handicap.

Voice 2: Holding the Negro down means the rest of us stay down with him. Keeping him unproductive hurts America. Keeping him in ill health hurts America and the cause of democracy for which we are fighting.[86]

Huntley expressly sought to assuage white concerns about the pace of racial change. Throughout the series, he repeatedly tried to reassure whites who were fearful of the impact of greater black freedom on the racial status quo. One show on the cultural contributions of blacks consisted entirely of a reading of Langston Hughes's poem "Freedom's Plow," which had been written for a National Urban League broadcast the previous year. Huntley used the poem's narrative of progress and patience to support and reinforce his own modest interpretation of African American goals for advancement. He told his listeners that "the Negro expects no overnight remedy, no instantaneous change, no sudden correction of discrimination by man-made laws."[87]

In fact, no other theme in the series received greater attention than the fears of whites about a time when African Americans would consider themselves their social equals. The series repeatedly cautioned African American listeners to accept new opportunities without overstepping existing social and political boundaries too quickly. In the first broadcast, for example, a white character declared: "I don't want an 'uppity,' smart-aleck black boy here. The way I see it, is that the Negro has to give way in this whole thing as much as the white man. I don't mean I want him to stand around with his hat in his hand and call me mister, but I do want him to take his rights in stride and just go on and work as though nothing much had happened." Consider this conversation between Doc, who was white, and Jim, who was black:

> Doc: But what about the Negro who gets a taste of equality and then gets a little out of line?
> Jim: He's a problem, and every Negro leader of any merit isn't overlooking that reaction.
> Doc: I know . . . we can explain it, Jim. He's known Jim Crow laws so long that when they're gone, he just goes too far, is all.
> Jim: We Negroes must take our advancement in dignity and majesty . . . not in bowing and scraping nor in obtrusive, aggressive celebration.

In another episode, a character advised African Americans who got good jobs to "accept [their] improved economic status with reserve and dignity. Over aggressiveness or brash conduct wipes out any advancement."[88]

Huntley returned to the familiar immigrant melting pot paradigm as he tried to create a calming vision of a future in which African Americans were accorded full rights as Americans and accepted as equal members of the body politic. The message of the final broadcast in the series was a throwback to the rhetoric of the 1938 series *Americans All, Immigrants All:* "Fellow Americans! Take pride in what has happened here! They came to this land as bohunks, wops, spicks, greasers, cockneys, cousin-jacks, micks, chinks, slaves and kikes. But read the honor roll of Americans and note the names: Patrick Henry, Albert Einstein, General 'Ike' Eisenhower, Arturo Toscanini, Pisudski [*sic*], Lin Tutang [*sic*], Booker T. Washington, and so on down the immortal list. We're a bit of every race and every people on earth and we have reason to believe that it's been a good idea."[89] The show continued the theme of an earlier broadcast devoted to a choral rendition of "Ballad for Americans," first popularized by Paul Robeson in the late 1930s.[90]

Because *These Are Americans* was a western regional broadcast rather than a national one, the show's creators were not working in a racial frame of reference that was dominated by worries about an imagined southern white reaction, as were the creators of many coast-to-coast shows. Overall, the show spent relatively little time justifying the extension of full rights to African Americans and devoted most of its attention to the issue of how to manage the resulting transition in race relations. Huntley was therefore able to address more directly white fears about black progress and the inevitable change in social relations that would follow. This is most obvious in the recurring cautions to blacks not to gloat or overstep new racial boundaries. But even this appeal for racial gradualism envisioned exactly the kind of future that southern congressmen had protested so vehemently in their 1943 showdown over the OWI pamphlet *Negroes and the War*. Those white men recognized as clearly as Huntley did the broader social implications of full opportunity; they had recoiled from such images just as quickly as black Americans had embraced them. Since it did not have to dodge imagined southern white fears, this western regional radio series was able to offer a melting pot vision that included "the Negro," but not without encountering resistance from some white listeners who, although they were not southern, shared many of the same views. Indeed, this is one reason why the series had as its central mission the reassurance of whites fearful of changes in the racial status quo. Huntley revealed in a later interview that during this period a man wrote to ask him the definitive racial inquiry: "[W]ould you want your daughter to marry a Negro?"[91]

Wherever they lived, most white Americans harbored deep fears about changes in the existing racial regime, and the specter of equal social and sexual relations with blacks was at the root of many of these fears. Subsumed under the rubric of social equality was a nexus of fears about the implications of political and economic equality—a loss of value for the currency of American whiteness. When the radio networks entered the terrain of racial equality, those fears governed the content of their broadcasts. With few exceptions, national radio programming remained under the control of powerful whites who, even if they were opposed to racial injustice, remained unwilling to permit African Americans to speak freely for themselves on the medium.

The National Urban League's appeal to the networks to grant it broadcast time to respond to the riots was overshadowed by what CBS had already done. League officials instead turned their attention to urging local affiliates to call special meetings of local leaders from labor, politics, and civic groups to "prepare counter offensive of public opinion" to prevent further violent outbreaks. The league decided to dedicate its upcoming October national annual conference to the theme "Victory through Unity" and to make it a publicity centerpiece for interracial unity.[92]

In preparation for the conference, Ann Tanneyhill approached officials at NBC and asked for national airtime for the league's use during the annual meeting, having worked with the network earlier that year when it broadcast "Freedom's Plow." NBC officials, who had seen CBS's success with "Open Letter," accepted Tanneyhill's proposal and granted her fifteen minutes of airtime. Tanneyhill, who had a penchant for lining up popular entertainment figures, asked actor Edward G. Robinson if he would be willing to read a message about the league on the radio. Robinson agreed and asked Tanneyhill to write something suitable for him.[93]

Unlike the league's earlier shows, this one did not include music, entertainment, dramatizations, or interviews but simply featured Robinson speaking alone in conversational tones. Introduced by an announcer as the "great dramatic actor and distinguished American citizen," Robinson read his "special message," an understated plea for racial tolerance. Robinson's basic appeal rested on concerns that racial violence would prolong and weaken the war effort. Casting "morale" as a "force" and a "weapon" in modern warfare, Robinson held up racial intolerance as antithetical to national morale and ultimately to victory itself. He argued that if and when victory came, it "will have been the victory of Americans of every shade of color, political and religious belief. It will not be an Irish victory, nor a Czech victory, nor a black, nor a white, nor any other kind of victory —just an American victory." He addressed more subtly the issue at hand: prejudice and hostilities against African Americans, "a people who have knocked at the doors of industry and government for the chance to make their contributions."[94] After citing several compelling examples of black heroism during the war, Robinson made his most direct appeal to white listeners:

> The fellow whose skin is black, but to whose welfare you probably never gave too much thought, is on your side. Your battle is his battle! His battle must be yours! What is his battle? It's a battle for freedom of opportunity, security of living, happiness for his family, a future for his children. Well, these are the things that all of us want. The Negro American wants them just as deeply, and as rightfully as the rest of us. . . . Contrary to common belief, the Negro doesn't always sing his troubles away—nor does he always laugh in the face of adversity. Boil it down and you find that the Negro is not unlike the rest of us.[95]

In his closing remarks, Robinson emphasized that a lasting peace depended on the willingness of the American people to accept "that an unmolested life, an untrammeled liberty, and an unhampered pursuit of happiness are not just a dream, but a practical program."[96] This show made no direct reference to the spate of violent episodes that had precipitated both the broadcast and the expanded "Victory through Unity" campaign. But the racial violence created an atmosphere of urgency that permitted the directness of the political arguments Robinson made.[97]

League officials reserved their anger about the riots for the pages of the October 1943 *Opportunity*. The magazine's editorial page was packed with ardent political criticism unlike any heard in league broadcasts. League board member William Baldwin placed the riots in an international perspective. He noted that "Detroit, Harlem, Mobile and Beaumont are but thin strands in comparison with those of Burma and Malaya in the crazy quilt of the white man's pretensions to innate superiority in a world predominately colored and now awakened and on the move; but our local 'incidents' have a high visibility which cannot be concealed from the world at large, or even from us, by any techniques of camouflage or artificial blackout."[98] Once again, Lester Granger used the magazine as a forum for blasting federal leadership. He described the riots as the "inevitable outcome of a *laissez faire* policy followed by governmental leadership." Granger made this prediction: "Historians . . . will look back upon the present phase of our racial relationships with unbelieving wonder. They will find it hard to understand that a great nation, fighting for its very existence in the bloodiest war of all time, should have been forced to depend upon leaders so dismally incompetent in solving a fundamental problem of national unity and civilian morale."[99]

The league's contrasting timidity on radio reflected a pragmatic decision to use the medium primarily to draw general attention to its work because league officials knew that the national airwaves simply could not be used to advance political arguments like those found in *Opportunity*. League officials understood that retaining their "guest" status on national radio depended on the use of vaguely worded general appeals, as well as the crisis of war itself. Strongly worded sentiment had no place on the broadcasts the league made to a general national audience, but the recurring vivid contrast between the two levels of rhetoric does raise a question about whether the league's true political position rested at either extreme or somewhere in the middle.

The league shows broadcast in 1943 were the high point in the league's use of national radio during this period, although Tanneyhill and Granger continued to try to secure additional long-term commitments from the networks. After her success with "Heroines in Bronze," Tanneyhill had become more convinced than ever of the "wide possibilities for interracial education which can be developed through the radio," which led her to propose longer-running series about African Americans.[100] Although the riots had compelled CBS and eventually NBC to carry programs about the need for racial unity, that spirit of commitment was short-lived. Broadcasts about blacks and race relations seemed destined to remain restricted to the narrow confines of special programming; the idea of a longer-running series apparently exceeded the level of commitment the networks were willing to make to the league or to anyone else.

In 1944 and 1945, league officials produced two other special broadcasts, neither of which lived up to their expectations. In 1944, Tanneyhill envisioned a show that would "emphasize the contributions and sacrifices being made by Negroes in the war program, and the need for elimination of remaining racial discrimination that weakens our national morale, impedes our Victory effort, and endangers our hopes and aims in the post-war period."[101] The broadcast, however, did not achieve these goals. NBC carried the half-hour show, which returned to the comfortable formula of speeches interspersed with musical entertainment. That show did not elicit the volume of mail the league had received in response to previous shows.[102] It seems likely that the novelty of the league's national broadcasts and

*Radio commentator H. V. Kaltenborn, National Urban League official Ann Tanneyhill,
pianist Hazel Scott, and a member of the Charioteers preparing for the 1944 National
Urban League broadcast. (National Urban League Collection, Library of Congress)*

its timid radio message were wearing off. For 1945, Granger stressed
to ABC that the league wanted to make an annual broadcast "devoted to the vocational and economic development of Negro war
workers and veterans, and to interracial harmony." [103] Tanneyhill explained that the show must "point to the possibilities for continued
use of Negro labor in the post-war period," but she played no role
in writing the script, which was marked by incongruity and the virtual absence of a political message. [104] After the broadcast, Granger
thanked the network for airing "the type of message we wanted con-
veyed," although that hardly seemed to have been the case. [105] At the
same time, knowing that the war was drawing to an end, Granger
tried to secure a permanent place on national radio for the league.
He asked the networks to commit to a regular series of league broadcasts, a request that went unmet since the overriding rationale for
that kind of commitment—the feared impact of racial disunity—expired with the end of the war crisis. [106]

"The Story They'll Never Print"

The constraints under which the league sought national airtime became even more apparent after the war ended and when the organization attempted to use national radio to address one of its most pressing concerns: postwar employment opportunities for blacks. League officials wanted to raise public consciousness of the plight of black workers who were quickly being dismissed from jobs in the shrinking postwar economy. Rather than producing another show on its own, league officials in 1946 asked the respected writer Erik Barnouw to draft a script for a half-hour dramatic radio show appealing to large companies to hire blacks. CBS, the most reliable supporter of league broadcasts, agreed to carry the show.

Barnouw's script was a textured rendering of the preparations of a plant as it began the process of introducing blacks into skilled positions. In the script, a consultant from the league worked closely with plant officials and potential black employees to smooth the way for the employees' integration into the workplace. At the same time, plant officials attempted to assuage the fears of white employees. On the fateful day when blacks were to begin work at the plant, a newspaper reporter showed up to cover the expected violent confrontation between resistant white workers and the black newcomers. But no hostilities erupted, and the newspaper had no story: "Why should they? No blood was spilt. No bones broken. A reporter lounged about but nothing happened, so he went home. Nothing happened? No, that's not quite right. . . . Here, in a factory, mankind moved forward; it moved an inch, and did not slip back." With this ironic and optimistic ending, Barnouw concluded his script, entitling it "The Story They'll Never Print."[107]

After CBS officials reviewed it, "The Story They'll Never Print" became the show they would never carry. Barnouw later recalled that the script was among the best he had ever written and that officials at the network also liked it very much. But network officials felt that the script was too hard-hitting on the racial question for its national audience, and they told him they "didn't dare produce it."[108]

Quite surprisingly, however, the War Department approved the use of a recorded version of the play for the Armed Forces Radio Service. But Barnouw recalled that he learned in a later conversation with Samuel Newman, who had worked for the radio service during the war, what happened when it was time to air the broadcast: "So it went through all the clearances and it finally reached this point of a glass master from which the wax was going to be made. And Newman was called in to his superior who had the disc on his desk and said, 'Now listen, Sam, I'm not going to have any of this nigger-loving shit on this network.' And he took the glass and shattered it all over the desk. So it went down the drain 'for technical reasons.'" [109] This response is consistent with the army's decision the year before not to produce "The Glass" for broadcast on CBS because it intruded too deeply into civilian racial matters. In this case, a year later, both the network and the army drew lines on the race question. In both cases, what is most clear is that familiar wartime appeals for national unity had lost their usefulness and no substitute considered politically safe for national airing had emerged. Faced with the looming postwar question of how black loyalty and service were to be rewarded, both CBS and the army pulled back because they feared that the most obvious answers—integration and full opportunity—were simply too provocative to be advocated in a national broadcast.

Minority Opinion

As part of its overall media strategy during the postwar period, the league encouraged its local affiliates to pursue radio time on local stations, and many succeeded, with varying degrees of effectiveness. One example of an extraordinary local program was *Minority Opinion,* a series begun in 1945 by Sidney Williams, the energetic and aggressive director of the Cleveland Urban League. Carried on local ABC affiliate WJW, the show broke with the traditional formats of most other broadcasts about blacks, including those of the national league. It consisted of interviews with African American writers and political activists and provided in-depth analysis and commentary concerning problems facing the black community in the postwar period. The show began broadcasting in October 1945 and was heard at least through March 1946. The fact that this was a local rather than a national broadcast allowed Williams to "narrowcast" to a sympathetic northern urban audience that had grown in stature and size because of the wartime migration of African Americans. Although he was confined to a local station, Williams dedicated his time on the air almost exclusively to national rather than local political issues. [110]

Williams served as the show's moderator, interviewer, and political commentator. The first broadcast in the series was a vivid and imaginative narrative of the birth of the interracialism of the National Urban League itself, linking it to black migration, economic dislocation, and racial patterns during the period prior to World War I. Williams focused on particular problems facing blacks in urban areas like Cleveland, especially housing shortages. The first of two programs on the problem of housing shortages outlined the economic and racial causes of the shortages and the drastic social consequences of overcrowding. In the second show, Williams described permanent solutions to housing shortages for blacks, including pending federal housing legislation, the elimination of racially restrictive housing covenants, and an end to discriminatory lending practices.[111]

Williams also explored other pressing national issues of concern to African Americans through interviews with a wide variety of public figures, writers, and journalists.[112] One show featured an interview by Williams with writer and activist Carey McWilliams in which the two men discussed the responsibility of what Williams called "our makers of public opinion"—the press and the movie and radio industries —for improving intergroup relations. McWilliams announced that members of the screenwriters' and radio writers' guilds had signed an agreement "not to write scripts which portray Negro people and other groups in stereotyped situations." He concluded by predicting that the current period, 1945 and 1946, would be a time of testing in race relations: "We are going to have to decide whether we are going to preserve, consolidate, and extend the gains made in the war period, or whether we are going to revert back to the pre-war status-quo—as a matter of fact, I don't think we can adopt the latter policy because I do not think there is a status quo. The war undermined that—there is no racial status quo."[113]

Other shows emphasized the political ties between African Americans and other people of color around the world. Paul Robeson, then chairman of the Council of African Affairs, linked the struggles of African Americans with those of colonized people the world over, in "Africa, the West Indies, India, China."[114] This theme was repeated in a broadcast on the black press. There, journalist Horace Cayton noted the "world-wide scope of the Negro Press" and its concern "with subjected and subordinated people—and, especially non-white people all over the world." William Walker, the editor of the *Cleveland Call and Post*, argued that black Americans not only had a "deep community of interest with all minorities in the U.S. and throughout the world" but also had a special responsibility "to help build a world in which all the minorities can weave their peculiar contributions into a creative whole."[115]

Williams also occasionally explored political issues without the use of interviews. On the first anniversary of Franklin Roosevelt's death, his program "Jim Crow Is On the Run" provided an imaginative account of Jim Crow's birth and life. In it he argued, as C. Vann Woodward would in the 1950s, that legally sanctioned segregated practices were imposed well after the Civil War rather than being a continuation of antebellum racial practices. Williams also reminded his listeners that the South was not Jim Crow's only home:

> Jim Crowism spread throughout the economic and political and social life of America like a malignant cancer, eating away the cohesion and unity of our nation. Statutes and regulations were passed in the south requiring that negro and white people be separated. . . . [T]he intent was to make impossible social contacts between white and Negro people. . . . There is no such thing as "separate but equal." Let's be fair to the South and admit, with shame, that Jim Crow has invaded the North too — sub rosa. What other than Jim Crowism is our segregated housing? The quota systems in our colleges and universities? The discriminatory employment practices of our industries and businesses?

Williams ended the broadcast with a lively summary of the achievements of Roosevelt's New Deal and, particularly, the changes in racial policies that had come during the war. Roosevelt's spirit lived on, Williams argued, and could be memorialized best through continued efforts to make the Fair Employment Practices Commission permanent and abolish the poll tax.[116]

In these ways, Sidney Williams, working on an independent local station in Cleveland, broadcast programming that was more consonant with African Americans' views of the real questions of postwar race relations. He used a more serious and weighty radio format that included interviews, book reviews, and commentary, without any need to entertain or hedge. According to Ann Tanneyhill, Williams was seen by some at the national office as a "radical," but the views he solicited and expressed on *Minority Opinion* were far more consistent with those being advanced in the pages of *Opportunity* than were the views carried on the league's national broadcasts. Williams, operating on a local station, was able to say out loud on the air what officials of the National Urban League felt free only to write.

As Ann Tanneyhill remembers it, the networks "never told us what we could and could not do" and did not censor the league's scripts.[117] But incidents surrounding the rejection of "The Story They'll Never Print" remind us how well league officials knew and abided by the parameters that governed their wartime radio scripts and programs. Although Granger and others in the organization were involved in the ongoing campaign for racial reform, there is little to indicate that the organization, which was so deeply wedded to the notions of interracialism and corporate support, wanted to mount an overt challenge to the networks' conventional approach to racial discourse.

At the same time, league officials practiced what advertisers called "segmentation"—that is, they purposely targeted variations of the same basic message to different audiences, altering language and tone accordingly—albeit as much out of necessity as part of a general strategy. This is seen clearly in the pages of *Opportunity,* where Granger in particular wrote scathing editorials protesting the discriminatory practices of the federal government, the War Department, and unions and private industry. Yet league officials did not appear to be eager to push the boundaries of political discourse about race and endanger their limited free access to national radio. The language and tone of league national broadcasts actually changed very little from 1941 to 1945. The first show in 1941 had made a basic plea for fair play and equal opportunity, as did its final, aptly named 1945 broadcast, "Too Long America."

League officials tried to use their national radio programs mostly to draw attention to the organization and give it a more visible national presence among both blacks and whites. Broadcasting its name and its general goals to millions of listeners across the country did just that, as league officials stressed in a 1945 issue of *Opportunity* dedicated to a retrospective of the organization's first thirty-five years. Illustrated by two photographs from the broadcasts in 1944 and 1945, the accompanying text stressed that "nationally known news commentators and radio and motion picture stars have helped to bring the annual VOC to the attention of a nation-wide audience."[118]

Speaking half a century after her experience with the league's radio broadcasts, Tanneyhill emphasized what she saw as their enduring effects. Of most obvious significance was that the use of radio "drew attention to the Urban League and to the problems people were facing" and helped elevate the league's stature during the war period. Equally significant, although for different reasons, was its 1943 broadcast of "Heroines in Bronze." Never before had the achievements of African American women received the amount of national media attention that show and the simultaneous issue of *Opportunity* generated. Listeners and readers were presented images of black women they had never encountered in popular representations, images that would remain rare on radio and other media. Finally, she credits the success of shows with opening the way for African American entertainers in the radio industry as a whole.[119] The league's first show, "The Negro and National Defense," not only served as a model for many other wartime appeals featuring black talent but also brought to the air an all-black show with tremendous popular appeal, illustrating that the medium's virtual exclusion of black talent was shortsighted.

These shows also constituted part of the emerging public discourse on equal opportunity and race prejudice in the World War II era, although their exceptional nature means that their lasting impact on public opinion was limited. Yet these broadcasts and the stories of their production and reception give us a glimpse of radio's potential as a medium for "positive" racial propaganda.

CHAPTER 5

RADIO AND THE POLITICAL DISCOURSE OF RACIAL EQUALITY

Popular national political forums were among radio's most prominent features in the war era and one of its many gifts to television. All of the networks had some version of this public affairs format, and of these, the *University of Chicago Round Table* and *America's Town Meeting of the Air* were the most popular, respected, and influential.[1] Because of their continuity, these two shows are particularly valuable sites for observing how over the course of a decade the political subject of race, first deemed unspeakable, came to be aired and then rose to prominence as a national issue. Quite literally, these broadcasts chart the evolution of a permissible political discourse about racial oppression, a development that provides insights into the fashioning and limitations of the white liberal response to the emergence of the civil rights movement.

African Americans waged a mind war against the shameful paradox of a segregated democracy during this period, although it would take two decades of mass protests, litigation, and deaths to overcome virulent white resistance to dismantling its edifice. On a rhetorical level, the discourse of racial equality was challenged by a discourse of white resistance, a fight played out before a national listening audience.

The concerted assault by African Americans on the conceptual world of racial segregation and the airing of a new political narrative on race have been overshadowed by their legacy: the dramatic battles and victories of the 1950s and 1960s that would be carried not on radio but on television.

Airing the Race Problem

NBC's *University of Chicago Round Table* began in the early 1930s as a local program broadcast from the campus of the University of Chicago.[2] In 1937, the university hired William Benton to fill the new post of vice president for public relations, whose responsibilities included oversight of radio operations. Benton was the premier practitioner of radio advertising and founder of the advertising firm Benton and Bowles.[3] *Round Table* pioneered a format in which faculty members engaged national political figures and journalists in discussions about pressing national political and economic issues. The program built up a loyal nationwide listenership that it reinforced by distributing printed transcriptions and bibliographies for each of the weekly Sunday shows. *Round Table*'s audience grew rapidly, rising from 1.5 million in the 1930s to approximately 10 million by 1941. Although other programs competed with it, *Round Table* under Benton's leadership earned a reputation as the most stately of the panel discussion shows, becoming recognized as the "intellectual's radio paradise."[4]

Attempts to introduce the race question into this paradise repeatedly met with defeat, and the resistance to those attempts offers valuable insights into the volatility of the race question over the course of the decade. Sherman Dryer, director of radio for the university, believed that the program had a public responsibility to confront the race issue and that if it did not do so it would risk accusations that it was afraid to take on controversial contemporary topics. Under Dryer's leadership, the program staff approved, scheduled, and publicly announced a broadcast in 1939 with the provocatively simple title "Is the Negro Oppressed?" Answering that question would not be easy. A black newspaper announced the upcoming broadcast with a description of the show's conflicted intentions: "Lynchings, Jim Crow laws, and other evidence of discrimination against the Negro in the United States will be compared with the advances made by the American Negro with the assistance of government and individuals interested in more than legal emancipation of the Negro." Walter

White of the National Association for the Advancement of Colored People (NAACP) and University of Chicago sociologist Louis Wirth were expected to appear on the show. Wirth would argue, according to this report, that "Hitler has accused America of ignoring its own minority problem" but that progress for blacks had been made "principally by the efforts of the Negro himself." This meant, according to Wirth, that "the Negro is not one of our greatest problems."[5] The show never aired and was replaced by a special on the coal miners' strike.

Dryer persisted in trying to get a show about blacks on the air. After the cancellation, Dryer argued to Benton that, "purely from a public relations angle," the series had to include a show about blacks or it would become the target of a protest campaign from the "left-wingers." He explained: "Influences will begin seeping out from New York, through the Daily Record–Daily Worker on the one hand and through the National Association for the Advancement of Colored People on the other, the net effect of which will be to undermine the prestige of the Round Table. When I say undermine, I mean not only with the so-called liberals but the academic conservatives as well. If once our own Faculty feel (as some of them do already) that the Round Table had cut off a program because of external influences, even the most conservative members would tend to look at it askance." Dryer suggested ways that such a broadcast could be kept "innocuous," such as adding a white southerner to the show or limiting the discussion to "the economic position of the Negro in the Northern cities." He advised that the title be changed to the more neutral "Today's American Negro" instead of "Is the Negro Oppressed?," although he confessed that he liked the drawing power of the more controversial title, which "would have a million dials twisting our way."[6]

Benton's work as an advertising executive made him quite cautious about risking negative audience reaction to any direct discussion of blacks, a point of view shared by officials at NBC. This is richly ironic because Benton had been the person who had convinced Pepsodent to sponsor the local show *Amos 'n' Andy* for national broadcast on NBC, a decision that brought enormous profit to NBC and the two white men who played the black characters on the show.[7] But there was a difference between depicting African Americans and airing a show that actually discussed their place in American political life. Benton asked NBC president Niles Trammell for advice on the idea of airing a "Negro show," letting him know that the university had received quite a few letters protesting the cancellation and that "the Communist papers and other left wing papers have been nipping a bit at our heels."[8]

Judging by their reaction, NBC officials had been unaware that any show on blacks was even under consideration. NBC vice president John Royal advised Trammell to prevent the show from being broadcast: "Today's American Negro, Today's American Jew, Today's American Catholic, Today's American Irish, would all be difficult and dangerous subjects to be discussed by the Round Table. The fact that the Communist papers and other left-wing papers have been nipping at his heels is not reason why it should be done, but perhaps a very good reason why it shouldn't be. Anyone who knows what's going on in this country realizes that the Communists are making a very strong play to arouse the Negroes in America."[9] Royal disparaged Dryer, describing him as "a young man with radical or at least 'broad minded' tendencies," and vowed to keep him from maneuvering "us into some embarrassing positions." The question of "the Negro," Royal told Trammell, simply had to be left alone: "They would *like* to have a Southerner on, but you, as a Southerner, know more than anyone else that you cannot *discuss* the nigger question." Royal recommended that the network cancel *Round Table* altogether if it decided to go forward with the show.[10] Trammell agreed with Royal's assessment. He called Benton directly to make the network's position clear, later informing Royal that Benton "understands our position and I don't think you will hear from this matter again."[11]

Silence on the race matter was enforced for another three years until American entry into the war emboldened Dryer in May 1942 to raise once again the idea of a "Negro Round Table." Dryer explained that Archibald MacLeish at the Office of Facts and Figures "now smiles upon a discussion of the Negro." Indeed, MacLeish had made a closed-circuit address to the radio industry in which he spoke of his fears of the dangers to national security of continued low black morale.[12] Taking MacLeish's remarks as federal imprimatur for radio to act, Dryer argued that as the most popular show of its type, *Round Table* had "a patriotic obligation to treat this topic" because "we can do more with one broadcast on 'The Negro' than probably a score of certain other 'national' programs, more than a hundred local programs." Dryer emphasized that the declaration of war had changed everything: "[T]he Negro problem today is not, as it was before

December 7, a Southern problem. It is now a nationwide problem, which the government has officially recognized." Certain precautions and "practical considerations" would be necessary, he warned. One participant had to be a white southerner, and no panelist should be black, "for if a Negro is on the program, whatever good things we have to say about the Negro race will be construed by a lot of people as something we couldn't avoid because of the Negro's presence."[13] Implicit here is Dryer's fear that plans to include an African American panelist would automatically doom the broadcast just as surely as ignoring the southern white perspective would.

Encapsulated here once again was the struggle about who was to speak on the "Negro problem," who was to serve as an expert at a time when African Americans were excluded from the symbolically equalizing formality of political discourse. Southern politicians and journalists still claimed sole legitimate authority concerning "the Negro" and most other white Americans deferred to this claim. When fears had surfaced among federal officials at the Office of War Information (OWI) and the War Department about their broadcasts about race relations, the specific object of concern was southern congressional control of key appropriations committees. But the immediate concern of *Round Table* was to avoid offending southern white listeners' sensibilities, to which it ceded veto powers.

Dryer sought advice from federal officials on how to proceed with his proposed show in June 1942. He turned to his liberal friend, Charles Siepman, in the Radio Division of the Office of Facts and Figures at the OWI. As discussed in chapter 3, Siepman was battling to convince his agency to sponsor broadcasts on the race question. Although he and Dryer were in agreement on the need for radio to act, he cautioned Dryer about the risks of talking about a problem that had no political solution:

> It is obviously a delicate subject, the action necessary to the solution of the problem being as yet absent. An airing of the issues involved may prove the most helpful contribution that can be made over the radio as long as this is done in terms that do not provoke hotter feelings than at present exist. Any decent person will sympathize with the negroes [*sic*] aspirations. Many will recognize and with distress, the anomaly of his position under the law and in a free democracy. But many too, will realize what deep seated prejudice and what a long tradition lies behind this unhappy story. . . .

To arouse false hopes would be as dangerous as to inflame violent passions.[14]

Siepman's warning was unnecessary since Dryer's efforts to bring a show on African Americans to the air were for naught. Benton again had sent a copy of Dryer's suggestions to Trammell, asking him once more for guidance but without taking any position himself: "Does Dryer make his case or is it your feeling that we'd better pass up this subject?" Two months later, Benton reported to Dryer that Trammell had told him that "he doesn't want to face this issue. He doesn't want us to face it." Trammell feared "all kinds of trouble and tribulation in the future around this issue." Benton asked Dryer, for the sake of the series, to "drop the matter," predicting that "if we decide to press this further, we'll have a major issue with NBC."[15]

The subject might have been dropped, but it would not go away. In August, Edwin Embree of the Chicago-based Rosenwald Fund called Benton to suggest that the series include a show about the poll tax, with Senator Claude Pepper as a participant. Benton replied that the subject would need to be broadened to include all "interferences with the democratic franchise," presumably to keep its focus away from the South and the race question. When he passed the information on to his colleagues at *Round Table,* the reaction was negative, based on the realization that "any broadcast on this subject will be in substantial part a discussion of the Negro question," which remained an issue to be avoided.[16]

Finally, after four years of internal conflict and delay, *Round Table* could no longer ignore the continued escalation of white fears concerning black activism and the steady migration of African Americans out of the Deep South. In 1943, faced with the eruption of racial violence in several major cities, the show publicly acknowledged the existence of African Americans, albeit only as a problem. Within a four-month period, *Round Table* broadcast two discussions on the "Negro question," the first in April and the second in July. Despite the decision to go forward, there was still serious trepidation about how to frame the issue and how it would be received. For example, to be on the safe side and to avoid any preshow objections, Benton cleverly suggested that the announcement for the first show include no reference to blacks but that it be given the covert title "Minorities," even though, as Dryer would later point out, "two-thirds of this program was devoted to the Negroes."[17]

The broadcast on minorities featured none, having as its guests Avery Craven of the university's history department; Robert Redfield, an anthropologist and the school's dean of social sciences; and Ralph McGill, editor of the *Atlanta Constitution*.[18] In many ways, the discussion played itself out as representational performance, with each panelist speaking on behalf of an imagined constituency. Although he was chosen to represent the South, McGill took on the role of the southern race moderate. A more conservative southern position on race was expressed by Craven, who was a southerner by birth as well as a historian of the South and the Civil War. That left Redfield to advocate more liberal views. What was missing from the performance, although not yet noticeable to the actors, was the voice of the subject itself.

Although they knew the show's purpose was to raise the race issue, the three men alternately approached and avoided the subject. They began with vague introductory comments about the importance of "minorities" to the country's development, with brief mention of Jews, Mexican Americans, Japanese Americans, and "the Negro." After the discussion skirted specifics, Redfield boldly suggested the obvious—that "the Negro" was the "number-one minority" problem and the one they needed most to discuss. Craven agreed and described the "problem" this way: "The Negro does . . . represent the minority group to the nth degree. . . . If ever there is a problem under the sun that had something of history and tradition back of it, it is the Negro problem. If I can judge, there are the other two things—restlessness on the part of the Negro as never before, demanding his rights and recognition as an individual, and there is also a stimulation of fears on the part of the dominant group. The majority, in other words, are as much disturbed over the race question as the Negro himself."[19]

It was this duality—black restlessness and white fear—that had finally rendered the issue of race too urgent to be ignored. McGill confirmed Craven's definition, but he added that African Americans were dissatisfied because they wanted their rights and that they should have them. Craven and McGill quickly agreed that the "Negro problem" was a national matter and no longer simply a southern concern, as if to protect themselves (and the show) from charges that they were "interfering" in the South's business. Furthermore, Redfield added that "what we do with reference to the Negro is attended to by persons in all parts of the world today, including our own allies of a different color."[20] This laying claim to the race problem marked a shift in thinking among some whites as they embraced the idea long advanced by African Americans that the national deference that had been accorded the South on the race issue was harmful to national interests, both domestic and foreign.

Redfield dared to raise the obvious question: "[S]omething has to be done about the Negro—is this the time to do it, and what is it we are to do?" When Craven argued that the war's needs came first, Redfield countered that the two problems—the war and "the Negro"— were intertwined and that solving the "Negro problem" was "also a strong step toward winning the war and the peace to come after."[21] Redfield concluded that "the race problem has now become so important to the security of the nation that the national government must, in some form or another, declare its interest in solving it and implement that interest by appropriate legislation." But when Redfield suggested that enforcement of such legislation be left to the localities, both Craven and McGill predicted failure, with Craven citing Reconstruction as a particularly strong case in point.

The show concluded with agreement that progress on the issue in the South would be most likely to come from the leadership of southern race moderates. In summation, Redfield laid out this consensus statement: "We seem to agree that this racial discrimination is a great evil; that it is in conflict with our democratic principles; that that conflict was never more dangerous, perhaps, than it is today when we are so seriously at war; and that some solution must be found. The difficulty we have is that difficulty of making a reasonable-enough progress in the direction of granting to the minority groups the rights which they should have without, at the same time, endangering public safety by stirring up reactionary resentment and perhaps violence."[22]

The consensus then, it seemed, was that this was a moral and political problem for the nation as a whole and that the war required that something *might* need to be done, perhaps by the federal government, provided it did not create reactionary violence among whites. That, of course, was the crux of the political problem of "the Negro"—what should be done, under whose leadership, at what pace, and with what resistance. The long-avoided *Round Table* discussion about "the Negro," mounted under the rubric of "minorities," had reached the heart of the racial dilemma in 1943.

Even with all of its limitations, this representational drama cap-

tured well certain shifts in white thinking on race relations that occurred during the war, in large part because of the increasing political
visibility of African Americans. One was the idea that the war itself
had nationalized the issue of race relations, especially in the eyes
of the international community. The other was the notion that the
federal government, and not just white southern leaders, had the authority and obligation to address the race issue. The political narrative
on the race question was being reopened, although its ending was
still far from being rewritten. On this broadcast, the solution to the
question was represented as resting in the hands of southern white
moderates; the power of Supreme Court intervention or the potential force of white resistance or that of the "Negro" race itself was
missing both from this drama and from its expected denouement.[23]

Round Table staff members had not misjudged the amount of interest a program about racial matters would draw, even one masked
under the bland title "Minorities." Dryer reported that the volume of
mail sent in response to the broadcast had been extraordinarily heavy,
about 500 letters a day. To reassure his superiors, he emphasized that
"only an infinitesimal percentage of the letters" had been critical.[24]
Most listeners had voiced enthusiastic approval, he explained, and he
noted with some surprise that a number of the letters had come from
black listeners.

The idea that African Americans also were listening to a political discussion show had not occurred to many people at *Round Table,*
and this broadcast about the "Negro problem" had the ironic effect
of making black listeners visible and enlarging the scope of the
show's "imagined community" of listeners.[25] Two prominent African
Americans, Ira De A. Reid of Atlanta University and Claude Barnett of the Chicago-based Associated Negro Publishers, contacted
McGill directly. Both recognized the political significance of the
broadcast, but neither was satisfied with its content. Barnett thanked
McGill for saying as much as he could to a broad national audience: "You must have known that many Negroes, as well as whites,
north and south, some of whom are inclined to be apprehensive of
the southern liberal's attitudes these days, were vitally interested in
your pronunciations. On the other hand, they must have known that
southern conservatives and demagogs were listening with equal attention and appreciated the necessity for careful statement on your
part." Reid was more blunt about the show's weaknesses, concluding
that the discussion about a remedy was "not very fruitful" because

it was so limited, both in content and in duration. He urged McGill to "give this thing further 'airing'" in his daily *Atlanta Constitution* column.[26] As Siepman at the OWI had warned Dryer, discussing a problem that had no current solution carried the risk of criticism, in this case from African Americans who found the program lacking in that regard.

The eruption of racial violence later in 1943 helped propel *Round Table*'s return to the issue of race on a July 4 broadcast. Efforts to avoid controversy for fear of inciting racial tensions seemed moot at that point, so this discussion, unlike the earlier one, was more urgent and, as a result, more daring in tone and content. Under the title "Race Tensions," the show featured as panelists black sociologist E. Franklin Frazier and Carey McWilliams, a leading activist writer on issues of race and ethnicity. Robert Redfield, who had espoused the most liberal views on the previous program, made a return appearance, this time in company more in line with his own views.

Despite the generally favorable reactions to the earlier show on the "Negro question," officials at NBC were no less nervous about airing this second broadcast. Indeed, they were more anxious because of the decision to include a black panelist. On the Friday before the scheduled Sunday broadcast, Judith Waller, director of public service for NBC's Central Division in Chicago, phoned Dryer to warn him that many stations, mostly southern, had notified the network in New York City that they would not carry the program because they objected to the topic or "the participation of a Negro" or both. Waller said that "the South was 'irrational,'" but she insisted that a white southerner be added to the program or that the program be canceled. Dryer suggested that a southerner "introduce" the panel and the topic, an arrangement that was acceptable to NBC. NBC also approved the choice of University of North Carolina sociologist Howard Odum for that role.[27]

Once again the white South was represented on *Round Table* by a racial moderate. In brief remarks broadcast from North Carolina, Odum opened the show by asserting "that our problem of the Negro in America is a southern and national problem." Odum urged "that America, and the South in particular, declare a moratorium on all violence." He ended with this plea: "[O]ur immediate problem here now is to covenant together for some new high morale on the part of all the people everywhere and for a master strategy for the better ordering of race relations in the war and the post-war period."[28] It is difficult to imagine how Odum's remarks about a "better ordering of race relations" could have been intended to achieve the show's goal of appeasing the imagined southern listener. Presumably Odum's representational status as a southerner slurred the starker political implications of his words.

The discussion in Chicago then began in the same way that other shows that reacted to the race riots had begun: by tantalizingly drawing attention to the fact that the subject matter was considered "hot" and taboo. Redfield opened by admitting that some *Round Table* listeners had advised against broaching the topic, saying it was "too dangerous" to discuss (avoiding any reference to the fact that NBC held a similar view). Thus cued, McWilliams and Frazier reassuringly responded that the times demanded the attention and that, in Frazier's words, "an intelligent understanding of the situation is necessary for intelligent action." Redfield insisted that the "objective atmosphere of a university group" was indeed the best place for that discussion, laying claim to the show's special qualifications as an objective site and protecting it from attacks for broadcasting the issue.[29]

McWilliams and Frazier both came prepared with demographic data about various minority groups, including their relative poverty in terms of health care, education, employment, and housing. Both also emphasized the impediments of the color line and, as Frazier described it, a "melting pot" that had "excluded the dark ingredients." McWilliams sought to set the record straight on the "zoot-suit riots," stressing that they were the result of attacks by white soldiers and sailors on Mexican Americans and African Americans and not the reverse. All of the panelists agreed that small steps toward advancement had come with the war but that it was that progress—and the potential for more—that was stirring up racial violence against blacks.[30]

The panel advanced the idea that changes were necessary to address the problem of racial inequality. Frazier insisted that an end to segregation and the sanctioned inferior status of blacks was the first step. McWilliams and Redfield quickly agreed, and McWilliams argued that the Fair Employment Practices Commission (FEPC) be strengthened and a new federal civil rights statute be passed. But the question of how to bring about the changes in public attitudes necessary for federal action to end segregation stumped and silenced this panel of opinionated men.[31] Indeed, this silence was telling, as the issue of segregation would continue to be an issue that divided blacks and white race liberals, especially southern moderate whites.[32] Although most blacks, like Frazier, saw segregation as the primary issue, whites who considered themselves southern race liberals did not follow suit.

After the show had aired, officials at *Round Table* congratulated themselves for having balanced the concerns of network officials and "southerners" sufficiently to get the show on the air. But as Judith Waller had feared, some local stations had refused to broadcast the show: only 86 of the 100 to 120 affiliate stations that normally carried *Round Table* aired the "Race Tensions" show.[33] Like the response to the earlier broadcast on minorities, however, the volume of mail sent in response to the show was heavy, about 1,700 letters, and again Dryer reported that most letters were approving. Dryer as usual tried to cast as positive a spin as possible on the overall response, but even he admitted that many less supportive listeners also had written in, almost all of whom were from the South. After reading a random selection of these letters, Dryer was shocked to find that almost all criticized "the Negro" because "he was a menace to white women." That topic certainly had not been discussed or alluded to in any way, but it was clearly on many southerners' minds as the unspoken and unspeakable implication of the changes in the racial order envisioned by the panelists. The mere advocacy of ending segregation was heard by these listeners as the equivalent to arguing for intermarriage.

This reading of sexual transgression was related to the political transgression of having a black man on the show as an equal participant with white men in advancing political solutions to the race problem. Indeed, the Tennessee father of one of Benton's neighbors registered a protest about the broadcast that captured just that sentiment. The man objected vehemently because "the colored participant addressed the other two participants by their last names without a 'Mr.'"[34] Southern racial etiquette had been violated not only by Frazier but also by the two white men who allied themselves with him as coequals.

Frazier's presence on the show was as much an affront to prevailing racial etiquette as it was a divergence from familiar aural representations of African Americans; he was, after all, no Amos, Andy, or Rochester. Vigilant white listeners, ever on the lookout for blacks' assertions of social equality, heard such a claim in Frazier's voice, his erudition, and the assumption of his authority to speak forcefully on behalf of African Americans against segregation and discrimination.

This was not a misreading or an overreaction on the part of those listeners, for they understood well that Frazier's presence represented a powerful symbolic shift. His arguments signaled that the prevailing political narrative about race was losing some of its potency and that African Americans, although long rendered voiceless, were laying claim to a place in that process. One supportive listener wrote Frazier that he had been listening to the show for many years, hoping to hear just such a discussion. He thanked Frazier for being such a "capable representative" and for his "profound and enlightening presentation," sentiments he also expressed in a letter to *Round Table*.[35]

Dryer tried to protect Frazier from the racist remarks in some listeners' letters about the show. He thanked Frazier for his "excellent participation," but he was purposely vague in describing listener reaction, telling him only that the response had been extraordinary and that "most" listeners agreed that the show was among the most "stimulating and socially important." Dryer shared with Frazier his hope that "not too long a time will pass before you'll face our microphone again."[36]

This hope would go unrealized. After these two shows in 1943, *Round Table* was almost completely silent on the subject of American race relations, despite the extraordinary public attention the subject received in this period.[37] Indeed even the appearance of Gunnar Myrdal's *American Dilemma* in 1944 failed to motivate this "intellectual's radio paradise" to revisit the questions that had been so eloquently raised by panelists on its two broadcasts in 1943 and that were now encapsulated in Myrdal's encyclopedic treatise. Myrdal's book, like the two shows, had the effect of airing, describing, and validating the race problem, but both the book and the shows faltered when the discussion shifted to solutions.

Although it was no longer taboo or controversial to admit that race was a national political issue, the question of remedy or intervention remained as elusive and volatile as ever. Only when the federal government acknowledged in 1947 the need for a remedy would this series be drawn back to search for solutions in the political minefield of American race relations.

America's Town Meeting of the Air

America's Town Meeting of the Air, a New York City–based town hall–style political discussion program, had a much livelier and less pretentious tone than the staid academic atmosphere of *Round Table*. A descendent of the suffragist-founded League for Political Education, the town hall discussion meeting originated in 1921 and was brought to radio in 1935. Hosted by George Denny, a former drama teacher and professional actor, *Town Meeting* was intended to be a nation-wide version of the old New England town meeting. Guests on the show debated controversial issues in front of audiences of over 1,000 people who were allowed to ask the panelists questions. As in *Round Table*, listeners could obtain weekly transcriptions of the program, which *Town Meeting* used in an aggressive public outreach campaign, actively promoting the use of its broadcasts and transcriptions in schools and the hundreds of listening and discussion clubs that gathered during the show's weekly broadcasts. The show also took to the road for half of the year, broadcasting live from cities around the nation. Although both *Round Table* and *Town Meeting* were broadcast initially on NBC, *Town Meeting* seemed to have enjoyed far greater independence from the network than *Round Table*, and for that reason, it followed a different path to the racial issue.

Panelists first debated the issue of racial inequality in 1941 on a show broadcast live from Birmingham, Alabama, on southern economic problems. Although Mark Ethridge, the moderate editor of the *Louisville Courier-Journal*, appeared on the show to cast the North as a colonizer of the South and to blame northern economic oppression for part of the South's problems, he also emphasized that he refused to defend the South for "the KKK, lynchings, floggings of union organizers, violations of civil rights," for poll taxes and white primaries, or for "imitating" Hitler. Columnist John Temple Graves admitted that the South had "sinned against the Negro," but he argued that the "number one problem of the Negro" was the "man across the sea," whom he called "the greatest race hater in history, the Jim Crow of all ages." This was as aptly ironic a description of Hitler as any. As argued earlier, Hitler's brand of white supremacy conjured up resonating visions of American white supremacy—for African Americans and here some southerners as well. Ethridge's reference to "imitating" Hitler and Graves's attempt to compare Hitler to Jim Crow only begged the question of why this American referent was so fitting in the first place.[38]

An editorial in a Birmingham newspaper pointed to the broadcast with pride, calling it a "bloodless battle" of free expression and apparently attaching much significance to the fact that southern whites were willing to disagree with each other in front of a national listening audience.[39] Similar conclusions were reached in an editorial in the newspaper at Birmingham Southern College, where the debate was held. In its extolling of the value of free debate, it barely hinted at the discussion of the economic plight of southern blacks, referred to obliquely as a "few local ills."[40] The on-air discussion and these editorials showed how narrow the southern views on the race question were, even if they alluded to differences of opinion. That narrowness and those differences would drive southern political responses to the questions of racial inequality for the duration of the war and decades to follow. Most significant by its absence, however, was the voice of African Americans themselves, a point lost entirely on the editorial writers in their rush to celebrate the value to democracy of southern white free speech.

Attempts to move *Town Meeting* into a more direct confrontation with the race issue met with little success, and as on *Round Table*, radio officials looked to the federal government for guidance. In 1942, Denny proposed to Archibald MacLeish at the OWI that *Town Meeting* broadcast a program on "the Negro in the defense problem." Theodore Berry, the black lawyer working inside the agency to get radio programming about African Americans on the air, strongly advised MacLeish to encourage Denny.[41] But *Town Meeting* never presented an episode on that subject, and indeed it is difficult to imagine how the series could have addressed such a volatile issue. Fairness would have required the presence of an African American and a government representative, and the broadcast could not have avoided confronting the paradox of fighting a segregated war for democracy. This was too much too soon for *Town Meeting,* as we have seen was the case for the duration of the war for the OWI.

Figuring out how to openly confront the race issue was a puzzle for *Town Meeting,* as it had been for *Round Table. Town Meeting*'s initial foray into the question of race also relied on the tactic of exploring a seemingly neutral subject, but the show was set in a more daring symbolic space. In May 1942, *Town Meeting* aired a show from the chapel of the premier black academic institution of the day, Howard University. The show's guests were all black Howard faculty members: philosopher Alain Locke; Howard's president Mordecai Johnson; Leon Ransom, dean of the Law School; and Doxey Wilkerson, professor of education. In introducing these representatives of the black intelligentsia, Denny hastened to emphasize that although the panelists were all African Americans, they had been asked to deal not with the "race problem" but with the broader philosophical question, "Is there a basis for spiritual unity in the world today?" Despite the designated topic, to those eager to hear the race question aired, the site selection alone served as clue and cue enough, as it did for the panelists.[42]

Locke and his colleagues took the show as an opportunity not only to discuss the philosophy of religion, which they did with vigor, but also to portray racism as an international ethical problem. Locke, for example, characterized as "poor seedbeds for world unity and world order" what he called the "superciliously self-appointed superior races aspiring to impose their preferred culture, self-righteous creeds and religions expounding monopolies on ways of life and salvation." Wilkerson was even more blunt, noting that "in this war in which colonial people play such an important role, the traditional relations of master and subject peoples are being altered. The chain of imperialist slavery tends definitely to weaken." Taking the point further, Ransom asked, "[H]ave we, Negroes and whites in this country, for instance, achieved any sort of spiritual unity? Are we not still enslaved by the idea that one must be dominant and the other the subservient group?"[43]

The audience's questions generated responses from the panelists that were more wide-ranging than the initial discussion. One person asked: "[D]o you agree that the Negro has made his progress in America because of cooperation rather than through his opposed struggles?" Ransom's answer drew hearty applause: "[B]eing a realist, I am afraid that I must say that the Negro has made his progress in America *in spite* of the majority group."[44]

The broadcast from Howard put African American intellectuals on display, where they embraced a cultural and political role, not just through the logic of their arguments but also through their aural presence as articulate, thoughtful representatives of the race. Operating in an educational forum, they could engage in a relatively free and protected level of political discourse. One paradoxical characteristic of the broadcasts from Birmingham and Howard was that the subject of race was being discussed by a group of whites and a group of blacks separately, although both groups were speaking to an integrated radio listening audience. Segregated arguments were being made but with no dialogue or dialectic. For *Town Meeting,* the subject still remained too volatile to be discussed by a mixed-race panel in any setting.

Although it had not yet broadcast a show devoted solely to the race question, *Town Meeting* had not shied away entirely from related controversial issues. For example, whereas *Round Table* resisted a request to deal with the poll tax issue, *Town Meeting* broadcast not one but two shows on the issue, including one with Claude Pepper, whom *Round Table* officials had specifically rejected. The series also aired discussions about problems facing other minority groups, such as the detention of Japanese Americans, the continued restrictions on Chinese immigration, and the prospect of Japanese assimilation.[45] Plainly, the creators of this series viewed controversy as a way to sustain and build audience interest. Taboo subjects stir listener interest, and racial equality was still at the top of the list of such subjects, although that fact alone was not enough to overcome deep fears about how to address the issue more directly.

As was the case with *Round Table,* the escalation of racial tensions and growing political attention directed at African Americans finally drove *Town Meeting* to abandon its caution. Departing from its usual practice of presenting the week's debate topic in the form of a question, *Town Meeting* aired a show from New York City in early 1944 with the imperative title: "Let's Face the Race Question." Following the seemingly standard format for introducing shows on race at the time, Denny opened the broadcast by warning that "[t]onight we're going to discuss a question that is considered by some timid souls to be dangerous—the race question, more specifically, the Negro question." He predicted that there would be "no disagreement among our speakers that we have a race problem. The difference in opinion lies in the way it should be approached." Adding to the air of danger, Denny took the very unusual cautionary step of asking the audience "to refrain from applause or demonstrations of any kind during the program."[46] Special care had also been taken to balance the presentation and debate. The show's panelists were well-known African American poet and writer Langston Hughes; Carey McWilliams, an effective progressive radio presence throughout the decade; John Temple Graves, again representing a white southern point of view; and James Shepard, the president of North Carolina College for Negroes, who expressed a more conservative black southern stance.

E. Franklin Frazier had broken the race barrier on *Round Table* the year before, but on the more relaxed and freewheeling *Town Meeting,* Langston Hughes launched a frontal attack on the race problem unlike anything heard on national radio before. He accused the country of treating black soldiers shabbily, of being "unwilling to provide more than inadequate Jim Crow cars or back seats in buses south of Washington for our own colored soldiers," and of undermining "the morale of Negro soldiers by segregating them in our armed forces and by continuing to Jim Crow them and their civilian brothers in public places." Hughes blasted opposition to social equality as a smoke screen for a profound fear of intermarriage, "as if permitting Negroes to vote in the poll-tax states would immediately cause Whites and Negroes to rush to the altar." That conception of equality, he concluded, had "nothing to do with the broad problem of civil, legal, labor, and suffrage rights for all Americans." What was needed was an "over-all federal program protecting the rights of all minorities and educating all Americans to that effect."[47]

The voice of the South then rose to answer Hughes. John Temple Graves began his rebuttal of Hughes's performance with some drama of his own, silencing the audience to offer a prayer "that nothing tonight will increase the sum total of race hate in America." Graves argued again, as he had on the 1941 broadcast from Birmingham, that states should be left alone to deal with the race problem because "not all the laws this nation can pass, not all the excitement this Nation's race leaders can create, not all the federal bureaus laid end to end, can force 30 million white people in the South to do what they are passionately and deeply resolved not to do in race relationships."[48]

Carey McWilliams took on Graves, countering that Americans "cannot solve this question even in their own communities until it is solved nationally, for the question has now become national in scope and effect, and it now falls full square within the field of federal action." His reading of the war's import contrasted starkly with Graves's plea for the status quo. McWilliams used Popular Front rhetoric to argue that the war was a "world revolution" that had "profoundly altered the relationships and factors involved in what we call the race question."[49]

Members of the live audience eagerly pushed this frank discussion even further. One person asked Graves, "How can you expect the states down south to handle the Negro problem when these states are in the hands of men who don't represent the people?" Denny called that question unfair and excused Graves from answering it, but the very vocal New York City audience insisted on an answer. Graves did not deny that democracy had historically been restrained by race in the South, but he said that such restraints were necessary because blacks so outnumbered whites. McWilliams pointed out that "the Negro minority in the South is declining decade by decade" and asked Graves how much it would have to decrease to satisfy southern whites.[50]

The broadcast generated a large volume of letters and would remain among *Town Meeting*'s most popular shows ever by that measure. The staff seemed relieved that there were so few negative responses to the program, credit for which rested with McWilliams and Hughes, both of whom had amicable styles that softened the political meanings of their arguments for some white listeners. After all, radio listeners heard tone as well as content in these discussions, and one could override the other. Hughes and McWilliams both had managed to project a nonthreatening tone even as they made fairly radical arguments in substance. Indeed, most listeners complimented the show for its fair discussion and the absence of bitterness.[51]

Hughes's appearance sparked an outpouring of personal support from many listeners who valued his message and his tone. They wrote him directly rather than through the network to thank him and to commend his political courage. "Not only did you ably tell how and why Federal action would be more effective than the states' in attacking the race problem, but you so intelligently discussed that delicate aspect of the problem—'social equality,'" one listener wrote; "[u]nlike some Negroes would have done, you did not evade that issue, you faced it and upheld it."[52] Perhaps the letter that best captured the meaning of the broadcast for many black listeners came from a group of students at Spelman College: "Thousands and thousands of thanks. . . . As all of us students . . . huddled around the radio in our various dormitories here on campus tonight, we rallied and cheered you as you so frankly and beautifully spoke the truth on the 'race question.' Thanks a million for your wisdom in treating and combing out the kinks in our dear Mr. Graves' approach. . . . The questions of the audience certainly did 'stick him up'—It was so amusing."[53] The managing editor of a black newspaper in Kansas City wrote him that "you did a swell job and I just wanted you to know that we out here in the Middle West enjoyed it very much." She also asked the question that may have been on many minds: "What percentage of the audience was colored and how many of those who asked questions were colored? We couldn't tell over the air."[54]

Several white listeners also commended Hughes, one of whom thanked him for his "fine contribution towards a better understanding of one of America's greatest problems."[55] A recent white immigrant from England congratulated Hughes and observed "that towards the negro question the Southerners have a blind spot, to overcome which nothing avails, neither argument nor logic, neither appeal to Christian principles nor appeal to national or self interest."[56]

Hughes knew the power of radio and had repeatedly sought access to it, although with much disappointment. He had written poems and dramatic plays for radio but as a black writer had faced difficulties in getting his work aired. Indeed, in a 1943 *Chicago Defender* column, Hughes wrote a letter to "Southern White Folks" in which he subverted the usual "Negro problem" imagery to make a point about radio's refusal to broadcast more of his work: "I tell you, you are really a problem to me. I, as a writer, might have had many scripts performed on the radio if it were not for you. The radio stations look at a script about Negro life that I write and tell me, 'Well, you see, our programs are heard down South, and the South might not like this.' You keep big Negro stars like Ethel Waters and Duke Ellington off commercial programs, because the sponsors are afraid the South might not buy their products if Negro artists appear regularly on their series."[57] Hughes recognized that the imagined southern listener was not the only reason or perhaps even the real reason radio executives were so reluctant to air more serious programming about race. Several weeks after his *Defender* column, he observed that during the war radio had become "fairly receptive" to presenting material about the "positive achievement" of particular African Americans, like George Washington Carver and navy hero Dorie Miller, but was still unwilling to air anything "setting forth the difficulties of the Jim Crow military set-up, segregation in war industries, etc., and what people of good will can do about it." The fact that radio had "censored out any real dramatic approach to the actual problems of the Negro people" rendered the radio industry "almost as bad as Hollywood." African Americans, he wrote, continued to hold a deep disdain for radio's presentation of what he called " 'handkerchief head' sketches" in which black stars usually were featured.[58]

Fueled by his anger over radio's failure to treat the race issue, Hughes seized the opportunity to appear on *Town Meeting* to present his own political views. His appearance on this national broadcast also opened the way for him to undertake an extremely successful speaking tour that included sizable white audiences. As a result of the broadcast, he became the first African American to be booked on a national tour by Feakins, the country's most well respected speakers' bureau. On tour for three months after the show, Hughes made over fifty appearances throughout the Midwest and the Southwest, addressing a variety of enthusiastic audiences, black and white.[59]

Hughes's experience with the power of radio only fed his anger and disappointment over radio's failures on the race issue. "Considering the seriousness of the race problem in our country," he wrote in 1945, "I do not feel that radio is serving the public interest in that regard very well. And it continues to keep alive the stereotype of the dialect-speaking amiably-moronic Negro servant as the chief representative of our racial group on the air." Recounting that "liberal" network executives lacked the political resolve to air a dramatic series about African Americans that he had repeatedly proposed to them, Hughes concluded: "I DO NOT LIKE RADIO, and I feel that it is almost as far from being a free medium of expression for Negro writers as Hitler's airlanes are for the Jews."[60]

Despite Hughes's continued disappointment in radio's treatment of the race issue, his appearance on *Town Meeting* had brought listeners face to face with the race question. The scarcity of listener protest eased the way for *Town Meeting* to tackle the more difficult issue of what to do about racial inequality. A discussion of the provocative question, "[S]hould government guarantee job equality for all races?," was aired in reaction to the ongoing campaign to make the FEPC a permanent agency.[61]

What remains most remarkable about this 1944 debate is the fact that point for point the arguments made against a government role in helping African Americans obtain fair access to employment were exactly the same as those directed at federal affirmative action programs decades later. Opponents blamed affirmative intervention for creating the very bitterness and racial hatred that mandated the measures in the first place, as if race prejudice, discrimination, and segregation had no prior independent or enduring existence. For example, Texas congressman Clark Fisher claimed to support equal opportunity in principle but said he opposed any federal role in furthering it because it will "stir up race consciousness, bitterness, and intolerance." He preferred the current system, which allowed "the poorest boy in the poorest family if he will work, if he has the ability and the initiative, to lift himself to the very top." Journalist Ray Thomas Tucker made a corollary and very creative argument that the creation of a permanent FEPC would prevent African Americans from following the traditional difficult path to success of immigrants and therefore "will breed bitterness and racial hatred."[62] Lillian Smith, the controversial writer and liberal activist, appeared in support of the FEPC, arguing that "it is the Government's job to protect the individual against those people who would endanger his basic right to work, just as the Government protects our safety on the streets and our health in epidemics."[63]

This exchange captured well public disagreement about the role the federal government should play in protecting and furthering the access of African Americans to employment. But as in many of the *Town Meeting* broadcasts, the debate expanded when questions were taken from the 1,500 people in the audience and telegrams sent in anticipation of the broadcast. One leading question came from Mary McLeod Bethune, who wrote, "[C]onsidering the increased industrialization of the USSR, China, and India, will the US be able to successfully compete for the postwar world trade without guaranteeing job equality for Americans of every race?" A person in the studio audience pointed out that blacks had already proven themselves capable of hard and ardent labor and yet they were still discriminated against in employment and therefore needed government help.[64] Once again, *Town Meeting* staff were surprised by the degree to which white listeners wrote long "dissertations on their personal feelings about the Negro question," exhibiting "emotional reactions" that far exceeded the issue of employment.[65]

Having faced one aspect of racial discrimination, *Town Meeting* turned to the broader question of racial injustice in a May 1945 broadcast entitled "Are We Solving America's Race Problem?" This topic generated passionate expressions of white resistance to the very idea of raising it for public discussion. Indeed, this would be one of *Town Meeting*'s most controversial and tumultuous broadcasts. The mere announcement of the topic drew letters of protest from white listeners, even before the show was aired. Many fear-filled letters came from outside the South, evidence in part that wartime migrations of African Americans had nationalized the race problem in many whites' minds.

These fears may have been amplified by Franklin Roosevelt's death and the growing anticipation that the war would soon end, although such concerns were not given direct voice in the letters. Several writers warned that the show was "playing with dynamite" and would only encourage more racial strife.[66] Some listeners earnestly suggested remedies to the race problem, including the often-repeated

TOWN HALL
AND
THE BLUE NETWORK

PRESENT THE 371ST BROADCAST OF

America's Town Meeting

MAY 24, 1945

SUBJECT:

Are We Solving America's Race Problem?

SPEAKERS:

Affirmative:

IRVING IVES
Majority leader of N. Y. State Assembly and co-author of the Ives-Quinn Anti-Discrimination Bill

ELMER CARTER
Former editor of the magazine "Opportunity" and member of N. Y. State Unemployment Appeal Board

Negative:

Representative JERRY VOORHIS
Democrat of California

RICHARD WRIGHT
Author of "Native Son" and "Black Boy"

Moderator: **GEORGE V. DENNY, JR.**

PRE-MEETING: 8:00 P.M.   BROADCAST: 8:30 TO 9:30 P.M.

Announcement of an America's Town Meeting of the Air *broadcast,*
*"Are We Solving America's Race Problem?" (*America's Town Meeting of the Air
Collection, New York Public Library*)*

idea that all blacks be relocated to reservations or separate cities, regions, or states of their own, with one writer suggesting that this
be done "in the same spirit as Zionism."[67] Another wanted blacks
to have completely equal opportunities with whites in employment,
education, and housing, provided that this could be done in a way
that would keep "ALL OF THOSE THINGS SEPARATE."[68] The most
colorful description of the race problem came from a man in Seattle
who may have mixed metaphors but captured well the fears held to
some degree by many people outside the South: "The Negro popu-

lation, is like the Sahara Desert, advancing every year about a mile, with overwhelming and irresistible force. Only one thing can stop the Desert, by drowning or letting in the sea. . . . But you cannot drown America's no. 1 problem, the negro. We are saturated with an incurable cancer. It has been allowed to go on so long, to operate now is impossible." [69]

When the show was aired, these prebroadcast responses prompted Denny to spread the responsibility for the choice of the topic, reminding his listeners that their votes and letters "had put this subject near the top of the list of America's major domestic problems." [70] Richard Wright, one of the country's most powerful black writers, and Elmer Carter, the black former editor of the National Urban League's *Opportunity* magazine, took opposing points of view on whether the race problem was being solved. Carter was paired with Irving Ives, the majority leader of the New York State Assembly; on Wright's side was liberal congressman Jerry Voorhis of California. [71] Carter offered the more conservative black position that the country was making progress toward racial equality, contrasting the record of the treatment of blacks in World War I with that in World War II and also noting that "the lynching record has almost been eradicated." [72]

In sharp contrast to Carter's voice of moderation, Wright launched an aggressive and unrelenting attack on racism and its effects, overstepping the bounds of politically acceptable discourse much further than had Langston Hughes the year before. Wright's extraordinary use of language not only overpowered Carter's arguments but also allowed him to dominate the program in a way that was utterly beyond the moderator's ability to control. Wright essentially reframed the entire debate and took over the show by asking:

What do we mean by a solution of the race problem? It means a nation in which there will exist no residential segregation, no Jim Crow Army, no Jim Crow Navy, no Jim Crow Red Cross Blood Bank, no Negro institutions, no laws prohibiting intermarriage, no customs assigning Negroes to inferior positions. . . . Racial segregation is our national policy, a part of our culture, tradition, and morality. . . . We see reflections of it in our films and hear it over our radios. . . . Gradual solutions are out of date. . . . [H]ere is the truth, whites can no longer regard Negroes as a passive, obedient minority. Whether we have a violent or peaceful solution of this problem depends upon the degree to which white Americans can purge their minds of the illusions that they own and know Negroes.[73]

Taking his argument a step further, Wright told his listeners that the "Negro has a sacred obligation and a moral duty to bring before the people of this country again and again and again the meaning of his problem," but, he added, "the fundamental problem rests upon whites and I believe that Negro protests, Negro agitation, should increase and become intense."[74] In replying to a question about intermarriage prohibitions, Wright insisted that such laws should be abandoned because they were meaningless: "I was down in Mississippi in 1940 and I saw the streets thronged with Mulattoes in a state where you have an airtight anti intermarriage law."[75]

Wright's call for black agitation and his comments on intermarriage jolted white listeners across the country. Denny, who had been unable to harness Wright on the air, feared that a negative response might ensue. In an unusual step, the day after the broadcast he asked for daily verbal reports on letters received rather than waiting for the normal weekly written tabulation and summary. His fears were well founded. Not only did the show generate an extraordinary volume of mail, but it drew long, passionate letters from well-educated white listeners who heard Wright's spirited argument as a threat to the racial world as they knew it, regardless of whether they lived inside or outside the South. According to an internal report, listeners were "highly critical of Richard Wright's attitude" and deplored the airing of the discussion of intermarriage.[76] A closer look at a sample of the mail reveals that this was an understated summary of the audience's reaction. Furthermore, these letters demonstrate the high levels to which white preoccupation with and fears about the race problem had risen nationwide by 1945. Again, this was a time when many whites were eager for normalcy after a war period marked both by southern black migration into areas previously without a visible black presence and by increasing expressions of black bitterness and anger, whether in city streets or under the sanction of radio forums like *Town Meeting*.

Wright's remarks about intermarriage sparked outrage, especially among white women. The year before, when Hughes had raised the issue, he had reassured whites that blacks did not want intermarriage but just wanted equal rights and equal opportunity. Wright inverted the entire question and ridiculed white men for their hypocrisy by citing his Mississippi example. White women attacked Wright for his obvious implication that mulattoes were the result of liaisons between white men and black women. One woman referred to the show as "revolting," and another reported that she had been "appalled" by what she described as Wright's demand for a "hybrid cesspool."[77] Other women called Wright's comments a "disgrace" and warned that they would lead to lynchings and encourage the resurgence of the Ku Klux Klan.[78] Indeed, white women listeners who wrote in seemed most concerned about defending the honor of their husbands, brothers, and fathers. Moreover, these women apparently did not oppose intermarriage as a way of defending themselves from imagined black suitors, as white men felt compelled to do on their behalf. Rather, in criticizing Wright's accusation that mulattoes were evidence of white male desire for black women, these white women revealed deep fears of sexual competition from black females, which might increase, they believed, if unions between white men and black women were sanctioned by law. In defense of white men, one woman from Detroit asserted that "I have never heard yet of a white man raping a colored woman."[79] White men, of course, also had been angered by Wright's comments, and their letters were even less polite. One particularly vehement man from Houston accused all black men of wanting to rape white women and referred to Wright repeatedly as "that buck negro" or "that ignorant negro buck."[80]

Many listeners, like those who had written letters before the show aired, offered as a solution to the race problem the idea of sending blacks away or somehow physically roping them off from whites. Some people earnestly suggested that African Americans be granted a homeland in the United States, be given a portion of the Pacific Northwest to settle, or be returned to Africa. One anonymous writer thought the only solution was to send all black Americans to Europe and to "exchange them for whites who would appreciate the advantage given them here, and eliminate these eternal race riots."[81] Many whites still searched for a solution to the "Negro problem" that would not upset the racial status quo; they simply wanted the problem to go away, as had whites who had embraced similar schemes throughout American history.[82]

While some listeners offered solutions to the race problem, others eagerly denied that there was a problem. One writer from Chicago explained, without intended irony, that the only problem was that African Americans had been exposed to too much "propaganda employing such words as freedom and equality."[83] More predictably, some white southerners insisted that the problem was northern agitators themselves, dubbed by one person as "noisy mouthed reformers in the North" who were "broadcasting their views" and "trying to stir up unhappiness and discontent among our colored citizens."[84]

Among the most interesting letters about the broadcast were those that revealed whites' anxieties that many blacks were no longer as deferential in their interactions with them as they used to be or ought to be. Some writers offered specific examples of increasing black arrogance and transgressions of racial etiquette, especially in southern border states and midwestern cities. One of the most telling letters came from a listener in Oklahoma who detailed what he called the "overbearing" ways in which blacks had begun to "push white people around." He complained that there was an "organized effort" among blacks "to make one day of the week a sort of 'push day,' on which the colored women of the town throng the places of business, and the sidewalks, just to shove white folks about." He warned that "once the war is over," blacks would be forced to "desist" from all such activities.[85] A writer from Chicago reported that he saw blacks on streetcars and buses acting as if "they are better than the white." Another listener from Cincinnati complained of blacks' new "overbearing attitude toward white people."[86]

These reactions of white Americans mirror descriptions of everyday acts of resistance waged by black working-class men and women in crowded and contested public spaces and in other interactions with white people during this period. They also represent white fears about any acts that appeared to be out of line with white expectations of black positionality, as was apparent in the spate of rumors of organized black resistance. "Race rebels" like Wright employed discursive and ideological tactics with the same effect in their intellectual encounters with white audiences.[87] Wright's arguments on the broadcast served as further confirmation for whites that these acts of racial rebellion were not isolated but were likely to increase, and Wright himself provided a frightful personification of this change, all adding further fuel to white fears.

Many white listeners channeled their fury about Wright's arguments into an attack on *Town Meeting* for allowing him to be on a "nation-wide radio hook-up," permitting him such free expression, and not having a southerner to defend the "white" point of view or at least "some one well acquainted with the negro faults and shortcomings."[88] One California woman complained that "the white man's mistreatment of the negro" was "not good material for radio comment," chiding Wright for even mentioning intermarriage, which she thought only worked to close the minds of the "millions of people" who were listening.[89] One listener chastised Denny for his polite handling of Wright: "He should have been cut off the air— with apology to the listening audience."[90]

Some considered Denny personally complicit in Wright's racial transgressions, specifically the fact that he referred to Wright as "MISTER"—a complaint similar to the one lodged against *Round Table* because Frazier addressed the white panelists without calling them "mister."[91] When an established radio forum lent its credibility, respect, and reach to black intellectuals like Frazier, Hughes, and now Wright, many white listeners deeply resented the division in white ranks it represented and the breach in the sanctioned silence on racial inequalities they desperately desired.

On the other hand, some listeners, mostly black, wrote in support of airing the issue in such a forthright manner, many writing to Wright directly or through his publisher rather than through the network.[92] The president of a black women's club in Mt. Vernon, New York, wrote: "I have never heard anything as well done as your expressions of last night at Town Hall. It was amazing and very much to the point."[93] Among the most emphatic responses was a letter written to Wright on behalf of the black men assigned to the army air force's 477th Bombardment Group stationed in Kentucky: "All radios of this group were tuned in on the program, so keen is the interest. Especially did we enjoy the way you handled the $64 question. It always comes up and we were glad to hear you handle it as you did. From all of us thanks a million. That personifies our outlook. We do not *ask* for democracy we *demand* it. In order to make democracy work it must work for all not just a few 'Uncle Tom' leaders."[94]

Other listeners also allied themselves with the thrust of Wright's overall views, often basing their arguments on contemporary examples. A black listener in Richmond asked: "[W]hy do Americans go 1,000 miles across the ocean to defend Democracy against the same evils as they are tolerating here upon our race?" The writer attached clippings about the police beating of a black soldier in Mississippi, inquiring, "[W]hy is it that the Secretary of War does not give our Negro in uniform the protection from white police officers and civilians wherever they may be?" Another particularly poignant letter asked: "How can we fight for the minorities abroad and keep our own in virtual slavery? If it is not corrected our boys will have died in vain." This writer recounted an incident in which Tuskegee airmen had to be partitioned off by a screen before they were allowed to eat in a public space shared by whites.[95]

Others who agreed with Wright made their case on moral grounds. "In every important event in our American History," one listener wrote, "the negro has been present, taking part regardless of danger for his white countryman and country—and you can't over look a people like that and still think you are right in doing so."[96] A few white listeners wrote in asking how they could support African Americans in their struggle for racial justice. One writer wanted to know what groups she could join to help, and another asked, "[W]hat is there that we can do?"[97]

Wright's controversial appearance on *Town Meeting* demonstrates once again the crucial cultural and political role African American intellectuals played in this period. Wright, Locke, and Hughes used their limited guest privileges on these political discussion shows to advance arguments too daring for most political figures to make, especially politicians who would have been featured on these programs. They offered a new representation of African Americans and their abilities, arguing point for point with whites and sparring as equals in the arena of political debate. On a medium that was ideal for the skilled use of language and oratory, these accomplished African American writers took on the duty of becoming public intellectuals, serving the race by fighting the battle of ideas that was essential to bringing about shifts in public opinion.

These men and others argued eloquently for an end to discrimination and segregation, but that goal was still not even rhetorically acceptable to the majority of white Americans in 1945. For that majority, the solution was to simply send the problem away or to continue to cordon it off. Until that view changed, there was little more that was politically safe to say. Silence on the issue of racial inequality set in again at *America's Town Meeting of the Air*, just as it had at the *University of Chicago Round Table*. Both of these important national political forums would confront the question of fashioning a remedy for racial discrimination and segregation only after the end of the war and after the insertion of a federal voice on the issue.

To Secure These Rights

The end of World War II brought with it an even bolder assertion by African Americans that their claims for an end to racial inequalities were now more timely than ever and could no longer be excused or postponed by fears of disunity during the war crisis. Several factors combined to make the postwar racial landscape as volatile as it had been during the war. Violence against returning black soldiers escalated. African Americans who had migrated to the North during the war transformed themselves into powerful urban voting blocs as they began their crossover into the Democratic Party. To a lesser degree, the Supreme Court's 1944 decision outlawing all-white primaries opened the way in some southern states for significant increases in black voting strength, although other locally administered impediments to voter registration continued to thwart most southern black citizens. Nonetheless, in the elections of 1944, 1946, and, most emphatically, 1948, the northern urban black vote became an important factor in national politics, as Harry Truman learned.[98] By 1947, the country was engaged in another war, and one that would span four decades: the Cold War against Soviet aggression and domestic communism.

Harry Truman's assumption of the office of president in 1945 coincided with this period of racial turmoil and competing political demands. Truman responded in several ways to these pressures, taking bold rhetorical and symbolic steps but offering limited action on behalf of African Americans, with one extremely significant exception: his executive order desegregating the military. On the symbolic level of national political discourse, Truman renewed the promise of national intervention on behalf of African Americans, a commitment his successors would be left to implement with varying degrees of success in the next two decades.

Under pressure from African Americans, Truman in 1946 created a committee to investigate the subject of civil rights and expand public awareness of the topic and the need to address it.[99] During the year the committee conducted its work, Truman took several steps aimed at reassuring black citizens and their white allies about his own commitment to equal opportunity. At the urging of Walter White, the executive director of the NAACP, the president accepted an invitation to speak at the Lincoln Memorial at a mass meeting to be held during the NAACP's annual conference in June 1947. White convinced Truman to come by arguing in part that the Soviet Union was using continuing evidence of racism to reduce the United States' international stature and that some reassurance from the president was necessary. This line of argument would be exploited as the Cold War picked up its pace. Acting on an earlier suggestion from W. E. B. Du Bois, the NAACP in October 1947 filed a petition with the newly formed United Nations Human Rights Commission detailing racial injustices against African Americans. Although action on the petition was blocked by American opposition, the petition drew much international press attention. It also attracted considerable notice and official complaint in the United States.[100]

White, a skilled publicist, worked to ensure the maximum amount of press coverage for the event. He paid special attention to radio, helping to arrange coverage not only by all four networks but also by most of the independent radio stations in major markets. The State Department agreed to carry the speech via shortwave for worldwide broadcast. White hoped that 400,000–500,000 people would attend local NAACP meetings at the time of the speech's broadcast "to form one gigantic mass meeting linked together by radio," making this in White's eyes possibly the largest mass meeting in the nation's history.[101]

White House officials were not unaware of the political and historical significance of this occasion for Truman, who would become the first president ever to deliver a live address to the NAACP.[102] Preceded at the microphone by Eleanor Roosevelt and Walter White, Truman spoke from the steps of the Lincoln Memorial to an audience estimated at about 10,000 people.[103] He asserted that "new concepts of civil rights" meant "not protection of the people *against* the Government, but protection of the people *by* the Government." He explained: "There is no justifiable reason for discrimination because of ancestry, or religion, or race, or color. . . . Every man should have the right to a decent home, the right to an education, the right to adequate medical care, the right to a worthwhile job, the right to an equal share in making the public decisions through the ballot, and the right to a fair trial in a fair court." Truman also made clear that the federal government would defend these rights and override recalcitrance at the state and local levels.[104]

For Walter White and other African Americans, this nationally broadcast address was the culmination of a decade of requesting a presidential radio appeal on racial issues. As early as 1938, White had

President Harry S. Truman addressing the 1947 NAACP convention from the steps of the Lincoln Memorial. (NAACP Collection, Library of Congress; the author wishes to thank the National Association for the Advancement of Colored People for the use of this NAACP material)

urged Roosevelt to devote a radio "fireside chat" to race relations, but he never did. In 1941, Ambrose Caliver secured Roosevelt's commitment to appear on the closing episode of *Freedom's People,* only to have the president withdraw at the last minute and submit a bland written statement instead. In the aftermath of the riots of 1943, black activists had renewed their pleas for the president to broadcast a statement

The Political Discourse of Racial Equality : 225

against racial violence, again without success. Truman's 1947 radio address to the NAACP was a long-awaited and long-overdue public display of presidential support for the general principle of equal opportunity and the expansion of the federal government's role in ensuring that opportunity.

Many African Americans saw the speech for what it was: a significant symbolic step, but one lacking in specific political proposals or commitments.[105] But for White, who believed deeply in the power of the media to change public opinion, the most significant aspect of the president's speech was that it had been broadcast nationally and internationally and that it produced, in White's words, "by far the largest single audience in history to hear the story of the fight for freedom for the Negro in the United States."[106]

Truman's reference to "new concepts of civil rights" marked the public introduction of an expanded view of the federal government's assumption of responsibility to protect citizens from the tyrannical acts of states, localities, and, eventually, private actors. Although the novelty of the term "civil rights" may have shielded the president's remarks from greater scrutiny by his critics, its meaning was not lost on other listeners among the "several hundred million" people across the globe estimated by White to have heard it. A group of black American soldiers who listened to the program via shortwave on the remote Pacific island of Tinian were so moved by the speech that they took up a collection and sent it to the NAACP to support its work.[107] The president's speech had not gone as far as most African Americans wanted, but its symbolic importance was not lost either, for it sounded like the beginning of something new to eager listeners like those on the tiny island of Tinian.

When the President's Committee on Civil Rights issued its report in the fall of 1947, the expansive nature of its recommendations exceeded most expectations. *To Secure These Rights* was a detailed blueprint for remedying sanctioned racial injustices in every aspect of American life. The report explicitly rejected the "separate but equal" doctrine as a failure and placed the responsibility for securing basic civil rights on the federal government itself, quoting Truman's remarks to the NAACP that the government must become the "friendly, vigilant defender of the rights and equalities of all Americans."[108] The report concluded with a legislative and legal manifesto for securing the litany of specific rights to which all Americans were entitled. It also called for an end to all discrimination and segregation in the armed services.[109] Writing before the report was released, the NAACP lawyer Charles Hamilton Houston had urged the committee to approach its work with "a sense of political maturity" that would require a willingness from the white committee members to write "a report for an entirely different world from that in which they were born."[110] In many ways, the report's recommendations outlined a plan to begin building a new world of racial ordering.

With its sweeping indictment, expansive recommendations, and extensive circulation, the report became a big news event.[111] Many of the overarching political justifications for the recommendations in *To Secure These Rights* were not new but had been articulated incessantly by African Americans and encapsulated in a body of intellectual work by blacks and others, including Gunnar Myrdal. Indeed, in the aptly titled 1944 volume, *What the Negro Wants,* over a dozen African American intellectuals and activists sounded a chorus of unanimity that what "the Negro" wanted were "the same rights, opportunities and privileges" extended to all other Americans in all aspects of public and civic life.[112] An end to legalized segregation, discrimination, and disenfranchisement was at the core of that claim, as they would be in *To Secure These Rights*. But this report generated much attention because it was well timed, coming after the war rather than during it, and most important, because it bore the stamp of federal approval and endorsement.

Truman followed the report by delivering a historic February 1948 special message to Congress devoted exclusively to the subject of civil rights in which he laid out a ten-point legislative agenda that mirrored many of the committee's suggestions. Most significant, he did not mention or endorse in any way the committee's general attack on segregation, but he did promise executive orders against discrimination in federal employment. Of special importance was his promise to end discrimination in the armed forces "as rapidly as possible."[113]

The president's special message to Congress provided more ammunition for African Americans, who were clamoring for Truman to back his rhetoric with action, especially in areas such as federal employment and the military where he could assert his presidential powers without waiting for legislative approval, which could be withheld indefinitely. A month after this civil rights speech, A. Philip Randolph advised young black men to refuse to comply with a proposed peacetime draft until segregation in the military was eliminated. That fall, after he had secured his party's nomination, Truman charged a committee with the task of planning for desegregating the military.[114] This was a significant victory in one of the longest and hardest fought battles in the struggle for racial equality by African Americans. Randolph had threatened an action, like his proposed 1941 March on Washington, that forced Truman to issue the executive order his predecessor had refused to issue during the course of World War II.

Once the president laid out his broad set of specific civil rights proposals, vehement opposition coalesced. Within weeks of the speech, Governor Strom Thurmond of South Carolina was urging the president to withdraw all of his legislative proposals or risk a southern rebellion against Truman and the Democratic Party.[115] Other pressures also were exerted on the president, including the insurgent presidential candidacy of the racially progressive Henry Wallace, who had been unceremoniously dethroned as vice president in 1944 and replaced on the Roosevelt ticket by Truman. In 1947, a few weeks after the civil rights report was issued, Wallace began a historic tour through the South and used the report to argue against racial segregation and discrimination. Significantly, Wallace defied local traditions by only speaking to nonsegregated audiences, a symbolic step that won approval from most African Americans whether they supported him or not. Southern whites who had identified themselves throughout the decade as race moderates attacked Wallace, his southern tour, and, most important, his call to end segregation. Segregation had become the defining issue for moderate southerners, ultimately ending any hope for a potential alliance between African Americans, northern liberals, and southern moderates on the race issue.[116]

Deeply intertwined with these issues was the Cold War itself, which began in this same period. Truman's administration set up the apparatus for waging a Cold War that was both foreign and domestic. Early in 1947, the president endorsed congressional attempts to ferret out subversives in federal positions and instituted an extensive "loyalty program" to do just that.[117] This bolstered attacks on political activists in general, especially those associated with liberal causes like racial equality and labor reforms, reproducing the 1930s era of smear tactics, innuendo, and purges. When Wallace announced in December 1947 that he would run for president the next year on a third-party ticket, one of his primary motivations was his virulent opposition to Truman's Cold War attacks.

It is an understatement to say that electoral politics in 1948 were deeply affected by these contradictory and intermeshed political developments. At the Democratic National Convention, a young Hubert Humphrey helped pass a strong civil rights plank, evidence of the growing potency of the northern black vote.[118] A contingent of southerners formed a Dixiecrat ticket headed by Strom Thurmond, which the majority of voters in four southern states later supported. Again, it was only after the convention, with his nomination assured, that Truman issued the executive order integrating the military that he had promised Randolph earlier that year.

In contrast, Wallace's bid for the presidency against Truman enthusiastically embraced the most liberal proposals on ending racial inequality advanced by black and white activists who were engaged in an uncompromising attack on segregation. Wallace's candidacy was an act of political courage for him and his black and white supporters in the South, as he returned once again to tour that hostile region. Subjected to harsh attacks and threats of physical violence, Wallace preached passionately against segregation and poll taxes, but his speeches and appearances were disrupted or ended by hecklers and, on occasion, by eggs and tomatoes.[119] Faced with this opposition, Wallace depended on cooperative local radio stations to get his message across in the region, presenting long speeches or interviews from the sanctuary of radio studios in Birmingham, at a black college campus in Mississippi, and in Shreveport. After Wallace was refused a speaking forum in Little Rock, in the name of free speech a local newspaper provided him with free airtime on a local radio station, where he was interviewed by moderate journalist Harry Ashmore.[120] Wallace's third-party effort sputtered along through the campaign and carried no states in the end, although it diverted some votes that otherwise would have gone to the Democrats in key northern states.

The strength of the urban black vote helped counter the loss of votes in normally Democratic states as well as the defection of a significant portion of the southern white vote, contributing to Truman's shocking victory over Thomas Dewey. In January 1949, the reelected President Truman renewed his request that Congress implement his 1948 civil rights proposals, although he knew as well as anyone that his proposals were once again dead on arrival, in part because he lacked the political strength to overcome massive white resistance in Congress, in the South, and across the nation. Although it would take another two decades of violence and struggle to move the nation to implement them, the goals of the civil rights movement lay encased in the report, *To Secure These Rights*.

Radio and Civil Rights

The political events of 1947 and 1948 brought the *University of Chicago Round Table* and *America's Town Meeting of the Air* back to the question that had silenced them earlier in the decade: how to remedy the problem of racial inequality. These shows returned to the issue not only with a newfound air of confidence but also with an eagerness to help rewrite the political narrative of race. Broadcasts at the end of the 1940s also reveal the evolution of a style of political engagement by radio that blurred the distinctions between educating, reporting, and editorializing, foreshadowing a fusion of functions that television would embrace.

After four years, *Round Table* began its return engagement on the race issue in 1947 with a broadcast called "Civil Rights and Loyalty" devoted to an endorsement of *To Secure These Rights*. Continuing its standard three-panelist format, the show featured the NAACP's Walter White, University of Chicago sociologist Louis Wirth, and historian Arthur Schlesinger Jr. The panelists abandoned the pretense of objectivity by firmly supporting the report's conclusions and recommendations without even debating the merits and political difficulties of the proposals. Schlesinger was especially effective at granting the report an air of legitimacy by characterizing it as a natural next step in the trajectory of American history and as being entirely consistent with the country's founding political philosophy.[121]

This show functioned as both an editorial on and an advertisement for the president's plans. Wirth explained on the air that one mission of this particular broadcast was to help create the "mass climate of opinion and information" necessary to put the proposals into effect. The show itself became a part of the massive public education program the report generated; the published transcription of the broadcast included a reprint of the committee's findings and recommendations.[122] "Whether America will be a more moral nation, or a more orderly and just nation, and a leader in the world," Wirth pleaded on the air, "depends upon what you and I and our listeners do to make these promises come true, by doing the things that the President's committee report recommends." The show ended as White chimed in "Perfect" and Schlesinger seconded with "Agreed." [123] "Perfect" was exactly the right word from White, who had been either relieved of the burden or denied the opportunity of arguing for the proposals since Schlesinger and Wirth had assumed that role, with White serving merely as political authentication. In all of these ways, *Round Table* was able to lend its own prestige to the findings of the politically charged report.

Cold War politics also may have played a significant role in this shift toward the embrace of the report's recommendations. At the time of the broadcast, for example, Schlesinger was firmly identified as a liberal anti-Communist. Six months before the broadcast, he had written a *Life* magazine article that warned of Communist influence in liberal circles, from labor to Hollywood to Washington, D.C., and urged progressives to purge themselves of that influence or risk the future of liberalism itself.[124] Unstated but perhaps lurking underneath this broadcast's enthusiasm for Truman's proposals were fears of the more progressive proposals of his rival, Wallace, and the more conservative southern position that threatened to disrupt the Democratic Party.

But African Americans also had played a major role in shaping the idea, now espoused by Schlesinger and others, that equal rights for African Americans were essential to the nation's global ambitions. That idea had been reinforced throughout the war by the "Double V" campaign and by the exertion of political pressure by Randolph, White, and many others that forced the production of *To Secure These Rights*. One Cold War casualty, however, was aptly captured in this show's title, "Civil Rights and Loyalty," as the cause of black american rights required the merger of the double political consciousness described in W. E. B. Du Bois's *Souls of Black Folk* and embodied in the "Double V" campaign. It was their claim to Americanness on which African Americans now rested their case for civil rights, which in turn required, at least initially, loyalty to American policies, domestic and foreign, and, as some have argued, a lessening of emphasis on anti-imperialist and anticolonialist struggles elsewhere. But African Americans' identification with other oppressed people of color never disappeared completely, even in this period when political exigencies required a full embrace of American identity as part of the price for structuring a targeted appeal for basic political rights at home.[125]

Radio's assumption of some civic responsibility on the race issue in this period reflects a significant transformation both in the influence of racial moderates on radio's political discussion shows and in the nature of radio as a political medium. Radio's emergence as a political agent in the 1940s added a new dimension to the world of political symbolism in which national and international politics operated. Roosevelt knew this well and exploited it fully as a politician and as the leader of a nation at war. The model of cooperative symbiosis that existed between radio and the federal government during the war crisis did not end when the war ended but was replicated here through the open endorsement of the president's positions.

Throughout the war period, members of the radio industry had looked to the federal government for political leadership and protective cover on how to discuss the race problem, if at all, repeatedly turning without much success to the OWI, for example. When no guidance was forthcoming, staff at radio programs like *Round Table* approached the issue with great caution and fear, and when faced with the unavoidable question of a racial remedy, they retreated for years into official silence. The concrete set of proposals in the report on civil rights finally opened the way for that discussion, and the stamp of official imprimatur allowed racial moderates to endorse its tenets under the guise of "educating" the public. Truman's open rhetorical embrace of the central claims of African American activists carried enormous symbolic power in the national discourse of the politics of race, in which radio played an important role. This was the case despite the fact that Truman's words far exceeded his actions, a shortcoming that for African Americans nullified much of the political symbolism.

Other *Round Table* discussions during this period starkly demonstrate two related developments: the abandonment of a deference to white southern politicians as representatives of a monolithic "white" point of view and the emergence of the voices of white southerners who were less rabidly opposed to alleviating some forms of racial injustice. This shift can be seen on *Round Table* discussions in which southern politicians were challenged and rendered largely ineffectual as they gave rote performances mouthing the same old arguments over and over again. In a 1948 show about the southern revolt in the Democratic Party, Senator John Sparkman of Alabama, who had helped lead the movement out of the party, reiterated that states' rights were under siege and that the South, if freed from its own economic disadvantages, could "solve" its own race problem. University of Chicago historian Avery Craven hosted this panel. In 1943, Craven, a southerner by birth and a historian of the Civil War and the South, had defended the conservative southern viewpoint on *Round Table*'s first treatment of the race issue. But on this show five years later, Craven not only openly disagreed with that view but also urged southern politicians to enact the president's proposals at the state level or risk massive federal intervention. Craven had recently made the same arguments in an opening address to the Conference on Civil Rights held at Atlanta University. As if to endorse Craven's legitimacy as a southern voice for moderation, *Round Table* reprinted the entire address as part of the transcription of the show.[126]

Not only was deference to white southern politicians abandoned, but in some cases, these men were held up for subtle ridicule, especially after the 1948 election. After Truman renewed his civil rights proposal in his 1949 State of the Union Address, *Round Table* responded with a show featuring the newly elected Senator Hubert Humphrey, who had led the fight for civil rights at the convention. Humphrey credited the Democratic victory to the votes in key states of people "who would be directly affected by an active civil rights program"—a veiled but obvious reference to the northern urban black vote. Senator Allen Ellender of Louisiana argued against Humphrey's interpretation of the election and against all of the president's proposals. Louis Wirth, who was moderating the discussion, baited Ellender by asking him if he was opposed to white supremacy. Trapped in a "yes or no" question, Ellender answered: "The Negro himself cannot make progress unless he has white leadership. If you call that 'supremacy,' why suit yourself. But I say that the Negro race as a whole, if permitted to go to itself, will invariably go back to barbaric lunacy."[127] In these and other ways, more liberal whites on *Round Table* rendered Ellender and many of his southern colleagues simple caricatures of the powerful politicians they had embodied in earlier appearances on radio forums. By 1948 and 1949, this group of politicians continued to voice positions that were offensive not only to African Americans, as always had been the case, but also to increasing numbers of whites—otherwise, this show would not have battled so openly with men like Ellender. The shift depended on an overly optimistic reading of general public opinion by Wirth and others, however. For example, on another show, Wirth read the 1948 election as a reaffirmation of domestic progress on the race issue, concluding that by "hopelessly" outvoting the Dixiecrats, American voters had emphatically rejected white supremacy, which was plainly not the case.[128]

Round Table broadcasts during this period continued to link the race issue at home with the country's ability to exert world leadership and counter Russian claims about American racism. Throughout the early Cold War period, this line of argument was one of the most commonly voiced rationales for adopting the president's proposals, a position espoused during the war by African Americans but now embraced by some whites as well.[129] After Ellender's assertion of black inferiority in the 1949 show featuring Humphrey, Wirth admonished his listeners that "the eyes of the world are upon us, and our deeds will demonstrate more than our words whether we mean democracy genuinely."[130] *Round Table* also linked the need for American racial reform to developments in India and South Africa. Wirth began a *Round Table* discussion on international race relations by asking, "[W]hat is the changing position of the white man in the United States, in Asia, in Africa, and elsewhere?" Philips Talbot, a political science professor at the University of Chicago, answered quite frankly that "the white man does not even recognize himself today compared to where he stood ten years ago."[131] Although this probably would have been heard as bad news in many quarters, the comment was offered here more as an observation rather than as a call to action.

The ten-year period that Talbot mentioned also spanned a marked shift in the way that racial inequality was discussed on national radio. The significance of rhetorical shifts should not be overstated, for, after all, words are no substitute for actions, but they also should not be overlooked or casually discounted. Ten years earlier, network officials had warned the staff at *Round Table* to abandon attempts to "discuss the nigger question" or risk cancellation. Even when the war and racial unrest forced the question onto the air, it was treated gingerly, after which it was ignored once again. By 1948, resident moderators like Louis Wirth were able to use *Round Table* to advance the new political rhetoric on racial inequality that had developed over the course of the decade, first among African Americans and then increasingly among a minority of like-minded whites. Wirth and other liberal academics used the radio forum to promote their own evolving political stance on the race problem, usually abandoning the cloak of academic objectivity. Instead, Wirth functioned as a playmaker, handing off assists to some white panelists, like Humphrey, and teasing others into foolish moves, like Ellender.

One notably ironic and extremely telling consistency throughout the decade, however, was the scarcity of African American voices on these shows. Although what was being said changed dramatically during this period, who said it did not, as these deliberations about the race issue continued to be held primarily by white men. Although there was still an attempt to include representative white southerners on these panels, white academics and intellectuals who had laid claim to civil rights ideology apparently believed they were better qualified than their black counterparts to advance that cause. The need to address racial inequality may have been settled as a matter of rhetoric, but accepting the fact that African Americans could and should speak for themselves on this powerful and protected medium was not.

Some of the patterns seen on *Round Table* also held true for *Town Meeting* broadcasts on civil rights during this period, but *Town Meeting*'s dedication to debate and its emphasis on audience and listener response provide a more nuanced reading of the political reality and resistance that met the president's proposals. Also, by this time, the show was reaching 20 million listeners over 225 local stations, and its growth and stature had outpaced *Round Table*'s.[132] White listeners talked back to the radio during and after these *Town Meeting* broadcasts, demanding to be heard. These listeners sensed not only that the debate was almost over but also that the South's position was being silenced in defeat.

Although *Town Meeting* also worked to ensure that southerners were represented on its broadcasts, as on *Round Table,* their performances became unconvincing redundancies as they refused to offer any new responses to the changing political landscape. After Truman's civil rights message to Congress in 1948, the show featured a debate on the narrow question of whether the president's proposals ought to be adopted. On opposing sides were Senator John Sparkman of Alabama, who had already declared his opposition to Truman's renomination, and Senator Wayne Morse of Oregon and Roger Baldwin of the American Civil Liberties Union.[133] Baldwin argued, as many others did during this period, that the president's plan was needed to strengthen democracy's appeal to the peoples of the world tempted by communism, which was the liberal anti-Communist stance.[134] Sparkman could only call the Truman program "unconstitutional, unwise, and unworkable," dismissing it as "a political football" put into play to win the black vote but offering no alternative response to the reality of racial inequality.[135]

It would be a mistake, however, to assume that the declining on-air effectiveness of these politicians rendered them unrepresentative of the racial sentiments of most white southerners or white Americans in general. On the contrary, *Town Meeting* broadcasts on the race issue uniformly generated long, deeply emotional letters from white listeners, largely but not exclusively from the South, who passionately supported the positions espoused by white southern politicians. For example, the show that included Sparkman brought in nearly 4,000 letters, most of which made the same well-worn arguments against racial equality: that any move toward political and economic equality would bring social equality, intermarriage, and mongrelization; that no one had the right to dictate policies to white southerners; and that the FEPC would discriminate against whites. According to *Town Meeting*'s analysis of its mail, the minority of listeners who admitted the need for fairness for "the Negro" also always included a "conditional 'BUT'" usually "contingent upon the Negro 'remembering his place.'" In contrast to these emotionally charged responses, *Town Meeting* staff noted that reactions from outside the South were far less engaged, expressing "neither enthusiasm for nor opposition" to the president's proposals. Those who supported them did so because they believed that racial injustice rendered the nation vulnerable to Cold War propaganda about the issue.[136]

The voice of African Americans was still largely absent from these *Town Meeting* discussions, the exception being that of Walter White. Both *Round Table* and *Town Meeting* relied on White to represent the African American position during this period. White appeared on *Town Meeting* in October 1947, a month before he would appear on *Round Table*. One shift during this period was that White and other members of the NAACP, who had been considered too politically risky for most radio broadcasts earlier in the decade, were now seen as acceptable and necessary participants on occasion. Other African American leaders, most notably A. Philip Randolph, would rarely be accorded that status, a measure both of fears of Randolph's political prowess and independence and of the growing legitimacy of the NAACP among white race moderates, including those in the broadcast industry.

On a show about how to improve race and religious relationships,

White participated in a discussion with Charles Taft, head of the Federal Council of Churches; former congresswoman and journalist Clare Boothe Luce; and Max Lerner, editorial writer for the leftist *PM* newspaper. White cast the race question in Cold War terms, arguing that Americans could not "talk of freedom and democracy" as long as African Americans were "scorned, disfranchised, segregated, denied education and jobs, tortured, even lynched."[137] As noted earlier, African Americans in this period portrayed segregation as weakening American claims to international leadership, especially vis-à-vis the Soviet Union. The efficacy of this appeal to moderate white listeners had not been lost on White and others who recognized that the threat of Communist gains might be more frightening to some Americans than racial equality.

White felt compelled to confront the question that loomed over all discussions of racial equality, just as Langston Hughes and Richard Wright had done in their appearances on earlier *Town Meeting* broadcasts: "Now let's face the bugaboo of social equality and intermarriage: the $64 question that always comes up—'How would you like your daughter to marry a Negro?'" He answered, as Hughes had, that there was no concerted campaign among blacks to marry white people in America, but at the same time, he acknowledged, as Wright had, that the law against intermarriage "placed a premium on bastardy and illicit sexual relations."[138] White added that antimiscegenation laws "deprive women of legal protection of their persons," meaning, although he did not say it, that the laws deprived African American women of protection from white men. This new argument in favor of lifting the ban on interracial marriage aimed squarely, as Wright had in 1945, at the hypocrisy of white men who supported the law as it applied to black men and white women but not as it applied to themselves and black women.

The mention of the intermarriage issue drew an angry response once again from white listeners, but the broader topic of race and religious relationships yielded letters that attacked not just White but other issues and panelists as well. Internal mail reports characterized one-third of the comments as "either anti-Negro, anti-Semite or anti-Catholic." Letters expressed a broad range of concerns, from fears of the Vatican to fears of "Jewish financiers." Some wrote to protest Lerner, who sounded, and was, much more liberal than White. "Several listeners deplored Max Lerner's exhortation to minorities to struggle to escape their caste," the mail report explained, because they heard his remarks as a call to violence for African Americans. Not all of the responses to the broadcast were negative. A minority praised the show for "having brought into the open the pettiness, the hypocrisy, and the bigotry."[139]

Truman's surprising reelection in 1948 emboldened *Town Meeting* to finally confront the question that remained politically untouchable on the air: "What should we do about race segregation?" As veteran radio moderator George Denny searched for a way to introduce this discussion, he fumbled through a familiar but ill-fitting paradigm: "Our melting pot—the great American melting pot—has still some lumps in it. What should we do about them? What is being done about them? Does the pot need more heat, or is the temperature about right? Will more stirring help? What should we do about race segregation in America today? One of the planks in President Truman's platform was the enactment of a civil rights program on a nationwide basis. Was the election a mandate to the Congress to pass this legislation?"

The first speaker to try to answer Denny's question was Ray Sprigle, a white journalist who had disguised himself as a "Negro," traveled throughout the South, and written a series of syndicated articles about his experience. Sprigle spoke as if still in his assumed identity, taking the liberty of talking "from the standpoint of the Southern Negro." He described segregation as part of "the whole vicious and evil fabric of discrimination, oppression, cruelty, exploitation, denial of simple justice, denial of the rights of full citizenship and the right to an education."[140]

By the time of this broadcast, *Town Meeting* had begun to be carried on television as well as radio. For that reason, when Denny introduced Walter White, he alerted his viewing audience that White was a "Negro," although they would not "recognize him as such."[141] So this show had the odd pairing of Sprigle, who temporarily had turned himself into a "Negro," and White, who looked as white as Sprigle but identified himself as black. Southern journalists Harry Ashmore and Hodding Carter also appeared on the show, and Ashmore found much humor in the fact that White "seemed the most conspicuous Aryan among us, while the swarthy Carter's skin was dark enough to prompt a Mississippi theater usher to direct him to the balcony. The makeup man was instructed to darken down White and lighten up Hodding."[142] The quirkiness of the politics of racial representation was never more visible.

Once again, White led with his strongest appeal to white Americans that without racial reforms, the country was a vulnerable target of Russian propaganda and subject to international shame. "Our enemies today," he said, "broadcast to the world that we in the United States talk about democracy but we separate and discriminate against our own citizens because of race, color, or creed."[143] He characterized segregation as antithetical to equality and offered the civil rights report as "one of many proofs that decent Americans want segregation abolished and they want it abolished now."[144]

Ashmore, the racially moderate executive editor of the *Arkansas Gazette,* reframed the discussion by arguing that the problem was not what to do about segregation "but what to do about those injustices and inequalities that have accompanied it," essentially rejecting White's claim that inequality and segregation were inextricably linked.[145] In the studio after the broadcast, White confronted Ashmore and Carter, asking them why they would not admit that segregation was "morally indefensible," a question neither could answer privately, much less publicly.[146]

This discussion about ending segregation reveals once again how those with access to the national airwaves tried to manipulate public opinion on this crucial question, although they met with limited success. Whereas White was predicting that decent people were ready and willing to end segregation, Ashmore, as had Wirth, was reading the 1948 election results as evidence that white southerners had already declared race no longer a political issue. All three men were engaging in rhetorical hyperbole, wishing for what was not in hopes of making it so. Panelists like White and Ashmore tried to prod public opinion by telling the audience to believe things that were not yet true; as a result, their glowing pronouncements had a hollow ring to them. If nothing else was clear, it certainly was true that the question of what to do about segregation remained unresolved and unresolvable in national politics, despite the harbingers of change in 1947 and 1948.

Still, the reemergence of race as a national political concern not only eased the way for shows like *Town Meeting* to face the segregation issue but also encouraged some moderate white southern listeners to raise their voices. Response to the discussion about segregation surprised the *Town Meeting* staff, who reported that the letters showed a "beginning of change in the attitude of Southerners," that the relative number of letters expressing a "deep hatred" of blacks had diminished, and that there was growing support for the extension of basic citizenship rights to blacks. Still, the show's staff concluded, these more racially moderate listeners did not think segregation could or should be ended immediately but thought it should be eliminated gradually without outside intervention.

Obviously, these letters did not represent the sea change in the white southern point of view that the *Town Meeting* staff reported them to be, but they were an indication that a racially moderate white minority had become convinced that change was on its way and that outright resistance offered them no opportunity to direct or control that change, although they remained deeply opposed to ending segregation. Rabidly racist responses to the question of segregation were still plentiful, but many who subscribed to those views had begun to feel silenced by the rhetorical alliance between African Americans like White, white northerners like Sprigle, and southern white moderates like Ashmore. And it was that silence that created the illusion among New York City–based *Town Meeting* officials that the mass of white southerners might be changing their point of view, which was clearly not so.

Not only was this not the case, but many of the listeners most opposed to the attack on segregation turned their anger against the act of discussing segregation and against the broadcast for fostering that discussion. Several letter writers argued that it was "dangerous" to discuss segregation and that the series should broadcast "no future programs" on the issue. A Memphis station reported that it had received "over 50 protest calls" during the broadcast. One listener urged that Sprigle and White be "kept off the air entirely," and another said she would "never listen again to your dreadful programs of hate," referring to hatred against white southerners. To other listeners, the decision to air shows about race relations was the underlying problem: "[T]he colored people are happy, but this stirring up of the question makes them unhappy and dissatisfied. They are negroes by the hand of God, and they cannot blame that on anyone."[147]

Town Meeting continued to debate the political consequences of racial inequality early into the next decade, at a time when there was both a lull and a stalemate on the issue. Early in 1950, the show returned to the question of whether Truman's civil rights program should be adopted, again pairing Humphrey with a southern Senate colleague, this time John Stennis of Mississippi. Stennis cast himself as the defender of the "plain old average common-garden variety of American" who "doesn't belong to any of those minority groups" and who opposed the president's proposals. Humphrey argued passionately, as White and other African Americans had argued repeatedly, that "our moral standing, our political standing in the eyes of the free world hinge pretty much on whether or not we pass civil rights legislation." Stennis replied that people the world over not only envied this country but were clamoring to emigrate here.[148] Although Humphrey and Stennis sparred to a rhetorical draw, southern strength to filibuster continued to thwart those in Congress like Humphrey who favored the president's proposals. The political narrative on race may have been shifting, but the effectiveness of political resistance to change continued in full force. More moderate voices might be winning the rhetorical battle, but the political power still remained lodged in the hands of racial conservatives and their constituencies.

Officials at *Town Meeting* did not escape that resistance as it played itself out in daily practice. Langston Hughes, Richard Wright, and Walter White would remain the only black voices to be heard debating the race question on these broadcasts. For each of White's appearances, he was paid an honorarium of $100, for which he expressed great appreciation. Little did he know that his fellow panelists were all paid at least twice as much and often much more than that. Sprigle, who had made a name for himself by pretending to be black, requested and was paid $400 for his appearance alongside White. Although Sprigle had assumed the racial status of the "Negro" and wanted to speak on behalf of the race, apparently he had no desire to suffer the economic consequences of having that identity, which the show repeatedly visited upon the unsuspecting White.[149] Representations of race remained market driven, primarily by the currency of racial category itself.

In a poetic end to its run of programs about race during this period, *Town Meeting* broadcast a show in 1950 to celebrate the fortieth anniversary of the *Pittsburgh Courier,* the largest black newspaper in the country and the one most closely identified with African American activism during World War II. The newspaper invited *Town Meeting* to broadcast live from Pittsburgh and paid the usual $1,000 fee for the privilege of hosting a debate on the question, "What effect do our race relations have on our foreign policy?" This was as leading a question as any the show could ask about racial inequality in the Cold

War era. Sociologist Charles Johnson, the first black president of Fisk University, wasted no time in arguing that the country's "racial system" was the "Achilles heel of both our domestic and foreign policy." World War II, he contended, had been fought to end the "arbitrary brutalities of a master race." A questioner from the audience picked up Johnson's argument by asking how a segregated country could ever criticize "an inclusive communism." The other panelist, Congressman Brooks Hays of Arkansas, accused Johnson of emphasizing "imperfections" while ignoring the "tremendous progress" that had been made on the race issue during the previous decade. Johnson closed the show by holding up segregated schools as the clearest "indication of an incomplete democracy," a claim that was already working its way through the legal system.[150]

Listener response to the show once again provided a vivid picture of the continuing deep division over the implications of discussing the race issue at the beginning of the 1950s. One writer praised the show as one of the "finest" *Town Meeting* broadcasts ever, calling Johnson "his own best argument for justice for the American Negro." But another listener wrote: "I wonder why you have so much discussion on the Negro question. It's terribly irritating to white people. . . . [F]or God's sake and white America, cut out the (Negro) question."[151]

National politics have always operated in the dual realms of actions and symbols. With the advent of a mass communications system like radio, the symbolic realm assumed an even greater authority as the performative aspect of politics found its natural audience: a body politic of millions of listeners. In this new, expanded public sphere, the manipulation of language as political imagery became more important than ever. Without visual images and the elixir of music, political meanings on these panel discussion shows had to be spoken or left unspoken. The contest over what could and could not be said took on paramount importance and mirrored the struggle over real political boundaries and limitations. The crisis of language in turn signified the crisis in the racial order. This transition period in American race relations, with all of its promise and its limitations, played itself out eloquently and paradoxically on these political discussion programs. Radio became a perfect carrier for the performative discourse of the politics of race.

These broadcasts also captured the shifting dialectic between actions and words as race riots, black migration, and black protests pushed the boundaries of the political worlds of actions and symbols. Truman's engagement in the symbolic and real politics of civil rights was one manifestation of that dialectic. The intellectual, moral, and legal potency of the claims that African Americans made during this decade of war, riots, and peace—as well as their emergence as a potent northern urban voting block—led Truman to a rhetorical embrace of the fundamental principles of the modern civil rights movement.

The World War II era brought a consolidation in African American discourse about the claim for black freedom. The politics of inclusion that African Americans had argued for at the beginning of the decade had silenced earlier black nationalist claims for a separate economic and political realm and had muffled anti-imperialist and anticolonialist critiques. Instead, the dominant black discourse of the postwar era called for full American rights and full access to the nation's institutions and privileges. Although this goal was conservative on its face, achieving it would require an aggressive and unified claim for freedom by African Americans and a willingness to engage in the struggle necessary to attain it.

Although this way of articulating black claims may have united African Americans across class, regional, and political lines, to white listeners, it sounded heretical. The mere act of discussing the race question on the air was a rite of legitimation for African American arguments for freedom, and fear and extreme caution among whites accompanied this change. African Americans also repeatedly challenged the censorship that excluded them from the groups and places that spoke with authority to the body politic about themselves, their history, and their needs. Radio was one of those places. When they were able to breach the censorship, African American intellectuals as varied as Alain Locke, Franklin Frazier, Langston Hughes, and Richard Wright presented and represented an image of "the Negro" that challenged the language of authority, even as more insistent black voices remained excluded. Nonetheless, allowing African Americans to enter the realm of performative political magic was heard by the listening white public as another threat to their entire universe of social, political, economic, and sexual relations. And it was. If American racism was the national religion, the liturgical conditions were shifting, and both its adherents and its heretics recognized as much.[152]

Visceral listener reactions to these broadcasts captured the cognitive and political shock that the idea of ending racial inequality stirred in most white Americans. Despite unifying wartime rhetoric, most white Americans regardless of region adhered to a "politics of exclusion" that depended on the continuation of racial discrimination and various forms of segregation. Indeed, this was no longer a southern problem, as many northern and western cities had responded to black migration by hardening rather than erasing the lines of racial exclusion. The majority of white Americans plainly feared a new racial frontier in which black Americans would escalate their struggle to secure the rights that democracy promised. Continued struggle would be necessary not only because of massive white resistance but also because of the limitations of white liberalism. Despite their claims to enlightenment, white race liberals were unable at the time to embrace the idea that politically sanctioned segregation and discrimination were antithetical to democracy and would need to be dismantled if "the Negro" was indeed now also an "American."

African Americans were kept on the margins of radio's public discourse about race relations, the rationale being that broader popular sentiment—especially southern views—required this restriction. Lurking beneath that claim was a certain hypocrisy. The postwar sparring on the radio between white liberals and white southern conservatives obscures the fact that white liberals engaged in the airing of the race question were themselves afraid of the full meaning of African American freedom. These whites who considered themselves liberals on race worked to distinguish themselves from southern segregationists, but they remained uncertain about any remedy-oriented definition of the race issue. Nonetheless, they projected confidence about their own "take" on race. Even when they invited a few black voices, they also felt sure that whites and only whites could lead on the race issue.

For all of these reasons, the politically permissible discourse that emerged in the late 1940s maintained a very narrow approach to ending segregation and race discrimination. That approach, as we have seen, was consensus oriented, casting the race problem primarily as a question of the nation living up to its founding principles but with no engagement of the actual mechanics and fundamental restructuring essential to advancing racial equality in a society that also was founded on the principle of racial inequality. "Americans who profess to believe in democracy will have to face the dilemma of cooperating," the African American historian Rayford Logan had warned in 1944, "or of limiting their ideals to white Americans only."[153] That was the dilemma that faced most white Americans at the end of the 1940s as the politics of racial exclusion and limitation prevailed.

CHAPTER 6

NEW WORLD A'COMING AND
DESTINATION FREEDOM

National radio's treatment of the race issue is a tale of caution and restriction. Network officials carefully scrutinized the political risks associated with federally sponsored programming, even as the federal agencies themselves adhered to their own strict tests of what was "speakable" about African Americans. The National Urban League bridled its rhetoric in deference to the industry's implied political boundaries, and when the league overstepped them, the networks simply refused to cooperate. Fears of audience disapproval haunted radio's political panel discussion shows as they alternately approached and avoided the race question, emboldened only in the postwar period after President Harry Truman's embrace of the rhetoric of racial equality. The exigencies of war and African American protests had pushed the issue onto the public airwaves, but the radio industry's cooperation was contingent on the observance of a political etiquette of race.

The full force of African American political thought rarely pierced the national airwaves as accepted political discourse about race. Two local urban radio programs aimed at more sympathetic and politically progressive audiences, *New*

World A'Coming and *Destination Freedom,* were far more consistent in tone and content with the claims and aspirations of African Americans in this era. This kind of broadcast was possible because of black migration and the formation of an urban market of working-class and middle-class African Americans. These local shows owed their existence to coalitions of African American and racially liberal white activists, intellectuals, entertainers, and artists. The fullest realization of radio's use as a forum for voicing African American political thought came in northern urban radio markets such as New York City and Chicago, where the specter of an offended white southern listener faded, as did the need to honor a nationally acceptable political discourse. The unfettered political voices of African Americans more fully entered the radio debate about their future only through "narrowcasting" to northern urban audiences.

New World A'Coming

Although the radio airwaves were dominated by the national networks at the beginning of the 1940s, by the end of the decade some of that power shifted as changes in the industry and in national demographics allowed for the emergence of stronger independent urban radio stations and more politically daring broadcasts. This was the case with WMCA, an independent station in New York City and a pioneer in forging a civic role for local broadcasting. Some of WMCA's programming embodied contemporary African American political thought, made possible by a coalition of African Americans and racially progressive whites.

The radio series *New World A'Coming,* first broadcast by WMCA in 1944, was the collaborative project of black writer Roi Ottley; Nathan and Helen Straus, the wealthy owners and operators of WMCA who dedicated their independent station to liberal causes; and the City-Wide Citizens Committee on Harlem, an interracial civic and political group that cosponsored the series. In every respect, *New World A'Coming* was a different kind of radio series on a very different kind of radio station in a city with a large black and politically progressive listening public.

A native New Yorker, Ottley had worked simultaneously as a social worker in Harlem and a reporter for the New York City black newspaper, the *Amsterdam News.* He served as publicity director for the National Congress of Industrial Organizations War Relief Committee and became the first black journalist to win a Rosenwald Fel-

Roi Ottley. (Photographs and Prints Division, Schomburg Center for Research in Black Culture, The New York Public Library, Astor, Lenox and Tilden Foundations)

lowship, which he used to study the war relief needs of working-class minorities in Allied nations. During his travels abroad, Ottley became a foreign correspondent for the liberal *PM* newspaper and the *Pittsburgh Courier*. He also broadcast radio news reports for CBS and the BBC during his travels.[1] Through these activities, Ottley became a well-respected and nationally known writer and journalist.

Ottley's acclaimed 1943 book, *New World A-Coming,* gave the series not only its title but much of its content for its first two years. Ottley's book focused on Harlem and more generally on the economic and political plight of African Americans. Its topical appeal and format were particularly well suited for radio adaptation.[2] Ottley saw his book, which included narratives about historical and contemporary black political figures, as "a study of black nationalism." He explained: "I have explored its ramifications in Negro life, its progress in very recent years, its vagaries, and its effect upon the Negro's thinking as he views the future cast of the world—waiting for the new world a-coming."[3]

Although New York City was the nation's radio capital, access to local airwaves was complicated by the fact that the most popular local stations also served as the flagship stations for the major radio networks rather than as true local affiliates. However, shifts in federal regulations brought changes in the broadcast industry and in local radio in the city.[4] Those changes allowed Nathan Straus Jr. to buy and transform WMCA into a beacon of liberalism. Earlier, Straus had left his work as a journalist with the *New York Globe* to represent Manhattan in the New York State Senate, where in his five years of service he became an expert on housing problems affecting the poor. In 1937, President Franklin Roosevelt appointed Straus to serve as the first head of the Federal Housing Authority, which had been created to eliminate slums and provide low-cost housing for the poor. But in January 1942, Straus resigned from that position as a result of a protracted battle with southern conservative congressmen over funds for public housing.[5]

After his resignation from the Roosevelt administration, Straus returned to New York City. When he bought WMCA the following year, he stressed his commitment to operate it as an independent station, without network affiliation, and to do so without undue influence from commercial considerations. In remarks reminiscent of early thinking about radio, Straus emphasized that "radio is one of the great factors in molding public opinion and in a democracy public opinion makes the laws." He hoped that WMCA would "help to promote the development of an informed public opinion concerning the great problems and issues of our troubled times."[6] Helen Straus joined her husband in fashioning WMCA into a station whose programming, although commercially supported, reflected her own strong belief in the efficacy of radio for educational and public service purposes. Helen and Nathan Straus shared the view that local radio had a different mission—and commercial niche—from network radio. To them, local stations had a special obligation to provide programming or information that could not be presented by a national network.[7] It was Helen Straus who read Roi Ottley's book and saw in it material about race relations that easily could be dramatized for radio.[8] Based on her idea, the station began the work of bringing Ottley's book to life on the radio.

When station officials announced in January 1944 that they were planning "a series of programs on Harlem and Negro culture," they indicated that the series would be presented in cooperation with the City-Wide Citizens Committee on Harlem. Chaired by Algernon Black and Adam Clayton Powell Sr., the committee had worked for several years to raise public awareness of and improve the conditions facing blacks in the city.[9] In addition to focusing on economic and political problems, the committee had a deep appreciation of the role of the media in shaping New York City's race relations. Part of that interest in media came from Schomburg Center for Research in Black Culture curator L. D. Reddick, who cochaired the group's subcommittee on education and broadened its agenda to include the mass media. He worked to change the image of blacks presented in school textbooks and the media, arguing for "an end to caricaturing and slander in the newspapers, and on the screen, stage and radio." [10] Reddick had worked as a consultant on *Freedom's People* in 1941 and 1942. He also had made local radio appearances in which he discussed black political and historical issues and had written a seminal 1944 article on the influence of media on race relations.[11] Working on a radio series about black life and culture provided the committee with an unusually rich opportunity to address issues on its own political agenda. To facilitate its work, Nathan Straus took the unusual step of endowing the committee with the funds to sponsor the broadcast.[12] This arrangement elevated the visibility of the City-Wide Citizens Committee on Harlem, affiliating Straus and his station with it and its political issues, and set a unique example of public service cooperation.

For a young station striving to build a reputation for public service, this was a bold step. Despite the station's public service commitment, in the end it still depended on advertising revenue for support. As an independent station, WMCA could not rely on programming from a national network to build its listenership. Creating an audience niche in New York City's competitive radio market would require as much work for WMCA as it would for any independent commercial station, and the Straus family took imaginative steps toward that goal. For instance, one particularly effective way that the station began the process of building audience identification was by airing a five-hour Christmas program in 1943 in which members of the armed services serving abroad could speak with their relatives in New York City.[13]

The radio series *New World A'Coming* premiered before a live audience for half an hour on a Sunday afternoon at "three past three" in March 1944, less than six months after Straus had acquired the station.[14] From the opening credits to sign-off, this program was an extraordinary divergence from national radio's timid and cautious approach to discussions about racial inequality. The series was laced with an unabashed political message and offered a prime example of the use of radio programming as a political medium. The early scripts for the show were based on Ottley's writing, sometimes repeating his prose word for word. Ottley himself was often out of the country during the early years of the program, but the shows reflected the tenor and tone of his book as well as the influence of his reporting from abroad.[15] Canada Lee, the respected black actor who had narrated most of the National Urban League's radio programming, was chosen to moderate many of the early broadcasts in the series.[16]

In its initial episodes, *New World A'Coming* cast itself as a program built on black hopes for the future. After a musical introduction written by Duke Ellington, the show's standard sign-on was: "With the sweep and fury of the resurrection . . . there's a new world a-coming! . . . a series of vivid programs dramatizing the inner meanings of Negro life." Lee spoke in modulated and restrained tones that shielded some of the "heat" of the scripts' political content. His description of the "new world" was taken virtually verbatim from Ottley's book: "The Negro stands at the door of a fretful future. What his future will be no man can say. There are no blue prints. The Negro may not be able to predict his future, but he knows what he wants — liberty and peace, and an enriched life, free of want, oppression, violence and presumption. In a word, he wants democracy — cleansed and refreshed. He wants to be able to feel, see, and smell, and get his teeth into it."[17] If there were any questions about the program's intentions, Lee answered them: "The great American dream is still to be realized. The whip of intolerance has been felt by the Jew and the Catholic, the Immigrant. . . . But it is the Negro who feels the hand of intolerance most heavily. . . . No attempt will be made to approach radio as a wailing wall, no lamentations will be offered. Only the sort of tales you never read about in your newspapers."[18] A representative from the City-Wide Citizens Committee on Harlem emphasized on the same broadcast that the series was designed to "help create a living democracy" and to "help the new world to be born."[19]

New World A'Coming featured a heavy dose of programming built around the war and the issues it raised for blacks. But unlike earlier radio programs, it did not use the war merely as a backdrop for black heroism or as a basis for moral pleas for black rights. Instead, the war served to fuel the show's determined assertion that "Negroes feel that the day for just talking has passed."[20] In one episode, a character said: "We're the most loyal race in the country yet we's always gettin' kicked around—even the soldiers."[21] Here and elsewhere the show reported or dramatized actual instances of black soldiers in uniform being subjected to Jim Crow treatment at home. One episode told the true story of black soldiers who were denied access to a movie theater in New Jersey, and another dramatized an incident in which black servicemen were refused food service at a train station where Nazi prisoners of war were being served, a commonly reported complaint from black soldiers. On a later show, listeners heard the story of Captain Hugh Malzac, the black commander of the USS *Booker T. Washington* who was denied the right to buy the house of his choice.[22] All of these accounts were used as examples of continued glaring injustices against black servicemen, as well as reminders of their service to the country.

Despite its often critical tone, the series also included moving accounts of black patriotism, but with an edge. Channing Tobias, a black official with the national Young Men's Christian Association (YMCA) and a member of the National Selective Service Board, concluded the show's D-Day broadcast with a stirring vision of the war's end: "There will be the usual parade of the survivors of the victorious armies up Fifth Avenue. Let me express the hope that it will differ from previous parades of returning victors in one respect: namely, that instead of white, yellow and red Americans forming one unit, and black Americans another, there will be one American army of men rendered color blind by the fires of common suffering marching together to one tune and under one flag." The show also included remarks by Walter White, who contrasted the unified D-Day offensive with strikes at home against hiring black workers at defense plants.[23]

The series also reminded listeners that American racism weakened the country's standing abroad, a common argument against racial injustice. In one commentary, the narrator observed: "Treatment of the Negro soldier is propaganda meat for the Japanese, who would persuade the dark races of the world that Japan is fighting a war to liberate the oppressed colored people."[24] A subsequent program raised the subject again: "[The] Japanese are acutely aware that the problem of color is one of the pivotal questions in the war. They know that there are millions of colored peoples who want to see the last of the white man's rule—witness India, Singapore, Burma, and Malaysia."[25]

New World A'Coming also tackled political and racial issues at home with unusual candor. Discrimination in employment was a key and recurring subject on the series during its first year. One show mixed a demand with a gentle plea: "Essentially, the masses of Negroes are concerned only with jobs, for they believe that fundamentally their problem is an economic one. This belief accounts for the depths of passion that underlie their hopes and sacrifices for the new world a'coming. They want the 'For whites only' signs torn from American jobs. We believe it can be done. Don't you?"[26] A broadcast about residential segregation and Ku Klux Klan activity against blacks who had moved into a white Queens neighborhood included this commentary: "Mixed housing—where Negro and white live together in harmony—should be encouraged. If—as is declared daily—we are fighting a war to extend democracy to all people we must fall into step and plan for a new and democratic era. The ghetto in American life must go."[27] The series also covered racial disparities in access to health care.[28]

Jim Crow conditions received special attention, although sometimes they were treated in a peculiarly optimistic way, with one show concluding that "today the solid South is cracking. More and more, there are white people in the South who are raising the banners of democracy in the Negro's behalf."[29] Here, as elsewhere, the show pointed to increasing numbers of progressive whites who were prepared to work with blacks to end discrimination, although it was clear that the majority of whites were not. As one character expressed it: "[T]here are white men in America who would rather lose the war—even their own freedom—than see any change in the racial status quo."[30]

One recurring theme in the scripts was that black urban areas were tense, expectant "hot spots" susceptible to rumors that could become the "spark which awakened the Negro's deep seated sense of wrong, denial and frustration . . . [as] daily victims of insult, violence, and discrimination in a nation talking loudly about expanding democracy."[31] In an early dramatization, an old southern black character said: "When an old Negro was insulted, he shed a tear; but when these young 'uns is insulted, they sheds blood."[32]

The series's treatment of black performers also strayed from the pattern of earlier shows. When it paid attention to black entertainers, it did so only to illustrate particular Jim Crowisms or broader cultural arguments. For example, in a show featuring black pianist Hazel Scott, a character remarked: "[Whites are] willing to laugh and enjoy Negro performers, but they won't let you eat in the same restaurant, live in the same building, or ride in the same car with them." A program on W. C. Handy concluded with a dramatization of an incident in which his wife was denied hospital emergency room care. Billie Holiday was a featured performer on one broadcast, but its focus was on the African origins of black American music.[33]

The first broadcast of the program's second season came in October 1944. The show's creators were so enthusiastic about the favorable reception that had greeted its first season that they began the broadcast with an imaginative recap of the series's rave reviews:

"A hard hitting show that is certain to create a furor" (*The Daily News*); "It's a straight from the shoulder approach to the problem of racial prejudice and intolerance" (*Variety*); "[I]t gives dignity to the Negro and states his case" (*The People's Voice*); "[I]t is a project of the first importance. It is a factual recital. It asks only that the listener listen with an open mind and having listened, rely on his intelligence and his conscience. It is a public invitation to decent thinking. . . . This is really public service" (*The New York Times*); and, "It's a powerful and important program. . . . [It] does not mince words. . . . It's true to life. It's honest" (*Billboard*).[34]

Indeed, hyperbole was unnecessary. The show's initial run of broadcasts had been an astounding success, in large part because it was such a dramatic departure from other programs of the day. The *New York Times* contrasted the traditional image of an "Uncle Tom bowing and scraping under magnolia trees" to the image of "the Negro" on *New World A'Coming*:

He is a human being like any other, who thinks and talks like the actual person he is free of the subtly degrading burnt-cork clichés that have been fastened upon him by stage, screen and radio for lo! these many years. The impact of this simple reality is at once great and refreshing, and if the series accomplished no more than the sweeping away of these delusions it would serve its purpose. It has, of course, a great deal more to say, and to one listener it seems that it has thus far said it with a skill altogether worthy of the design.[35]

The *Pittsburgh Courier* also gave special notice to the show, even though it was only broadcast in New York City. Like the *New York Times*, it emphasized that *New World A'Coming* was "being handled without the old embellishments of Uncle Tom and the other typed characterizations so often utilized by producers to portray the race on the American scene." The *Courier* credited the collaboration between Ottley and Straus with fashioning a new kind of radio series about black life and concerns.[36]

Emboldened by its own success, the series opened its second season with "The Negro and the 1944 Elections." It attacked the poll tax and other barriers to black suffrage, declared that the primary issues of the election were "jobs, job equality, discrimination," and emphasized that the black vote was large enough in seventeen key states to swing the national election. It concluded with specific voter registration information and a plea for black voters to support those "who will support win the war and win the peace measures." This was not the first show in the series to raise electoral politics, but it was the first to do so explicitly. An earlier show, broadcast live from Abyssinian Baptist Church, interpreted the history of the black church as an expression of political self-sufficiency. It concluded with a reminder that "[t]oday—the Negro church—to all practical purposes —is a powerful, cohesive group, forming a mobile political bloc."[37] The sentiment expressed on the show about the importance of the 1944 elections was shared not only by Ottley but by Straus as well. In a speech in September, Straus had called for the election of public officials who will "strike the shackles of economic disability from our Negro brethren." He was emphatic in calling for "equal and exact justice" for all: "It means equality of economic opportunity, equality of educational privileges, equality at the ballot box. It means the end of the poll tax and the death of Jim Crow."[38]

As the end of the war approached, episodes in the series argued consistently that the war could only lead to world peace if democratic nations extended respect and cooperation to the nonwhite world and that before the United States could do that, it had to address racial inequality at home. Ottley, who had just returned from six months of covering the war overseas, hosted the program's 1944 Christmas Eve broadcast. He set the tone for the program, which looked ahead to the war's end: "One of the crucial questions which must be answered openly and truthfully *before* the war is ended concerns the Negro's place in the future peace of the world."[39] Other broadcasts repeatedly linked the struggle for black rights in the United States with those of other oppressed people of color fighting white domination throughout the world. Channing Tobias spoke with great conviction on the subject:

> Any plan for the future peace of the world that does not include racial equality as a major consideration is not only lacking in realism but destined to failure from the beginning. When Woodrow Wilson rejected the proposal for a racial equality clause in the covenant of the League of Nations, it was tantamount to saying that the victorious nations in World War I would continue to hold unbroken the ring of white dominance that then encircled the darker peoples of the world. World War II is in part the result of that decision. The question now is—will the victorious nations in this war heed the lesson of history, or must we have another war? There can be no middle ground.[40]

African American women and issues of special importance to them were not ignored by the series. An early Ottley dramatization told the story of a black domestic worker who quit her job in protest over her white employer's disrespectful treatment of her soldier son.[41] Black army nurses "fighting the color line" were featured as the subject of a radio play in one show. The D-Day broadcast specifically acknowledged the contributions of black women in uniform and on the assembly line.[42] The series also commended white women writers whose works protested racial inequality, such as Lillian Smith and Dorothy Parker.[43]

On occasion, the series used radio plays with dramatic roles for African American actors and actresses to make its political arguments. Ottley explained why in introducing a dramatization of a story by Dorothy Parker: "Until now, Negroes in radio have been mainly caricatures. They are usually hysterical servants frightened at ghosts—lazy buffoons—charmingly naive children or menials rapturously enchanted by their white masters. Rarely have the dramatic intensities of the Negro experience been portrayed." At first glance, the plot and message of Parker's story appear conservative compared with the stridency of the series's narratives and reporting. The story showed the reactions of white society women attending a party for a black male classical music star. One of the white women, after much tortured soul-searching and awkwardness, for the first time in her life "calls a Negro—*mister*." A narrative interlude from Ottley emphasized a broader message for a social world in racial transition—one aimed at the white listener: "Today we are approaching a third phase in race relations. For white and blacks are meeting more and more as equals—in business, progressive movements, and often socially. This meeting is still tentative, uncertain, and awkward on both sides. For the races have much to learn about each other. . . . Today's presentation was offered as a 'what not to do.'"[44]

In its second season, as in its first, *New World A'Coming* received accolades from the broadcast industry and from organizations dedicated to improving race relations. The prestigious Institute for Education by Radio unanimously named Straus as the recipient of its top award in 1945 for sponsoring "the radio program or series doing the most to further democracy in America." The series also was described as being "fearless and socially responsible" and as setting an example of "what can be done by any independent radio station." Indeed, in its first year, the series won seven awards, a number that would have made any of the major networks proud but was unprecedented for an independent station. Belatedly, in 1948, the series would also win a coveted Peabody Award, making it the first show on an independent station to do so.[45]

To Straus, these awards were a sad reminder of how limited the world of radio remained in addressing the issue of race. In a *New York Times* essay, he explained that he had hoped that radio, unlike older, more tradition-bound industries, would be freer from conventional trappings and beliefs. Instead, Straus had found that radio was as conservative as the rest on the topic of race relations: "On the airwaves, the Negro has been portrayed as a stock character—the lovable fool, the illiterate rascal or the old family retainer. It hardly seems open to dispute that this has been a trend with divisive implications all too obvious to need detailed recital. . . . It gives pause for thought that a program which adheres basically only to honesty and truthful portrayal of a vital problem should be so widely regarded in radio as 'new' and 'unusual.' "[46]

Although *New World A'Coming* challenged radio's traditional approach to discussions about race, it was only one of several "unconventional" ideas that Straus put in place at WMCA, all of which reflected ideas about radio's responsibility as a player in the political sphere.[47] The station openly challenged a National Association of Broadcasters rule that banned the sale of airtime for the discussion of controversial issues, a rule used effectively to bar the airing of political views by labor unions and other groups who had no other means of access to the airwaves. Straus implemented a policy that permitted the sale of time for the discussion of a controversial issue provided that another entity with an opposing view also bought time.[48] In his public appearances and writings, Straus also repeatedly warned of the dangers of radio's dependence on the advertising industry.[49] WMCA also had the important distinction of being the first major station in the country to hire an African American staff announcer. In February 1945, it hired as an announcer Gordon Heath, an actor and director well known in New York City's theater and radio circles for his work with Owen Dodson on several Negro Freedom Rally shows and with the American Negro Theatre.[50]

Despite Straus's deep commitment to the cause of African American racial equality, *New World A'Coming* maintained its sole focus on African Americans only for its first two seasons. Ottley's involvement with the series ended around the end of World War II.[51] In late 1945, the show broadened its coverage to include problems affecting other marginalized groups. This required a change in the wording of its original introduction from "dramatizing the inner meanings of Negro life" to "dramatizing the contributions of minority groups to the strengthening of American democracy."[52] The range of issues, now presented exclusively through radio plays, expanded to include protests against the mistreatment of Filipinos, European refugees, and Italian immigrants. But the series also continued to broadcast shows about African Americans. For example, the opening episode of the fall 1948 season focused on the impact of restrictive covenants in housing on the creation and perpetuation of ghettos, particularly Harlem.[53] The series remained on the air through 1957, by which time it also probed "discrimination against Puerto Ricans in New York

City, the plight of war refugees still living in European camps, and the nature of apartheid in South Africa."[54]

Special circumstances brought *New World A'Coming* to the air at WMCA. Local political support and supportive owners provided Ottley's political arguments with a radio forum. In many ways, WMCA embodied the civic vision of radio that the medium's early supporters had hoped for, a mission lost in the process of national commercial domination. This show stands in stark contrast in tone and content to the national broadcasts about African Americans and race relations, including those that specifically responded to racial violence. The history of the series provides a glimpse of the factors necessary for such a break with national radio's traditions: greater African American authorial imprint, local political support, cooperative radio owners, and a narrowed urban audience of African Americans and sympathetic whites. Sidney Williams's local Urban League interview program in Cleveland, *Minority Opinion,* depended on some of the same conditions. Such developments opened the way for the postwar emergence of "black radio," a genre that depended on and advanced the popularity of African American music but whose political implications still remain largely unexplored.

Unrestrained by fears of offending southern whites or national advertisers, *New World A'Coming* more fully reflected the rhetorical stance of black activism as the postwar period approached. The show owed its existence not only to Ottley's writing but also to a model of urban interracial cooperation that characterized the early stages of the modern civil rights movement and was reflected in the goals and operation of the New York City station WMCA. What Nathan and Helen Straus were able to achieve was not only exemplary but also exceptional because so few people had the financial resources and personal stature necessary to build such a successful independent station in a major radio market. Straus was a New Dealer with progressive views on race who also had the financial freedom to defy radio's conventions on race and other politically controversial issues. That combination of circumstances was rare in independent radio, and for that reason, the success of both the series and WMCA remained unparalleled.

New World A'Coming also depended on collaborations between progressive writers, artists, and musicians; civic groups; and sympathetic members of the radio and entertainment industry. This followed a pattern first established by federally sponsored radio programs like *Americans All, Immigrants All* and *Freedom's People*, as well as shows produced by the National Urban League. Interracial partnerships brought all of these broadcasts to life and to the air. The cultural apparatus of the New Deal state had been transformed to meet the needs of morale building in time of war. Racial unity had become a permanent part of that campaign, and along the way, it also was becoming a core tenet of urban liberalism.

Destination Freedom

By the end of the 1940s, the radio and advertising industries realized that many major urban areas had been transformed by the massive black migration of the war years. Many of the country's largest cities now had newly enlarged and more visible communities of African Americans, including a black working and middle class that the booming war economy had expanded. Even though many of the jobs that were opened to African Americans during the war had ended, the relatively strong postwar economy helped sustain a new black consumer class. It was this new black urban audience that local radio stations—and local advertisers—were beginning to "see" for the first time in the postwar period.[55] At the same time, the early stages of television's emergence as the new national advertising medium transformed radio into a locally oriented medium with targeted markets. Network radio's strongest days were nearing an end.

For African Americans, the late 1940s were a period of transition, as frustration continued around the issues of inequality, segregation, and discrimination. The more aggressive style of protest of the war period, albeit for the goal of being fully included and accepted in American democracy, did not disappear but continued to gather support. That postwar position found poignant expression on a Chicago radio program aptly titled *Destination Freedom*. Aired for two years beginning in 1948, this series had a consistency of authorial imprint, control, and relative independence that made it by far the single most effective use of radio to teach black history and to make political arguments on behalf of the black quest for freedom.

Destination Freedom was conceived and written by Richard Durham, a black journalist and experienced radio scriptwriter who had a gift for searing language and aural drama and a deep belief in the political and redemptive value of black history. Born in Mississippi but brought up in Chicago, Durham had worked with the

Works Progress Administration Writers Project, the *Chicago Defender,* and *Ebony* magazine, which the Chicago-based Johnson Publishing Company had introduced in 1945. A versatile radio writer, Durham had gained experience writing scripts for another local Chicago radio show, *Democracy U.S.A.* That experience, from 1946 to 1948, was valuable for Durham, but he also criticized the show for its patronizing attitude about black historical personalities. His dramatic skills as a scriptwriter also benefited from his work on a local black soap opera titled *Here Comes Tomorrow.*[56] The proliferation of local radio programming aimed at African Americans is another indication that the black urban radio market had been discovered and was being targeted.

Although still only in his twenties, armed with experience, Durham conceived the idea of writing and airing his own series about black history and contemporary political issues. He had a clearly articulated vision for his radio series and its focus on black lives:

[S]omewhere in this ocean of Negro life, with its cross-current and under-currents, lies the very soul of America. It lies there regardless of the camouflage of crack-pots and hypocrites—false liberals and false leaders—of radio's Beulahs and Amos and Andys and Hollywood's Stepin Fetchits and its masturbation with self-flattering dramas of "passing for white" such as "Pinky" and "Imitation of Life." It lies there because the real-life story of a single Negro in Alabama walking into a voting booth across a Ku Klux Klan line has more drama and world implications than all the stereotypes Hollywood or radio can turn out in a thousand years.[57]

He specifically chose a dramatic format that could exploit "the infinite store of material from the history and current struggles for freedom" and spotlight "new types of human beings who have never strutted across a stage or been portrayed on a radio program or in a novel."[58] Durham also saw his work as both educational and political in nature: "A good many white people have cushioned themselves into dreaming that Negroes are not self-assertive, confident, and never leave the realm of fear or subservience—to portray them as they are will give a greater education (to the audience) than a dozen lectures."[59]

As a ready resource, Durham could draw on a pool of talented, politically astute black dramatic actors and actresses affiliated with the progressive Du Bois Theater Guild in Chicago. They shared with

Richard Durham. (Courtesy of Clarice Durham)

Durham a deep commitment to their work and their political obligation as artists to the struggle for black freedom. From *Freedom's People* forward, this linkage between vocation and political expression continued a pattern of commitment among African American intellectuals, writers, and artists that had made possible national radio programming about black history and black contributions to the war effort. Relying on his own connections as a journalist, Durham persuaded the *Chicago Defender* to buy the time needed to broadcast the series.[60] He was able to convince WMAQ, the local NBC-owned affiliate, to carry the show, under the oversight of Judith Waller, a veteran NBC employee with twenty years of experience in educational and public affairs radio programming.[61] The *Defender* ended its sponsorship of the program during the first year when Oscar Brown Jr., a leading actor on the show, became a candidate for a political position in competition with a person endorsed by the paper. Although the Chicago Negro Insurance Association had offered to help cosponsor the program, WMAQ itself assumed the entire cost of the show. In early 1950, the Chicago Urban League helped sponsor a few of the episodes, but for most of its two-year life, the series had no outside funding.[62] The postwar economy now supported some black-owned businesses with the financial resources to buy the access on local stations that African Americans had been unable to secure on national radio. Even the local station itself, eager to draw black audiences, was willing to donate the time for black-oriented public service programming in much the same way that the national networks had used public affairs broadcasts to attract coast-to-coast audiences.

Destination Freedom premiered in June 1948. The series relied on a conventional format of half-hour radio dramatizations of the lives of black historical figures, transformed by Durham's eloquence and explicit political stance into powerfully effective racial protest. Durham coupled his clear sense of mission with good historical research. In a 1975 interview, Oscar Brown Jr., the program's lead actor and narrator, recalled the long hours Durham spent in public libraries poring over monographs, autobiographies, and other historical materials as he wrote scripts. In fact, in many cases, the words spoken by historical figures in the sketches were taken from their biographies, speeches, or writings.[63] *Destination Freedom* told the stories of leaders and activists like Crispus Attucks, Frederick Douglass, Toussaint L'Ouverture, Harriet Tubman, and Mary Church Terrell;[64] writers and poets like Richard Wright and Gwendolyn Brooks;[65] performing artists such as Katherine Dunham and Canada Lee;[66] scholars;[67] scientists and explorers;[68] and black cultural legends such as Stackalee and John Henry.[69]

Durham did not intend simply to illustrate black capabilities and heroism. He provided political lessons by implication, as had programs like *Freedom's People* and those sponsored by the National Urban League. Durham's shows dramatized how individuals confronted and overcame racial injustice through resourcefulness and principled resistance, characteristics Durham described as enduring aspects of black history. This black quest for freedom was explicit in everything about the series, from its title to the recurring use of the spiritual "Oh, Freedom":

Oh freedom, oh freedom
Oh freedom over me
And before I'd be a slave
I'd be buried in my grave
And go home to my Lord and be free.

On one show, historian Herbert Aptheker was quoted as saying that "the desire for freedom is the central theme, the motivating force in the history of the American Negro people."[70]

The show's essence resided in its emphasis on the interrelatedness of freedom, education, and historical knowledge. Education was not only a tonic against prejudice and hatred but also a route to black empowerment and progress, as indicated by the persistent black belief in education's emancipatory possibilities. A thirst for education and the personal struggles to satisfy it were central features in the lives portrayed in the series. For example, Richard Wright's story included an incident in which he persuaded an Irish coworker to check out books for him at a public library that refused service to blacks. One program traced the grass-roots campaign by Atlanta blacks for better schools and included Walter White saying that "education should mean equal opportunity for all children, Negro and white."[71]

The series also underscored the political responsibilities accompanying the privilege of education. In the dramatization of the life of Mary McLeod Bethune, she urged her students "to use education to liberate mankind . . . and free not just themselves, but those around us" by going "into the homes of every man and woman who has the right to vote and to escort him to the polls." The show on Bethune also contained an account of a true incident in which a Ku Klux Klan cross was burned at the campus of Bethune's fledgling Bethune-Cookman College. She had ignored white accusations that her "radical classes" were "teachin' the good colored folks down here the Bill of Rights 'n the Constitution an' all them un-American attitudes towards the state segregation laws."[72]

Many broadcasts stressed the importance of learning the history of African Americans in order, as Carter G. Woodson was quoted as saying, "to uncover the treasure of Negro life, so that America's goal of equality and justice may be strengthened by the knowledge of their struggle for freedom in the past."[73] Although the same belief in the value of black history was implicit in *Freedom's People* and *New World A'Coming, Destination Freedom* expressed a more profound faith in the power of historical truth, as revealed in these remarks by a character on the episode about Woodson: "[T]hey don't dare print the truth about the Negroes' role in American history. If they did, in one generation school children would grow up hating segregation and race discrimination, and those who profit by prejudice would have the ground shaken under them like an earthquake."[74]

Black women leaders were a prominent and distinguishing feature of the history lessons on *Destination Freedom*. Durham fought with his director at WMAQ to make sure that the black women characters the series featured were accorded the respect he believed they deserved: "[To] present Harriet Tubman as a sort of refined version of Aunt Jemima would be criminal. To present her as a sort of religious fanatic would be far-fetched. To present her as so many Negro women are, dauntless, determined, who have a healthy contempt for people who live by race prejudice and who are quick to recognize and extend a warm hand to other humans would be an honest, but for radio, a radical approach."[75]

This "radical" approach was embodied in the stories not only of Harriet Tubman and Mary McLeod Bethune but also of activists like Ida B. Wells and Mary Church Terrell and artists like classical singer Marian Anderson and choreographer Katherine Dunham. Both women and men were heard challenging the dual restrictions of race and gender, but the women characters spoke with a special passion. Terrell's character, for instance, spoke in a calm, determined voice: "I'll devote my life to being 'out of place,' to convincing most people that only when woman stands with man and demands full rights for all is she really—in her place."[76] She also spoke directly to white women about the racist actions of white men: "Women—since when have we needed cowards in bedsheets and masks and shotguns to safeguard our person and our homes. The only protecting women need is protection by equality under the law. Equality of opportunities and the right to share the benefits of this land, alongside men. . . . In the right to vote and the right to work—will freedom be found—for once a white woman bows down before white masculinism—she is ready for slavery!"[77]

More generally, episodes in the series repeatedly attacked race prejudice, Jim Crow, and injustice. One broadcast literally put a character named Race Prejudice on trial: "He's been found guilty in every civilized court in the world. What is your verdict? Will it be to exterminate him?" Durham often used a technique of giving objects and concepts human personalities and voices. For example, in an award-winning show about the two black surgeons who perfected human heart suture procedures, the heart narrated the story, eerily punctuated by the sound of a heartbeat. A broadcast about poor housing conditions featured Slums as a sinister character proud of his origins in racial covenants and other restrictive real estate practices. On the lighter side, the story of Louis Armstrong was told in part by his trumpet.[78]

Not only did the series assail prejudice, but it repeatedly linked segregation and economic profit. Poor housing conditions and residential segregation received special attention. One episode was devoted to the report of the National Commission against Segregation in the Nation's Capital. In addition to advocating the need to make "trade in racial prejudice as illegal as trade in opium and other habit-forming drugs," the broadcast's narrator concluded that "segregation of the Negro is planned as a matter of business."[79] On another show, a black Chicago alderman was dramatized as saying: "The slums are the cemeteries of the living. Jim Crow is the undertaker. Both must disappear from American life."[80]

The words of Ida B. Wells were used to connect racial violence with economic oppression: "The real motive behind all lynching was not the 'moral' issue pretended—but underneath it was a matter of murder for money and jobs. The base of all race terror . . . [is] a weapon to enslave a people at the bottom of the economic scale—and the 'moral' charges were just the envelope of the letter."[81] Similar sentiment was expressed in a drama about Walter White, who explained lynching by saying that "segregation and Jim Crow encourage lawlessness." The actor portraying White's father observed that "as long as there's segregation, and profit to be made outta prejudice, it'll rise up again." The show also included a scene in which White's dying father said to his son: "I didn't kill Jim Crow, and it's killing me. It's not my chance any more, it's yours."[82]

The series sometimes framed its arguments for lifting racial restrictions and expanding black economic and political opportunities in ways that appealed directly to white self-interest in freedom, the national interest in economic progress, and the country's pride in itself as an emerging world leader. An actor portraying Charles Caldwell, the black Reconstruction era Mississippi state senator, quoted Lincoln: "'Liberty is the heritage of all men. Destroy it, and you plant the seeds of despotism at your own door. Get familiar with hatred and put chains around others, and you prepare your own limbs to wear chains.'" [83] In a dramatized speech, Caldwell combined an appeal to white ethnic groups with a reminder that the world was peopled mostly by nonwhites: "If they say these rights are not for Negroes, what's to stop them from saying it's not for Catholics, Jews, Irish, the foreign born? If the right to rule is for white men only, what will you say to the majority of the world which is not white?" [84] The actor portraying William Lloyd Garrison spoke directly to the national economic interests when he said, "I know that she [the nation] can neither be truly happy nor prosperous while she continues to manacle and brutalize every sixth child born on her soil—the Negro people." [85]

Durham's extraordinary talents and intellectual versatility were reflected in the canon of African American history that he broadcast on *Destination Freedom*. But he still had less than absolute control over the series. He battled frequently with WMAQ, which ultimately had final approval over his scripts. He wanted to do shows on the lives of Nat Turner and Paul Robeson, but both were rejected by WMAQ as being too controversial. The political verve of some of his scripts was diminished by the editing of station officials. In a script on segregation in Washington, D.C., Durham equated segregation with a "master race" philosophy, but this characterization was eliminated from the show. Station officials probably had reason to be nervous. Durham himself later recalled that the station received complaints from the American Legion about several programs in the series. [86]

Despite Durham's disagreements with WMAQ, the station remained committed to the series, no doubt because of the praise and commendation Durham's work garnered. In 1949, the show won a first-place award from the respected Institute for Education by Radio for "its vital compelling use of radio technique in presenting contributions of Negroes to the development of democratic traditions and the American way of life." On the first anniversary episode of the series, Governor Adlai Stevenson commended the show for its work in reducing racial intolerance and educating the public about African American contributions. [87]

As we have seen, his success and the local nature of his program did not protect him from pressures to make his broadcasts conform

Cast of Destination Freedom.
(Courtesy of Clarice Durham)

to certain political expectations. Meanwhile, Durham's friend Langston Hughes continued his own confrontations with radio's ethos of race at the network level. Hughes had been hired by NBC in 1948 to rewrite a proposed black variety show titled *Modern Minstrels* whose cancellation had been urged by Walter White of the National Association for the Advancement of Colored People. According to Arnold Rampersad, Hughes was brought in "to lift its tone—although with blacks involved, NBC did not want it lifted too high." Hughes took on the well-paying job but criticized the entertainment industry for "not doing anything with the racial taboos surrounding either Hollywood or the radio." Hughes eliminated the minstrel angle and recast the show as *Swing Time at the Savoy,* which was broadcast for five shows and enjoyed some success. But Hughes was never able to achieve his full potential as a radio writer or secure more permanent radio work because, as his biographer Arnold Rampersad notes, "the doors to the radio networks were as tightly guarded as ever." [88]

In a letter to Hughes in November 1949, Durham described with great prescience the Cold War political shifts that would eventually doom his local show and his involvement with it:

Destination Freedom is still going strong, although censorship drops down on me rather unexpectedly at times. Recently I took the tactic of point-blank refusing to blue-pencil anything that I felt was healthy for inspiring Negroes to a more militant struggle. They threatened and cajoled but I held pat. Then they said I "must be Red." Can you imagine that? White folks will call you "Red" in a minute, won't they? Anyhow—I still put words like this in the mouths of some of my colored women characters (when they're giving advice to white women): "any white woman who accepts white supremacy is getting ready to sleep with fascism." . . . However, I admit that I'm forced to throw in some neutral character now and then.[89]

In 1950, Durham's relationship with *Destination Freedom* ended when, in the face of rising anti-Communist conservatism in radio and the arts, WMAQ discontinued his series. Another show continued under the same title but was hosted by "Paul Revere," who extolled white patriots like Nathan Hale and Dwight Eisenhower.[90]

After leaving the show, Durham reported to Hughes that it "seems like I've squeezed Chicago dry of anymore writing jobs, if you hear of anything in New York, please let me know."[91] But Durham stayed on in Chicago, where he secured work in the 1950s as national program director of the United Packinghouse Workers of America. His tenure there was not without controversy, as he tried to move the union toward making black advancement a priority, advocacy that resulted in his resignation under pressure in 1957.[92] Following that, he edited Elijah Muhammad's publication *Muhammad Speaks* in the 1960s, created and wrote scripts for the award-winning Chicago public television series *Bird of the Iron Feather* in the 1970s, and helped Muhammad Ali write *The Greatest*, Ali's life story, which was published in 1977.[93]

Durham's *Destination Freedom* produced a body of intellectual work that stands apart from all other attempts to bring black history and culture to a radio audience. The strident, undisguised messages in Durham's scripts would not have been acceptable on a national broadcast. In many ways, *Destination Freedom* was the culmination of efforts to use radio to teach African American history and to use that history to make political arguments in favor of contemporary struggles for racial reform. In 1941, the Office of Education show *Freedom's People* tried to do that but under dramatically different circumstances. The programs sponsored by the National Urban League relied on a softer and much narrower political appeal and only alluded to the extraordinary history of oppression and resistance that Durham so eloquently brought to the air. Only the local New York City show *New World A'Coming* took as bold an approach as Durham, but that show's focus soon shifted away from African Americans to embrace a broader appeal against oppression at home and abroad. Under Durham's hand, the focus of *Destination Freedom* never blurred and remained crisply centered on black history and black people.

The entertainment and communications industries were deeply influenced by and embedded in American racism and had profited from perpetuating racially restrictive portrayals. In the war era, a few members of the radio industry acknowledged the medium's complicity in and responsibility for shaping public attitudes about African Americans and about their place in American history and culture. All of the programs discussed in earlier chapters reflect a slow and subtle change in the way radio and the nation viewed race relations. These local shows in New York City and Chicago also offer a glimpse of radio's enormous potential as a medium for cultural and political education, as shown by the extraordinarily creative ways that Durham molded black history into living political argument. The tentative tone and content of the national broadcasts, whether produced by the federal government, the National Urban League, or the networks themselves, are most clear when those shows are compared with *New World A'Coming* and *Destination Freedom,* which were intended for black and liberal white audiences in northern cities. Black writers Roi Ottley and Richard Durham helped craft powerful and evocative programming with political and social messages that complemented and bolstered the arguments for racial equality that black leaders and organizations were then advancing. This radio programming also expressed a consolidation in black political discourse that occurred in the war period and prepared the way for the campaign against racial discrimination — through legal action, civil disobedience, and economic boycotts — that shook the nation in the 1950s and 1960s. But at the end of the 1940s, that new world had not come, and freedom still remained a destination.

Writing in 1944, L. D. Reddick urged "national Negro improvement associations" to begin "drying up the stream of anti-Negro propaganda" by developing "concrete plans" to control the "instruments of mass communication for the broad social purpose of bettering Negro-white relations." Reddick chided African American organizations for holding on to strategies directed at " 'converting *individuals* to right thinking' " rather than developing methods to intervene with the media, "which reach virtually every citizen and present their message to him so often and in such forms that he is powerless to escape it." Radio was one of the forms of mass communication Reddick had in mind. What distinguished this medium, he observed, was that "radio programs come directly into the home. This gives them a convenience and an intimacy that no other communication agency has." It was his hope, he wrote, that "such a powerful instrument as radio may be won to the cause of democracy in race relations."[1]

This book charts the efforts of African Americans in the 1930s and 1940s to win national radio to that cause. At a time when the medium denied them access, except in stereotypical and restricted roles, African American intellectuals, activists, federal officials, journalists, and entertainers, actors, and musicians welcomed the introduction of special programming about black history and culture. That programming allowed a new black voice to be heard on the radio, a voice that challenged accepted portrayals of black abilities and placed African American contributions and culture at the heart of American history. Perhaps more important, these radio shows helped to redefine and expand the concepts of Americanness and freedom in a decade that ended with the banning of discrimination in federal employment, the phasing out of segregation and discrimination in the armed forces, and the delineation of a litany of rights that would serve as the map of discovery for the new racial frontier pioneered in the modern civil rights movement of the 1950s and 1960s.

By the late 1930s, radio was the dominant and most pervasive form of national mass communication. The federal government's own ambiguous authority over the public airwaves created an enigmatic relationship with the nascent radio broadcast industry that accorded the state certain privileges of access. At the same time, the industry had made race and ethnicity "visible" on radio largely through entertainment programs that depended on a shared need for stereo-typical depictions. It is significant then that radio also was enlisted to help with the project of creating a new national historical narrative that included ethnic groups, Jews, and African Americans. That was precisely the mission of *Americans All, Immigrants All,* which brought the notion of intercultural education to national radio, only to see it abandoned and thwarted by members of Congress who sought to eliminate the vestiges of the New Deal cultural apparatus, including its heavy reliance on radio.

Only the threat of war and internal domestic disturbance kept federal officials in the business of producing radio programs about race relations. Persistent black protests over existing racial policies—especially in the federal government and the military—and the fear of violence and disloyalty by African Americans convinced those officials that they needed to speak to and about black citizens. A show like *Freedom's People* benefited from the use by black intellectu-als and federal officials of black culture and history to make broader although somewhat truncated arguments for black political and eco-nomic rights, under the guise of providing "educational" materials for blacks and whites. Franklin Roosevelt's tepid written remarks read on that show marked the emergence of a consensus view—for fair play and equal opportunity for blacks—that was yet too tentative and too weak to be given the weight of the voice of the president most associated with radio's emotive and political powers.

The fragile nature of that newly articulated plea for racial tolerance is best illustrated in struggles at the Office of War Information and the War Department over the development of radio programming suitable for lifting African American morale. Faced with the inherent contradictions of federal acquiescence to segregation and discrimi-nation, federal officials fought among themselves over what could be said about "the Negro's" place and plight, who spoke or quali-fied as an expert on the race question, and, ultimately, the impossible task of raising black morale without embracing African Americans' clamors for equality. For all of these reasons, the amount of fed-eral wartime radio programming on the issue of race was relatively limited compared to the enormous amount of airtime the networks devoted to federal messages about the war in general. The content of the programs that did emerge reflected the nature of the compro-mises that were being made within the federal government among a complex set of actors that included "Negro" advisers inside the government and racial reform advocates outside it; New Deal hold-overs now charged with doing the domestic cultural work of war; and those charged with carrying on the project of war itself.

Just as the New Deal state represented a new level of federal infra-structure and intervention, the federal government in World War II expanded its own authority and domain in order to wage a war of massive scale. Black activists in the late 1930s were challenging lynch-ing, the poll tax, and discriminatory practices in public places. The war against fascism became a splendid symbol of American racial hy-pocrisy. A. Philip Randolph and many others highlighted that con-flict incessantly with an eye toward the country's enemies and allies abroad. They were aided in their cause by the U.S. government's obsession with fears that Japanese and German propaganda would exploit the country's obvious racial contradictions.

Southern conservatives in Congress tried to use the war as one more excuse for delaying any changes in national racial policies. With the power of the appropriations purse under their control, their presence overshadowed attempts to construct radio programming designed to create even the *illusion* of racial unity. That illusion would have required a new imagining of race relations and rights for blacks, a racial future that many whites in the South and elsewhere resisted mightily.

The nationalization of the race question was an important by-product of the war. Most often, this metaphorical migration of the "Negro problem" is attributed to the migration of large numbers of blacks out of the South during the war. That is not the whole story, however, for something far more complex also was going on. Black activists and the black press held the country's policies of discrimina-tion and segregation up to shame in front of a world engaged in a war against a racial tyrant. The federal government and its armed forces emerged as a site for the "nationalization" of the South's prescrip-tions on race. Under the glare of an international spotlight, federal officials became uncomfortable with the notion that the South knew best how to manage the race problem. The need to look elsewhere for racial guidance became even more apparent after well-publicized incidents of black soldiers being attacked and lynched occurred dur-ing the war and in the immediate postwar period. The race problem became nationalized when federal officials no longer wanted acts and policies that were morally indefensible and ultimately damaging internationally attributed to them—and to the nation as a whole—as a new war, the Cold War, neared. It was only then that the national deference shown the South on how to handle the "Negro problem" began to be questioned.

The federal government served as a stage for these internal disputes about race, as evidenced by decisions of officials about what to air and what not to air. But throughout this period, radio programming from private sources also confronted the question of racial inequality. Although the national radio networks remained cautious about engaging the political issue of race for most of the 1940s, some exceptions at least permitted the introduction of that issue into political discourse on the air. The race riots of 1943 provided the most impetus for that change, as 1943 and 1944 brought the most radio programming on race relations of the decade to the air, including special broadcasts produced by the networks and the National Urban League. Another round of programming grew out of the need to discuss the political disruptions that accompanied President Truman's racial reform proposals in 1947 and 1948. That turmoil was embodied in the 1948 presidential campaign, which was marked both by southern defection and by the emergence of a northern urban black voting bloc.

Programs such as *America's Town Meeting of the Air* and the *University of Chicago Round Table* struggled throughout this decade with the question of how to acknowledge and treat race as a political issue suitable for open public discussion. These programs began with the belief that the "nigger question" could not be discussed, but the political events of the decade demanded that the question be addressed. By 1948, these shows had introduced a new white political discourse about race relations in which the more moderate position on the issue most often was expressed by the resident and guest intellectuals who shaped the broadcasts and the contrary view was expressed by white southern politicians. But black voices on these influential national radio forums were still few and far between, as white men continued to treat the race issue as one they could solve alone. More often than not, these broadcasts also revealed a discourse of white resistance as white listeners responded vociferously to the mere mention of the concept of ending segregation and discrimination.

Whites on these broadcasts who considered themselves moderates on racial issues stopped short of advocating remedies designed to end segregation, as that issue continued to stand as the dividing line between black and white agreement on a strategy for attacking racial inequality. In these ways, whereas many African Americans were advocating for racial equality by promoting an end to discrimination and segregation, many sympathetic white Americans were simply hoping for an end to the dangers of race hatred. During the war, those dangers included the threat of racial violence and national disunity, and in the Cold War period, the danger also included a weakened international position in the eyes of the rest of the world. Whites engaged in a propaganda of unity in this period, whereas African Americans engaged in a propaganda of equality, believing that unity without equality was impossible. Many whites recognized the necessity of having a common cause with African Americans, but whites and blacks nonetheless held quite conflicting political goals.

The spirit of black protest and militancy of the 1940s was best heard not on national radio but on two local radio shows whose titles also captured a prevailing sense of optimism and hope: *New World A'Coming* and *Destination Freedom*. On stations in New York City and Chicago, black writers Roi Ottley and Richard Durham captured the essence of what many African Americans hoped the war had brought: a redefinition of who could live as truly free Americans and an assumption by the federal government of the responsibility to help enforce that redefinition.

But even for those involved with these exceptional and progressive local shows, the idea persisted that radio was failing in its civic responsibility to address racial issues and in employing black actors, musicians, and technicians. Canada Lee, the veteran stage actor and political activist who had appeared on *New World A'Coming,* gave a scathing critique of radio's failures in a speech delivered in 1949. "The plain fact is that a virtual 'Iron Curtain' exists against the entire Negro people as far as radio is concerned," he said. Lee criticized both the paucity of African Americans on national radio and the nature of depictions of them, "for, with rare exceptions, it is the cannibal, the lazy gambler, the shiftless-thieving razor-wielding Negro, that has come to represent the totality of Negro life." In a call to arms against the broadcast industry, Lee concluded that "all the evidence of our lives bears testimony to the bitter seeds broadcast by the day and the hour and the minute to sixty million American radios." [2]

More broadly then, this book underscores the widespread recognition among African American activists and intellectuals that the mass media was essential to the fight for racial equality, a battle that could not be won by litigation alone but required shifts in public opinion that in turn determined changes in legislated public policy. On national radio, the emphasis on the spoken word limited what could be said about racial injustice and, as we have seen, who could say it. In the 1950s and 1960s, local radio stations, including those with black audiences as their target, would come to power as national radio's strength was diminished by television's coast-to-coast reach (and its easy appeal for corporate advertisers). Local black-oriented radio opened a new channel of communication within and between black communities, establishing itself as a new political organizing tool and ironically imperiling the livelihood of the black press.

Just as national radio was a forum in the political debates about race in the 1940s, national television functioned in a similar way for the civil rights struggles in the 1950s and 1960s. On television, African Americans could show the camera images of themselves engaged in the heroic work of seeking the right to sit at the front of the bus, waiting patiently in line to be denied once again the right to vote, or being attacked by the full power of Bull Connor's dogs and hoses. African Americans had no greater control over television than they had over national radio, but these were the images that television needed, and only they could provide them: images of themselves so at odds with traditional media depictions of "the Negro" that the images themselves were newsworthy. The codependency between the national media, especially television, and the civil rights movement of the 1950s and 1960s would have disadvantages, but the movement's relationship with the media would be one of its defining characteristics. The awareness of the potential power of such an alliance is one of the legacies of the exceptional programs highlighted in this book. The ideological framing for that movement was already in place at the end of the 1940s. During the war, African Americans had used democracy's rhetoric against itself. In the years to follow, they would add to that a Cold War rhetoric designed to hold the nation up to international shame. The effectiveness of that strategy would require compelling public demonstrations, sophisticated interventions with the mass media, and coalitions with cooperative whites, patterns already present by the 1940s.

The political, cultural, and intellectual history of the twentieth century is inextricably intertwined with the history of the modern mass media, particularly the mass electronic media of radio and television. There is no better example of the relationship between media, politics, and culture than in the history of race relations, a nexus evidenced in the politicized racial imagery of popular culture in both the entertainment and the news industries (a distinction now blurring) in all of their forms. Certainly, if race is a construction, as some have argued, then in the twentieth century, the mass media is its primary building site.

APPENDIX

Radio Programs Discussed in the Text

Program	Dates	Network or Station	Sponsor
Americans All, Immigrants All	Nov. 13, 1938– Apr. 30, 1939	CBS	Office of Education; Service Bureau for Intercultural Education; Carnegie Foundation
"America's Negro Soldiers"	Aug. 12, 1941	NBC	War Department
America's Town Meeting of the Air			
"Are We a United People?"	Feb. 20, 1941	NBC	*Town Meeting*
"Are We Solving America's Race Problem?"	May 24, 1945	ABC	*Town Meeting*
"Is There a Basis for Spiritual Unity in the World Today?"	May 28, 1942	NBC	*Town Meeting*
"Let's Face the Race Question"	Feb. 17, 1944	ABC	*Town Meeting*
"Should Government Guarantee Job Equality for All Races?"	Dec. 7, 1944	ABC	*Town Meeting*
"Should the Anti–Poll Tax Bill Be Passed?"	May 18, 1944	ABC	*Town Meeting*
"Should the Poll Tax Be Abolished?"	Oct. 20, 1942	NBC	*Town Meeting*
"Should the President's Civil Rights Program Be Adopted?"	Mar. 23, 1948	ABC	*Town Meeting*
"What Can We Do to Improve Race and Religious	Oct. 7, 1947	ABC	*Town Meeting*

Program	Dates	Network or Station	Sponsor
Relationships in America?"			
"What Should We Do about Race Segregation?"	Nov. 8, 1948	ABC	*Town Meeting*
Destination Freedom	June 1948–1950	WMAQ–Chicago	*Chicago Defender;* Chicago Urban League; WMAQ
Freedom's People	Sept. 21, 1941–Apr. 19, 1942	NBC	Office of Education; Rosenwald Fund; Southern Education Foundation
"Freedom's Plow"	Mar. 15, 1943	NBC	National Urban League
"Heroines in Bronze"	Mar. 20, 1943	CBS	National Urban League
"Judgment Day"	Sept. 27, 1942	NBC	War Department
Minority Opinion	Oct. 22, 1945–Mar. 1946	WJW–Cleveland	Cleveland Urban League
My People	Oct. 11, 1942, Feb. 13–Mar. 6, 1943	Mutual	OWI; G. Lake Imes
"The Negro and National Defense"	Mar. 30, 1941	CBS	National Urban League
New World A'Coming	1944–57	WMCA–New York	WMCA; City-Wide Citizens Committee on Harlem
"Open Letter on Race Hatred"	July 24, 1943	CBS	Entertainment Industry Emergency Committee; CBS
"Opportunity to Serve"	Mar. 20, 1943	Mutual	National Urban League
"Salute to Freedom"	Mar. 18, 1944	NBC	National Urban League
These Are Americans	Jan. 29–Mar. 3, 1944	CBS–Pacific Network	Hollywood Writers' Mobilization; CBS Department of Education, Western Division
"Too Long America"	Mar. 14, 1945	ABC	National Urban League

Program	Dates	Network or Station	Sponsor
University of Chicago Round Table			
"Civil Rights and Loyalty"	Nov. 23, 1947	NBC	*Round Table*
"Minorities"	Mar. 28, 1943	NBC	*Round Table*
"Race Relations around the World"	Dec. 5, 1948	NBC	*Round Table*
"Race Tensions"	July 4, 1943	NBC	*Round Table*
"Should We Adopt President Truman's Civil Rights Program?"	Feb. 6, 1949	NBC	*Round Table*
"The South and the Democratic Convention"	June 13, 1948	NBC	*Round Table*
"What Does the Election Mean?"	Nov. 7, 1948	NBC	*Round Table*
"What Do We Know about Prejudice?"	May 2, 1948	NBC	*Round Table*
"Victory through Unity"	Oct. 2, 1943	NBC	National Urban League

NOTES

ABBREVIATIONS
The following abbreviations are used for frequently cited sources throughout the notes.

ATMA Collection
  America's Town Meeting of the Air Collection, New York Public Library, New York, New York
Barnouw Papers
  Erik Barnouw Papers, Mass Communications Research Center, State Historical Society of Wisconsin, University of Wisconsin, Madison, Wisconsin
Benton Papers
  William Benton Papers, Records, Office of the Vice President, William Benton, 1937–46, Special Collections, University of Chicago Library, Chicago, Illinois
Caliver Papers
  Ambrose Caliver Papers, Moorland-Spingarn Research Center, Howard University, Washington, D.C.
FDRL
  Franklin D. Roosevelt Library, Hyde Park, New York
Hughes Papers
  Langston Hughes Papers, James Weldon Johnson Collection, Beinecke Rare Book Library, Yale University, New Haven, Connecticut
LC
  Library of Congress, Washington, D.C.
Locke Papers
  Alain Locke Papers, Moorland-Spingarn Research Center, Howard University, Washington, D.C.
MacLeish Papers
  Archibald MacLeish Papers, Manuscripts Division, Library of Congress, Washington, D.C.
MCRC
  Mass Communications Research Center, State Historical Society of Wisconsin, University of Wisconsin, Madison, Wisconsin
MSRC
  Moorland-Spingarn Research Center, Howard University, Washington, D.C.
NAACP Papers
  Papers of the National Association for the Advancement of Colored People, part 1, Meetings of the Board of Directors, 1909–50, and part 9, Discrimination in the United States Armed Forces, 1918–55, series A, General Office Files on Armed Forces Affairs, Microfilm Collections, Manuscripts and Archives, Sterling Library, Yale University, New Haven, Connecticut

NBC Collection
: NBC Collection, Mass Communications Research Center, State Historical Society of Wisconsin, University of Wisconsin, Madison, Wisconsin

NUL Papers
: National Urban League Papers, Manuscripts Division, Library of Congress, Washington, D.C.

PAL
: Performing Arts Library at Lincoln Center, New York Public Library, New York, New York

RBB
: Records of the Bureau of the Budget, Record Group 51, National Archives, Washington, D.C.

RCASW
: Records of the Civilian Aide to the Secretary of War (William Hastie–Truman Gibson Papers), Record Group 107, National Archives, Washington, D.C.

R. D. DuBois Papers
: Rachel Davis DuBois Papers, Immigration History Research Center, University of Minnesota, Minneapolis, Minnesota

ROE
: Records of the Office of Education, Record Group 12, National Archives, Washington, D.C.

ROWI
: Records of the Office of War Information, Record Group 208, National Archives, Washington, D.C.

RSC
: Radio Script Collection, Schomburg Center for Research in Black Culture, New York Public Library, New York, New York

SCR
: Schomburg Center for Research in Black Culture, New York Public Library, New York, New York

Straus Papers
: Nathan Straus Papers, Franklin D. Roosevelt Library, Hyde Park, New York

W. E. B. Du Bois Papers
: W. E. B. Du Bois Papers, Microfilm Collections, Manuscripts and Archives, Sterling Library, Yale University, New Haven, Connecticut

Wright Papers
: Richard Wright Papers, James Weldon Johnson Collection, Beinecke Rare Book Library, Yale University, New Haven, Connecticut

INTRODUCTION

1 Although few books treat the African American experience in the 1930s and the 1940s as a whole, a number of works address specific aspects of that period. For example, on the war years, see Wynn, *Afro-American and the Second World War,* and Blum, *V Was for Victory.* On political developments, see Dalfiume, *Desegregation of the U.S. Armed Forces;* Finkle, *Forum for Protest;* Washburn, *Question of Sedition;* Kesselman, *Social Politics of FEPC;* and Plummer,

Rising Wind. The magnitude and implications of the wartime migration remain largely unaddressed, although some works illuminate important aspects of that movement. See, for example, Lewis, *In Their Own Interests*, and Lemke-Santangelo, *Abiding Courage*. More general histories that cover early parts of this period include Weiss, *Farewell to the Party of Lincoln*, and Kirby, *Black Americans in the Roosevelt Era*. On the development of the predecessor cases to *Brown*, see Tushnet, *NAACP's Legal Strategy*, and on legal challenges to the electoral process, see Hine, *Black Victory*.

2 Among these were books as varied and provocative as Anne Petrey's *Street*; Richard Wright's *Native Son*, *Black Boy*, and *Twelve Million Black Voices*; W. E. B. Du Bois's *Dusk of Dawn* and *Color and Democracy*; Horace Cayton and St. Claire Drake's *Black Metropolis*; Gunnar Myrdal's *American Dilemma*; and Rayford Logan's *What the Negro Wants*, to name but a few.

3 See, for example, Payne, *I've Got the Light of Freedom*; Fairclough, *Race and Democracy*; Reed, *Simple Decency and Common Sense*; and Norrell, *Reaping the Whirlwind*.

4 Kelley, *Hammer and Hoe* and "'We Are Not What We Seem'"; Sullivan, *Days of Hope*; Von Eschen, *Race against Empire*; Denning, *Cultural Front*.

5 This also applies to studies of media treatments of African Americans, in which discussions of film and television also predominate. See, for example, Cripps, *Making Movies Black*; Bogle, *Toms, Coons, Mulattos*; Snead, *White Screens, Black Images*; Guerrero, *Framing Blackness*; MacDonald, *Blacks and White TV*; and Gray, *Watching Race*.

6 There are some exceptions. See, for example, MacDonald, *Don't Touch That Dial!*, a general history of radio programming that also includes an important chapter on the history of African Americans and radio, and Ely, *Adventures of Amos 'n' Andy*. More recently, some historians have discussed radio in their treatments of broader subjects, including Brinkley, *Voices of Protest*, and Cohen, *Making a New Deal*, which explores radio's impact on urban working-class community formation. Rubin, *Making of Middlebrow Culture*, includes a chapter on radio book-discussion shows. An important recent study of radio's importance as a national cultural force is Hilmes, *Radio Voices*. The early history of radio has received considerable attention but not the period under study here. See Douglas, *Inventing American Broadcasting*; Smulyan, *Selling Radio*; McChesney, *Telecommunications, Mass Media, and Democracy*; and Bergreen, *Look Now, Pay Later*. Broader treatments of mass media that include some discussion of radio include Czitrom, *Media and the American Mind*; Barnouw, *Tower in Babel* and *Golden Web*; and Hilmes, *Hollywood and Broadcasting*.

7 See, for example, Morrison and Lacour, *Birth of a Nation 'hood*, and Morrison, *Race-ing Justice, En-gendering Power*.

8 U.S. Department of Commerce, Bureau of the Census, *Sixteenth Census of the United States, 1940: Housing*, vol. 2, *General Characteristics*, table 10, "Persons per Room and Radio for Occupied Units, by Tenure and Color of Occupants, for the United States, by Regions, Urban and Rural, 1940," 38–39; Sterling and Kittross, *Stay Tuned*, 81, appendix C, table 8, "Ownership of Radio Receivers, 1922–1988," 656.

9 Pells, *Radical Visions and American Dreams,* 41; MacDonald, *Don't Touch That Dial!,* 60–61; *Fortune* 17 (January 1938): 88.

10 Radio's arrival as a source of hard news reporting was delayed until the mid-1930s due to a conflict between the networks and newspaper publishers over whether the wire services for newspapers should be allowed to offer their reports to radio stations. Czitrom, *Media and the American Mind,* 86–87. See also Lott, "Press Radio War," 275–86, and Jackaway, *Media at War.*

11 NBC beat out its competitor CBS by paying $100,000 to air the show nationally for fifteen minutes every evening of the week except Sunday. The show also had several marketing spin-offs, including a film, a daily comic strip, and phonograph records of the characters. MacDonald, *Don't Touch That Dial!,* 26–28; Barnouw, *Tower in Babel,* 229; Bergreen, *Look Now, Pay Later,* 70, 73–74; Czitrom, *Media and the American Mind,* 84. For detailed biographical information on Gosden and Correll, see Ely, *Adventures of Amos 'n' Andy.*

12 Ely, *Adventures of Amos 'n' Andy,* 63. For the most insightful treatment of the racialized basis of this appeal, see Hilmes, *Radio Voices,* 85–90. Media historian Erik Barnouw has characterized *Amos 'n' Andy* as "the ghetto-keeper's fantasy picture of the inside of the ghetto." Barnouw, *Tower in Babel,* 276. Czitrom argues that Gosden and Correll, themselves the products of the minstrel and carnival circuit, simply transformed and expanded the minstrel tradition for radio. Czitrom, *Media and the American Mind,* 83–84.

13 Hilmes, "Invisible Men," 309–10, 313, and *Radio Voices,* 89–90. Indeed, blackface on the radio may have been the ultimate "ventriloquized cultural form," a term used more generally in Lott, "Love and Theft," 38.

14 Two studies that point to the importance of *Amos 'n' Andy* in the lives of working-class Americans make no mention of the show's racialized content. In their exceptional study of southern cotton mill workers, Jacquelyn Dowd Hall and her colleagues conclude that radio programming actually dictated the reorganization of workers' recreational time, using *Amos 'n' Andy* as a prime example. Hall et al., *Like a Family,* 258. In her discussion of the importance of radio in bringing together white industrial workers in this period, Lizabeth Cohen also relies on several examples of the popularity of *Amos 'n' Andy* to help make that argument but again with no attention to the show's racial content. Cohen, *Making a New Deal,* 325, 328–29. Lawrence Levine's discussion of the show recognizes its roots in minstrelsy but credits its success to its portrayal of an ethic of hard work and achievement that had universal appeal to listeners, rendering the racial aspect of the show less relevant. Levine, *Unpredictable Past,* 220.

15 Bishop W. J. Walls, quoted in "Handicap of *Amos 'n' Andy* Is Related," *Pittsburgh Courier,* May 23, 1931. For a chronicle of the protests against the radio show and its 1950s television successor, see Ely, *Adventures of Amos 'n' Andy.*

16 Reddick, "Educational Programs," 367–89. Nannie Burroughs claimed credit for starting the fight against the show in 1929 and chided the *Courier* for not helping her earlier. Nannie H. Burroughs, "Negroes Fighting All Over the Place," *Pittsburgh Courier,* September 12, 1931.

17 *Pittsburgh Courier,* September 5, 1931. Walls also had commented on the por-

trayal of black women: "If a woman is not a tool she becomes a senseless, bossy wife, or a tyrannizing vampire." Walls, quoted in "Handicap of *Amos 'n' Andy* Is Related." For other examples of concern about the portrayal of black women, see *Pittsburgh Courier,* May 30, June 13, 1931.

18 *Pittsburgh Courier,* May 30, June 6, 20, 1931.

19 Ibid., June 20, 1931.

20 Ibid., June 6, 1931.

21 Ibid., October 24, 1931.

22 Ibid., May 23, 1931.

23 Ibid.

24 Ely, *Adventures of Amos 'n' Andy,* 171. According to Ely, "Ultimately, the *Amos 'n' Andy* controversy stirred new debate of an old but abiding issue: Which aspects of black life and culture should Afro-Americans display in public and which—if any—should they keep among themselves or abandon altogether."

25 Quoted in Bay, *White Image in the Black Mind.* For histories of the evolution of the black press, see Suggs, *P. B. Young, Newspaperman, Black Press in the South,* and *Black Press in the Middle West.*

26 For a discussion of African American protests against the film, see Cripps, *Slow Fade to Black,* 41–69.

27 MacDonald, *Don't Touch That Dial!,* 328–32, 27–28; Bergreen, *Look Now, Pay Later,* 70, 73–74; Czitrom, *Media and the American Mind,* 84; L. S. Cottrell, quoted in MacDonald, *Don't Touch That Dial!,* 328. Hilmes also concludes that representations of African Americans were "far more central to the overall discourse created by broadcasting than standard histories, through omission, have implied." Hilmes, *Radio Voices,* 33, 77–78.

28 MacDonald, *Don't Touch That Dial!,* 365–69.

29 See, for example, the attention given black consumers in *Sponsor,* a radio trade magazine: "The Forgotten 15,000,000: Ten Billion a Year Negro Market Is Largely Ignored by National Advertisers," *Sponsor,* October 10, 1949, and "The Negro Market: $15,000,000,000 to Spend," *Sponsor,* July 28, 1952.

30 Both NBC and CBS and their trade association, the National Association of Broadcasters (NAB), openly solicited the president's presence on the air. The day after his inauguration, the networks and the NAB promised that all broadcasting stations would grant him immediate access "on a moment's notice." Roosevelt took them up on that offer, making fifty-one network broadcasts in his first year of office. He soon learned firsthand the power of the medium: his first "fireside chat" drew an estimated 60 million listeners. The broadcast averted a run on the banks that had been anticipated the next day. For a discussion of the overall New Deal public relations operation, see Winfield, *FDR and the News Media,* 79–102. Of special value on the relationship between Roosevelt and all media, including radio, see Steele, *Propaganda in an Open Society,* 20; McChesney, *Telecommunications, Mass Media, and Democracy,* 182; Leuchtenburg, *Franklin D. Roosevelt and the New Deal,* 44–45; and Cantril and Allport, *Psychology of Radio,* 32. On Eleanor Roosevelt, see Beasley, *Eleanor Roosevelt and the Media* and *White House Press Conferences of Eleanor Roosevelt.*

31 For several examples of how federal officials took advantage of radio's avail-

ability, see Sayre, *Analysis of the Radiobroadcasting Activities of Federal Agencies,* 13–15, and Smulyan, *Selling Radio,* 21–22. Louisiana governor Huey Long's reliance on radio also was an attempt to circumvent the press. The "lying newspapers," as he referred to them, led him to make many of his appeals via radio. Brinkley, *Voices of Protest,* 71, 79.

32 For example, in 1935 CBS broadcast a two-hour program, "Of the People, by the People, for the People," that commemorated the second anniversary of the New Deal. The show, which the federal Office of Education recommended for classroom use, featured dramatizations of key administration achievements, followed by discussions of the events by New Deal officials. One widely circulated editorial blamed the broadcasters for having "succeeded in over-selling the administration on the advantages of radio advertising." Steele, *Propaganda in an Open Society,* 23.

33 When the president decided to use his 1936 State of the Union Address to kick off his reelection campaign, he insisted that the address be given before an evening session of Congress in order to draw a large listening audience, an unprecedented action that deeply angered the Republicans. Later, the 1936 Democratic National Convention adjusted its schedule and program to take advantage of radio coverage, prompting one writer to refer to it as a "radio convention." So extensive was the president's use of radio that on at least one occasion during the campaign, his opponent, Alf Landon, was proudly introduced as "the candidate who was no 'radio crooner.'" Schlesinger, *Age of Roosevelt,* 503, 574, 588, 590, 612; Chester, *Radio, Television, and American Politics,* 38, 40; Nimmo, *Political Persuaders,* 133.

34 In addition to their outreach to black voters, the Democrats focused on a dozen key cities and broadcast appeals in Italian, German, Polish, Hungarian, and Greek. Chester, *Radio, Television, and American Politics,* 39. Roosevelt's radio appeals to industrial workers also paid off in the 1936 election. Lizabeth Cohen observes that "radio helped rally industrial workers for Roosevelt." One Chicago steelworker reported that listening to the president on the radio made it seem like "FDR was on our side, organizing with us." Cohen, *Making a New Deal,* 332.

35 Harold Ickes, "Readin', Ritin', and Radio," opening address at the First National Conference on Educational Broadcasting, December 1936, in Marsh, *Educational Broadcasting,* 8–9. The Office of Education, in conjunction with an unusually broad coalition of government officials, network officials, educators, and civic groups, sponsored two national conferences on educational broadcasting, the first in 1936 in Washington, D.C., and the second the next year in Chicago. Over 700 people attended the 1936 conference, which also attracted much coverage in the national, trade, and education press.

36 The Interior Department building constructed in 1938 included the first federal broadcasting studio. Entry 182, "Radio Education Project, Audience Preparation Division, Office File of Ben Brodinsky, 1936–40," box 7, folder, "Speeches," ROE; John W. Studebaker, U.S. commissioner of education, "Radio in the Service of Education," speech, in Marsh, *Educational Broadcasting,* 28–29, 31.

1 DuBois and Okorodudu, *All This and Something More,* 34–41.

2 Ibid., 47–48, 51–53, 59, 65–69, 81–83.

3 Interview with Rachel Davis DuBois, November 17, 1984, in Crispin, "Rachel Davis DuBois." The idea of a radio program had been raised by members of DuBois's Advisory Committee at the liberal Progressive Education Association, with which the Service Bureau was briefly affiliated. DuBois and Okorodudu, *All This and Something More,* 86; Crispin, "Rachel Davis DuBois," 29, 32.

4 DuBois and Okorodudu, *All This and Something More,* 86; Montalto, *History of the Intercultural Education Movement,* 149–50.

5 William Boutwell to Rachel Davis DuBois, August 26, 1938, series V, box 22, folder 8, "Correspondence, John W. Studebaker," R. D. DuBois Papers.

6 Illinois governor Henry Horner also brought the idea directly to the attention of RCA's (NBC's parent company) president, David Sarnoff, who was himself the child of Jewish immigrants. Avinere Toigo to David Sarnoff, October 14, 1938, box 54, folder 49, *"Americans All, Immigrants All,"* NBC Collection. Although Angell had no experience in radio, NBC considered his association with the network a public relations coup of great magnitude and orchestrated an elaborate publicity campaign around his hiring. Prior to his sixteen years as president of Yale, Angell had been on the faculty and then the acting president of the University of Chicago. He also had headed the Carnegie Corporation. John F. Royal, vice president for programming, to Lenox Lohr, NBC president, March 8, 1937; "Dr. Angell Takes Full-Time Post as Radio Education Counselor; Retired Yale President, Vigorous at 68, Sees Chance of Rendering a Real Public Service," *New York Herald-Tribune,* June 28, 1937; "Dr. Angell to Receive $25,000 Radio Salary; NBC President Says He Will Have Free Hand to Devise Ways to Serve Listeners," *New York Times,* June 29, 1937; and "The Place of Radio in Education," NBC publication commending Angell's appointment, box 92, folder 64, "Department Files, Program Educational Counselor, J. R. Angell, 1937," NBC Collection.

7 Officials at NBC's Chicago affiliate WMAQ encouraged Angell to meet personally with Toigo in New York City to discuss what they called "the Melting Pot program." James R. Angell, educational counselor, NBC, to John F. Royal, NBC, January 24, 1938; Judith Waller, NBC, Chicago, to Angell, February 8, 1938; Angell to Philips Carlin, NBC, February 10, 1938; and Avinere Toigo to Angell, February 21, 1938, box 58, folder 48, *"Americans All, Immigrants All,"* NBC Collection.

8 In the handbook for listeners published after the series was broadcast, the Office of Education credited Toigo as "one of the most enthusiastic advocates and supporters of a comprehensive radio presentation of the immigrants' contribution to American life." Jones, *Americans All . . . Immigrants All,* iv.

9 William Boutwell to Franklin Dunham, educational director, NBC, March 7, 1938, and James R. Angell to Dunham, March 11, 1938, box 58, folder 48, *"Americans All, Immigrants All,"* NBC Collection.

10 Unfortunately, CBS corporate records for this period are unavailable, so it is impossible to be more precise about internal decision making concerning

this series and others covered in this book. Where such information is presented, it has been gleaned from other archival sources as noted.

11 Television broadcasting and the manufacture of television sets had begun in the 1930s, but availability was stalled first by the depression and then, more significantly, by World War II. Smith, *In All His Glory*, 186–88. This period in television history is referred to as its "false dawn." Sterling and Kittross, *Stay Tuned*, 151.

12 DuBois also proposed a program of active listener follow-up through study groups and local activities built around the concept of learning about different cultural values. Rachel Davis DuBois to John W. Studebaker, August 16, September 26, 1938, series V, box 22, folder 8, "John W. Studebaker, ca. 1938–39," R. D. DuBois Papers.

13 Seldes's 1936 book, *Mainland*, according to Richard Pells, captured well the "love affair with the past" and with "the people" that Seldes and other Popular Front intellectuals embraced after 1935 when they abandoned their earlier more radical visions. Pells, *Radical Visions and American Dreams*, 316–17. In his biography of Seldes, Michael Kammen characterizes Seldes as the person who "launched" the radio series, but that conclusion overstates his role and ignores the indispensable work of DuBois and others at the Office of Education in creating and molding this series. Kammen, *Lively Arts*, 261.

14 Gilbert Seldes, "General Approach to Writing of Script for *Immigrants All—Americans All*," n.d., entry 174, "Radio Education Project, Office of the Director, Records Relating to Radio Programs, 1935–41," box 1, folder, "Prospecti, *Americans All, Immigrants All*," ROE.

15 Novick, *That Noble Dream*, 224–39.

16 Minutes, Advisory Committee meeting, September 28, 1938, entry 174, "Radio Education Project, Office of the Director, Records Relating to Radio Programs, 1935–41," box 1, folder, "Prospecti, *Americans All, Immigrants All*," ROE.

17 President Franklin Roosevelt, "All of Us, and You and I Especially, Are Descended from Immigrants and Revolutionists," speech to Daughters of the American Revolution, April 21, 1938, Washington, D.C., in Rosenman, *Public Papers and Addresses of Franklin D. Roosevelt*, 258–60; "Report of Trip to New York, September 8th to 10th Inclusive, 1938," entry 174, "Radio Education Project, Office of the Director, Records Relating to Radio Programs, 1935–41," box 1, folder, "*Americans All*, Planning," ROE; Rachel Davis DuBois to Gilbert Seldes, John W. Studebaker, and William Boutwell, September 26, 1938, series V, box 22, folder 7, "Correspondence, Gilbert Seldes, ca. 1938–39," R. D. DuBois Papers.

18 Philip Cohen to William Boutwell, October 14, 1938, entry 170, "Radio Education Project, Office of the Director, Subject File, 1936–40," box 5, folder, "Cohen, Phil, Jan. 1, 1938–Dec. 31, 1938," ROE.

19 John W. Studebaker to William Boutwell, October 29, 1938, entry 174, "Radio Education Project, Office of the Director, Records Relating to Radio Programs, 1935–41," box 1, folder, "*Americans All*, Planning," ROE. The recordings were an expensive innovation, costing at least $10,000 to produce. Funds

for this purpose were provided by the Carnegie Corporation Foundation Committee on Scientific Teaching Aids. Boutwell to Gilbert Seldes, January 13, 1939, entry 170, "Radio Education Project, Office of the Director, Subject File, 1936–40," box 9, folder, "Gilbert Seldes," ROE.

20 Coverage of the Nazi takeover of Austria in the spring of 1938 was broadcast via live CBS pickups. The Munich crisis in September 1938 precipitated a CBS innovation of daily European "roundup" broadcasts, with radio correspondents reporting from Prague, London, Paris, Rome, Berlin, and Munich. Edward R. Murrow and William Shirer came to fame because of their evocative coverage of these events. Moreover, public interest in war coverage rose dramatically thanks to the recent development of portable radio sets. Hosley, *As Good As Any,* 51, 53, 58–59; William Boutwell to Rachel Davis DuBois, August 26, 1938, series V, box 22, folder 8, "Correspondence, John W. Studebaker," R. D. DuBois Papers. On Boutwell's entreaties to CBS, see Boutwell to Sterling Fisher, education director, CBS, September 23, 1938; Boutwell to Edward R. Murrow, CBS correspondent, September 24, 1938; Boutwell to Harry C. Butcher, vice president, CBS, November 19, 1938; and Boutwell to Fisher, November 21, 1938, entry 174, "Radio Education Project, Office of the Director, Records Relating to Radio Programs, 1935–41," box 3, folder, "Correspondence, William Boutwell, July–December 1938," ROE.

21 The number of national and racial groups to be highlighted in individual shows was reduced to thirteen (from the thirty-one groups suggested by DuBois). Another thirteen episodes interspersed the broader themes Seldes promoted about immigration in the frontier and colonial eras and the contributions of many groups in the American Revolution and in crosscutting fields such as science, industry, social progress, and the arts. Some groups were eliminated altogether (e.g., Puerto Ricans, East Indians, and West Indians), and many others were combined (e.g., Latin Americans, Mexicans, and Spanish into "Our Hispanic Heritage"). Instead of several shows featuring different eastern European nationalities, two episodes were devoted to "Slavic peoples." In order of broadcast, separate episodes were dedicated to the following national and racial groups: English, Hispanic, Scotch-Irish and Welsh, "Negro," French-speaking, Irish, German-speaking, Scandinavian, Jewish, Slavic, Oriental, Italian, and Near Eastern. "Revised Outline (Based on Suggestions of Committee on Sept. 28th), *Immigrants All—Americans All,* Columbia Broadcasting System, November 14–May 15, 1938 [*sic*], Gilbert Seldes, Script Writer, Rachel Davis-DuBois, Consultant on Intercultural Education, Ruth E. Davis, Research, Radio Division, U.S. Office of Education," entry 174, "Radio Education Project, Office of the Director, Records Relating to Radio Programs, 1935–41," box 1, folder, "Prospecti, *Americans All, Immigrants All,*" ROE.

22 Rachel Davis DuBois to William Boutwell, December 21, 1938, series V, box 22, folder 4, "Correspondence, William D. Boutwell, 1937–39," R. D. DuBois Papers. Seldes insisted that the show emphasize that "life in the United States is the achievement of many kinds of people" and that any plea for tolerance "strikes me as exhortation" and should not be included. Memorandum,

Boutwell to Phil Cohen, January 3, 1939; Gilbert Seldes to Boutwell, January 12, 1939; Boutwell to Seldes, January 13, 1939; Seldes to Cohen, January 1939; and Boutwell to John W. Studebaker, January 20, 1939, entry 170, "Radio Education Project, Office of the Director, Subject File, 1936–40," box 9, folder, "Gilbert Seldes," ROE. Scripts for *Americans All, Immigrants All* are located at PAL. Phonograph recordings of the broadcasts are available at the National Archives, Audiovisual Collections, Washington, D.C., and in the R. D. DuBois Papers.

23 The Office of Education attempted to exert some control over the project by preparing a detailed written assignment of duties for the 2,000 hours of work it estimated each half hour broadcast would require. Publicity and promotion for the show would come from the Office of Education, the Service Bureau, and CBS, each assuming responsibility for different audiences and outlets. Listening aids would be prepared by DuBois, and audience mail was to be handled by the federal government. The scripts would be subject to review by the Office of Education, the Office of Immigration and Naturalization, and CBS, as well as by the advisory committees suggested by DuBois. CBS would transcribe the shows for public use; distribution of the transcriptions would be handled by the Office of Education. Memorandum, "Three Partners," n.d., entry 174, "Radio Education Project, Office of the Director, Records Relating to Radio Programs, 1935–41," box 1, folder, "Prospecti, *Americans All, Immigrants All*," ROE.

24 In the series, the English were lauded for constructing all of the nation's basic legal and political infrastructure. The episode about the French descendants was based on the premise that there was "not a moment in the life of the United States in which their influence could not be traced." "Our English Heritage," script 2, November 20, 1938, 6, and "The French and Netherlanders," script 7, December 25, 1938, 5, PAL; memorandum, Laura Vitray to William Boutwell and Philip Green, December 9, 1938, entry 174, "Radio Education Project, Office of the Director, Records Relating to Radio Programs, 1935–41," box 1, folder, "*Americans All*, Planning," ROE. In a weak attempt to link the English to a broader group of Americans, the script argued unconvincingly that their inspiring notion of rebellion against tyranny was taken up "by rich men and poor men, by landowners and bondsmen, by gentlemen and by ex-convicts, by Episcopalians, Catholics, Protestants, and Jews." This seemed a very taut stretch of the chain of immigrant connection and collaboration. The series built a more prototypical immigrant experience around the Scots, who were portrayed as invited workers and indentured servants who arrived seeking to practice their religion freely and then rose to stability. "The Scots, Scotch-Irish, and Welsh," script 4, December 4, 1938, PAL.

25 Rachel Davis DuBois to Gilbert Seldes, October 17, 1938, series V, box 22, folder 7, "Correspondence, Gilbert Seldes, ca. 1938–39," R. D. DuBois Papers.

26 "The Slavs: Russians, Ukrainians, Yugoslavs," script 14, February 12, 1939, 5, and "The Slavs: Poles, Czechs, and Slovaks," script 15, February 19, 1939, 28,

PAL; Laura Vitray to Gilbert Seldes, January 28, 1939, entry 174, "Radio Education Project, Office of the Director, Records Relating to Radio Programs, 1935–41," box 1, folder, "*Americans All,* Planning," ROE.

27 "The Irish," script 9, January 8, 1939, 5, PAL.

28 George Fredrickson points to the existence of "romantic racialism" among white liberals in the 1930s, although he does so specifically in reference to the practice of arguing that African Americans contributed their "gifts" of joy, expressiveness, music, and songs to the United States. Fredrickson, *Black Image in the White Mind,* 328–29. See also Ignatiev, *How the Irish Became White.*

29 "The Italians," script 17, March 5, 1939, 10, 12, 16, PAL. In a review session on a script, one participant suggested to Office of Education official Philips Carlin that many immigrants had a much harder time than was depicted in the script. Carlin dismissed the comment, defending the narrower focus on achievement and success as a matter of selection. "Notes from March 9, 1939, Reading of Script No. 19," "Other Groups," entry 174, "Radio Education Project, Office of the Director, Records Relating to Radio Programs, 1935–41," box 1, folder, "*Americans All,* Planning," ROE.

30 "Our Hispanic Heritage," script 3, November 3, 1938, PAL.

31 "Orientals in America," script 16, February 26, 1939, 6–9, PAL; Laura Vitray to Gilbert Seldes and Philip Green, February 14, 1939, entry 174, "Radio Education Project, Records Relating to Radio Programs, 1935–41," box 1, folder, "*Americans All,* Planning," ROE.

32 "Orientals in America," 14, 25, 27–28.

33 Quoted in Seelye, "Broadcasting Tolerance," 54–55. Seelye's thesis is a detailed analysis of a sample of the mail. She also includes the results of a questionnaire sent to a small sample of listeners. Unfortunately, the actual letters and cards on which Seelye relied in 1941 cannot be located in the textual holdings about the series at the National Archives, Washington, D.C. However, her thesis contains many detailed excerpts of the letters grouped by content. Her references to the letters only include the state from which they came and do not allow one to discern the gender of the writer. Furthermore, her analysis does not provide a breakdown of the mail by the show to which it responded, although that can often be surmised.

34 Ibid., 94, 80–84.

35 Ibid., 70–71, 85.

36 Ibid., 73–78.

37 Ibid., 53, 56.

38 Ibid., 67.

39 Ibid., 93.

40 On *Birth of a Nation,* see Cripps, *Slow Fade to Black.* For a discussion of the NAACP's campaign against radio programming, including *Amos 'n' Andy,* see Archer, *Black Images in American Theater,* 225–62. On the *Pittsburgh Courier*'s petition drive against *Amos 'n' Andy,* see Reddick, "Educational Programs," 367–89. For details on recurring campaigns against *Amos 'n' Andy,* see Ely, *Adventures of Amos 'n' Andy.*

41 *Opportunity* 7 (July 1928): 218.

42 Rachel Davis DuBois to Alain Locke, September 19, 1938; DuBois to John W. Studebaker, October 26, 1938; Studebaker to DuBois, October 29, 1938; and DuBois to Locke, November 1, 1938, box 164-23, file 43, "Davis-DuBois, Rachel," Locke Papers.

43 Rachel DuBois admired Locke and had come to know him through a circle of intellectuals and artists in Harlem. DuBois also had a personal and professional relationship with W. E. B. Du Bois, as well as a lively correspondence. John W. Studebaker to Alain Locke, November 17, 1938, box 164-87, file 40, "Studebaker, J. W.," Locke Papers; Studebaker to W. E. B. Du Bois, November 1, 1938; Du Bois to Studebaker, November 4, 7, 1938; and Philip Green to Du Bois, November 29, 1938, reel 49, W. E. B. Du Bois Papers; DuBois and Okorodudu, *All This and Something More,* 48–49.

44 Locke, *New Negro;* Meier and Rudwick, *Black History and the Historical Profession,* 8, 47, 79; Logan and Winston, *Dictionary of American Negro Biography,* 398–404; Locke and Stern, *When Peoples Meet.*

45 Huggins, *W. E. B. Du Bois,* 1297–99, 1063.

46 Rachel Davis DuBois to Alain Locke, n.d., and Locke to DuBois, n.d., box 164-23, folder 43, "Davis-DuBois, Rachel," Locke Papers; Rachel Davis DuBois to W. E. B. Du Bois, n.d., and Rachel Davis DuBois to W. E. B. Du Bois, November 30, 1938, reel 48, W. E. B. Du Bois Papers. Rachel DuBois also solicited comments on the script from New York City friends, actor Alvin Childress and writer Marion Cuthbert. DuBois to Alvin Childress, November 29, 1938, and DuBois to Marion Cuthbert, November 29, 1938, series V, box 22, folder 18, "Research and Resources, Program 6," R. D. DuBois Papers. For example, Seldes's early writings elevated "white" jazz to superior status, a distinction he later regretted. Kammen concludes that by the 1930s Seldes was an "ardent advocate of African Americans" and "a cultural negrophile," although he offers little evidence to support those claims. It appears more likely, however, that Seldes still harbored mixed feelings about the place of African American culture in American culture. Kammen reports that in 1939, the same period as this radio broadcast, Seldes lamented that the United States had "no folk song," ruling out black spirituals "because they are not the product of the dominant people." Kammen, *Lively Arts,* 58, 98, 303.

47 DuBois to Locke, December 6, 1938, box 164-23, folder 43, "Davis-DuBois, Rachel," Locke Papers. Rachel DuBois also consulted W. E. B. Du Bois on "The Upsurge of Democracy," script 8, passing his suggestions on the topics of slavery and Frederick Douglass to Seldes. W. E. B. Du Bois to Rachel Davis DuBois, December 12, 1938, and DuBois to Seldes, December 19, 1938, box 22, folder 18, R. D. DuBois Papers.

48 W. E. B. Du Bois to Philip Cohen, November 30, 1938, with enclosure, "Criticism of *Americans All—Immigrants All,* Episode Six, 'The Negro,'" reel 48, W. E. B. Du Bois Papers.

49 Alain Locke to Rachel Davis DuBois, November 30, 1938, box 164-23, folder 43, "Davis-DuBois, Rachel," Locke Papers.

50 Historical assessments of slavery and the Reconstruction period were under some dispute at the time of this series, and the murkiness here may have

reflected that, as well as political considerations. Novick, *That Noble Dream,* 224–39. The interpretation Du Bois had offered in his own *Black Reconstruction* (1935) was not yet accepted by the historical profession, a change that would take at least another half a century.

51 Script, *"Americans All—Immigrants All,* Episode 6, 'The Negro,'" entry 187, "Radio Education Project, Script Writers and Editors, Office Files, 1936–40," box 4, folder, *"Americans All—Immigrants All,"* ROE. The previous version of the script is in box 164-106, folder 6, *"Americans All—Immigrants All* Script," Locke Papers. The final version of the script and all others in the series are at the PAL. I was unable to locate the earliest Seldes draft, to which specific criticisms were submitted, in the National Archives, Washington, D.C.; the Locke Papers; the R. D. DuBois Papers; or the W. E. B. Du Bois Papers. My conclusions about that version of the script are based on reading backward from the criticism and the subsequent drafts.

52 Memorandum, George G. Murphy Jr. and Roy Wilkins, NAACP, December 9, 1938, series V, box 22, folder 18, R. D. DuBois Papers. There is no indication of who provided Murphy and Wilkins with the script. It also is unclear which version of the script they worked from, but their general criticisms would have held for all versions, including the final one.

53 Rachel DuBois had informed Boutwell that White was "very eager" to publicize the show through local NAACP branches; he also had requested 600 copies of the program brochure to be distributed to black youth and college groups. Overall publicity for the episode was channeled through the NAACP's local chapters and the black press. Rachel Davis DuBois to William Boutwell, November 15, 1938, series V, box 22, folder 4, R. D. DuBois Papers.

54 Press release, "Nation to Hear Broadcast of Negro Contributions to American Life, Sunday; JULES BLEDSOE TO SING," entry 182, "Radio Education Project, Audience Preparation Division, Office File of Ben Brodinsky, 1936–40," box 7, folder, "Service Bureau for Intercultural Education," ROE.

55 Rachel DuBois also had organized script-reading sessions in New York City to ferret out any material that might offend or be objectionable to African Americans. Rachel Davis DuBois, invitation notes, December 6, 1938, and undated handwritten notes, series V, box 22, folder 18, R. D. DuBois Papers. I have found no specific reactions to the draft scripts of this episode from Office of Education officials; the script came relatively early in the series (it was the sixth), and as such it preceded, for example, Chief Script Editor Laura Vitray's involvement with the series.

56 Rachel Davis DuBois to Rudolph Schram, December 13, 1938, ibid.

57 Rachel Davis DuBois to Alain Locke, December 21, 1938, box 164-23, folder 43, "Davis-DuBois, Rachel," Locke Papers.

58 Alain Locke to Rachel Davis DuBois, December 19, 1938, box 164-21, folder 38, "Columbia Broadcasting Company," Locke Papers.

59 Alain Locke to Rachel Davis DuBois, December 19, 1938, box 164-21, folder 38, "Columbia Broadcasting Company," and DuBois to Locke, December 21, 1938, box 164-23, folder 43, "Davis-DuBois, Rachel," Locke Papers.

60 Alain Locke to CBS, n.d., box 164-21, folder 38, "Columbia Broadcasting

Company," Locke Papers. CBS vice president for broadcasting W. B. Lewis responded to Locke, indicating that he was referring his letter to the Office of Education, "under whose auspices the program was presented." Despite this passing of the blame, officials at CBS also were disappointed by the broadcast, viewing it as further evidence that the overall series was weak. W. B. Lewis to Locke, January 10, 1939, ibid.

61 Alain Locke to Rachel Davis DuBois, December 19, 1938, box 164-21, folder 38, "Columbia Broadcasting Company," Locke Papers.

62 Philip Cohen to William Boutwell, December 27, 1938, box 5, folder, "Cohen, Phil, Jan. 1, 1938–Dec. 31, 1938"; Cohen to Boutwell, January 9, 1939; Boutwell to Cohen, March 3, 1939; and Cohen to Boutwell, March 15, 1939, entry 170, "Radio Education Project, Office of the Director, Subject File, 1936–40," box 6, folder, "Cohen, Phil, Jan. 1, 1939–Aug. 31, 1939," ROE.

63 Memorandum, William Boutwell to Philip Cohen, March 3, 1939, entry 170, "Radio Education Project, Office of the Director, Subject File, 1936–40," box 6, folder, "Cohen, Phil, Jan. 1, 1939–Aug. 31, 1939," ROE. Caliver, who held a Ph.D. from Columbia University, had been appointed as a "senior specialist in education of Negroes" at the Office of Education in 1930, the first African American to hold a professional position there. Caliver created an Office of Education radio series in 1941 that focused exclusively on African Americans. That show, *Freedom's People,* and Caliver are the subjects of chapter 2. "Nomination for Distinguished Service Award: Ambrose Caliver, Ph.D.," entry 17, "Records of the Office of the Commissioner, Office File of Ambrose Caliver, 1946–62," box 3, folder, "Dr. Caliver, Biographic," ROE.

64 W. E. B. Du Bois to Rachel Davis DuBois, January 4, 1939, reel 50, W. E. B. Du Bois Papers.

65 Script, "The Negro," 25, PAL; Rachel Davis DuBois to Gilbert Seldes, November 29, 1938, series V, box 22, folder 7, "Correspondence, Gilbert Seldes, ca. 1938–39," R. D. DuBois Papers.

66 Walter White to John W. Studebaker, December 23, 1938, series V, box 26, folder 2, R. D. DuBois Papers.

67 "Mail Tabulations: *Americans All—Immigrants All,*" January 28, 1939, series V, box 26, folder 2, R. D. DuBois Papers; Alain Locke to Rachel Davis DuBois, December 19, 1938, box 164-21, folder 38, "Columbia Broadcasting Company," Locke Papers.

68 Quoted in Seelye, "Broadcasting Tolerance," 73–78.

69 Ibid., 71.

70 Ibid., 71, 80–81, 83.

71 DuBois and Okorodudu, *All This and Something More,* 72. DuBois had faced similar questions from members of other ethnic groups. Ibid., 68. DuBois and Schweppe, *Jews in American Life.* The book was one of two completed in a planned series called Building American Culture that was never realized for lack of funding.

72 William Boutwell to Sidney Wallach, American Jewish Committee, June 24, 1938, series V, box 22, folder 4, "Correspondence, William D. Boutwell, 1937–39," R. D. DuBois Papers; "Report of Trip to New York, Septem-

ber 8th to 10th Inclusive, 1938," entry 174, "Radio Education Project, Office of Director, Records Relating to Radio Programs, 1936–41," box 1, folder, "*Americans All,* Planning," ROE; Advisory Board of the Service Bureau to John W. Studebaker, October 5, 1938, series V, box 22, folder 8, R. D. DuBois Papers; Montalto, *History of the Intercultural Education Movement,* 184; DuBois and Okorodudu, *All This and Something More,* 72.

73 Montalto, *History of the Intercultural Education Movement,* 156–57; minutes, Advisory Committee meeting, September 28, 1938, entry 174, "Radio Education Project, Office of Director, Records Relating to Radio Programs, 1936–41," box 1, folder, "*Americans All,* Planning," ROE. On the refugee question, see Wyman, *Paper Walls,* and Friedman, *No Haven for the Oppressed.*

74 Minutes, Advisory Committee meeting, September 28, 1938, entry 174, "Radio Education Project, Office of Director, Records Relating to Radio Programs, 1936–41," box 1, folder, "*Americans All,* Planning," ROE. DuBois recalls that the matter was actually put to a vote, although no vote was recorded in the minutes of this particular meeting. DuBois and Okorodudu, *All This and Something More,* 87.

75 Minutes, Advisory Committee meeting, September 28, 1938, entry 174, "Radio Education Project, Office of Director, Records Relating to Radio Programs, 1936–41," box 1, folder, "*Americans All,* Planning," ROE.

76 On Seldes's own views and the anti-Semitism directed at him, see Kammen, *Lively Arts,* 19.

77 Montalto, *History of the Intercultural Education Movement,* 211; Theresa Mayer Durlach, vice chair, Service Bureau for Intercultural Education, to John W. Studebaker, October 27, 1938, series V, box 22, folder 13, R. D. DuBois Papers.

78 Entry 175, "Radio Education Project, Office of the Director, Supplemental Material for Radio Programs, 1935–41," box 1, folder, "Teaching Tolerance a Major Problem in 1939," ROE.

79 Derounian, who had escaped the Turkish massacre in Armenia as a child in 1915, had been a student of DuBois's at New York University. Under an assumed name, he later infiltrated the American Nazi movement on behalf of the federal government and wrote a book in 1943 detailing the extent of Nazi influence in the United States. DuBois and Okorodudu, *All This and Something More,* 83; Crispin, "Rachel Davis DuBois," 26. Derounian's book, *Under Cover,* was published under his pseudonym, John Roy Carlson.

80 Derounian built on the bureau's lists of foreign-language newspapers, black newspapers, religious news services, leading churches and denominations, YMCAs and YWCAs, and a wide range of other civic and social organizations. He also made personal appeals to prominent columnists, publishers, and radio editors in key cities with large immigrant populations. He buttressed all of this activity with personalized mailings to group leaders and newspapers serving the particular national or racial group to be highlighted in the upcoming episode. Arthur Derounian to Philip Cohen, January 17, 1939, entry 187, "Radio Education Project, Script Writers and Editors, Office Files, 1936–40," box 4, folder, "Plans for Promotion," ROE.

81 Ben Brodinsky, head of the Office of Education's Audience Preparation Division, essentially adapted Derounian's tactics when he instituted the use of extensive mailings and contacts with Jewish news services and organizations with memberships totaling nearly 1 million Jews. The targeted groups included forty-three mostly national organizations, such as organizations serving synagogues and rabbis, B'nai B'rith, Hadassah, and the Council of Jewish Federations, to cite but a few. Ben Brodinsky to William Boutwell, "Why We Should Get 50,000 Letters on the Feb. 5 Program," n.d., entry 187, "Radio Education Project, Script Writers and Editors, Office Files, 1936–40," box 4, folder, "Plans for Promotion," ROE. The Jewish War Veterans and the National Council of Jewish Women were among many organizations that sent the Office of Education letters of support prior to the broadcast. Entry 187, "Radio Education Project, Script Writers and Editors, Office Files, 1936–40," box 12, folder, "Jews," ROE.

82 Hadassah to Ben Brodinsky, Office of Education, January 14, 1939, entry 187, "Radio Education Project, Script Writers and Editors, Office Files, 1936–40," box 12, folder, "Jews," ROE. The Office of Education also made direct contact with local branches of Jewish organizations in key cities. For instance, the Washington, D.C., United Jewish Appeal made its mailing list of 4,500 names available to Brodinsky through his brother Joseph, a prominent Washington, D.C., lawyer. Rabbi Isadore Breslau to Joseph Brodinsky, January 3, 1938, ibid. The agency also sent special releases to the general press in thirteen major cities and to Jewish news organizations such as the Jewish Telegraphic Agency.

83 "Tolerance on the Air," *Hebrew Union College Monthly* 26 (January 1939), clipping in ibid.

84 Ibid.

85 "The Jews in the United States," script 13, February 5, 1939, 5, 8–15, 17, 21, 23, 25, 27, 29–32, PAL.

86 Ibid., 33–34.

87 In less than two weeks of the broadcast, the Office of Education had received over 9,000 letters and cards seeking more information. "Mail Report for *Americans All—Immigrants All*," January 20, 1939, box 26, folder 2, R. D. DuBois Papers; William Boutwell to Gilbert Seldes, February 18, 1939, entry 170, "Radio Education Project, Office of the Director, Subject File, 1936–40," box 9, folder, "NBC, J. W. Studebaker, 1937," ROE.

88 Quoted in Seelye, "Broadcasting Tolerance," 72, 58.

89 Ibid., 63.

90 Philip Cohen wrote Boutwell the day after the broadcast that he and the New York City staff "were all very much thrilled" by it and that it was "the best to date." The next day, he reported that Commissioner Studebaker had "again expressed his enthusiasm" for the broadcast. Philip Cohen to William Boutwell, February 6, 7, 1939, entry 170, "Radio Education Project, Office of the Director, Subject File, 1936–40," box 6, folder, "Cohen, Phil, Jan. 1, 1939–Aug. 31, 1939," ROE.

91 Edward Ashley Bayne, Service Bureau for Intercultural Education, to Jeanette

Sayre, February 16, 1939, and Sayre to Bayne, February 18, 1939, series V, box 26, folder 2, R. D. DuBois Papers.

92 *Social Justice,* November 27, 1939, quoted in Seelye, "Broadcasting Tolerance," 4.

93 "Radio Mail Returns, Educational Sustaining Programs, December 1938" and "Mail Tabulation, *Americans All—Immigrants All,*" January 28, 1939, series V, box 26, folder 2, and William Boutwell to Edward Bayne, March 2, 1939, series V, box 22, folder 4, R. D. DuBois Papers; John W. Studebaker to Senator Robert Reynolds, March 27, 1939, entry 170, "Radio Education Project, Office of the Director, Subject File, 1936–40," box 10, folder, "Studebaker Letters, Sept. 1938–Aug. 1939," ROE; Boutwell to Keith Tyler, September 1, 1939, entry 170, "Radio Education Project, Office of the Director, Subject File, 1936–40," box 4, folder, "1939 Corres., William Boutwell, June–Dec.," ROE.

94 Rachel DuBois also complained repeatedly about the delays, arguing that "publicity is valueless unless there is an adequate follow-through." Entry 187, "Radio Education Project, Script Writers and Editors, 1936–40," box 4, folder, "Requests," ROE; William Boutwell to Edward Bayne, March 14, 1939, and Rachel Davis DuBois to Boutwell, April 4, 1939, series V, box 22, folder 4, R. D. DuBois Papers. Eventually, funds provided by the Carnegie Foundation enabled the office to make available the most comprehensive work from the series: a glossy 120-page study guide booklet, Jones, *Americans All . . . Immigrants All,* that included narrative histories, bibliographies, and suggested discussion questions for each of the groups covered in the series. Rachel DuBois helped draft the booklet. Listeners who bought the phonograph recordings would also receive the booklet and a manual outlining more than 100 ways to use the recording in classrooms, assemblies, night schools, and adult education. Entry 187, "Radio Education Project, Script Writers and Editors, 1936–40," box 4, folder, "Two Highest Awards," ROE; Franklin Dunham to James Angell, November 20, 1939, box 66, folder 9, "*Americans All, Immigrants All,* CBS, 1939," NBC Collection. (NBC officials followed every aspect of this CBS series very closely.) There were few orders from individuals for recordings; as expected, they were bought primarily by schools, teachers, and educational groups. Entry 175, "Radio Education Project, Office of the Director, Supplemental Material for Radio Programs, 1935–41," box 1, folder, "*Americans All* Recordings," ROE. However, some of the records were purchased by ethnic and civic organizations. For example, the International Workers Order of New York City, a group self-described as having 165,000 members representing twelve nationalities, used the recordings in "lodge broadcasts." Peter Chaunt, International Workers Order, to Office of Education, September 14, 1939, entry 187, "Radio Education Project, Script Writers and Editors, 1936–40," box 3, folder, "Booklet, *Americans All,* Correspondence," ROE. The Service Bureau, credited at the end of each broadcast, was inundated with requests for additional information from all over the country; however, it was unable to meet the demand from even New York City alone after the New York City Board of Education responded to the series by requiring that all high schools teach "tolerance"

as part of the curriculum. DuBois to John W. Studebaker, December 21, 1938, March 2, 22, 1939, and Louis Posner to Studebaker, January 3, 1939, series V, box 22, folder 8; DuBois to Gordon Studebaker, Radio Script Exchange, March 24, 28, 1939, series V, box 26, folder 3; and Bayne to Boutwell, n.d., and Boutwell to Stanley Walker, Service Bureau, April 4, 1939, series V, box 22, folder 4, R. D. DuBois Papers. DuBois solicited funds from private groups and individuals to pay for mimeographing scripts about their group. Robert Valeur, French Information Center, to DuBois, April 4, 1939; DuBois to Valeur, April 4, 1939; DuBois to Italian Welfare League, April 11, 1939; and Italian Welfare League to DuBois, April 10, 1939, with $40 check, series V, box 26, folder 3, R. D. DuBois Papers; Rachel Davis DuBois to W. E. B. Du Bois, April 12, 1939, and W. E. B. Du Bois to Rachel Davis DuBois, April 14, 1939, with $10, reel 50, W. E. B. Du Bois Papers.

95 The annual American Legion Auxiliary Radio Award went to the series. The episode on Jewish Americans won an honorable mention from the prestigious Institute for Education by Radio, as well as an education citation from the National Council of Jewish Women. Elsewhere the show's renown continued to increase, generating one inquiry about the possibility of making a film based on *Americans All*. "Immigrant Radio Program Wins Award in Annual Women's Poll; Series Sponsored by U.S. Named as Leading Work of Year; Democracy Theme Is Found Popular on the Air," *New York Times*, April 20, 1939; entry 187, "Radio Education Project, Script Writers and Editors, 1936–40," box 4, folders, "Plans for Promotion" and "Two Highest Awards for *American All*," ROE; Edward Bayne to William Boutwell, April 24, 1939, box 22, folder 4, R. D. DuBois Papers; Boutwell to Jack Leighter, Paul Kohner, Inc., May 16, 1939, entry 170, "Radio Education Project, Office of the Director, Subject File, 1936–40," box 4, folder, "1939 Corres., William Boutwell, Jan.–June," ROE.

96 As soon as the series was announced, Avinere Toigo, who had first offered the idea to NBC, wrote RCA president David Sarnoff and NBC educational programs director James Angell letters in which he gloated about CBS's acceptance of the show. At that point, NBC vice president John Royal asked his staff to keep a close eye on the series. Avinere Toigo to David Sarnoff, October 14, 1938; Toigo to James Angell, October 14, 1938; John Royal to Wayne Randall, October 27, 1938; and Royal to Sarnoff, November 21, 1938, box 58, folder 48, "*Americans All, Immigrants All*," NBC Collection. Royal tried to denigrate the awards themselves and implied that John Studebaker had manipulated the judges, especially the women. Royal to Lenox Lohr, president, NBC, April 24, 1939, box 66, folder 9, "*Americans All, Immigrants All*, CBS, 1939," NBC Collection; "Corwin Play Honored by Air Teachers; CBS Far Ahead in Radio Programs for Education," *New York Post*, May 5, 1939; Sarnoff to Lohr, May 6, 1939, box 66, folder 9, "*Americans All, Immigrants All*, CBS, 1939," NBC Collection.

97 James Angell to Lenox Lohr, May 8, 1939, box 66, folder 9, "*Americans All, Immigrants All*, CBS, 1939," NBC Collection. In a postscript, Angell further ridiculed the awards process: "You probably know that the unfortunate

judges spent three days listening to 245 records. Only fragments of intelligence could have survived such an ordeal." Angell to Niles Trammell, NBC, May 24, 1939, and Trammell to John Royal, May 12, 1939, ibid.

98 Rachel Davis DuBois to John W. Studebaker, Gilbert Seldes, William Boutwell, Philip Cohen, Louis Adamic, and Frank Tager, AJC, March 1939, entry 174, "Radio Education Project," box 1, folder, "*Americans All,* Corres. with Service Bureau," ROE; Boutwell to DuBois, March 16, 1939, box 22, folder 4; Studebaker to DuBois, May 25, 1939, and DuBois to Studebaker, June 1, 1939, box 22, folder 8, R. D. DuBois Papers.

99 Sayre, *Analysis of the Radiobroadcasting Activities of Federal Agencies,* 27–28, 31, 91.

100 John W. Studebaker to Senator Robert Reynolds, March 27, 1939, and Interior Assistant Secretary Oscar Chapman to Senator Tom Connally, February 16, 1939, entry 170, "Radio Education Project, Office of the Director, Subject File, 1936–40," box 10, folder, "Studebaker Letters, Sept. 1938–Aug. 1939," ROE.

101 Sayre, *Analysis of the Radiobroadcasting Activities of Federal Agencies,* 91.

102 Jay Clark Waldron, station relations director, Office of Education, Federal Security Agency, to J. W. Somerville, assistant to Eleanor Roosevelt, July 1940, file 6G, "Interior Department, Department of Education, 1940," President's Official File, Franklin D. Roosevelt Papers, FDRL. Eleanor Roosevelt also donated $500 to Rachel DuBois in support of her work in the New York City schools, according to DuBois. DuBois and Okorodudu, *All This and Something More,* 91; John W. Studebaker to President Franklin D. Roosevelt, July 19, 1940; Lowell Mellett, director, Office of Government Reports, to Roosevelt, July 26, 1940; and Roosevelt to Studebaker, July 26, 1940, file 6G, "Interior Department, Department of Education, 1940," President's Official File, Franklin D. Roosevelt Papers, FDRL.

103 DuBois and Okorodudu, *All This and Something More,* 91–94.

104 Ibid.

105 Kammen, *Lively Arts,* 262.

106 Dorothea Seelye to William Boutwell, n.d., entry 174, "Radio Education Project, Office of the Director, Records Relating to Radio Programs, 1935–41," box 1, folder, "*Americans All,* Promotion and Follow-up," ROE; Dorothea Seelye, quoted in Sayre, *Analysis of the Radiobroadcasting Activities of Federal Agencies,* 94; Seelye, "Broadcasting Tolerance," 103.

107 Seelye, "Broadcasting Tolerance," 104a–104b.

108 Gerstle, *Working-Class Americanism,* 290.

109 For a comparison of assimilationist strategies of African American and Jewish elites in an earlier period, see Lewis, "Parallels and Divergences," 543–64. For a general treatment of the involvement of Jewish Americans in the civil rights efforts of African Americans in an earlier period, see Diner, *In the Almost Promised Land.*

110 For example, psychologist and radio analyst Hadley Cantril concluded that a widespread fear of war or internal revolution was a precipitating factor in the mass panic that followed the radio broadcast of *The War of the Worlds* in October 1938. Heyward Hale Broun observed at the time that if the invasion from

Mars broadcast had occurred six months earlier, no panic would have ensued. Daily radio reports about Hitler's actions had, in Broun's view, given Americans a general case of the "jitters." Cantril, *Invasion from Mars,* 159, 202–3.

111 On whiteness in the nineteenth century, see Saxton, *Rise and Fall of the White Republic,* 10, 18, 296, 314, 388, and Roediger, *Wages of Whiteness,* 12, 20, 140. Similarly, George Fredrickson has argued for the existence in the mid-nineteenth century of a "racial basis of American citizenship" that was anti-assimilationist against nonwhites and an embodiment of a kind of white nationalism. Fredrickson, *Black Image in the White Mind,* 137; Fields, "Slavery, Race, and Ideology," 111, 115; Gary Gerstle, quoted in Kazal, "Revisiting Assimilation," 470.

112 Morrison, *Playing in the Dark,* 47. Recent discussions of ethnicity continue to wrestle with the relationship between race and ethnicity, most often attempting to avoid contaminating one with the other. See, for example, Sollars, *Beyond Ethnicity,* 39, 36–37: "I think it is most helpful not to be confused by the heavily charged term 'race' and to keep looking at race as one aspect of ethnicity" and "The term 'ethnicity' here is thus a broadly conceived term. This choice does not represent an attempt to gloss over the special legacy of slavery and racism in America."

113 Huggins, "Deforming Mirror of Truth," 27.

114 Duberman, *Paul Robeson,* 236–37. See also MacDonald, *Don't Touch That Dial!,* 346–47.

115 This pun would have worked with his listeners since Robeson also imitated the dialect used in *Amos 'n' Andy,* which is what caught my ear in listening to a recorded version. *Check and Double Check* also had been the title of a 1930 RKO film spin-off from the radio series. Cripps, *Slow Fade,* 269. L. D. Reddick noted that "for a while expressions like, 'check and double check' . . . were so popular that teachers of English were disturbed over the corruption of their students." *New York World Telegram,* December 1, 1933, cited in Reddick, "Educational Programs," 383.

CHAPTER TWO

1 Johnson, "Present Status of Race Relations," 323.

2 "Nomination for Distinguished Service Award," entry 17, "Office File of Ambrose Caliver," box 3, folder, "Dr. Caliver, Biographic," ROE.

3 In 1929, President Herbert Hoover appointed an Advisory Committee to determine the appropriate federal role in education. Black members of the committee submitted a minority report that urged the federal government to address the issue of racially disparate spending. In response, the Office of Education created a Division of Special Problems and designated funds for a "specialist in Negro education," Caliver's position. Hutchinson, "Marginal Man with a Marginal Mission," 104–8.

4 Caliver's reports included *Background Study of Negro College Students: Rural Elementary Education among Negroes under Jeanes Supervising Teachers* (1933); *The Education of Negro Teachers* (1933); *Secondary Education of Negroes* (1933); *Availability*

of Education to Negroes in Rural Communities (1936); and *Vocational and Education Guidance of Negroes* (1938).

5 John W. Studebaker to NBC, December 8, 1938, box 65, folder 8, "U.S. Dept. of Interior, Genl.," NBC Collection; Hutchinson, "Marginal Man with a Marginal Mission," 186. Among the positions Caliver had held at Fisk University was that of publicity director. "Ambrose Caliver," in Logan and Winston, *Dictionary of American Negro Biography*, 85–86.

6 Ambrose Caliver to Harold Ickes, January 22, 1934, and copy of *Fundamentals in the Education of Negroes*, Office of Secretary, Central Classified File, 1907–36, file 6-61, Negro Education, Records of the Department of Interior, Record Group 48, National Archives, Washington, D.C.; Sitkoff, *New Deal for Blacks*, 201.

7 Hutchinson, "Marginal Man with a Marginal Mission," 137–38, 154, 187–88. He apparently also served sixteen years without any upgrade in his civil service rank. When he was finally promoted, it was not to the level for which his length of service would have qualified him.

8 Caliver also was among those whom W. E. B. Du Bois asked to serve on the editorial board of his ill-fated *Encyclopedia of the Negro* project. W. E. B. Du Bois to Ambrose Caliver, March 22, 1938, and Caliver to Du Bois, April 2, 1938, reel 48, W. E. B. Du Bois Papers.

9 For discussions of aspects of African American history in the 1930s, see generally Kelley, *Hammer and Hoe*; Sullivan, *Days of Hope*; Weiss, *Farewell to the Party of Lincoln*; Sitkoff, *New Deal for Blacks*; and Kirby, *Black Americans in the Roosevelt Era*.

10 See Dalfiume, *Desegregation of the U.S. Armed Forces*; Davis, "The Negro in the United States Navy, Marine Corps, and Coastguard," 3; and Wynn, *Afro-American and the Second World War*, 22.

11 Anderson, "Last Hired, First Fired," 82–97; Jones, *Labor of Love, Labor of Sorrow*, 238. See also Franklin, *From Slavery to Freedom*, 425–26.

12 Franklin, *From Slavery to Freedom*, 442; Wynn, *Afro-American and the Second World War*, 61.

13 MacDonald, *Don't Touch That Dial!*, 60–61, 65, 302–4, 307–12; Barnouw, *Golden Web*, 140–42, 147–52.

14 Spaulding, "History of Black Oriented Radio in Chicago," iii, iv, v.

15 For a discussion of the NAACP's campaigns against specific radio programming, including *Amos 'n' Andy*, and for increased employment opportunities for blacks in radio, see Archer, *Black Images in American Theater*, 225–62. On the *Pittsburgh Courier*'s petition drive against *Amos 'n' Andy*, see Reddick, "Educational Programs," 384. For details on recurring black campaigns against *Amos 'n' Andy*, see Ely, *Adventures of Amos 'n' Andy*.

16 It appears that Caliver's only direct involvement with *Americans All* came after the Jules Bledsoe incident on the show about African Americans, when he was asked to review the final revised script that was used to reenact the show for the purpose of making a phonograph recording.

17 Unsigned handwritten notes, August 23, 1940, box 83C, folder 22, *"Freedom's*

People," NBC Collection. Despite Caliver's long career at the Office of Education, the National Archives, Washington, D.C., contains only a scanty collection of materials from his office files. I am grateful to Jewel Caliver Terrell, who led me to a collection of her father's papers at the MSRC. I am especially indebted to Joellen P. El-Bashir, senior manuscript librarian at the MSRC, who retrieved and granted me special access to unprocessed materials that have subsequently been processed and cataloged. Because my research preceded that processing, my references to the material do not include specific locations within the collection, although I have examined the processed collection to ensure that I have not overlooked any relevant material.

18 Ambrose Caliver to John W. Studebaker, September 5, 1940, Caliver Papers.

19 In a 1939 article, Caliver argued that black students needed to accept the values of American culture at the same time that they needed to be taught the worthiness of their own race and its culture and values. They also needed to be taught, he cautioned, that "the problems which have resulted from their minority group status cannot be solved overnight." Ambrose Caliver, "Education of Negroes," *School Life* (June 1939), quoted in Hutchinson, "Marginal Man with a Marginal Mission," 155. See also Caliver, "Power Within."

20 "*We, Too, Are Americans:* Education Radio Series and Transcriptions on Contributions of Negroes to American Life: Prospectus, October 1940," Caliver Papers.

21 William Boutwell to John W. Studebaker, November 5, 1940, ibid.

22 Bess Goodykoontz, Office of Education, to Edwin Embree, Rosenwald Fund, November 7, 1940; John W. Studebaker to Goodykoontz, November 11, 1940; Embree to Goodykoontz, November 13, 1940; Studebaker to Embree, December 16, 1940; Embree to Studebaker, January 15, 1941; and Studebaker to Embree, January 27, 1941, ibid. In January 1941, the Rosenwald Fund agreed to provide the $5,200 that Caliver estimated he would need. These funds were funneled through the National Education Association. The $5,200 budget was divided largely between the scriptwriter's salary ($2,000) and the cost of recording the phonographs ($2,000). Caliver expected that the research would be done by a voluntary advisory group and that he would be able to work on the project as needed. It appears that additional funds were needed after the American Federation of Musicians notified NBC that although the services of the musicians would be donated for the live broadcast, the musicians would expect to be paid a fee for the recorded versions of the shows. James Petrillo to William Burke Miller, NBC, July 28, 1941, box 83C, folder 22, "*Freedom's People,*" NBC Collection. The Southern Education Foundation added $2,000 to the $5,200 that Rosenwald had contributed. "Annual Report of the President to the Southern Education Foundation, Inc., Arthur D. Wright, President," January 8, 1942, Southern Education Foundation Archives, Robert W. Woodruff Library, Atlanta University Center, Atlanta, Georgia; Studebaker to Arthur D. Wright, September 30, 1941, Caliver Papers.

23 John W. Studebaker to Niles Trammell, January 31, 1941; Ambrose Caliver to Philips Carlin, February 5, 1941; Trammell to Studebaker, February 11, 1941;

Studebaker to Trammell, February 18, 1941; I. McGeary, NBC, to Walter Preston, manager, Public Service Division, NBC, February 21, 1941; Preston to Caliver, February 24, 1941; H. B. Summers, NBC, to Preston, February 28, 1941; James Angell to Trammell, February 7, 1941; and Angell to Preston, February 17, 1941, box 83C, folder 22, *"Freedom's People,"* NBC Collection.

24 *"Freedom's People,* Educational Radio Series and Transcriptions, Tentative Outline of Programs," March 15, 1941, Caliver Papers; Ambrose Caliver to Walter Preston, March 1, 1941, and R. W. Friedheim to H. B. Summers, March 11, 1941, box 83C, folder 22, *"Freedom's People,"* NBC Collection.

25 H. B. Summers to Ambrose Caliver, March 26, 1941, and Caliver to Summers, April 10, 1941, Caliver Papers; Caliver, "Developing Racial Tolerance in America," 447–58.

26 Blum, *V Was for Victory,* 185–88.

27 Finkle, *Forum for Protest,* 116, 111.

28 The army itself was investigating black press activities extensively throughout 1941, and a year earlier, FBI agents had begun to visit the *Pittsburgh Courier.* Washburn, *Question of Sedition,* 33.

29 Ambrose Caliver to H. B. Summers, July 17, 1941, and Summers to Caliver, July 25, 1941, box 83C, folder 22, *"Freedom's People,"* NBC Collection. This program is discussed in chapter 3, which focuses on the War Department and the Office of War Information.

30 T. D. Rishworth, NBC, to H. B. Summers, June 2, 1941, and John W. Studebaker to Niles Trammell, September 6, 1941, box 83C, folder 22, *"Freedom's People,"* NBC Collection; Summers to Ambrose Caliver, August 19, 20, 1941, Caliver Papers.

31 The FCC in June had issued formal regulations based on its antitrust investigations of the broadcast industry. Among other provisions, the FCC prohibited duopoly—the ownership of two stations in the same area by one licensee—and broadcasting company ownership of more than one network. Both provisions would drastically affect NBC's current operations: the network would likely be forced not only to divest one of its two networks (dubbed NBC-Red and NBC-Blue) but also to sell off one of the two stations it operated in each of the lucrative New York City, Chicago, Washington, D.C., and San Francisco markets. In October 1941, NBC sued the federal government to block the FCC's action, triggering litigation that would take two years and a Supreme Court decision to resolve. Sterling and Kittross, *Stay Tuned,* 189–91. In reference to these developments, Summers told Caliver that "as long as the whole network situation remains so uncertain, it's almost impossible for us to plan anything definite—even some of the program series which we have had on the air for years have not yet been booked for the fall season." H. B. Summers to Ambrose Caliver, July 25, 1941, Caliver Papers.

32 To reduce program costs, Caliver convinced the American Federation of Radio Actors and the American Federation of Musicians to authorize their members to perform on the show without compensation. Later, Caliver also persuaded officers of the American Society of Composers, Authors, and Publishers (ASCAP) to waive members' rights to fees for the songs used on

the show, an exception they granted only on the condition that the show be defense related. NBC in turn accepted the waiver purely on the basis that the show was a "morale building defense program." ASCAP licenses to the networks had expired in December 1940, and apparently ASCAP and the networks were still at odds about a new agreement. In the interim, ASCAP members had allowed their music to be used only on programs strictly relating to national defense. In granting their approval, ASCAP officials also emphasized their appreciation of black contributions to American music as part of their rationale. John W. Studebaker to Niles Trammell, September 6, 1941; Gene Buck, president, ASCAP, to Studebaker, September 8, 1941; Trammell to Studebaker, September 8, 1941; Ambrose Caliver to H. B. Summers, NBC, September 9, 1941; Summers to Sidney Strotz, NBC, September 11, 1941; Strotz to Studebaker, September 11, 1941; Studebaker to Trammell, September 11, 1941; Summers to William Boutwell, September 12, 1941; and Studebaker to Strotz, September 13, 1941, box 83C, folder 22, *"Freedom's People,"* NBC Collection.

33 See, for example, Ambrose Caliver to Ethel Waters, April 23, 1941, Caliver Papers.

34 *Freedom's People* Scripts, box A58758, James Weldon Johnson Collection, Beinecke Rare Book Library, Yale University, New Haven, Connecticut. This collection includes scripts, press releases, and bibliographical materials for each program, as well as the original printed publicity circular. All quotations from *Freedom's People* scripts are found in this collection. Caliver sent the scripts and background materials to Carl Van Vechten specifically for the Johnson Collection at Van Vechten's request. Ambrose Caliver to Carl Van Vechten, May 14, 1942. The following are the scripts by title and date: "Music," September 21, 1941; "Science and Discovery," October 19, 1941; "Sports," November 23, 1941; "Military Service," December 21, 1941; "The Negro Worker," January 18, 1942; "The Education of the Negro," February 15, 1942; "Creative Art," March 15, 1942; and "The Negro and Christian Democracy," April 19, 1942. A partial collection of the scripts can also be found in the RSC. Taped recordings of the radio broadcasts are available in the Recorded Sound Division, LC. Complete program recordings for the September 1941 and January, March, and April 1942 broadcasts and half of each of the October, November, and December 1941 broadcasts are available. The recording for February 1942 is not available. Additional textual materials about the series are at the MSRC and the NBC Collection. I have been unable to locate any material on *Freedom's People* at the National Archives, Washington, D.C., in the records of the Office of Education, the Department of Interior, or the Federal Security Agency or in Caliver's files there.

35 "Meeting of the Advisory Committee in Connection with the Proposed Radio Broadcasts on the Negro," June 11, 1941; "Minutes of Meeting of the Advisory Committee of the Radio Project," September 16, 1941; and "Meeting of the Advisory Committee in Connection with the Negro Radio Project," October 7, 1941, Caliver Papers. Derricotte's comment shows that although television was not yet available, awareness of the technology was widespread

even though its full development had been delayed first by the depression and then by the war itself. Sterling and Kittross, *Stay Tuned*, 151.

36 For example, Wesley wanted to correct the misconception that there was no free black population during the time of slavery, a change that Tunick incorporated into the program's standard introduction. "Meeting of the Advisory Committee in Connection with the Proposed Radio Broadcasts on the Negro," June 11, 1941, Caliver Papers.

37 Choral music for the programs was performed live by the Leonard De Paur Chorus. The NBC Orchestra provided the accompaniment.

38 W. C. Handy emphasized that the blues, like spirituals and work songs, had "passed out of the hands of the Negro into the folk-lore of our great country." Noble Sissle's band followed with a snappy rendition of "St. Louis Blues."

39 For example, each show was accompanied by a brief bibliography on a particular subject, which Caliver himself prepared. Organizers also had planned to distribute to civic and church groups, schools, and the general public study guides on each subject along with the scripts and the bibliographies. The study guides, if they were produced, were not included in the materials Caliver collected and forwarded to Carl Van Vechten for inclusion in the James Weldon Johnson Collection at Yale, nor are they among Caliver's own papers.

40 "Eleanor Roosevelt's 'My Day': New Radio Program to Aid Negro Cause," September 18, 1941, Caliver Papers. Caliver responded to her column by writing to thank her and to request a meeting to discuss his future plans for the series. Ambrose Caliver to Eleanor Roosevelt, September 18, 1941, and secretary to Mrs. Roosevelt to Caliver, September 19, 1941, Correspondence Files, 170, C–D, 1941, Eleanor Roosevelt Papers, FDRL.

41 John W. Studebaker to Franklin Roosevelt, September 19, 1941, file 6G, "1941–42," President's Official File, Franklin D. Roosevelt Papers, FDRL.

42 See, for example, "Radio Drama on Negroes to Bolster National Unity," *Chicago Defender*, September 27, 1941; "Nation Acclaims Robeson on *Freedom's People* Broadcast," *Pittsburgh Courier*, September 27, 1941; "*Freedom's People* to Air Monthly," *Minnesota Spokesman*, September 19, 1941; and "Freedom's Program," *Louisville Defender*, September 27, 1941, Caliver Papers.

43 Form letter, John W. Studebaker, "Dear Fellow-Workers," n.d., ibid.

44 *New York Times*, September 21, 1941, ibid.

45 See, for example, letters from Ellison M. Smith, director, Division of Elementary Education, University of South Carolina, October 16, 1941, and R. A. Carter, Association of Colleges and Secondary Schools for Negroes, State Agricultural and Mechanical Institute, Alabama, October 17, 1941, ibid.

46 Letter from Agnes J. Tallent, principal, P.S. 196, Brooklyn, n.d., ibid.

47 See, for example, letters from Robert P. Daniel, Shaw University, September 22, 1941; Ralph Davies, Department of Records and Research, Tuskegee Institute, September 26, 1996; R. G. Reynolds, Faculty Committee on Public Relations, Teachers College, October 8, 1941; Charles L. Trabert, Department of Education, Newberry College, October 16, 1941; and Jean Kircher, Manhattanville College of the Sacred Heart, September 22, 1941, ibid.

48 See, for example, letters from Fay Collier, educational director, Young Com-

munist League of New York State, September 24, 1941; Florence Owens, YWCA, Fargo, North Dakota, September 22, 1941; Florence Coles, City Federation of Colored Women's Clubs, Charlottesville, Virginia, September 30, 1941; Lucy J. Harris, Franklin, Kentucky, October 1, 1941; Mrs. Reginald V. Bovill, Dearborn, Michigan, October 2, 1941; Clifford Burr, Department of Education, State of California, October 3, 1941; Elizabeth Brown, Board of Education of the Methodist Church, Nashville, Tennessee, October 14, 1941; and W. A. McKinney, principal, Uvalde Elementary School, Uvalde, Texas, October 20, 1941, ibid.

49 "Radio Reviews," *Variety,* September 24, 1941, ibid.

50 "For Native Sons," *Time,* September 29, 1941, ibid.

51 David Bradford, "A New Series of Radio Programs Will Help to Publicize the Achievement of the Negro," *Louisville Courier-Journal,* September 30, 1941, ibid.

52 The writer also urged his readers to write NBC and ask that the show be given a weekly spot. "*Freedom's People:* Paul Robeson, W. C. Handy, Josh White, and Others in a New Radio Program," *New Masses,* October 6, 1941, ibid.

53 "Radio Drama on Negroes to Bolster National Unity," *Chicago Defender,* September 27, 1941; "Nation Acclaims Robeson on *Freedom's People* Broadcast," *Pittsburgh Courier,* September 27, 1941; "Freedom's People Speak: Tribute to Nation's Progress," *Washington Afro-American,* October 11, 1941; George Schuyler, "The World Today," *Pittsburgh Courier,* October 25, 1941; "Billy Rowe's Notebook," *Pittsburgh Courier,* September 27, 1941; and John Lloyd to Ambrose Caliver, October 4, 1941, with clipping from unknown newspaper, n.d., ibid.

54 See, for example, letters from H. H. Hamilton, YMCA, Dallas, Moorland Branch, September 22, 1941, and Cecelia Cabaniss Sanders, executive secretary, YWCA, New York City, September 22, 1941, ibid.

55 Letters from Mary McLeod Bethune, September 23, 1941; William J. Trent Jr., racial relations officer, Federal Works Administration, September 23, 1941; and Channing Tobias, September 22, 1941, ibid.

56 "Meeting of the Advisory Committee in Connection with the Negro Radio Project," October 7, 1941, ibid.

57 John W. Studebaker to Ambrose Caliver, September 23, 1941, ibid.

58 John W. Studebaker to NBC officials Niles Trammell, Sidney Strotz, and Clarence Menser, September 25, 1941, and Ambrose Caliver to Menser, Walter Preston, and Strotz, September 26, 1941, box 83C, folder 22, "*Freedom's People,*" NBC Collection.

59 Ambrose Caliver to W. W. Alexander, Office of Personnel Management, October 25, 1941, box 164-19, folder 17, "Corres., Caliver, Ambrose," Locke Papers.

60 H. Wheelan, WSMB, New Orleans, to John W. Studebaker, September 18, 1941, Caliver Papers.

61 Ted Thompson, NBC, to William Boutwell, October 14, 1941, and "NBC Public Service Program Suggestions, *Freedom's People,*" bulletin 62, October 1941, ibid.

62 Ambrose Caliver to Malvina C. Thompson, White House, October 7, 1941,

and Thompson to Caliver, October 9, 1941, Correspondence Files, 30.1, A–F, 1941, and Caliver to Eleanor Roosevelt, October 14, 1941, Correspondence Files, 95, A–C, 1941, Eleanor Roosevelt Papers; and Caliver to Eleanor Roosevelt, February 20, 1942, box 7974-8032, file 8026, *"Freedom's People* Broadcasts," President's Personal File, Franklin D. Roosevelt Papers, FDRL; "My Day," quoted in *Federal Radio Education Committee Service Bulletin* 3, no. 11 (November 1941), and Caliver to Frank Wilson, October 16, 1941, Caliver Papers. Caliver to W. E. B. Du Bois, October 15, 1941, reel 53, W. E. B. Du Bois Papers, describes the visit and reports that the Roosevelts and their guests "expressed great interest and commendation" about the show. It appears that Caliver sent identical letters to all of the Advisory Committee members.

63 For a discussion of this idea, see Dates and Barlow, *Split Image.*

64 Henry Louis Gates has observed that "at least since 1600, Europeans had wondered aloud whether or not the African 'species of men' . . . could ever create formal literature, could ever master the 'arts and sciences.'" Gates, *Loose Canons,* 53. The question of black humanity seemed sufficiently unsettled in white minds in the 1940s that African Americans still felt they had to offer this proof.

65 Mary Church Terrell to John W. Studebaker, March 30, 1941, Caliver Papers.

66 "Meeting of the Advisory Committee in Connection with the Proposed Radio Broadcast on the Negro," June 11, 1941, ibid.

67 Locke, *Negro Art;* George Hall, "Alain Locke and the Honest Propaganda of Truth and Beauty," in Linnemann, *Alain Locke,* 91–99; Huggins, *Voices from the Harlem Renaissance;* Du Bois, "Criteria of Negro Art."

68 As a young man, Caliver had earned a certificate in cabinet making from Tuskegee, where he became friends with Carver. Caliver requested his appearance on the show.

69 George Washington Carver to Ambrose Caliver, October 22, 1941; J. T. Gordon to Carver, n.d.; Carter G. Woodson to Caliver, November 3, 1941; "Meeting of the Advisory Committee in Connection with the Proposed Radio Broadcasts on the Negro," June 11, 1941; and "Minutes of Meeting of the Radio Project Advisory Committee," November 14, 1941, Caliver Papers.

70 Letters from Arthur Wright, Southern Education Foundation, October 23, 1941; Marie McIver, supervisor, Colored Elementary Schools, Raleigh, North Carolina, October 21, 1941; F. M. Wood, Baltimore Colored Schools, November 3, 1941; and Lillian P. Falls, Ogden Park Civic Committee, Chicago, November 3, 1941; "Meeting of the Advisory Committee in Connection with the Proposed Radio Broadcasts on the Negro," June 11, 1941; and "Minutes of Meeting of the Radio Project Advisory Committee," November 14, 1941, ibid.

71 Niles Trammell to John W. Studebaker, October 22, 1941; Philip Cohen, chief of Radio Research Project, Library of Congress, to Ambrose Caliver, November 15, 1941; "Meeting of the Advisory Committee in Connection with the Proposed Radio Broadcasts on the Negro," June 11, 1941; and "Minutes of Meeting of the Radio Project Advisory Committee," November 14, 1941, ibid.

72 "Minutes of Meeting of Radio Project Advisory Committee," November 11, 1941, ibid.

73 This refusal to accept black blood and then acceptance of black blood only if it was kept segregated particularly galled blacks because Charles Drew, who had helped develop the technique for storing blood plasma, was a black man. Finkle, *Forum for Protest*, 105–6.

74 Ibid., 106–7.

75 Washburn, *Question of Sedition*, 54, 59.

76 Finkle, *Forum for Protest*, 100.

77 Ibid., 112; Washburn, *Question of Sedition*, 55–56.

78 Washburn, *Question of Sedition*, 59.

79 Ambrose Caliver to L. D. Reddick, December 3, 1941, Caliver Papers.

80 "Broadcast Speech for January 18, 1942, on *Freedom's People* Program: Negro's Contribution to Economic Development of the United States of America," box 34, folder, "Speeches and Writings, June 25, 1941–June 26, 1942," A. Philip Randolph Papers, Manuscripts Division, LC.

81 Letter from Samuel Spritzer, Bronx, December 21, 1941; "Doing Our Part," *Boston Guardian,* January 24, 1942; letters from Emory B. Smith, Washington, D.C., January 19, 1942; M. H. Powell, St. Paul, Minnesota, n.d.; and Charles Handy, New York City, December 22, 1941, Caliver Papers. For a discussion of the meeting of black leaders, see Finkle, *Forum for Protest,* 107.

82 "Minutes of Meeting of Radio Project Advisory Committee," March 9, 1942, Caliver Papers. Caliver served in leadership roles in his church most of his adult life. The series's normal broadcast had been from 12:30 to 1:00 P.M on Sundays, which brought complaints from some black listeners because many black churches were still holding services at the time of the broadcast. That may have given Caliver the idea of turning the time to his advantage.

83 Benjamin Mays, president, Morehouse College, to Ambrose Caliver, September 20, 1941, ibid.

84 Caliver, "Developing Racial Tolerance in America," 452; "Minutes of Meeting of Radio Project Advisory Committee," March 9, 1942, Caliver Papers.

85 "Minutes of Meeting of Radio Project Advisory Committee," March 9, 1942, Caliver Papers.

86 The language appears to be derived from Galatians 3:28.

87 For background on Jernagin and the council, see Sawyer, "Fraternal Council of Negro Churches."

88 Stephen Early to Ambrose Caliver, March 10, 1942, file 8026, *"Freedom's People* Broadcast," President's Personal File, Franklin D. Roosevelt Papers, FDRL.

89 This was a widely held view. See, for example, George E. Haynes, Department of Race Relations, Federal Council of Churches of Christ, to Ambrose Caliver, March 30, 1942, Caliver Papers.

90 Ambrose Caliver to Eleanor Roosevelt, February 20, 26, 1942; Franklin Roosevelt to William Hassett, White House, February 26, 1942; Edwin Embree, Rosenwald Fund, to Franklin Roosevelt, April 1, 1942; W. H. Jernagin to Franklin Roosevelt, April 7, 1942; telegram, Alain Locke, Charles Wesley, Sterling Brown, Joseph Houchins, Arthur Wright, Campbell Johnson, and Mary McLeod Bethune to Franklin Roosevelt, April 8, 1942; William Hassett, White House, to Aubrey Williams, National Youth Administration,

April 6, 1942; Williams to Hassett, April 10, 1942; Franklin Roosevelt to Caliver, April 15, 1942; and Stephen Early to Locke, Embree, and Jernagin, April 16, 1942, file 8026, *"Freedom's People* Broadcast," President's Personal File, Franklin D. Roosevelt Papers, FDRL. Aubrey Williams drafted the letter for the president's approval.

91 Charles A. Schenck Jr., BBC, New York City, to Ambrose Caliver, January 22, 1942, Caliver Papers.

92 In the Advisory Committee meeting about this show, Caliver seemed confident that the president would agree to appear despite Tunick's doubts. Caliver also believed that if the president did not agree to appear, Mrs. Roosevelt would, but I have found nothing that indicates that such a request was ever made, perhaps because of the lateness of the president's final answer. "Minutes of Meeting of Radio Project Advisory Committee," March 9, 1942, ibid.

93 *Variety,* n.d., ibid.

94 Alfred Edgar Smith, Federal Works Agency, to Ambrose Caliver, April 21, 1942; Dutton Ferguson, Federal Works Agency, to Caliver, April 21, 1942; and Campbell C. Johnson, Selective Service System, to John W. Studebaker, May 4, 1942, ibid. The entire series won a place on the prestigious annual Honor Roll of Race Relations, which was based on a nationwide poll conducted by the SCR. "Minutes of Meeting of Radio Advisory Committee," March 9, 1942, ibid.; Ambrose Caliver, "To All Members Who Have Contributed to *Freedom's People,*" n.d., box 164-19, folder 17, "Corres., Caliver, Ambrose," Locke Papers.

95 "Comments on *Freedom's People* Broadcasts," n.d., based on letters, evaluation forms, and comments sent to Caliver after the final show from Marguerite White, Bethel A.M.E. Church, San Francisco; Maude Winston, Southampton County, Virginia; Helen Marshall, Canton, Mississippi; Odessa Wilson, Farmerville, Louisiana; Dr. Carmichael, Canton, Mississippi; and F. D. Parrish, Canton, Mississippi; and letters from James A. Jackson, New York City, April 20, 1942; Josephine Schuyler, New York City, April 24, 1942; and Jessie Treichler, Equal Rights Committee, Antioch College, April 10, 1942, Caliver Papers.

96 Letters from O. R. Pope, principal, Booker T. Washington High School, Rocky Mount, North Carolina, April 19, 1942, and Jane White, New York City, May 6, 1942, ibid.

97 As part of his plan to broadcast the show live in churches, Caliver relied heavily on black teachers to persuade local ministers to cooperate. His efforts met with mixed success, but as part of his evaluation process, Caliver sent out "utilization reports" on the last broadcasts to gather specific information on audience size and reactions. Responses came primarily from southern states, including Alabama, Louisiana, Mississippi, North Carolina, Oklahoma, Texas, and Virginia, as well as from California and Michigan. "Utilization Report of the *Freedom's People* Broadcast of April 19, 1942," ibid.

98 John W. Studebaker to Sterling Brown, Elise Derricotte, Alain Locke, Carter G. Woodson, Arthur D. Wright, Campbell A. Johnson, and Joseph R. Houchins, April 28, 1942, ibid.

99 Ambrose Caliver, "Memorandum for the Commissioner re *Freedom's People*," n.d., ibid.

100 "Radio and National Morale," in "This Is the NBC: Program Service of the NBC," March 1942, ibid.

101 Thomas D. Rishworth, director of public service programs, NBC, to John W. Studebaker, May 19, 1942, ibid.

102 Hilmes, *Radio Voices*, 263.

103 Ambrose Caliver, "Tentative Outline for Proposed Negro Study and Music Hour of the Southern Education Foundation," April 29, 1943, Caliver Papers.

104 Hutchinson, "Marginal Man with a Marginal Mission," 145–46.

CHAPTER THREE

1 Sterling and Kittross, *Stay Tuned*, 213–23, 239–43; MacDonald, *Don't Touch That Dial!*, 65–70.

2 Weinberg, "What to Tell America," 80–81; Blum, *V Was for Victory*, 26.

3 Weinberg, "What to Tell America," 73–74, 77; Winkler, *Politics of Propaganda*, 20–23.

4 Davis was a nightly news commentator for CBS radio, with an audience of over 12 million daily listeners. Shulman, *Voice of America*, 35; Weinberg, "What to Tell America," 81–82. Although the agency was charged with coordinating war information, it was never fully empowered to do that, nor did it receive acquiescence to its leadership from the Departments of State, War, and the Navy and the myriad other federal agencies involved in the war effort. Winkler, *Politics of Propaganda*, 6–7. See also Jones, "United States Office of War Information."

5 Henry Wallace, quoted in Shulman, *Voice of America*, 65.

6 Box 46, folder, "Speech, National Urban League," MacLeish Papers. This idea spread elsewhere at his agency. For example, the Office of Facts and Figures Radio Guide, April 27, 1942, lists as one of its principal goals the effort to show "that we are fighting for freedom and against slavery." Entry E-93, box 599, folder, "Radio War Guides," ROWI.

7 R. Keith Kane to Archibald MacLeish, February 14, 1942, entry 5, box 3, folder 002.11, "Special Groups, Negroes, Jan.–Mar. 1942," ROWI.

8 Caroline Blake, New York City, to Archibald MacLeish, February 14, 1942, and MacLeish to Blake, February 28, 1942, ibid. Blake's letter referred to a series of well-publicized incidents in early 1942, such as the Sikeston, Missouri, lynching and burning of a black man and mounting controversy in Detroit over wartime housing for black migrants seeking jobs. MacLeish might have been referring to several OWI intelligence studies about racial unrest: Division of Surveys, Office of Government Reports, "Survey of Intelligence Materials, No. 13," March 9, 1942, and "Preliminary Appraisal of the Present Negro Situation," March 9, 1942, cited in Jones, "United States Office of War Information," 351, 388. See generally Sitkoff, "Racial Militancy," 661–65.

9 Moss offered MacLeish use of his production, "Salute to Negro Troops," for this purpose, but MacLeish refused the offer. Carlton Moss, production

manager, Council on Negro Culture, to Archibald MacLeish, February 1942, entry 5, box 3, folder 002.11, "Special Groups, Negroes," ROWI. Moss was not alone in offering ideas for steps to boost black morale. During the spring of 1942, several other private black enterprises and organizations sent similar suggestions to MacLeish, but he declined them all. Alton Davis of the popular gospel radio show *Wings over Jordan* proposed a touring morale caravan, composed of Joe Louis and Dorie Miller, funded by the federal government. Theodore M. Berry to Ulric Bell, May 13, 25, 1942, and John Kelly to George Barnes, May 16, 1942, ibid. John Sengstacke, publisher of the *Chicago Defender*, offered to devote a special edition of his newspaper to OFF use, but the offer was declined by MacLeish. John Sengstacke to Bell and MacLeish, April 11, 1942, and MacLeish to Sengstacke, April 28, 1942, ibid. Claude Barnett, director of the Associated Negro Press, sent a detailed proposal on black morale to Vice President Wallace, who in turn referred it to MacLeish. Claude A. Barnett to Vice President Henry Wallace, April 20, 1942, and Wallace to MacLeish, May 4, 1942, ibid. The Fraternal Council of Negro Churches also offered its assistance to MacLeish. The organization's annual meeting included a session entitled "How Can the Negro Church Work for Victory?" R. R. Wright Jr. to MacLeish, April 18, 1942, and MacLeish to Wright, April 22, 1942, ibid.

10 Examples include "Reports from the Special Service Division Submitted April 23, 1942: Negro Organizations and the War Effort"; Cornelius Golightly, "Negro Morale in Boston," Special Services Division Report 7, May 19, 1942; and "Negro Conference at Lincoln University," Special Services Division Report 5, May 15, 1942, cited in Dalfiume, *Desegregation of the U.S. Armed Forces*, 127.

11 Berry was an assistant prosecuting attorney in Hamlin County, Ohio. He also previously had served as director of a Works Progress Administration survey project in Ohio for the U.S. Department of Interior. *Who's Who in Colored America*, 588.

12 T. M. Berry, "Blue Print of Program for Strengthening Negro Morale in War Effort," March 4, 1942, entry 1, box 8, folder, "Racial Relations," ROWI.

13 Ibid., 2–3.

14 Archibald MacLeish to Dean R. O'Hara Lainer, Hampton Institute, March 9, 1942, entry 5, box 3, folder 002.11, "Special Groups, Negroes, Jan.–Mar. 1942," ROWI.

15 Archibald MacLeish to Walter White, March 5, 1942, and Roy Wilkins to White, March 23, 1942, reel 8, NAACP Papers, part 9.

16 Minutes, board meeting, March 12, 17, 19, April 4, 7, 8, 1942, box 52, folder, "Office of Facts and Figures, Minutes of Meetings," MacLeish Papers.

17 R. Keith Kane to Archibald MacLeish, February 14, 1942, entry 5, box 3, folder 002.11, "Special Groups, Negroes, Jan.–Mar. 1942," ROWI.

18 The OFF commissioned the National Opinion Research Center to study the attitudes of New York City blacks toward racial discrimination and toward the war itself. The methodology for the study was curious at best: 1,008 blacks in New York City were interviewed, half by white interviewers and

half by black interviewers, presumably to neutralize or to assess the impact of cross-racial interviewing. Although an economic cross section of blacks were interviewed, for comparative purposes, the survey designers decided to have a white interviewer survey 500 poor whites. It is unclear whether the decision to focus on New York City was driven by convenience, special concerns about racial tensions in the city, or assumptions of New York City's typicality. Extensive Surveys Division, Bureau of Intelligence, Office of Facts and Figures, "The Negro Looks at the War: Attitudes of New York Negroes toward Discrimination against Negroes and a Comparison of Negro and Poor White Attitudes toward War-Related Issues," Report 21, May 19, 1942, Microfilm Collections, Manuscripts and Archives, Sterling Library, Yale University, New Haven, Connecticut.

19 Ibid., 7–8.

20 Ibid., 14–16.

21 Ibid., 4–5. The OWI received at least one report that supported the idea that some blacks identified with the Japanese as other people of color but in a very complicated way. A memo to Elmer Davis about racial tensions in Detroit reported this incident: "A short time ago a propaganda poster was erected in the Ford Motor Car Company where they are manufacturing bombers. The poster depicted a Jap soldier with his bayonet poised above a white woman. Five negro workmen left their machines, tore down the posters and destroyed them. When questioned, one of the [men] replied 'The Japs are colored people. So are we. We are not fighting colored people. We are fighting for democracy.' No further explanation was given. A committee of white workmen urged that the negroes be dismissed . . . but the negroes were not dismissed." It is also possible that these black men were especially sensitive to the poster's use of the image of a white woman under attack by a "colored" man, in this case Japanese. "Negro Problem in Detroit," C. M. Vanderburg, Office of Emergency Management, to Elmer Davis, OWI, September 19, 1942, entry 1, box 8, folder, "Racial Relations," ROWI. Earlier that year, officials at the OWI had received information from the State Department that some blacks on the West Coast were frightened by the internment of Japanese Americans and were "asking whether they would be the next to be moved from their homes and interned." Memorandum of conversation, State Department, Division of Near Eastern Affairs, "Racial Problem on the West Coast," May 27, 1942, entry 5, box 3, folder 002.11, "Special Groups, Negroes, June 1942," ROWI.

22 "The Negro Looks at the War," 6–7.

23 For a valuable review of African American political dissent against war, see Gill, "Afro-American Opposition."

24 Examples include Division of Surveys, Bureau of Intelligence, Office of Facts and Figures, "Current Problems of Negro Morale," Special Report 10, May 16, 1942, cited in Gosnell, "Obstacles to Domestic Pamphleteering," 364; Office of Facts and Figures, "Preliminary Appraisal of the Present Negro Situation," Survey 7, March 7, 1942; and "Negro Attitudes toward the War as

Reported by Correspondents of the Special Services Division," May 19, 1942, cited in Jones, "United States Office of War Information," 146, 388.

25 Archibald MacLeish to the secretary of war, April 7, 1942, box 226, folder, "Office of Facts and Figures," RCASW; MacLeish to General Philip Fleming, administrator, Federal Works Agency, April 7, 1942, entry 5, box 3, folder 002.11, "Special Groups, Negroes, April 1942," ROWI.

26 Robert Huse to Elmer Davis, Milton Eisenhower, Gardner Cowles, and Archibald MacLeish, including minutes, board meeting, November 14, 1942, box 52, folder, "Office of Facts and Figures, Minutes of Meetings," MacLeish Papers.

27 For a description of the creation of new images of white women for wartime use, see Honey, *Creating Rosie the Riveter*, 33, 48, 49, 54, 98, 212.

28 It appears that Starr volunteered his time to the federal government, making him what was called a "$1 a year man" and signifying that he was given token compensation but full status as a federal adviser nonetheless.

29 Milton Starr, "Report on Negro Morale," n.d., Microfilm Collections, Manuscripts and Archives, Sterling Library, Yale University, New Haven, Connecticut. Although this report is undated, textual references reveal that it was written in either July or August 1942.

30 Ibid., 1–5. Starr also urged that William Hastie's speeches be subject to censorship because of his public criticism of the War Department's policies affecting African American soldiers. Hastie, a black lawyer, had been appointed by the president in 1940 as an aide to the secretary of war. Milton Starr to Ulric Bell, June 1, 1942, entry 5, box 3, folder 002.11, "Special Groups, Negroes, June 1942," ROWI.

31 Starr, "Report on Negro Morale," 9, 12–29.

32 Berry returned to Cincinnati and in the 1950s became active in electoral politics, serving on the city council from 1950 to 1957. He returned to Washington, D.C., during President Lyndon Johnson's administration, serving from 1965 to 1969 as assistant director of the Office of Economic Opportunity. From 1972 to 1975, he was mayor of Cincinnati. Low and Clift, *Encyclopedia of Black America*, 173; Phelps, *Who's Who among African Americans*, 95.

33 "Summary and Suggestions on Negro Morale Problems," Theodore Berry to Elmer Davis, July 24, 1942, entry 1, box 8, folder, "Racial Relations," ROWI. Berry continued to argue that the campaign must include messages directed at whites. Berry's memorandum included a June 23 organization chart for staffing and implementing his plan: Theodore M. Berry, "Proposed Organization Chart of Division Relating to Negro Activities in Office of War Information," June 23, 1942, ibid. Berry had generated repeated memoranda on this question and argued that an official policy was needed if "Negroes in the government" rather than "outside forces" were to establish leadership on the issue. In his view, such a policy was necessary to show "Negro citizens at home that government is taking the lead without pressure and with the advice of Negro leadership." Berry to Ulric Bell, May 25, 1942; "Negro Morale: Suggested Policy," Berry to George Barnes and Bell, April 1, 1942; and Berry

to Bell, June 1, 1942, entry 5, box 3, folder 002.11, "Special Groups, Negroes, April, May, June 1942," ROWI.

34 Theodore Berry to Elmer Davis, July 24, 1942, 4–5, and Davis to Gardner Cowles, n.d., entry 1, box 8, folder, "Racial Relations," ROWI. During the summer of 1942, various proposals for a consolidated approach to the "Negro problem" at the federal level were offered. For example, Lawrence Cramer of the Fair Employment Practices Committee suggested that the committee be made the central agency for the problem. Lawrence Cramer to Judge James Landis, Office of Civilian Defense, May 22, 1942, ibid. Some officials proposed that the number of black advisers be increased or that a bureau dedicated to African Americans be created; others made what was referred to as the "perennial proposal"—the appointment of a "Negro affairs" expert at the White House. Ironically, several OWI officials believed that if a centralized "Negro agency" was to be established, it ought to be at the OWI, despite that agency's inability to construct an African American morale policy. George A. Barnes to Ulric Bell, June 5, 1942, and Barry Bingham to Elmer Davis, July 21, 1942, entry 5, box 3, folder 002.11, "Special Groups, Negroes, June 1942," ROWI. Officials at the Office of Civilian Defense considered creating a Division of American Unity to deal with black morale but abandoned the idea after deciding that the topic was " 'too hot a potato.' " Dalfiume, *Desegregation of the U.S. Armed Forces,* 126, n. 80.

35 William B. Lewis, CBS, "Radio's Vital Role in the Time of Crisis," *CBS Student Guide,* July 1941, entry E-93, box 601, folder, "CBS," ROWI; Jones, "United States Office of War Information," 109–10, 328; Winkler, *Politics of Propaganda,* 60–63. The OWI coordinated programmatic matters as well. The Radio Bureau received and passed on to network programming officials war messages from federal agencies that were to be incorporated into popular radio series like *The Lone Ranger, Fibber McGee and Molly,* and *The Jack Benny Show.* The bureau organized a series of one-minute talks by popular radio commentators and, in conjunction with the Bureau of Campaigns, used radio to enlist support for various home front drives, including buying bonds, collecting salvage, and saving fuel. Radio Bureau staff also read and approved scripts from various federal agencies and private entities for radio programming or specials about the war. Winkler, *Politics of Propaganda,* 60–63; MacDonald, *Don't Touch That Dial!,* 65, 69–70. For more on the relationship between the OWI and radio, see Jones, "United States Office of War Information," 107, 142, 328–33.

36 "Summary and Suggestions on Negro Morale Problems," Theodore Berry to Elmer Davis, July 24, 1942, entry 1, box 8, folder, "Racial Relations," ROWI. Apparently, Starr also had made his suggestion early in 1942, but no action had been taken on it. Starr, "Report on Negro Morale," 17–18.

37 Jones, "United States Office of War Information," 361.

38 "The Negro Problem," Charles Siepman to W. B. Lewis, August 6, 1942, 1–2, entry 1, box 8, folder, "Racial Relations," ROWI.

39 Bureau of Intelligence, OWI, Intelligence Report 36, August 14, 1942, 11–13, box 53, folder, "OWI Intelligence Reports," MacLeish Papers.

40 As early as 1938, Walter White had written President Roosevelt to request that an upcoming presidential broadcast "include specific reference to the campaign of bigotry now current in the country against the Negro as manifested in the wholesale discharge of the Negro under the wages-and-hours act, and in the record, to date, of six lynchings since Congress adjourned in no one of which has there been even a single arrest." Walter White to President Franklin Roosevelt, October 31, 1938, box 2, file 93, "Colored Matters," President's Official File, Franklin D. Roosevelt Papers, FDRL. Roosevelt had promised to make a statement on racial matters on the final episode of the *Freedom's People* radio series in 1942 but substituted a weak written statement instead, as discussed in chapter 2.

41 Elmer Davis to William Hassett, August 27, 1942, entry 1, box 8, folder, "Racial Relations," ROWI.

42 Milton Eisenhower, associate administrator at the OWI, had asked his deputies to propose a way to design some "over-all pattern" with respect to handling the information needs demanded by the "Negro situation." Although these deputies agreed that a position should be created at the OWI to coordinate efforts to raise black morale, they balked at the idea of establishing a separate bureau for such purposes, preferring instead to consult Will Alexander at the War Manpower Commission as needed. "Recommended Procedure on Negro-White Relations," Reginald Foster and Robert Huse to Elmer Davis, September 12, 1942, ibid.

43 George A. Barnes to Milton S. Eisenhower, September 28, 1942, ibid.

44 G. Lake Imes to L. D. Reddick, October 24, 1942, RSC; Sterling and Kittross, *Stay Tuned*, 157–58.

45 Script, "This Is Our War," *My People,* October 11, 1942, RSC.

46 Milton Starr to William Lewis, chief, Radio Bureau, October 22, 1942, and Lewis to Starr, December 2, 1942, entry 1, box 8, folder, "Racial Relations"; and Alvin M. Josephy Jr., Domestic Radio Bureau, OWI, to Tom Slater, Mutual Broadcasting System, January 28, 1943, entry 146, box 760, folder, "*My People,*" ROWI.

47 OWI files include scripts for broadcasts on February 13, 20, 27, and March 6, 1943. Records about the show in the ROWI are scanty; they contain what appear to be the draft and final scripts but little contextual information. Additionally, corporate records of WOR, the Mutual network's flagship New York City station, are held by the LC but are unavailable to researchers. The RSC holds the scripts for the broadcasts prior to the OWI's involvement with the show but not the subsequent ones. Recordings of the shows do not appear to be among those at the National Archives, Washington, D.C.

48 Script, draft and revised, *My People,* February 17, 1943, and Joseph Liss, script clearance editor, Domestic Radio Bureau, Office of Emergency Management, to G. Lake Imes, February 24, 1943, entry 146, box 760, folder, "*My People,*" ROWI. Available records do not reveal Imes's reactions to the OWI's cosponsorship or subsequent attempts to censor the show's content.

49 Script, *My People,* March 6, 1943, ibid.

50 Joe Liss to Douglas Meservey, February 18, 1943, ibid. Mutual network offi-

cials also wanted to carry the show through the duration of the series. Liss to Meservey, G. Zachary, and M. Starr, February 19, 1943, ibid.

51 Office of War Information, "Program for War Information to Negroes," n.d., Microfilm Collections, Manuscripts and Archives, Sterling Library, Yale University, New Haven, Connecticut. Textual references suggest that the document was probably written in April 1943.

52 Gill, "Afro-American Opposition," 38–40; Wynn, *The Afro-American and the Second World War,* 11; Harris, *Keeping the Faith,* 32–33.

53 The pamphlet the OWI eventually issued, *Negroes and the War,* met with such immediate controversy and criticism that the University of Chicago political scientist Harold Gosnell, then serving at the federal Budget Office, used it as a case study about the need for proper clearance procedures for federal publications. Owen's original pamphlet, Gosnell's report, and notes of his extensive interviews can be found in Series 41.3, Division of Administrative Management, War Records Section, 1941–47, box 15, folder 159B, "Case Study of the Clearance of an OWI Publication, *The Negroes and the War,*" RBB. For a brief account of the pamphlet's development, see Gosnell, "Obstacles to Domestic Pamphleteering," 364–69. See also Jones, "United States Office of War Information," 358–61. For brief mention of the pamphlet, see Winkler, *Politics of Propaganda,* 31, 56; Blum, *V Was for Victory,* 41, 194–96; Koppes and Black, "Blacks, Loyalty, and Motion Picture Propaganda," 360; and "Summary and Suggestions on Negro Morale Problems," Theodore M. Berry to Elmer Davis, July 24, 1942, entry 1, box 8, folder, "Racial Relations," ROWI.

54 Special Services Division, Office of Government Reports, "Developing Situations," August 12, 1942; R. Keith Kane, Bureau of Intelligence, OWI, "Report of Recent Factors Increasing Negro-White Tensions," November 2, 1942; and Bureau of Intelligence, Division of Surveys, Office of Facts and Figures, "March on Washington Movement among Negroes," May 12, 1942, cited in Jones, "United States Office of War Information," 389; Bureau of Intelligence, OWI, "Survey of Intelligence Materials, Supplement to Survey 26"; Division of Information, "Editorial Emphasis in the Negro Press, May 29–June 20," Special Intelligence Report 48; Service Materials Division, "Statements of Thought Leaders on the Negro Problem, March to June 1942" and supplementary report, June 24, 1942; and William Hastie to R. Keith Kane, chief, Bureau of Intelligence, July 24, 1942, entry 188, box 226, folder, "OWI," RCASW.

55 "The Government's Dilemma," *Pittsburgh Courier,* October 3, 1942.

56 William Hastie to Chandler Owen, May 2, 1942, reel 7, NAACP Papers, part 9. Phileo Nash at the OWI asked Leo Rosten, the deputy director of the Domestic Branch, to help him stop the publication. "The Pamphlet 'What Will Happen to the Negro If Hitler Wins?,'" Phileo Nash to Leo C. Rosten, November 17, 1942, RBB; Walter White to Archibald MacLeish, April 29, 1942, and MacLeish to White, April 30, 1942, entry 5, box 3, folder 002.11, "Special Groups, Negroes," ROWI; White to Frank Reeves, D.C. branch, NAACP, May 1, 1942, reel 8, and White to A. Philip Randolph, May 4, 1942;

Randolph to White, May 14, 1942; Henry F. Pringle to White, August 18, 1942; and White to Pringle, August 24, 1942, reel 7, NAACP Papers, part 9.

57 *Negroes and the War,* Selected Documents on World War II from the Franklin D. Roosevelt Library, Microfilm Collections, Manuscripts and Archives, Sterling Library, Yale University, New Haven, Connecticut. The quality of the pamphlet far exceeded that of any other OWI publication, as did its overall and per copy costs; the OWI spent $85,000 to produce 2.5 million copies of *Negroes and the War.* "Domestic Branch, Publications Issued, June 1942–March 1943," entry 1, box 8, folder, "Releases, Domestic, 1942–43," ROWI.

58 Blum, *V Was for Victory,* 195.

59 Walter White to Milton MacKaye, OWI, January 25, 1943; Carolyn Davenport, Philadelphia NAACP, to White, February 4, 1943; White to Davenport, February 5, 1943; and Davenport to White, February 6, 1943, reel 7, NAACP Papers, part 9.

60 Reportedly, Will Alexander explained that "nobody trusted Owen" because "Owen helped Randolph organize the Negro porters and then sold out to the Pullman Company." Some of Alexander's comments on Owen were reported to Gosnell by John Fleming; others came directly from Alexander. Ralph Bunche, then at the Office of Special Services, said that "it was a bad blunder" to assume that Owen was a "man of distinction" when the few people who knew him found him to be a "completely dishonest person." Franklin Frazier was not interviewed by Gosnell, but his reported comments were passed on by Phileo Nash to Gosnell. Frazier was reported to have said that "Owen was known as a 'chiseler.'" An OWI fieldworker in Chicago, where Owen resided, wrote a long report in which he concluded that "Owen had a reputation for being slick." Harold Gosnell, interviews with Donald Young, Special Service Division, U.S. Army, January 5, 25, 1943; Lieutenant William Bryant, Special Services Division, Bureau of Intelligence, OWI, January 22, 1943; Truman K. Gibson, assistant to the civilian aide, War Department, January 25, 28, 1943; Frances Williams, senior consumer relations officer, Department of Information, Office of Price Administration, January 28, February 1, 1943; Phileo Nash, Special Services Division, Organizations, Bureau of Intelligence, OWI, February 2, 1943; John Fleming, chief, Bureau of Publications and Graphics, Domestic Branch, OWI, February 2, 1943; Will Alexander, February 6, 1943; and Ralph Bunche, research analyst, British Empire Section, OSS, February 4, 1943, RBB.

61 Walter White to Elmer Davis, December 8, 1942; Theodore Berry to White, December 17, 1942; Davis to White, December 18, 1942; Milton Starr to George Barnes, December 17, 1942; William Hastie to White, December 8, 1942; White to Davis, January 15, 1943; Starr to Barnes, January 22, 1943; Davis to White, January 30, 1943; and White to Davis, February 3, 1943, entry 1, box 6, folder, "Racial Relations," ROWI; interview with George Barnes, January 23, 1943, RBB; Starr to Barry Bigham, June 30, 1942, entry 1, box 8, folder, "Racial Relations," ROWI.

62 Truman Gibson was among those who had suggested that whites needed to

see such materials, as was Will Alexander. Interviews with Donald Young, January 25, 1943; Will Alexander, February 6, 1943; and Truman K. Gibson, January 28, 1943, RBB.

63 Black readers specifically objected to photographs they thought were unrealistic or untruthful, such as one of nine white men and a lone black man celebrating a union election victory. Blacks said it was a "propaganda" photograph in which the black man was simply "stuck in" and that "they should have all Negroes join the union," not just one. *"Negroes and the War:* A Preliminary Test of an OWI Pamphlet," n.d., Selected Records of the Office of War Information, Microfilm Collections, Manuscripts and Archives, Sterling Library, Yale University, New Haven, Connecticut.

64 The survey was conducted by sociologist Paul Lazarsfeld's researchers at Columbia University. A seventeen-page memorandum summarized the results of interviews with seventy black New York City families about the pamphlet. Lazarsfeld, who had escaped Nazi-occupied Austria in the 1930s, was one of the pioneers in the relatively new field of opinion surveying. Although his work centered specifically on radio, apparently his methodology was thought by OWI officials to be suitable for measuring "audience reaction" to print production as well. For this study, black interviewers visited the families twice, once to leave a copy of the publication and the second time to ask a series of questions about their reactions to it. "General attitude questions" were asked before and after the respondents read the pamphlet; in a follow-up interview, they were queried about specific photographs "to see whether the respondents had understood the implications of these items." Ibid.

65 Sekula, "On the Invention of Photographic Meaning," 84.

66 *"Negroes and the War:* A Preliminary Test of an OWI Pamphlet."

67 Ibid.

68 Hall, "Reconstruction Work," 158.

69 For example, unaware that Owen's essay was going to open the pamphlet, Ralph Bunche had reviewed the photographic layout for it and had not found it objectionable. Interview with Ralph Bunche, February 4, 1943, RBB.

70 Gosnell, "A Case Study of the Clearance of an OWI Publication," 9; "Negroes and the War," Milton MacKaye to Mr. Guinzberg, April 6, 1943; and interview with Milton MacKaye, Division of Publications, Bureau of Publications and Graphics, Domestic Branch, OWI, January 26, 1943, RBB. Another OWI official confirmed that local NAACP branches were requesting additional copies of the pamphlet, in part because Walter White had stated that "it might do some good." Interview with Robert Martin, Schools and Colleges Division, Bureau of Special Operations, OWI, February 8, 1943, RBB.

71 Gill, "Afro-American Opposition," 267.

72 Interview with Robert Martin, OWI, February 8, 1943, RBB.

73 Sterling A. Brown, "Count Us In," in Sanders, *A Son's Return,* 72.

74 Alfred Palmer, quoted in Natanson, *Black Image in the New Deal,* 41. Palmer was an OWI photographer who worked with Poston in selecting the photographs.

75 "Negro Pictures," memorandum, Talbot Patrick to George Barnes, May 22,

1942, entry 5, box 3, folder 002.11, "Special Groups, Negroes," ROWI. Milton MacKaye, a magazine writer at the OWI, had primary responsibility for the layout and production of the pamphlet. Gosnell, "Obstacles to Domestic Pamphleteering," 364–65.

76 Rankin quoted in *New York Times,* March 14, 1943, in Jones, "United States Office of War Information," 458, 495; W. C. Hodges, Birmingham, Alabama, to Senator John Bankhead, March 16, 1943; Bankhead to Elmer Davis, March 18, 1943; Ralph Shikes, OWI, to James Allen, OWI, March 20, 1943; resolution, Shreveport Chamber of Commerce, April 8, 1943; Senator John H. Overton to Davis, April 15, 1943; and Davis to Overton, April 23, 1943, entry 1, box 8, folder, *"Negroes and the War,"* ROWI. In one reply, Davis explained that the pamphlet was designed specifically to "help counteract Japanese propaganda designed to foment racial discord in this country," a curious claim considering the pamphlet's sole focus on Hitler. Davis to Bankhead, March 30, 1943, ibid. Senator Harry F. Byrd of Virginia forwarded to Davis a letter from a white constituent in Richmond who called the publication a "mawkish, minority, glorifying hand-out" and a "rank waste of public money." C. F. Hatch, Richmond, to Senator Harry F. Byrd, June 19, 1943; Davis to Byrd, June 30, 1943, ibid.

77 *Congressional Record,* 78th Cong., 1st sess., vol. 89, pt. 5, June 18, 1943, 6133–37. The 1942 elections also yielded a more conservative Congress that was deeply suspicious of the administration's use of federal funds to advance its own popularity. Weinberg, "What to Tell America," 83; Shulman, *Voice of America,* 96.

78 He cited as examples "a very friendly statement to the Negro cause, by the President, along with his photograph" and another page that "portrays Mrs. Roosevelt in a role of great friendliness to the Negro race." Report of Hearings, Senate Subcommittee on Appropriations, 78th Cong., 1st sess., National War Agencies Appropriation Bill, 1944, June 26, 1943, 196–97.

79 Ibid., 171–78, 195–202, 246, 344–45.

80 Gosnell, "Obstacles to Domestic Pamphleteering," 368–69; Blum, *V Was for Victory,* 194–95; Weinberg, "What to Tell America," 83, n. 48. See also Koppes and Black, "Blacks, Loyalty, and Motion Picture Propaganda," 390, and Winkler, *Politics of Propaganda,* 56–57. *Negroes and the War* was not the only pamphlet under attack, although it was the primary one; two other pamphlets, one on taxation and one on inflation, also were criticized as pro–New Deal propaganda.

81 Shulman, *Voice of America,* 104; Jones, "United States Office of War Information," 466, 468.

82 Walter White, "People and Places: Slaughter on Capital Hill," *Chicago Defender,* July 10, 1943. White also linked Davis's troubles to a broader attack by congressional reactionaries against New Deal agencies, including the Farm Security Administration and the National Youth Administration.

83 Report to the president, Elmer Davis, "The Office of War Information, 13 June 1942–15 September 1945," Subject File, OWI, folder 2, Elmer Davis Papers, Manuscripts Division, LC; Shulman, *Voice of America,* 185.

84 Hall, "Reconstruction Work," 156.

85 For example, George Roeder's excellent book on visual images in World War II describes the reaction to a 1942 *Life* magazine article praising black contributions to the war: "[S]ome readers praised the article for 'building up confidence, morale and patriotism in our country,' while others found it odious. A reader from Kentucky [wrote,] 'Your Negro war article is inflammatory to the point of treason.'" Roeder, *Censored War,* 45.

86 Hall, "Reconstruction Work," 153, 152, 157; Sekula, "On the Invention of Photographic Meaning," 84.

87 With an eye toward gaining the black vote in the 1940 election and assuaging black concerns, Roosevelt made three war-related personnel decisions involving prominent black men. Aside from Hastie's appointment at the War Department, he assigned Major Campbell Johnson to be an assistant to the selective service director and promoted Colonel Benjamin Davis Sr. to brigadier general, making him the highest-ranking African American in the military. Roosevelt had appointed Hastie to the federal district court for the Virgin Islands in 1937. McQuire, *He, Too, Spoke for Democracy,* 9–13. For more about Hastie's long and distinguished career, see also Ware, *William Hastie.*

88 "Proposed Radio Broadcast Developing the Participation of Negroes in the United States Army," Truman K. Gibson to Major R. B. Lord, deputy director, Bureau of Public Relations, March 20, 1941, entry 188, box 183, folder, "Bureau of Public Relations, War Department," RCASW. In that same month, Hastie wrote a long internal critique of the army's proposed plans to limit black enlistment and to corral black troops into service and nontechnical positions. McQuire, *He, Too, Spoke for Democracy,* 25–27.

89 "Proposed Radio Broadcast Developing the Participation of Negroes in the United States Army," Truman K. Gibson to Major R. B. Lord, May 1, 1941, entry 188, box 183, folder, "Bureau of Public Relations, War Department," and Mr. Brechner, Department of War, to E. M. Kirby, chief, Radio Branch, Bureau of Public Relations, June 24, 1941, entry 188, box 240, folder, "Radio Broadcast, Negro in the Army," RCASW.

90 Mr. Brechner to E. M. Kirby, June 24, 1941, entry 188, box 240, folder, "Radio Broadcast, Negro in the Army," RCASW.

91 Dalfiume, *Desegregation of the U.S. Armed Forces,* 73, 74, 80.

92 Mr. Brechner to E. M. Kirby, June 24, 1941, entry 188, box 240, folder, "Radio Broadcast, Negro in the Army," RCASW.

93 Truman K. Gibson to press relations officers, July 3, 1941, ibid.

94 Some of the officers Gibson had contacted did appreciate the larger significance of his request and the underlying political issues at play. Several defensively sought to reassure Gibson about the status of black troops at their facilities. One wrote that "there is no distinction made in any way between the colored soldier and the white soldier here." Another officer explained that on his base "the colored soldiers have their own theaters, swimming pool, recreation buildings, club houses etc." The officer made a virtue of the fact that black soldiers had their "own" facilities without acknowledging that

segregation also demanded it. Glenn J. Jacoby, first lieutenant, Headquarters, Quartermaster Replacement Training Center, Fort Francis E. Warren, Wyoming, July 8, 1941; Victor L. Cary, second lieutenant, press relations officer, Quartermaster Replacement Center, Camp Lee, Virginia, July 19, 1941; and William W. Deaton, first lieutenant, Headquarters, Twenty-fourth U.S. Infantry, Fort Benning, Georgia, July 10, 1941, ibid.

95 Glenn J. Jacoby, July 8, 1941, ibid.

96 John R. Sutherlin, first lieutenant, press relations officer, 100th Coast Artillery, Camp Davis, North Carolina, July 15, 1941, ibid.

97 Truman K. Gibson to Edna Thomas, Negro Actors Guild of America, July 11, 1941, and Thomas to Gibson, July 18, 1941, ibid. However, Sissle and Gibson were unable to secure big-name stars like Marian Anderson, Ethel Waters, and Paul Robeson, apparently because of scheduling conflicts and the short lead time between the War Department's July decision to do the show and its actual broadcast the next month on NBC. Robeson expressed his regrets that he would be in Hollywood at the time of the broadcast. Gibson to Harold Gumm, July 30, 1941 (Ethel Waters); Hubert T. Delany to Gibson, August 1, 1941 (Marian Anderson); and Eslanda Goode Robeson to William Hastie, August 7, 1941, ibid. Scheduling was further complicated because the network twice delayed the broadcast, shifting it from July 27 to August 5 to the final broadcast date of August 12.

98 Truman K. Gibson to Edna Thomas, July 11, 1941, ibid.

99 H. B. Summers to Ambrose Caliver, July 25, 1941, box 83C, folder 22, "*Freedom's People,*" NBC Collection.

100 All references to "America's Negro Soldiers" in this chapter are to script, "America's Negro Soldiers," August 12, 1941, RSC.

101 William H. Hastie to acting director, Bureau of Public Relations, August 6, 1941, entry 188, box 240, folder, "Radio Broadcast, Negro in the Army," RCASW. Hastie did appear on a national CBS program on Saturday, September 20, 1941, commemorating the opening of a weekend recreational camp for black soldiers. War Department press release, "CBS Airs Program from 'Weekend Camp' for Negro Soldiers," September 11, 1941, and memorandum, Truman K. Gibson, September 15, 1941, ibid.

102 Patterson paid special attention to the Ninety-ninth Squadron, which, he announced, was preparing to relocate to the Tuskegee Air Field. Noting that the new base was designed by a black architect and was being built by a black contractor, he held up the squadron as a stellar example both of black achievement and of American democracy. The segregated treatment of the squadron and of black airmen would be the source of continuing criticism and eventually would play a major role in Hastie's decision to resign from the War Department. See, for example, Dalfiume, *Desegregation of the U.S. Armed Forces,* 84.

103 Patterson's complete remarks were issued as "The Negro and National Defense, Address by Honorable Robert P. Patterson, the Under Secretary of War, August 12, 1941, Press Release, War Department Bureau of Public Rela-

tions," box 17, series 6, "National Defense Program, War Department, 1941–42," NUL Papers.

104 Dalfiume, *Desegregation of the U.S. Armed Forces,* 72. Attributing political implications to the image of the black soldier was not new. Some had hoped, for example, that the presence of black Civil War soldiers in uniforms would combat claims of innate black inferiority and male docility. Fredrickson, *Black Image in the White Mind,* 168–69. For a recent discussion of hostilities directed at black troops who served in the Spanish-American War, see Gilmore, *Gender and Jim Crow,* 78–82. Not only were black soldiers subject to white rhetorical claims that they were incapable of being good soldiers, but when traveling on trains, they were sometimes the object of white gang attacks.

105 Sterling A. Brown, "Count Us In," in Sanders, *A Son's Return,* 78. A similar argument is made about the World War II period in Kelley, *Race Rebels,* 64, and " 'We Are Not What We Seem.' "

106 Truman K. Gibson to Earl Dickerson, August 20, 1941, entry 188, box 240, folder, "Radio Broadcast, Negro in the Army," RCASW. Dickerson, a member of the Chicago city council and of the President's Fair Employment Practices Committee, conducted the show's on-the-air interviews with black soldiers at the Illinois base.

107 "Radio's Greatest All-Colored Show for NBC; Stars to Take Part in Mammoth Program Next Tuesday Night," *Pittsburgh Courier,* August 9, 1941; "The Army's Radio 'Flop,' " *Pittsburgh Courier,* August 23, 1941.

108 "Program Requested by Mr. Truman Gibson," Brooks Watson to Lieutenant Colonel E. M. Kirby, Radio Branch, Bureau of Public Relations, January 22, 1942; Captain Bates Raney, assistant public relations officer, Military Intelligence, to Kirby, September 30, 1942; and Kirby to Raney, October 5, 1942, entry 188, box 240, folder, "Radio Broadcast, Negro in the Army #2 (1942)," RCASW.

109 Despite several press reports about the broadcast, I have been unable to locate either a script or a recording. Press release, Public Relations Section, New York City, Eastern Defense Command, n.d.; Jerry Franken, "Heard and Overheard," *PM,* September 29, 1942; "Listening In," *New York Daily News,* September 28, 1942; "Network Airing History of Negroes," *People's Voice,* September 26, 1942; *New York Times,* September 27, 1942; "History of Negro Troops on NBC Hookup Sunday," *Norfolk Journal and Guide,* September 26, 1942; and *New York World-Telegram,* September 25, 1942, ibid.

110 Lieutenant Colonel E. M. Kirby to Lieutenant Colonel Edward J. F. Glavin, public relations officer, Eastern Defense Command, n.d., ibid. NBC received over 120 telephone calls; the New York City affiliate where it originated, WJZ, counted 40 phone calls immediately following the broadcast. Calls from San Francisco and Hollywood listeners were also noted. The Chicago affiliate reported 60 calls, "an unusually high number for any show except the 'teaser' or 'give away' type." Captain Bates Raney to Kirby, September 30, 1942, and Kirby to Raney, October 5, 1942, ibid.

111 The national radio networks produced other programs about black wartime

contributions, including an August 1944 NBC coast-to-coast broadcast commemorating the Tuskegee Air Field and its black air corpsmen. NBC also sponsored *They Call Me Joe,* a series that spotlighted a representative ethnic soldier in each episode. One show profiled a black soldier. The Mutual network broadcast a national program in 1943 in which black soldiers described their ordeals in the war. A dramatization set aboard the all-black naval destroyer, the USS *Booker T. Washington,* was also broadcast in 1943. And in cooperation with the Negro Newspaper Publishers Association, black servicemen were featured in a nationwide Mutual broadcast from Hampton Institute. That show, "Fighting Men," included "first-hand descriptions of sea rescues" by an all-black Coast Guard station. Guzman, *Negro Yearbook,* 450, 497, 447; script, *They Call Me Joe,* September 30, 1944, RSC. See also MacDonald, *Don't Touch That Dial!,* 348.

112 Capra remained the producer in name for the film but delegated the film's direction to a younger colleague, Stuart Heisler. Cripps and Culbert, *"The Negro Soldier,"* 117–18, gives brief details on the army's planning for the film. It pays considerable attention to the film's reception but provides little content analysis. More recently, Cripps has written a detailed discussion of Moss's contribution to the film and its political context: *Making Movies Black,* 102–25. Aspects of the film's development and reception also are noted in Wynn, *The Afro-American and the Second World War,* 83–84. Cripps also includes a brief discussion of the film in *Black Film as Genre,* 108–14.

113 This church also stood in marked contrast to those depicted in films of that era about black religion such as *Green Pastures* and *Cabin in the Sky.*

114 Some critics have been less enthusiastic about the artistic aspects of the film even as they generally termed it a landmark film of its genre. For example, see Barnouw, *Documentary,* 162. While appreciating its significance, Barnouw derides the film for being "ultimately condescending" and "unrelenting in religiosity." Another critic thought the film was "impressive and moving" but found its narrative structure "unwieldy." Richard Dyer MacCann, "World War II: Armed Forces Documentary," in Barsam, *Nonfiction Film Theory and Criticism,* 156, n. 19.

115 Cripps, *Making Movies Black,* 112.

116 As previously noted, organized black protests against derogatory and one-dimensional black film characters had begun as early as the NAACP's 1915 campaign against *The Birth of a Nation.* Cripps, *Slow Fade to Black,* 41–69; Koppes and Black, "Blacks, Loyalty, and Motion Picture Propaganda," 391.

117 Cripps, *Making Movies Black,* 107.

118 Dalfiume, *Desegregation of the U.S. Armed Forces,* 88–89.

119 Cripps, *Making Movies Black,* 112, 114–15.

120 The film's greatest distribution came through a network of public libraries, schools, and colleges that screened it for civic and church groups. Cripps and Culbert, *"The Negro Soldier,"* 121–29.

121 Cripps, *Making Movies Black,* 112.

122 Cripps and Culbert, *"The Negro Soldier,"* 115, quoting a War Department offi-

cer's training manual, *Leadership and the Negro Soldier,* prepared by Donald Young and used as part of its orientation of white officers charged with leading black units.

123 Ibid., 130–31.

124 MacGregor, *Integration of the Armed Forces,* 50–57; Shulman, *Voice of America,* 111.

125 That hostility erupted publicly in 1945 after the War Department sent him to assess the performance of the all-black Ninety-second Division in Italy. Gibson confirmed that the unit had "'a rather dismal record'" and implied that its troops had behaved in a cowardly fashion in the heat of battle, which he attributed to their relatively low intelligence and educational levels. Stunned, the black press immediately began calling for Gibson's ouster. Labeling him a "'new enemy,' an 'Uncle Tom,'" and "'a liability,'" black editors heaped waves of criticism on Gibson and on his department's continued refusal to appraise fairly the contributions of black soldiers. Even Hastie, who had avoided any personal comments on Gibson since his resignation, condemned his remarks as a "'gross libel' against the race." Finkle, *Forum for Protest,* 185–88.

126 Actually, Gibson's letter had referred to the need for material on "returning Negro soldiers and Japanese American soldiers," although the remainder of his letter dealt solely with returning black soldiers. Truman K. Gibson to Robert Heller, vice president, CBS, June 13, 1945, entry 188, box 240, folder, "Race Relations, Radio Script," RCASW.

127 Ibid.

128 For a detailed discussion of the voluntary replacement plan, see Lee, *Employment of Negro Troops,* 688–705. For a briefer treatment, see Dalfiume, *Desegregation of the U.S. Armed Forces,* 99–100.

129 Script, "The Glass," *Assignment Home,* entry 188, box 240, folder, "Race Relations, Radio Script," RCASW.

130 Charles S. Johnson, Fisk University, to Truman K. Gibson, September 20, 1945, ibid.

131 Louis R. Lautier, administrative assistant, civilian aide to secretary of war, to Colonel Douglas Parmentier, News Division, August 25, 1945, and Lautier to E. B. Ray, theatrical editor, Afro American Newspapers, Baltimore, August 23, 1945, ibid.

132 Walter White to Henry Stimson, War Department, August 25, 1945; Luther Hill, public relations, War Department, to White, n.d.; and Roy Wilkins to Stimson, August 31, 1945, reel 10, NAACP Papers, part 9.

133 "Army KO's Negro Job Show," *PM,* August 23, 1945, and "Heard and Overheard: Action on All Fronts," *PM,* August 29, 1945, ibid.; John J. McCloy, assistant secretary of war, to Eleanor Roosevelt, n.d., entry 188, box 240, folder, "Race Relations, Radio Script," RCASW.

134 Letters from Dr. Luther S. Peck, president, Dusable Lodge 751, International Workers Organization, Chicago, Illinois, August 30, 1945; Arthur S. Harris, Worcester, Massachusetts, August 24, 1945; and Martha Weisman, Charlestown, Massachusetts, September 17, 1945; and Truman K. Gibson to Weis-

man, October 2, 1945, entry 188, box 240, folder, "Race Relations, Radio Script," RCASW.

135 Clifford Evans, director of news and special events, WLIB, New York City, to Truman K. Gibson, September 19, 1945, and Gibson to Evans, September 25, 1945, ibid.

136 Sterling A. Brown, "Count Us In," in Sanders, *A Son's Return*, 75.

CHAPTER FOUR

1 For a history of the early National Urban League, see Weiss, *National Urban League*. Weiss argues that the organization's philosophy mirrored that of Booker T. Washington. For a longer view of the organization's history, see Moore, *Search for Equality*, which emphasizes the league's social service orientation as well as the enlargement of its goals and strategies in the 1940s and 1950s. For a sympathetic insider's history of the organization, see Parris and Brooks, *Blacks in the City*. Parris was director of public relations at the league; his book, although unabashedly laudatory, is useful because it includes detailed observations about the league's programs in the late 1930s and 1940s.

2 *Opportunity* 7 (July 1928): 218.

3 Archer, *Black Images in American Theater*, 225; Reddick, "Educational Programs," 13.

4 Weiss, *National Urban League*, 255, 257, 258–59; Parris and Brooks, *Blacks in the City*, 263–64.

5 Weiss, *National Urban League*, 258.

6 Parris and Brooks, *Blacks in the City*, 212.

7 The National Urban League and its local affiliates sponsored annual weeklong activities advancing the vocational campaign from 1930 to 1933, suspended them for two years for lack of funds, and then resumed them from 1937 to 1964. Weiss, *National Urban League*, 259.

8 For example, one of the scripted programs entitled "Occupational Opportunity for Negroes" pitched league books and programs through a question-and-answer session involving fictionalized local board members and a white executive. Another was a dramatization of George Washington Carver's life. The scripts for the two 1939 programs were included as part of the league's manual for the 1940 campaign. National Urban League, Department of Industrial Relations, "Program Aids from the Vocational Opportunity Campaign," February 1, 1940, MSRC; "Vocational Guidance Week," *Opportunity* 16 (March 1938): 69.

9 Author's interview with Ann Tanneyhill, November 28, 1994, Mashpee, Massachusetts.

10 Davidson Taylor, vice president in charge of production, CBS, to Edward Lawson, National Urban League, February 11, 1941, box 30, series 7, "Vocational Opportunity Campaign Report, Unbound Miscellaneous Material, 1941," NUL Papers; interview with Ann Tanneyhill, August 11, 1978, in Hill, *Black Women Oral History Project*, 233.

11 Interview with Ann Tanneyhill, August 11, 1978, in Hill, *Black Women Oral History Project*, 234.

12 Ed Lawson to American Federation of Musicians, February 20, 1941, and George Heller, American Federation of Radio Artists, American Federation of Labor, to Lawson, February 21, 1941, box 30, series 7, "Vocational Opportunity Scrapbook, Report, Unbound Miscellaneous Material, 1941," NUL Papers. It also appears that a copyright fee waiver was sought and secured from the American Society of Composers, Authors, and Publishers for the songs that were to be used on the show. Lawson to Philip Cohen, CBS, February 25, 1941, and Local 802, American Federation of Musicians, New York City, to Lawson, February 25, 1941, ibid.

13 Hilmes, *Radio Voices*, 78.

14 Tanneyhill knew tap dancer Bill Robinson personally. Canada Lee, another close friend who was then acting on Broadway in the play *Native Son,* agreed to moderate the show. For a review of Canada Lee's pioneering career as a stage and screen actor, see Gill, "Canada Lee," 79–89.

15 Interview with Ann Tanneyhill, August 11, 1978, in Hill, *Black Women Oral History Project*, 234.

16 The show was broadcast on Sunday, March 30, 1941, at 5:00 P.M. Author's interview with Ann Tanneyhill, November 28, 1994, Mashpee, Massachusetts.

17 Script, "The Negro and National Defense," March 30, 1941, RSC. For a brief account of the broadcast, see "Negro Stars of Stage, Screen, and Radio on Urban League Program," *Opportunity* 19 (Spring 1941): 119.

18 Script, "The Negro and National Defense," March 30, 1941, RSC.

19 Reddick, "Educational Programs," 384.

20 "The Radio as a Propaganda Medium," *New York Age,* April 5, 1941, and Ann Tanneyhill, "9th Vocational Opportunity Campaign Report, 1941," vol. A, box 30, series 7, NUL Papers. The editorial writer also argued that black entertainment talent made radio "an excellent means of putting our cause before the public when an appeal can be mixed with such a program as that staged by the Urban League," indirectly confirming Reddick's comments on this strategy.

21 *Providence Chronicle,* April 5, 1941, and Ann Tanneyhill, "9th Vocational Opportunity Campaign Report, 1941," ibid.

22 *Michigan Chronicle,* April 5, 1941, and Ann Tanneyhill, "9th Vocational Opportunity Campaign Report, 1941," ibid.

23 *Philadelphia Tribune,* April 3, 1941, and Ann Tanneyhill, "9th Vocational Opportunity Campaign Report, 1941," ibid.

24 *Time,* April 7, 1941; *Philadelphia Tribune,* April 3, 1941; and Ann Tanneyhill, "9th Vocational Opportunity Campaign Report, 1941," ibid.

25 There were over 300 written comments. One of the unfavorable letters was from an Ohio man who wrote that "there must always be enough discrimination to prevent amalgamation" and said he supported segregation for that reason, although he did not "necessarily regard the Negro as inferior." Cyrus H. Eshleman, Lakewood, Ohio, April 9, 1941, and Ann Tanneyhill, "Report of

the Ninth Vocational Opportunity Campaign, March 16–23, 1941," vol. C, "Letters of Comment," box 30, series 7, NUL Papers.

26 Letters from social studies classes, Arts High School, Newark, New Jersey, April 2, 1941, and Abraham Lefkowitz, principal, Samuel Tilden High School, Brooklyn, March 31, 1941, and Ann Tanneyhill, "Report of the Ninth Vocational Opportunity Campaign, March 16–23, 1941," ibid.

27 Letters from Wallace Young, Torrington, Connecticut, March 31, 1941, and Joseph P. Cohen, April 2, 1941, and Ann Tanneyhill, "9th Vocational Opportunity Campaign Report, 1941," ibid.

28 Letter from Eileen Murphy, New York City, April 25, 1941, and Ann Tanneyhill, "9th Vocational Opportunity Campaign Report, 1941," ibid.

29 Roy Wilkins to Eugene Jones, March 31, 1941; Walter White to Lester Granger, March 31, 1941; Channing Tobias, YMCA, to Jones, March 31, 1941; Ambrose Caliver to Ann Tanneyhill, March 18, 1941; and Ann Tanneyhill, "9th Vocational Opportunity Campaign Report, 1941," ibid.

30 Regrettably, the unavailability of CBS records to researchers prevents a fuller understanding of the reaction of network officials there.

31 "The Right to Work: A Basic Liberty," March 20, 1941, WQXR, New York City, RSC.

32 "Negro Labor and National Defense," March 18, 1941, WEVD, New York City, RSC. Popular local radio commentator Tony Won also made a pitch for the league's goals. *Tony Won's Scrapbook*, March 18, 1941, WEAF, New York City, RSC.

33 Elmer Carter, "The Urban League," *Opportunity* 16 (December 1938): 355; "If War Comes," *Opportunity* 17 (April 1939): 4, 98.

34 "On Racial Prejudice at Home and Abroad," *Opportunity* 17 (January 1939): 2. See also "If War Comes," 98, which noted that "it seems strange that the great industrialists of America should be so indifferent and prejudiced against the employment of Negroes."

35 "Where Democracy Fails," *Opportunity* 19 (January 1941): 3; Lester Granger, "The Urban League Faces a New Year," *Opportunity* 19 (January 1941): 20. See also Lester Granger, "Report on Defense," *Opportunity* 19 (January 1941): 47.

36 "A Negro Pursuit Squadron," *Opportunity* 19 (April 1941): 98–99.

37 In declining, NBC officials referred to a speech in connection with National Negro Health Week, a dramatization on the Agriculture Department's *Farm and Home Hour* on farm opportunities for blacks, and a speech by Ed Lawson of the National Urban League on the regular series *On Your Job*. A. Philip Randolph to NBC, June 3, 1941; Warren Brown, executive secretary, National March on Washington movement, to William Burke Miller, NBC, May 27, 1941; and Miller to Brown, June 4, 1941, box 85, folder 60, "Corres., Negro March on Washington," NBC Collection.

38 A. Philip Randolph, "The Negro March on Washington," June 28, 1941, nationwide radio address, Mutual network, RSC. It does not appear that either CBS or NBC followed suit.

39 Lester Granger to Niles Trammell, president, NBC, February 17, 1942, and

William Burke Miller, NBC, to Mr. Hendricks, NBC, March 6, 1942, box 88, folder 22, "National Urban League," NBC Collection. Indeed, Dwight Herrick of NBC's Public Service Program Division passed along Granger's request to Ambrose Caliver and asked him to consider incorporating information about the National Urban League into one of the remaining episodes in his series because of the "similarity of purpose." Dwight Herrick to Ambrose Caliver, March 6, 1942, and Herrick to Granger, March 6, 1942, ibid.

40 Lester Granger to Dwight Herrick, March 16, 1942, and Thomas Rishworth to Granger, March 23, 1942, ibid.

41 Granger not only supported Randolph's efforts but also took partial credit for the early changes in federal racial policies that resulted from the threatened march. See, for example, Lester Granger, "The Negro Marches," *Opportunity* 19 (July 1941): 194, and "The President, the Negro, and Defense," *Opportunity* 19 (July 1941): 205–7. For an account of the March on Washington movement and the Fair Employment Practices Commission, see Anderson, *A. Philip Randolph*.

42 "The Fate of Democracy," *Opportunity* 20 (January 1942): 2; "The Red Cross, Alas," *Opportunity* 20 (January 1942): 3. See also "We Stand," *Opportunity* 20 (March 1942): 67.

43 "Vocational Guidance Week," *Opportunity* 20 (March 1942): 67.

44 Herrick wrote privately that Granger's response was simply evidence that there was a lack of cooperation between black groups. Dwight Herrick to William Burke Miller, March 19, 1942; Thomas Rishworth, director of public service programs, to Lester Granger, March 23, 1942; Granger to Rishworth, March 26, 1942; and Herrick to Rishworth and James Angell, NBC, April 2, 1942, box 88, folder 22, "National Urban League," NBC Collection.

45 Author's interview with Ann Tanneyhill, November 28, 1994, Mashpee, Massachusetts; interview with Ann Tanneyhill, August 11, 1978, in Hill, *Black Women Oral History Project*, 235. The veteran stage actor Paul Muni read the poem on the air. The show was given favorable notice in many newspapers, both black and white. So many requests for copies of the Hughes poem came in that it was reprinted separately and in *Opportunity* to facilitate its wide distribution to teachers, schools, and libraries. "Freedom's Plow," March 15, 1943, NBC, box 31, series 7, folder, "Vocational Opportunity Scrapbook Reports, Loose Material, 1943," NUL Papers. Articles about "Freedom's Plow" appeared, for example, in the *Norfolk Virginian Pilot,* March 14, 1943; *Springfield (Illinois) State Journal,* March 16, 1943; *Grand Rapids Herald,* March 15, 1943; *Bridgeport (Connecticut) Post,* March 14, 1943; and *Roanoke Times,* March 15, 1943, box 31, series 7, "Report of the 11th Vocational Opportunity Campaign, March 14–21, 1943," vol. V, "Newspaper Clippings," and vol. I, "Freedom's Plow," which includes letters requesting copies of the poem, NUL Papers.

46 Script, William Agar, acting president of Freedom House, "Opportunity to Serve," March 20, 1943, WOR and Mutual network, RSC.

47 Author's interview with Ann Tanneyhill, November 28, 1994, Mashpee, Massachusetts.

48 Honey, *Creating Rosie the Riveter,* 46–52, 82–83. These images of black women

as subservient or as mammy figures were consistent with the marginalized treatment of black women in radio in general. Hilmes, *Radio Voices,* 258.

49 Interview with Ann Tanneyhill, August 11, 1978, in Hill, *Black Women Oral History Project,* 235; "Script, 'Heroines in Bronze,'" box 14, series 12, folder, "Radio Scripts, Vocational Opportunity Campaign, 1941–44," NUL Papers; script, *The Spirit of '43,* March 20, 1943, RSC.

50 Tanneyhill attached to the script an extensive bibliography of the standard source materials from the SCR about each woman. For example, the list included narratives about Truth, Tubman, and Wheatley, as well as Wheatley's letters, poems, and other works. Tanneyhill also relied on general histories of African Americans, including those by Carter G. Woodson and Benjamin Brawley, as well as broader writings in black history by W. E. B. Du Bois, James Weldon Johnson, and Frederick Douglass. Ann Tanneyhill, "Source Material for Historical Portions of Script," "Script, 'Heroines in Bronze,'" box 14, series 12, folder, "Radio Scripts, Vocational Opportunity Campaign, 1941–44," NUL Papers; script, *The Spirit of '43,* March 20, 1943, RSC.

51 See Guy-Sheftall, *Words of Fire,* 1–22.

52 For an in-depth treatment of discrimination against black women nurses, including during World War II, see Hine, *Black Women in White.*

53 The songs included well-known spirituals like "Nobody Knows the Trouble I've Seen," "I Know the Lord Has Laid His Hands on Me," "No More" (which Paul Robeson had sung on *Freedom's People*), "Go Down Moses," "Walk Together Children," and "I Thank God I'm Free at Last." One notable exception was "My Man Has Gone," which the soprano Ann Brown, who was starring in *Porgy and Bess,* sang as a tribute to black men in the service.

54 For a discussion of the notion of "respectability," see Higginbotham, *Righteous Discontent.* On the images of black women, see White, *Arn't I a Woman?;* Turner, *Ceramic Uncles and Celluloid Mammies;* and Morton, *Disfigured Images.*

55 For various campaigns, Tanneyhill also encouraged league affiliates to help generate local publicity by asking the radio editors of local papers to list the programs and by sending out reminders of upcoming broadcasts during the week preceding the broadcasts. She enclosed a sample postcard for duplication; like a chain letter, the card instructed the recipient to send five copies to friends. Ann Tanneyhill, memoranda to affiliated organizations, December 14, 1942, January 7, 1943, and memorandum on national radio programs, 11th Vocational Opportunity Campaign, March 1, 1943, box 31, series 7, folder, "Form Letters, Memorandum, News Releases," NUL Papers.

56 Press releases, "Radio Programs to High-Light Job Campaign; Noted Entertainers to Aid Urban League Cause," February 20, 1943; "Urban League Campaign to Plan for Post-War Jobs," February 27, 1943; "Thousands Eagerly Await Urban League Coast-to-Coast Broadcast; Negro Women in War Zones to Participate; Fredi Washington, Canada Lee, Others Slated on National Hook-Up," March 13, 1943; "CBS Broadcast Dramatizes Negro Women in War; Urban League Reaches Nation-Wide Audience with Unusual Program," March 21, 1943; and "Biographical Information Regarding Principal Participants in 'Heroines in Bronze,'" n.d., ibid.

57 Franklin D. Roosevelt to Lester Granger, March 11, 1943; Thomas E. Dewey to Granger, March 13, 1943; "President Backs Negro Campaign," *New York World-Telegram,* March 17, 1943; and "FDR Endorses Job Campaign," *Pittsburgh Courier,* March 20, 1943, box 31, series 7, "Report of the 11th Vocational Opportunity Campaign, March 14–21, 1943," vol. V, "Newspaper Clippings," NUL Papers.

58 See, for example, "Opportunity Drive Opens; Campaign Designed to Increase Use of Negro Manpower," *New York Times,* March 16, 1943; "Radio to Highlight Job Campaign; Top Artists to Aid League's Cause," *Pittsburgh Courier,* February 27, 1943; "Radio Program to Highlight Job Campaign of Urban League," *People's Voice,* February 27, 1943; Ann Petrey, "The Lighter Side," *People's Voice,* March 20, 1943; "Unusual Air Program Planned by National Urban League via Columbia," *Pittsburgh Courier,* March 13, 1943; "Urban League Drive to Stress 'Womanpower,'" *St. Paul Pioneer Press,* March 13, 1943; and "Women's Part in War Efforts Highlight of Week's Broadcast; Workers Here and Abroad to Be Heard," *Baltimore Afro-American,* March 20, 1943, ibid.

59 "Brown American Womanpower Issue," *Opportunity* 21 (April 1943).

60 Honey, *Creating Rosie the Riveter,* 119.

61 Examples of the articles in the *Opportunity* 21 (April 1943) special issue include Mary Anderson, "Negro Women on the Production Front," 37–38; George DeMar, "Negro Women Are American Workers, Too," 41–43, 77; Estelle Massey Riddle, "The Negro Nurse and the War," 44–45, 92; "Women in the Defense Industry, with Some of Its Broader Implications," 46–49, 86; and "Negro Women and the WAAC," 54–55, 93.

62 Lester Granger, "Women Are Vital to Victory," *Opportunity* 21 (April 1943): 36.

63 "Loyalty of Colored America Told on Air Program; Tells of Woman Training White Pilots; Others in Various Defense Industries," *Baltimore Afro-American,* March 27, 1943; "The Listening Post" and "The Lighter Side," *People's Voice,* March 27, 1943; and "Notables Take Part in National Urban League Program," *Pittsburgh Courier,* March 27, 1943, box 31, series 7, "Report of the 11th Vocational Opportunity Campaign, March 14–21, 1943," vol. V, "Newspaper Clippings," NUL Papers.

64 Letters from Elmer Carter, n.d.; A. L. Foster, Chicago Urban League, March 20, 1943; Chas. W. Washington, executive secretary, Minneapolis Urban League, March 29, 1943; J. Harvey Kerns, executive secretary, Baltimore Urban League, March 30, 1943; William L. Evans, executive secretary, Buffalo Urban League, March 31, 1943; and Raymond Brown, executive secretary, Omaha Urban League, March 29, 1943, ibid., vol. III.

65 Letter from John T. Clark, executive secretary, St. Louis Urban League, March 26, 1943, ibid.

66 Letters from Ira De A. Reid, Atlanta University, March 28, 1943, and Mabel Staupers, National Association of Colored Graduate Nurses, March 26, 1943, ibid.

67 Guzman, *Negro Yearbook,* 448.

68 Letters from Herbert Barrett, New York City, March 24, 1943; George Towns, Atlanta, March 20, 1943; and Margaret Reilly, Jamaica, New York, March 20,

1943, box 31, series 7, "Report of the 11th Vocational Opportunity Campaign, March 14–21, 1943," vol. III, NUL Papers. The broadcast stimulated some listeners to write for information about National Urban League programs and memberships. See, for example, letter from Byron Hamlin, Dayton, Ohio, March 20, 1943, ibid.

69 For detailed treatment of the Detroit riots, see Buchanan, *Black Americans in World War II,* 46–51. See also Blum, *V Was for Victory,* 199–204, and Wynn, *The Afro-American and the Second World War,* 68–69.

70 Buchanan, *Black Americans in World War II,* 53–55; Blum, *V Was for Victory,* 206.

71 Brown, "The Negro Vote," 153. See also Blum, *V Was for Victory,* 204.

72 White, *A Man Called White,* 226, 231.

73 Ibid., 231. Although it is easy to compare the content of the programs NBC and CBS broadcast on this issue, the unavailability of CBS corporate records inhibits my ability to reach confident conclusions about internal attitudes at that network. More of NBC's records are available, making it possible to detect the fear and caution that drove NBC's decision making. But marked variations in particular programming decisions are an important reminder of the need to disaggregate the mass media. Although both networks worked within fairly restrictive boundaries on broadcasts about race, they were not absolutely consistent.

74 Robson had come to prominence as the director of the prestigious dramatic forum, the *Columbia Workshop,* as well as a number of other public affairs and documentary-style programs. Barnouw, *Radio Drama in Action,* 60; William Robson to Erik Barnouw, April 18, 1945, box 1, folder 24, "William N. Robson," Barnouw Papers.

75 Barnouw, *Radio Drama in Action,* 60; William Robson to Erik Barnouw, April 18, 1945, box 1, folder 24, "William N. Robson," Barnouw Papers.

76 Barnouw, *Radio Drama in Action,* 60; William Robson to Erik Barnouw, April 18, 1945, box 1, folder 24, "William N. Robson," Barnouw Papers.

77 Barnouw, *Radio Drama in Action,* 60; William Robson to Erik Barnouw, April 18, 1945, box 1, folder 24, "William N. Robson," Barnouw Papers.

78 Barnouw, *Radio Drama in Action,* 62. The script is reprinted in ibid., 60–77. It is also available at the RSC. A recording of the broadcast was provided by Professor J. Fred MacDonald, Department of History, Northeastern Illinois University, Chicago.

79 White, *A Man Called White,* 231. After Wilkie's defeat in 1940, he took on two important positions: special counsel to the NAACP and chair of the board of Twentieth Century Fox. Cripps, *Making Movies Black,* 35.

80 Barnouw, *Radio Drama in Action,* 76.

81 Ibid., 60.

82 Ibid.; Edmerson, "Descriptive Study of the American Negro in United States Professional Radio," 160–67.

83 Buchanan, *Black Americans in World War II,* 45–46. See also Blum, *V Was for Victory,* 205–6.

84 The two civic groups were the Committee of the Mayor of Los Angeles for Home Front Safety and the Citizen's Committee of the Council of Social Ser-

vices. A complete set of the scripts for the six broadcasts is contained in the RSC but without any contextual information. The Chet Huntley Papers at the MCRC and at the University of Montana do not appear to hold any additional information on his work on this series. As noted earlier, CBS's records are not available to the public. Although the federal government played no role in the broadcast, scripts from the series were made available for educational purposes through the federal government's free script loan service, a practice that continued as late as 1950. See, for example, Federal Security Agency, Office of Education, *Radio Script Catalog*.

85 Script, *These Are Americans,* January 29, 1944, RSC.

86 Ibid.

87 Ibid., February 19, 1944.

88 Ibid., January 29, February 12, 1944.

89 Ibid., March 4, 1944.

90 Ibid., February 26, 1944.

91 Interview with Chet Huntley, in Edmerson, "Descriptive Study of the American Negro in United States Professional Radio," 108.

92 Parris and Brooks, *Blacks in the City,* 299–303.

93 Author's interview with Ann Tanneyhill, November 28, 1994, Mashpee, Massachusetts; interview with Ann Tanneyhill, August 11, 1978, in Hill, *Black Women Oral History Project,* 234.

94 Script, "Victory through Unity," October 2, 1943, RSC.

95 Ibid.

96 Ibid.

97 Parris and Brooks, *Blacks in the City,* 304, 295.

98 William Baldwin, "The Common Good," *Opportunity* 22 (October 1943): 146.

99 Lester Granger, "Victory through Unity," *Opportunity* 22 (October 1943): 147.

100 In June 1943, Tanneyhill wrote Truman Gibson at the War Department with her idea for "a series of four to six dramatic broadcasts based on stories of heroism in World War II—stories of heroism of Negroes in the Army and Navy." She developed a prospectus for a show to be called "Gallant Black Eagle," which told the story of the all-black Ninety-ninth Fighter Squadron, but she was never able to place it with any network. It was also during this period that Tanneyhill enrolled in a summer radio scriptwriting course at New York University. Author's interview with Ann Tanneyhill, November 28, 1994, Mashpee, Massachusetts; Ann Tanneyhill to Truman K. Gibson, June 7, 1943; Gibson to Tanneyhill, June 28, 1943; Tanneyhill to Gibson, June 29, November 19, 1943; Gibson to Tanneyhill, November 25, 1943; and Tanneyhill to Gibson, 1943, entry 188, box 10, folders, "Radio, 1943" and "Radio Broadcast, Negro in the Army #2 (1942)," RCASW.

101 Ann Tanneyhill to Ted Poston, OWI, January 31, February 7, 14, 1944; Poston to Tanneyhill, February 16, 1944; Tanneyhill to Captain Daniel Day, Bureau of Public Relations, War Department, February 17, 1944; Day to Tanneyhill, March 4, 1944; and Truman Gibson to Tanneyhill, March 7, 1944, box 31, series 7, Vocational Opportunity Campaign Scrapbooks, folder, "'Salute to Freedom,' 1944," NUL Papers.

102 Letters from Edward Lewis, New York Urban League, March 28, 1944; William Ashby, Springfield Urban League, March 20, 1944; John T. Clark, St. Louis Urban League, March 23, 1944; "Salute Show Issues Plea for Justice," *New York Amsterdam News,* March 25, 1944; and "'Salute to Freedom' Nationwide Broadcast Climaxes Natl. Urban League Job Campaign," *New York Age,* March 25, 1944, box 31, series 7, "Report, Twelfth Annual Vocational Opportunity Campaign, March 12–19, 1944," NUL Papers; script, "Salute to Freedom," March 18, 1944, RSC.

103 Lester Granger to H. B. Summers, Blue network, December 1, 1944, box 32, series 7, folder, "'Too Long America,' 1945," NUL Papers. Although still called by its former NBC name, the Blue network would become ABC as a result of the federal antitrust action that forced NBC to sell one of its two networks.

104 Ann Tanneyhill to Lester Granger, February 6, 1945, ibid. Tanneyhill was assigned to gather and send research materials to ABC's scriptwriter. Memorandum, Tanneyhill to Ira Marion, February 13, 1945, ibid. Marion had been a scriptwriter for NBC's *Blue Playhouse* dramatic series. The league's press release for the show also emphasized that Marion had been the "producer of 'The Tree,' the first anti-lynch play to appear on Broadway." Press release, March 14, 1943, ibid.; "Too Long America," March 15, 1945, ABC, box 14, series 12, folder, "Radio Scripts, 1945," NUL Papers.

105 Lester Granger to Ira Marion, March 27, 1945, box 32, series 7, folder, "'Too Long America,' 1945," NUL Papers. It appears that the league specifically solicited reaction to the show from officials at local affiliates who had been urged to persuade members to express their appreciation for the show to local stations and the network. Letter from William V. Kelley, March 19, 1945, with attached "Postal Card Replies Commenting on Last Wednesday's Radio Program"; Los Angeles Urban League, "Comments, National Urban League's Broadcast," with brief reactions from over thirty listeners; Ann Tanneyhill to H. B. Summers, Blue network, March 26, 1945; and Tanneyhill to Ira Marion, March 26, 1945, ibid.

106 Lester Granger to H. B. Summers, Blue network, March 27, 1945, box 32, series 7, folder, "'Too Long America,' 1945," NUL Papers. Tanneyhill also was trying to arrange a variety show for national radio the following month as part of the league's new national fund-raising campaign. It does not appear that the league succeeded in getting additional airtime for either the series or the fund-raising campaign. Ann Tanneyhill to Floyd Covington, Los Angeles Urban League, March 26, 1945, ibid. The United Urban League Service Fund was established in 1945 under the leadership of pollster Elmo Roper with the expectation that a more professionalized fund-raising effort would lift the league's sagging budget. Overall, the effort met with little success, and the league's financial deficits continued unabated. Parris and Brooks, *Blacks in the City,* 346–52.

107 Script, Erik Barnouw, "The Story They'll Never Print," 1946, box 1, series 7, folder, "Barnouw, Erik: Film Proposals, 1947–57," NUL Papers.

108 The rejected script was eventually published in a collection of radio plays,

and as a result, many local and educational stations and university and ama-
teur theater groups produced it, including the American Negro Theatre.
Erik Barnouw oral history, 1975, 79–80, Oral History Collection, Columbia
University, New York, New York. Permission requests for the script were
received from the Theatre Department at Southern Methodist University,
April 24, 1948; B'nai B'rith Hillel, Ohio State University, November 25, 1948;
American Friends Service Committee, Philadelphia, June 4, 1948; Bloomington Public High Schools, Wisconsin, January 30, 1947; and Radio De-
partment, Northwestern University, April 2, 1948, box 1, series 7, folder,
"Barnouw, Erik: Film Proposals, 1947–57," NUL Papers. Barnouw proposed
to Parris that the script for the show be turned into a screenplay for a motion
picture, but that does not seem to have happened. Erik Barnouw to Guichard
Parris, National Urban League, December 6, 1949, ibid.

109 Culbert, "Education Unit in World War II," 279–80.

110 Scripts for *Minority Opinion* are available in the RSC. The program was first
broadcast on October 22, 1945, on WJW in Cleveland and was heard at
9:30 P.M. on Mondays at least through March 1946, the date of the last script
at the SCR. In a January 1946 script, the show was characterized as being
aired weekly, but a complete set of scripts is not available at the SCR.

111 Script, *Minority Opinion,* November 5, 19, 1945, RSC.

112 For example, he interviewed Charles Loeb, city editor of the black paper
the *Cleveland Call and Post* and a war correspondent for the Combined Negro
Press, and Lester Granger, who had recently returned from a tour of Pacific
navy bases. Both men argued that the desegregation of the military and the
use of mixed units was, in Loeb's words, the "necessary cog in the wheel of
national unity and inter-racial understanding." Ibid., December 10, 1945.

113 Ibid., December 17, 1945.

114 Ibid., January 14, 1946.

115 Ibid., March 4, 1946.

116 Ibid., April 8, 1946.

117 Author's interview with Ann Tanneyhill, November 28, 1994, Mashpee, Massachusetts.

118 "The Future Is Yours," *Opportunity* 23 (Fall 1945): 182–83.

119 Author's interview with Ann Tanneyhill, November 28, 1994, Mashpee, Massachusetts.

CHAPTER FIVE

1 Initially both *Round Table* and *Town Meeting* were carried by NBC, but when
ABC formed in the mid-1940s from one of the two networks that had com-
prised NBC, *Town Meeting* was broadcast by ABC from 1945 onward. The other
shows were *People's Platform* (CBS) and *American Forum of the Air* (Mutual).

2 When NBC agreed to carry the weekly show nationally, officials there were
skeptical about its ability to attract and sustain a national audience. Some at
the network suggested that the university was using the show simply to expand its fund-raising efforts, although it was also considered a program of
very high quality and the best of its kind. James R. Angell, NBC, to Lenox

Lohr, NBC, January 11, 1938; John F. Royal, NBC, to David Sarnoff, RCA, February 15, 1938; Angell to Martha McGrew, NBC, November 3, 1938; and Lohr to Robert Hutchins, president, University of Chicago, November 8, 1938, box 65, folder 21, "*U. of Chicago Roundtable, 1938*," NBC Collection.

3 *Current Biography*, 40–42.

4 Harry Schubert to William Benton, October 28, 1940, box 18, folder 13, "*Roundtable*," NBC Collection. Benton estimated that the show distributed 300,000 copies of the printed transcript per year. William Benton, "Radio Broadcasting: A Challenge to the Educator," March 27, 1941, box 86, folder 9, "*Chicago University Round Table*," NBC Collection; McCoy and Ruetten, *Quest and Response*, 93.

5 Sherman Dryer to William Benton, June 28, 1939, box 73, folder 32, "*U. of Chicago Roundtable, 1939*," NBC Collection; "The Oppressed Negro Will Be Discussed over Radio Sunday," *St. Louis Argus*, May 19, 1939, Clipping Files, "Radio Programs," MSRC.

6 Sherman Dryer to William Benton, June 28, 1939, box 73, folder 32, "*U. of Chicago Roundtable, 1939*," NBC Collection.

7 Mayer, *Robert Maynard Hutchins*, 298.

8 William Benton to Niles Trammell, June 30, 1939, box 73, folder 32, "*U. of Chicago Roundtable, 1939*," NBC Collection.

9 John F. Royal to Niles Trammell, July 6, 1939, ibid.

10 Ibid.

11 Niles Trammell to John F. Royal, July 17, 1939, and Royal to Trammell, July 6, 1939, ibid.

12 Sherman Dryer to William Benton, May 12, 22, 26, 1942, box 25, folder, "Dryer, May 1, 1942, to June 30, 1942," Benton Papers.

13 Sherman Dryer to William Benton, May 29, 1942, ibid.

14 Sherman Dryer to Charles Siepman, Radio Division, Office of Facts and Figures, June 1, 1942, and Siepman to Dryer, June 8, 1942, entry 5, box 3, folder 002.11, "Special Groups, Negroes, June 1942," ROWI.

15 William Benton to Niles Trammell, June 3, 1942, box 26, folder 20, "Trammell, Niles," and Benton to Sherman Dryer, August 7, 1942, box 32, folder, "Dryer, July 1–September 30, 1942," Benton Papers.

16 William Benton to Sherman Dryer, August 7, 1942, box 32, folder, "Dryer, July 1–September 30, 1942," Benton Papers.

17 William Benton to Sherman Dryer, April 9, 1943, box 32, folder, "Dryer, January 1, 1943–June 30, 1943," Benton Papers.

18 McGill was a last-minute substitution after Jonathan Daniels withdrew following his appointment as special assistant to President Franklin Roosevelt. Although *Round Table* discussions were not scripted or rehearsed as such, they were not spontaneous. The participants met in advance, prepared a topical outline for the discussion, and exchanged the views they expected to express.

19 "Minorities," *University of Chicago Round Table*, March 28, 1943, 7.

20 Ibid., 8–10.

21 Ibid.

22 Ibid., 12–13, 16.

23 The NAACP's success before the largely Roosevelt-appointed Supreme Court would soon be made clear. In 1944, in *Smith v. Allwright,* the Court outlawed the all-white primary, and in 1946, in *Morgan v. Virginia,* it barred racial segregation in interstate travel. Both cases opened the way for political battles that would be settled in the decades to follow. Sullivan, *Days of Hope,* 147–49, 194. See also Hine, *Black Victory.*

24 Sherman Dryer to William Benton, April 2, 1943, box 32, folder, "Dryer, January 1, 1943–June 30, 1943," Benton Papers.

25 Anderson, *Imagined Communities.*

26 Ira De A. Reid, Atlanta University, to Ralph McGill, March 28, 1943, and Claude Barnett, Associated Negro Publishers, to McGill, March 28, 1943, box 32, folder, "Dryer, January 1, 1943–June 30, 1943," Benton Papers.

27 On such late notice, both reliable southern moderate journalists Mark Ethridge and Ralph McGill were unavailable. Sherman Dryer to William Benton, July 7, 1943, and Benton to Dryer and John Howe, July 23, 1943, box 36, folder, "Dryer, July 1, 1943–November 30, 1943," Benton Papers.

28 "Race Tensions," *University of Chicago Round Table,* July 4, 1943, 1–2.

29 Ibid., 3.

30 Ibid., 3–6.

31 Ibid., 13–21.

32 Kneebone, *Southern Liberal Journalists,* 198, 201–8.

33 William Benton to Sherman Dryer, August 23, July 13, 1943, and Dryer to Benton, July 15, 20, 1943, box 36, folder, "Dryer, July 1, 1943–November 30, 1943," and Benton to Niles Trammell, July 26, 1943, box 37, folder, "Trammell, Niles," Benton Papers.

34 William Benton to Sherman Dryer, July 13, 1943, box 36, folder, "Dryer, July 1, 1943–November 30, 1943," ibid.

35 The race of the writer is not indicated in the letter, but it seems likely from its content that he was African American. Dr. J. B. Walker, Canton, Ohio, to Franklin Frazier, August 2, 1943, box 131-50, folder, "Speaking Engagements, 1942–43," E. Franklin Frazier Papers, MSRC. Franklin Frazier's papers at the MSRC do not appear to contain any additional information about his appearance on the show.

36 Sherman Dryer to Franklin Frazier, July 10, 1943, ibid.

37 Dryer, the strongest advocate of airing the shows on race, left *Round Table* in 1944 to assume a position at another local radio station, but so did the more cautious Benton, who left in 1945 to direct public relations for the State Department. The show continued to receive outside appeals that it address racial issues. For example, in 1945, John Harms of the Church Federation of Chicago suggested another broadcast on race relations but was told that the topic already had been covered. John Howe, *Round Table,* to John Harms, January 24, 1945, box 41, folder, "*Round Table,* General," Benton Papers. A 1944 program entitled "Peace as a World Race Problem" emphasized color-based discriminations in other countries and cultures, but Louis Adamic made observations about domestic racial issues on the show. The fact that almost all of the printed letters responding to the show included references to "the

Negro" and to domestic racial concerns indicates that listeners focused on those concerns. "Peace as a World Race Problem," *University of Chicago Round Table,* August 20, 1944, 8, 11, 19, 20. Other programs in which the issue of race drew some mention included "The People Say," *University of Chicago Round Table,* August 27, 1944, 26–27. A 1945 program entitled "We Hold These Truths" and a 1946 program called "What Is Equality?" made brief references to the "Negro problem," but both addressed broader, more philosophical political questions. "We Hold These Truths" and "What Is Equality?," *University of Chicago Round Table,* July 1, 1945, and February 10, 1946, respectively.

38 "Are We a United People?," *America's Town Meeting of the Air,* February 20, 1941, 5–6, 7–8, 16–19.

39 McClellan Van Der Veer, "Battle in Birmingham," *Birmingham Age-Herald,* February 24, 1941, reprinted in *America's Town Meeting of the Air,* March 3, 1941, 33–37. The show received prominent local news coverage: "Town Talkers Admit U.S. Rift, Split on Course," *Birmingham Age-Herald,* February 21, 1941, Tutwiler Collection, Birmingham Public Library, Birmingham, Alabama. It was the general view that broadcasting the show from Birmingham was good for the city's image and good for the image of southerners in general. John Temple Graves, "This Morning," *Birmingham Age-Herald,* February 25, 1941, ibid. Another editorial tried to smooth over differences between the panelists, although it did acknowledge novelist Erskine Caldwell's comments on the show about the "extensions of opportunity" to "the Negro and other handicapped Americans." "Great Stuff," *Birmingham Age-Herald,* February 22, 1941, ibid.

40 *"Town Meeting of the Air," Hilltop News,* February 28, 1941, Archives and History Division, Birmingham Southern College, Birmingham, Alabama.

41 Theodore M. Berry to Archibald MacLeish, March 17, 1942, entry 5, box 3, folder 002.11, "Special Correspondence, Negroes, January–March 1942," ROWI.

42 For example, in its prebroadcast publicity, the *Baltimore Afro-American* attached great significance to the fact that the show was being aired from the Howard University campus, despite the fact that the show's advertised title did not mention African Americans or race relations. *Baltimore Afro-American,* May 23, 1942.

43 "Is There a Basis for Spiritual Unity in the World Today?," *America's Town Meeting of the Air,* May 28, 1942, 8, 9, 11.

44 Ibid., 16.

45 Ibid.: "Should the Poll Tax Be Abolished?," October 20, 1942, with Senator Claude Pepper of Florida and Congressman Ed Cox of Alabama; "Should the Anti-Poll Tax Bill Be Passed?," May 18, 1944; "Should All Japanese Continue to Be Excluded from the West Coast for the Duration?," July 15, 1943, with Carey McWilliams; "Should We Repeal the Chinese Exclusion Laws Now?," September 2, 1943; and "Can the Japanese Be Assimilated?," August 3, 1944, with Carey McWilliams, broadcast from Sacramento.

46 "Let's Face the Race Question," *America's Town Meeting of the Air,* February 17, 1944, 4. In the period when the show began to consider race relations, it

was in transition between NBC's control and ABC's management, making attempts to examine relations between the show and its sponsoring network even more difficult.

47 Ibid., 5.

48 Ibid., 6–7. James Shepard, who was the least effective of the speakers, seconded Graves's argument that the perpetuation of democracy rested on personal religious beliefs and also commented that he trusted the generous and Christian attitudes of southerners like those he knew in North Carolina to make what he called "the needed social, political, and economic readjustments." Ibid., 10.

49 McWilliams suggested enforcing the Constitution, outlawing the poll tax, enacting an antilynching statute, making the Fair Employment Practices Commission permanent, amending labor laws to bar union discrimination, and guaranteeing open access to public health facilities, hospitals, libraries, schools and colleges, and housing. Ibid., 8–9.

50 Ibid., 14–15, 17–19.

51 Ruth Barash to George Denny, January 6, 1949, folder, "*America's Town Meeting* Reports, Mail and Ratings for ABC," and R. Huggins, "Preliminary Report on 2/17 Town Meeting," box 62, folder, "Preliminary Mail Reports (Jan. 44–Apr. 45)," *ATMA* Collection. Hughes's appearance did generate some negative response but very little compared to the outpouring of supportive letters. Among the negative responses were those of several listeners who urged Hughes to emigrate to Russia or Africa, and one writer referred to the broadcast as "that smelly program." Quoted in Rampersad, *Life of Langston Hughes*, 84.

52 Letter from Falba Ruth Conic, Jackson, Mississippi, February 2, 1944, box 183, folder, "*America's Town Meeting* (1944)," General Correspondence, Hughes Papers.

53 Letter from Savannah Ruth Ivory, Spelman College, February 17, 1944, ibid.

54 Letter from Lucile Buford, Kansas City, Missouri, February 24, 1944, ibid.

55 Letter from Bettye Steinberg, Louisiana, February 18, 1944, ibid. Another writer critiqued the panelists: "Your answers were 100 percent perfect. Carey McWilliams is good. Dr. Shepard was an ass as usual and of course Temple Graves was impossible." Letter from L. F. Coles, New York City, February 18, 1944.

56 Letter from Elinor Sundemeyer, Gettysburg, Pennsylvania, February 23, 1944, ibid.

57 Langston Hughes, "Letter to the South," *Chicago Defender*, July 10, 1943, in De Santis, *Langston Hughes and the "Chicago Defender*," 75–76.

58 Langston Hughes, Writers' War Board, to Katherine Seymour, July 29, 1943, box 169, folder, "Writers' War Board," General Correspondence, Hughes Papers. Hughes was writing in response to a request that he write scripts for a series of New York City broadcasts on racial and religious tolerance. He agreed to write the scripts but not without first venting his anger about radio's general treatment of racial issues. Seymour to Hughes, July 15, August 6, 19, September 3, 1943, and Hughes to Seymour, August 16, 20, 1943, ibid.

59 Rampersad, *Life of Langston Hughes,* 84–86.

60 Langston Hughes to Erik Barnouw, March 27, 1945, box 1, folder 10, "Langston Hughes," Barnouw Papers. Barnouw had written Hughes for permission to include Hughes's radio play "Booker T. Washington in Atlanta" in a collection of socially relevant radio dramas he was then editing, *Radio Drama in Action.* Barnouw's introduction to Hughes's play reiterates the criticisms in Hughes's response but not his angry tone or his reference to Hitler. Barnouw, *Radio Drama in Action,* 284–85.

61 "Should Government Guarantee Job Equality for All Races?," *America's Town Meeting of the Air,* December 7, 1944.

62 Ibid., 7, 9–10, 14.

63 Ibid., 18. Smith's 1944 novel about an interracial relationship, *Strange Fruit,* had been banned in Boston. Congressman Charles La Follette of Indiana also spoke on the affirmative side of the issue.

64 Tucker replied that even if the government could guarantee that equality, it would not benefit "the Negro" very much: "It's simply this—that we're an advanced race because we've had 2000 years of culture and education. The Negro has had only about 200." The transcription noted that his remarks were met with "shouts" from the audience. Ibid. African Americans listening at home also were angered by Tucker's attack on black culture. One writer offered evidence of early African achievements for Tucker's benefit. Kay Koch, "Confidential Audience Mail Report," December 15, 1944, box 15, folder, "Audience Mail Reports, 1944," *ATMA* Collection.

65 Members of the listening audience deeply disliked Lillian Smith, whom they characterized, perhaps with some sexist overtones, as "emotional and utopian rather than practical." Kay Koch, "Confidential Audience Mail Report," December 15, 1944, box 15, folder, "Audience Mail Reports, 1944," *ATMA* Collection.

66 Letters from Ira S. Dresbach, Tiffin, Ohio, May 13, 1945, and R. M. Halliburton, no address, May 18, 1945, box 26, folders, "May 24: Are We Solving America's Race Problem?," *ATMA* Collection.

67 Letters from W. A. Poyck, Wilkes-Barre, Pennsylvania, n.d.; V. S. Elliott, Birmingham, Alabama, May 11, 1945; Rebecca Carter, Cleveland, Ohio, May 20, 1945; and Nancy Hale, Deshaw, Massachusetts, May 12, 1945, ibid.

68 Letter from Ira S. Dresbach, Tiffin, Ohio, May 13, 1945, ibid.

69 Letter from W. H. Herrwood, Seattle, Washington, May 20, 1945, ibid.

70 "Are We Solving America's Race Problem?," *America's Town Meeting of the Air,* May 24, 1945, 4.

71 This was not Wright's first appearance on *Town Meeting.* In April 1939, Wright had been among the speakers on a show entitled "Can We Depend upon Youth to Follow the American Way?" Other speakers included the editors of *Vassar Magazine* and the *Yale Daily News.* Wright used his time to argue that black unemployment was unbearably high, that educational and professional opportunities for blacks were nonexistent, and that if the "American way" was to be preserved for youth, the test of its survival would be "embodied in the fate of the Negro in America." The moderator and the other panelists

completely ignored Wright and his comments, as did all of the questioners in the audience. Series 1, Writings, box 3, folder 26, "Can We Depend upon Youth to Follow the American Way?," Wright Papers. It is unclear why Wright was invited to join this panel, although he had been awarded a Guggenheim Fellowship a month before the broadcast. Rampersad, *Richard Wright—Later Works*, 852.

72 "Are We Solving America's Race Problem?," *America's Town Meeting of the Air*, May 24, 1945, 4–5.

73 Ibid., 6. By the time of his appearance in 1945, Wright had published *Native Son* (1940), *Twelve Million Black Voices* (1941), and, to enthusiastic reviews, *Black Boy* (1945). Rampersad, *Richard Wright—Later Works*, 857.

74 "Are We Solving America's Race Problem?," *America's Town Meeting of the Air*, May 24, 1945, 6–7.

75 Denny tried to shield Wright from a question about black anti-Semitism, but Wright insisted on answering it. He acknowledged that he had written or made recent remarks about Jewish-owned businesses in black communities but explained that his intentions were not anti-Semitic. Ibid., 18, 20. Several listeners protested that Wright's response was unsatisfactory and should have been probed further. Letters from Edith Dillion Stelling, New York City, May 24, 1945, and Shoshana Harr, May 24, 1945, box 26, folder, "Are We Solving America's Race Problem?," *ATMA* Collection. Some listeners wrote to Wright directly to express their disappointment with his explanation that black anti-Semitism grew out of the experience of African Americans with Jewish landlords and merchants. Letters to Wright from Beatrice Goldman, Brooklyn, New York, May 24, 1945; Ethel Goodman, New York City, n.d.; Stella Gosner, Chicago, May 27, 1945; Gertrude Howard, New York City, May 25, 1945; Jacob J. Leibson, Halcott Center, New York, May 25, 1945; and Joseph Kaufman, Brooklyn, New York, May 24, 1945, series 1, Writings, box 2, folder 8, "Are We Solving America's Race Problem?," Wright Papers.

76 Kay Koch to George Denny, May 29, 1945, box 62, folder, "Preliminary Mail Report (1945–46)," *ATMA* Collection.

77 Letters from Mrs. Henry Jay, Memphis, Tennessee, May 25, 1945, and Louise Smith, Kansas City, Missouri, May 24, 1945, box 26, folders, "May 24: Are We Solving America's Race Problem?," *ATMA* Collection.

78 Letters from Mrs. C. E. Fisher, Miami, Florida, May 24, 1945, and Louise Smith, Kansas City, Missouri, May 24, 1945, ibid.

79 Letter from anonymous, Detroit, Michigan, May 27, 1945, ibid.

80 Letters from L. C. Christian, Houston, Texas, May 25, 1945, and anonymous, Detroit, Michigan, May 27, 1945, ibid.

81 Letters from A. B. McAllister, Hinckley, Illinois, May 25, 1945; Louise Smith, Kansas City, Missouri, May 24, 1945; anonymous, Detroit, Michigan, May 27, 1945; H. C. Keller, Louisiana, May 29, 1945; and anonymous, May 25, 1945, ibid.

82 Fredrickson, *Black Image in the White Mind*. For a discussion of the response of nineteenth-century African Americans to colonization proposals, see Bay, *White Image in the Black Mind*.

83 Letter from "A. Nordic," Chicago, October 22, 1945, box 26, folders, "May 24: Are We Solving America's Race Problem?," *ATMA* Collection.

84 Ibid.; letter from Daisy Dean, Chattanooga, Tennessee, May 25, 1945, ibid.

85 Letter from Forde Harrison, McAlester, Oklahoma, May 24, 1945, ibid.

86 Letters from Chicago, name and date illegible, and anonymous, Cincinnati, Ohio, May 25, 1945, ibid. A listener in Washington, D.C., complained that "negroes are pushing themselves in all white sections." Letter from Rosa Cooley, Washington, D.C., May 17, 1945, ibid.

87 Kelley, "'We Are Not What We Seem,'" 76, 77, 102–10, and *Race Rebels,* 55–57, 67–68, 70–72.

88 Letter from H. E. Weinberger, Peoria, Illinois, May 25, 1945, box 26, folders, "May 24: Are We Solving America's Race Problem?," *ATMA* Collection. Others commenting about the absence of a white southerner included anonymous, Detroit, Michigan, May 27, 1945 ("Been wondering where all the 'white brains' were"); W. R. Thompson, Greenville, South Carolina, June 9, 1945 (next time get "well qualified Southerner"); Bob Noble, Alexandra, Louisiana, n.d. (not having a southerner was an "injustice"); William Estopinal, Gulf Port, Mississippi, May 18, 1945 (complained even before the broadcast that it included no "one from the deep south"), ibid.

89 Dorothy M. Kelly, Azuza, California, to Richard Wright, May 25, 1945, series 1, Writings, box 2, folder 8, "Are We Solving America's Race Problem?," Wright Papers.

90 Letter from Ira McBride, Cherryvalle, Kansas, n.d., box 26, folders, "May 24: Are We Solving America's Race Problem?," *ATMA* Collection.

91 Letter from L. C. Christian, Houston, Texas, May 25, 1945, ibid.

92 Letter from Ophelia Dudley Steed, Cleveland, Ohio, May 25, 1945, ibid.

93 Mrs. Thomas White, Mt. Vernon, New York, to Richard Wright, May 25, 1945, series 1, Writings, box 2, folder 8, "Are We Solving America's Race Problem?," Wright Papers.

94 Marsden A. Thompson, second lieutenant, Goodman Field, Kentucky, 477th Bombardment Group, to Richard Wright, May 24, 1945, ibid. For a general discussion of the 477th, see Lee, *Employment of Negro Troops,* 464, 466.

95 Letters from Richmond, name missing, May 27, 1945, and Grace Morrow, Raton, New Mexico, May 28, 1945, box 26, folders, "May 24: Are We Solving America's Race Problem?," *ATMA* Collection.

96 Letter from Robert B. Jones, Jamestown, New York, May 31, 1945, ibid.

97 Beatrice Goodman, Brooklyn, to Richard Wright, May 24, 1945, and Nicoline Mass, Berkeley, to Wright, May 24, 1945, series 1, Writings, box 2, folder 8, "Are We Solving America's Race Problem?," Wright Papers.

98 For a discussion of the shifting power of the black vote in this period, see Moon, *Balance of Power,* and Lawson, *Black Ballots.*

99 White, *A Man Called White,* 331–33; Bernstein, *Politics and Policies,* 277–79. See also Kluger, *Simple Justice,* 249–53.

100 White, *A Man Called White,* 347–48; McCoy and Ruetten, *Quest and Response,* 67–68; Bernstein, *Politics and Policies,* 279.

101 Walter White to independent radio stations, June 1947, with attached list of

stations; NAACP press release, "Pres. Truman to Speak at NAACP 38th Conference," May 30, 1947; and NAACP press release, "Largest Mass Meeting in Nation's History Planned by NAACP," June 6, 1947, reel 12, NAACP Papers, part 1; White, *A Man Called White*, 348.

102 Bernstein, *Politics and Policies*, 279.

103 McCullough, *Truman*, 569.

104 Harry S. Truman, "Address before the National Association for the Advancement of Colored People," June 29, 1947, in *Public Papers of the Presidents: Harry S. Truman*, 311–13.

105 *Pittsburgh Courier*, July 5, 1947; "Truman to the NAACP," *Crisis* 56 (August 1947): 233.

106 White, *A Man Called White*, 349.

107 Ibid. An NAACP member visiting Vienna wrote to White that the president's speech had received wide publicity there and had generated much discussion.

108 In each instance, the report cited current examples of the denial of those rights primarily to African Americans but also to Mexican Americans, Japanese Americans, and Native Americans. *To Secure These Rights*, 3–53, 79. The rationale for this new federal role was that states had failed to protect their citizens from the "most serious wrongs against individual rights" committed by both private actors and local public officials, that "the idealism and prestige of our whole people" were needed "to check the wayward tendencies of a part of them," that the country's bad civil rights record had international implications, and, finally, that the federal government itself should set an example as the country's largest employer and as the sole authority over the armed forces. Ibid., 99–103.

109 Among other things, the committee recommended an end to all discrimination in education, housing, employment, health care, public services, interstate transportation, and public accommodations. Ibid., 139–73.

110 Charles Hamilton Houston, in *Baltimore Afro-American*, August 23, 1947.

111 McCoy and Ruetten, *Quest and Response*, 92; Bernstein, *Politics and Policies*, 283. The committee coordinated a massive public education campaign "to inform the people of the civil rights to which they are entitled and which they owe to one another." Copies of its text were cheap and available either in pamphlet form or in serialized form in newspapers. The committee distributed 25,000 copies of the report to the press and a wide range of local and national civic, educational, business, labor, consumer, veterans, and women's organizations, to name only a few. Black newspapers like the *Pittsburgh Courier* and the *Baltimore Afro-American* serialized the report. *PM* included 160,000 copies in its Sunday edition and sold 230,000 reprints for 10 cents each. Simon and Schuster, with underwriting from the Congress of Industrial Organizations (CIO), printed a $1 hardcover edition of the report that became a best-seller within two weeks. Nearly 40,000 copies of the government edition of the report were distributed in three months. Private groups like the American Jewish Congress prepared and distributed versions of the report. Traditional liberal groups like the American Council on Race Relations, the American Civil Liberties Union, the CIO, the NAACP, the American Friends Service

Committee, and many others also helped publicize the report. McCoy and Ruetten, *Quest and Response*, 92–93.

112 First proposed by southern race moderates as a forum to refute Myrdal's claims of black dissatisfaction, the book was transformed by the African Americans who responded into an affirmation of their agreement on the need for an aggressive attack on segregation and discrimination. White southerners who considered themselves race moderates continued to defend segregation, ending the promise of a southern white and black alliance on the race issue. Logan, *What the Negro Wants*; Janken, *Rayford Logan*, 145–65.

113 Harry S. Truman, "Special Message to the Congress on Civil Rights," February 2, 1948, in *Public Papers of the Presidents: Harry S. Truman*, 121–26.

114 Earlier that year, Truman had issued an executive order barring discrimination in federal employment. Bernstein, *Politics and Policies*, 286, 290, 297; McCoy and Ruetten, *Quest and Response*, 106–8.

115 McCullough, *Truman*, 207–8.

116 White race moderates had been forced by southern blacks to face the segregation issue during the controversy over the publication of *What the Negro Wants*. Kneebone, *Southern Liberal Journalists*, 202.

117 Sullivan, *Days of Hope*, 237–38.

118 Ibid., 259.

119 Ibid., 250, 259–66.

120 Ibid., 266–68.

121 "Civil Rights and Loyalty," *University of Chicago Round Table*, November 23, 1947, 4–8.

122 Regrettably, records of the behind-the-scenes deliberations at *Round Table* and NBC are not available for this period. The NBC Collection at the University of Wisconsin is bereft of information and the NBC Manuscript Collection at the LC also does not appear to hold additional material on this show. William Benton's papers at the University of Chicago, which were the source of my earlier discussions about tensions around the race issue, end in 1946, the year he left the university. Sherman Dryer's departure from the show and the university preceded Benton's. As best as I can determine, the University of Chicago Library Special Collections do not hold any additional materials on the program, although some at the university suspect that additional material about the show exists somewhere on campus.

123 "Civil Rights and Loyalty," *University of Chicago Round Table*, November 23, 1947, 9. The civil rights report generated radio coverage in forums other than panel discussion shows. For example, in 1948, the Mutual network broadcast the national four-part series *To Secure These Rights*, a dramatization of the findings of President Truman's Committee on Civil Rights. Southern affiliates and politicians were outraged by the program despite the fact that an earlier version had been revised with their objections in mind. Eventually, Mutual granted the Conference of Southern Governors and twenty southern senators three hours of airtime in which to present their views on the committee's findings. MacDonald, *Don't Touch That Dial!*, 359–60. Records of the Mutual network and its flagship station WOR are in an unprocessed collection at

the LC and are currently unavailable to researchers. The local New York City show *New World A'Coming,* to be discussed in chapter 6, also devoted three episodes to dramatizing aspects of the report.

124 Sullivan, *Days of Hope,* 231–33.

125 In her useful and insightful analysis of the political currents of this period, Penny Von Eschen argues that the black press and civil rights organizations shifted dramatically away from anti-imperialist, anticolonial claims as part of their capitulation to Cold War politics and in favor of a focus on domestic racial concerns. However, that transition seems less dramatic and more unresolved as I also find a continuation of the pattern of identification with and repeated references to the struggles of other people of color during this time. Von Eschen cites as an example a 1949 *Town Meeting* broadcast with Walter White, whom she characterizes as "defending U.S. policy." Von Eschen, *Race against Empire,* 112. But a closer examination of White's comments on the show reveals that his position was much murkier, at least during this period. He attacked dictatorships of the left and the right and defended democracies in general, but he also warned that "democracies—particularly my own United States—must cleanse themselves of imperialism and racial arrogance," adding that he was "deeply aware of the grave shortcomings of my own country so far as democracy for minorities is concerned." He also urged that the Atlantic Pact and the European Recovery Program not be "achieved at the expense of Asia and Africa" and called the persistence of "master-race" theories tragic. "How Can We Advance Democracy in Asia?," *America's Town Meeting of the Air,* September 6, 1949, 11–13, 18–19. This was a rebroadcast of a program that originated in Karachi, Pakistan, on August 10, 1949.

126 "The South and the Democratic Convention," *University of Chicago Round Table,* June 13, 1948, 3, 9, 12–14.

127 "Should We Adopt President Truman's Civil Rights Program?," *University of Chicago Round Table,* February 6, 1949, 2–7.

128 "Race Relations around the World," *University of Chicago Round Table,* December 5, 1948, 1, 4–8, 12.

129 For example, Charles Hamilton Houston wrote in his regular column in the *Baltimore Afro-American* that "the greatest propaganda weapon which Russia has with the non-white countries is America's racial discrimination and segregation, particularly in the armed forces"; "[o]ne thing is certain: the United States cannot fight a war with Russia on a segregated basis." *Baltimore Afro-American,* April 3, 17, 1948.

130 "Should We Adopt President Truman's Civil Rights Program?," *University of Chicago Round Table,* February 6, 1949, 12, 14.

131 "Race Relations around the World," *University of Chicago Round Table,* December 5, 1948, 1, 4–8, 12.

132 George Denny to Charles Taft, September 18, 1947, box 37, folder, "October 7, 1947," *ATMA* Collection.

133 "Should the President's Civil Rights Program Be Adopted?," *America's Town Meeting of the Air,* March 23, 1948, 4.

134 Ibid.

135 Ibid., 6, 8.

136 Ruth Barash, *"America's Town Meeting* Audience Mail Report, Should the President's Civil Rights Program Be Adopted?," April 22, 1948, box 56, folder, "Should the President's Civil Rights Program Be Adopted?, March 23, 1948," *ATMA* Collection.

137 "What Can We Do to Improve Race and Religious Relationships in America?," *America's Town Meeting of the Air,* October 7, 1947, 4–6.

138 Ibid., 6.

139 Betty Cabana to George Denny, October 13, 1947, box 62, folder, "Preliminary Mail Reports (1945–46)," *ATMA* Collection.

140 "What Should We Do about Race Segregation?," *America's Town Meeting of the Air,* November 8, 1948, 3–4.

141 It appears that the show at that time was carried only on WJZ-TV in New York City and WFIL-TV in Philadelphia, probably because of the limitations of ABC's new television operations.

142 Harry Ashmore, quoted in Egerton, *Speak Now against the Day,* 527.

143 "What Should We Do about Race Segregation?," *America's Town Meeting of the Air,* November 8, 1948, 8.

144 Ibid., 10.

145 In the question-and-answer period, Ashmore also rejected the new definition of civil rights that had emerged from the committee's report and from some of Truman's own speeches. Ashmore asked "how the right to a job ever got to be a civil right anyway." He thought civil rights should be limited to "the right to the ballot, the right to freedom of person, the right of freedom of speech." Ibid., 11–13.

146 Egerton, *Speak Now against the Day,* 528.

147 *Town Meeting* received 2,580 letters in response to the show. Box 57, folder, "What Should We Do about Race Segregation?, November 11, 1948," *ATMA* Collection.

148 "Should President Truman's Civil Rights Program Be Adopted?," *America's Town Meeting of the Air,* January 31, 1950, 5–8.

149 In White's first appearance in 1947, for example, the other panelists, Clare Boothe Luce and Max Lerner, were each paid $200, which seemed to be the standard minimum, whereas Charles Taft, the president of the Federal Council of Churches, demanded and was paid $1,000. Taft also requested that Denny pay him in a way that allowed him to avoid paying a fee to his booking agent. George Denny to Charles Taft, September 18, 1947; *Town Hall* to Mrs. Henry Luce, May 12, 1947; Walter White to Marian Carter, *Town Meeting,* October 17, 1947; and Carter to Max Lerner, October 11, 1947, box 37, folder, "October 7, 1947"; and Denny to Ray Sprigle, November 23, 1948, and memorandum, Mr. Traum, *Town Meeting,* to Mr. Edwards, November 10, 1948, box 39, folder, "November 9, 1948," *ATMA* Collection. In 1944, Langston Hughes was paid $75, and Richard Wright was paid $150 in 1945. Anna Pascocello, Town Hall, to Langston Hughes, February 23, 1944, box 150, folder, "Town Hall Inc.," General Correspondence, Hughes Papers; Carter, Town Hall, to Richard Wright, May 12, 1945, series 2, Correspondence, box 93,

folder 1175, Wright Papers. I was unable to locate information to compare the fees paid to Hughes and Wright with those paid to their copanelists.

150 "What Effect Do Our Race Relations Have on Our Foreign Policy?," *America's Town Meeting of the Air,* April 18, 1950, 3–6, 13.

151 Quoted in Ruth Barash, *"America's Town Meeting* Audience Mail Report," May 4, 1950, box 40, folder, "April 18, 1950: What Effect Do Our Race Relations Have on Our Foreign Policy?," *ATMA* Collection.

152 This discussion was influenced in part by ideas in Bourdieu, *Language and Symbolic Power,* 112–13, 115–18, 126–31, 138, 142, 153, 223–24.

153 Logan, *What the Negro Wants,* vii–viii.

CHAPTER SIX

1 A graduate of St. Bonaventure College, where he was a track star, Ottley also spent time at the University of Michigan and St. John's University Law School. "Roi Ottley," in "News of Scribner Books and Authors," n.d.; "Roi Ottley's Prizes Are the Least of His Joys," unnamed source, n.d.; and "Negro Writer Wins Award for Tour," *Black Dispatch,* May 2, 1943, folder, "Roi Ottley," Clipping Files, MSRC; "Roi Ottley Dies; Wrote on Negro," *New York Times,* October 2, 1960. St. Bonaventure has some of Ottley's manuscripts, but unfortunately no publicly available general collection of his papers appears to exist.

2 Ottley, *New World A-Coming.* The book has twenty-three short chapters spanning Ottley's wide interests in black culture, politics, and history. Many chapters highlight special political concerns or historical moments, but most focus on the paramount issues of the day, particularly discrimination in employment and housing.

3 Ibid., vi.

4 When forced to divest itself of one of its two networks, RCA sold one of the networks to Edward Noble, who then created ABC. At the same time, FCC rules forced Noble to sell his local New York City station, WMCA. *Billboard,* September 25, 1943, Scrapbooks, Straus Papers; Sterling and Kittross, *Stay Tuned,* 210–11.

5 Members had threatened privately that no funds would be appropriated for housing until Straus resigned. Despite his stated personal regard for Straus, Roosevelt did not fight for his retention and accepted his resignation as part of the price of placating southern Democrats in the House of Representatives. "Straus Quits Post as Head of USHA," *New York Times,* January 6, 1942; "Capitol Termites Ate Housing Plan," *PM,* January 30, 1942; "Straus to Resign in Housing Fight," *New York Daily News,* January 5, 1942; information release, U.S. Housing Authority, "Letter to the President from Nathan Straus Tendering His Resignation as Administrator of the United States Housing Authority," n.d.; and "Memorandum of Conversation with President Roosevelt, 11:15 A.M., Tuesday, January 13, 1942," Scrapbooks, Straus Papers. One account placed racial considerations at the core of congressional opposition to Straus's ideas for funding public housing. Some members of Congress may have harbored lingering resentment against Straus's special assistant Robert

Weaver, who had reportedly stated in a speech two years earlier that "the two races can live harmoniously together in the same project." Alvin E. White, "Four Freedoms (Jim Crow)," *The Nation*, February 21, 1942, and "Southern Congressmen Force Negro's Friends from Fed. Post," *Tampa Bulletin*, January 24, 1942, ibid.

6 "Noble Sells WMCA to Nathan Straus," *Daily News*, September 30, 1943; "Straus Buy of WMCA from Noble Points Up Entry of Public-Minded into Field," *Billboard*, September 25, 1943; and "Who's Who in Radio: Nathan Straus," *Radio Daily*, October 13, 1943, ibid.

7 "Local Stations Still Have Far to Go to Meet Community Needs—Straus," *Variety*, January 19, 1944, ibid.

8 Author's telephone interview with Peter Straus, November 1994, New York City. Peter Straus worked at WMCA with his parents during much of the period when *New World A'Coming* was on the air.

9 "Report of Panel Discussion, City-Wide Harlem Week, May 28–June 2, 1945"; conference minutes, July 19, 1943; and "Statement of Program and Activities of the City-Wide Citizens' Committee on Harlem," May 22, 1943, in *City-Wide Citizens' Committee on Harlem*, n.d., SCR.

10 "Report of Panel Discussion, City-Wide Harlem Week, May 28–June 2, 1945"; conference minutes, July 19, 1943; and "Main Points in Radio Interview of Dr. Lawrence Reddick, Curator, Schomburg Collection of Negro Literature, New York Public Library, Station WQXR, Monday, May 24, [1943]," in ibid.

11 Reddick, "Educational Programs," 367–89.

12 Moreover, it appears that when Straus bought the rights to Ottley's book, he donated them to the committee for use on the series. "WMCA Endows Group for Negro-Life Project," *Radio Daily*, February 29, 1944, and "Books, Authors," *New York Times*, February 29, 1944, Scrapbooks, Straus Papers.

13 Although the networks had similar programming, WMCA's was on the air for a much longer period of time and, because it was limited to New Yorkers, permitted far more local families to actually get on the air. This in turn generated a larger local audience as well as local press coverage. "Troops and Kin Reunited by Radio," *New York Times*, December 28, 1943, and "Local Stations Still Have Far to Go to Meet Community Needs—Straus," *Variety*, January 19, 1944, ibid.

14 Scripts and tape cassettes through 1946 are available for the show in an unprocessed collection located in the Audiovisual Division at the SCR. A complete collection of the scripts from 1944 through 1953 and a sample from 1954 through 1957 are available in the unprocessed WMCA Collection at the MCRC. That collection also includes a limited amount of material on the station's administration, as well as some correspondence and background information on this particular series.

15 Examples of Ottley's varied reports from abroad include "Ottley Reports on Negro-White Troop Relations," *PM*, September 21, 1944; "There's No Race Problem in the Foxholes," *PM*, January 1, 1945; "No Self-Rule in Sight for World's Colonial Peoples," *PM*, January 2, 1945; "Fascists Used Jim Crow in Italy as Weapon against Allied Unity," *PM*, January 4, 1945; "Effects of Nazi

Brutality Linger in Liberated Areas," *PM*, January 5, 1945; "Interracial Mixing Common in France during Nazi Rule," *Pittsburgh Courier*, March 24, 1945; and "Roi Ottley Interviews Pope," *Pittsburgh Courier*, September 21, 1944, folder, "Roi Ottley," Clipping Files, MSRC. The actual scriptwriting and production rested with two white radio veterans, Mort Sklar and Mitchell Grayson, respectively. Grayson shared a belief in the political power of radio's images in the arena of race relations. According to a black press report, Grayson was "determined that the vaudeville type of Negro would never appear on New World A'Coming"; he also contended that "if a program does nothing but present Negroes as people it will be a major contribution to radio." *Chicago Defender*, March 3, 1945.

16 For a description of Canada Lee's groundbreaking work as a dramatic actor on stage and in film, see Gill, "Canada Lee," 79–89.

17 Script and cassette recording, *New World A'Coming*, March 12, 1944, SCR; Ottley, *New World A-Coming*, 343.

18 Script, *New World A'Coming*, March 5, 1944, SCR.

19 Ibid.

20 Script and cassette recording, ibid., March 12, 1944.

21 Ibid.

22 Ibid., March 5, 23, April 2, 1944.

23 Ibid., June 11, 1944.

24 Ibid., March 5, 1944.

25 Ibid., April 23, 1944.

26 Ibid., March 5, 1944.

27 Ibid., April 2, 1944.

28 Ibid., April 9, 1944.

29 Ibid., May 28, 1944. For other examples, see ibid., April 16, 1944, December 18, 1945, and April 30, 1946.

30 Script and cassette recording, ibid., April 23, 1944.

31 Script, ibid., March 12, 1944.

32 Ibid., March 5, 1944.

33 Ibid., November 19, 1944; script and cassette recording, June 25, 1944.

34 Script and cassette recording, ibid., October 8, 1944.

35 "This Is Service," *New York Times*, March 19, 1944, Scrapbooks, Straus Papers.

36 "Station Owner Goes All Out in Behalf of Race," *Pittsburgh Courier*, April 1, 1944, ibid.

37 Script and cassette recording, *New World A'Coming*, October 8, April 30, 1944, SCR.

38 "Must Remove Shackles from Negro—Straus," *Norfolk Virginia Journal and Guide*, September 30, 1944, Scrapbooks, Straus Papers.

39 Script, *New World A'Coming*, December 24, 1944, SCR.

40 Ibid. One broadcast featured Wendell Wilkie's controversial book *One World*, which also advanced these and other themes. Ibid., "In Memory of Wendell Wilkie," October 8, 1945.

41 Ibid., June 18, 1944, reprinted in Barnouw, *Radio Drama in Action*, 354–68.

42 Scripts, *New World A'Coming*, April 22, 1945, June 11, 1944, SCR.

43 One inadvertently amusing and ironic script in the series told the true story of a white female schoolteacher who was rushed to Harlem Hospital in an emergency and lived to tell the tale. One broadcast featured an account of the Colored Orphan Asylum, which had been organized by white Quaker women. Ibid., April 22, 1945, June 11, May 14, December 31, 1944. Additionally, on October 22, 1945, the series aired a dramatization of Lillian Smith's play, *There Are Things to Do,* about ways to fight racial prejudice, and on May 7, 1944, it aired a show based on a Dorothy Parker short story.

44 Ibid., "Arrangement in Black and White," May 7, 1944.

45 Not surprisingly, WMCA also won a place on the SCR's 1944 Honor Roll of Race Relations for broadcasting "the most forthright radio dramatization of Negro life and race relations on the air today." Straus credited the show's quality to Roi Ottley, producer Mitchell Grayson, narrator Canada Lee, and Helen Straus, who served as WMCA's director of educational and children's programming. Other awards were granted to the series by the National Conference of Christians and Jews and liberal groups like the Writers' War Board and the Interracial Film and Radio Guild. "Shrine to Be Scene of Gala IFRG Mass Meeting; Edward G. Robinson, M.C.," *Los Angeles Sentinel,* May 17, 1945, box 15, folder, "Interracial Film and Radio Guild," and "The 1948 George Peabody Foster Radio Awards," box 15, folder, "Peabody 1948," WMCA Collection, MCRC; "Eighteen Cited for Aid to Racial Unity," *New York Times,* February 11, 1945; "Americans of Good Will—Nathan Straus: Uses Radio against Bias," *Pittsburgh Courier,* February 24, 1945; "Bernays Award Is Presented to WMCA, Nathan Straus," *Radio Daily,* June 1, 1945; "Straus Receives Award," *New York Daily News,* June 11, 1945; and "Text of Greetings by Raymond Swing, Norman Corwin, and H. V. Kaltenborn on Bernays Award," June 10, 1945, Scrapbooks, Straus Papers.

46 Nathan Straus, "Those 'Don'ts': Mr. Straus Challenges a Few of Radio's More Prevalent Conventions," *New York Times,* June 3, 1945, and "Radio Inhibited by Convention, Lacks Daring Even Though Young, Says Straus," *Broadcasting,* June 11, 1945, Scrapbooks, Straus Papers.

47 Helen Straus pioneered popular daily educational programming for children. "Public Service," *Tide,* November 1, 1945, ibid.

48 However, judging what was and was not "controversial" was within the station's authority, so some subjects that the networks might have considered too touchy might not have been considered controversial by WMCA. In that case, time could be sold without allowing airtime for any counterpoint. "WMCA, N.Y., Lifts Controversy Ban," *Variety,* February 23, 1944; "Straus, WMCA, Defends Policy on Time Sales in Answer to NAB," *Variety,* March 15, 1944; "WMCA Adopts Policy of Time Sale to Both Sides of Controversy Groups," *Broadcasting,* February 28, 1944; and "WMCA Defends Policy in Replying to NAB," *Radio Daily,* March 15, 1944, ibid. The station also initiated a topflight radio news service. In conjunction with the *New York Times,* it ran news on the hour, every hour, fifteen hours a day and supplemented this heavy news coverage with a variety of news analysts and commentary. "Public Service," *Tide,* November 1, 1945, ibid. With the support of Senator

Claude Pepper of Florida, Straus also tried unsuccessfully to secure the rights to broadcast live debates from the floor of the U.S. Congress. Members of Congress objected, boldly citing their right to privacy as grounds for refusing to allow radio microphones in the chambers. Straus instead initiated *Halls of Congress,* a series of dramatized reenactments of key debates, relying on the *Congressional Record* for text. "Straus Would Air Congress on WMCA, Writes Sen. Pepper," *Radio Daily,* September 13, 1944, and "Congressional Airtime Okay, Straus to Pepper," *Variety,* September 13, 1944, ibid.

49 "Nathan Straus Says Radio Fails to Do Good Job," *New York Herald-Tribune,* May 4, 1946; "A Broadcaster's Advice to Radio," *St. Louis Post-Dispatch,* May 13, 1946; and "Radio Forum," *Tide,* May 17, 1946, ibid.

50 "Gordon Heath Vaults Racial Bar to Become WMCA Announcer," *Amsterdam News,* February 17, 1945, and "Heard and Overheard: WMCA Scores Again," *PM,* February 19, 1945, ibid. Although Heath's hiring was seen as a pioneering move by the station, he never saw the radio position as anything other than temporary since he remained deeply committed to pursuing a career as an actor. He later recalled that he enjoyed his time at WMCA "because they were such nice, casual guys who treated me like a fellow announcer and took me for granted as such." He left the position after six months to star in the Broadway play *Deep Are the Roots,* the story of a returning black soldier who rejects an interracial marriage in favor of working for black rights in the South. Heath also performed in the play's London production in 1947 and starred in 1955 in the BBC television production of *Othello.* In the late 1940s, Heath moved to Paris, where he was co-owner and operator of the nightclub L'Abbaye and continued to work as a director and actor until his death in 1991. Heath, *Deep Are the Roots,* 94–96, 98–105, 6–7. Thanks to Robin Kelley for bringing Heath's memoir to my attention.

51 Ottley continued his career as a journalist and author, writing three other books and moving to Chicago, where he worked as a reporter for the *Chicago Tribune* from 1953 until his death in 1960. He also hosted a radio interview program on the local *Tribune*-owned station, WGN. "Roi Ottley Dies," *New York Times,* October 2, 1960, and "Interesting People: Roi Ottley," *Post,* September 9, 1971, folder, "Roi Ottley," Clipping Files, MSRC.

52 Script, *New World A'Coming,* October 8, 1945, SCR.

53 "The Negro Problem Gets a Fair Airing," *New York Star,* November 23, 1948, folder, "Roi Ottley," Clipping Files, MSRC.

54 MacDonald, *Don't Touch That Dial!,* 355.

55 Newman, "Capturing the Fifteen Million Dollar Market."

56 For information on *Destination Freedom*'s origins and on Durham, see Hugh Cordier, "A History and Analysis of *Destination Freedom,*" seminar paper, Northwestern University, 1949, *Destination Freedom* Collection, SCR. See also MacDonald, "Radio's Black Heritage," and *Richard Durham's "Destination Freedom,"* which includes an introductory essay about Durham and the show as well as fifteen of the available ninety-seven scripts. MacDonald also discusses the series in *Don't Touch That Dial!,* 361–62.

57 Interview with Richard Durham, quoted in Cordier, "History and Analysis

of *Destination Freedom,*" 24–25. Durham refers to the two films *Pinky* and *Imitation of Life,* which portrayed black women trying to pass for white.

58 Ibid., 25–26.

59 Ibid., 28.

60 MacDonald, "Radio's Black Heritage," 67; Cordier, "History and Analysis of *Destination Freedom,*" 14.

61 Waller's experience spanned the development of network radio itself, especially in the fields of education and public affairs. She had helped bring the local Chicago predecessor of *Amos 'n' Andy* to the network in 1929 and had been one of the NBC officials who considered and rejected the Office of Education series *Americans All, Immigrants All* in 1938. She also was the network's liaison to the prestigious radio panel discussion show, *University of Chicago Round Table.*

62 Cordier, "History and Analysis of *Destination Freedom,*" 14; MacDonald, *Richard Durham's "Destination Freedom,"* 3.

63 Scripts and cassette recordings, *Destination Freedom,* June 27, 1948–July 23, 1950, *Destination Freedom* Collection, Audiovisual Division, SCR; Oscar Brown Jr., quoted in MacDonald, "Radio's Black Heritage," 69.

64 Scripts, *Destination Freedom:* "The Knock-Kneed Man," June 27, 1948 (Crispus Attucks); "Railway to Freedom," July 4, 1948 (Harriet Tubman); "The Making of a Man" and "The Key to Freedom," July 25, August 1, 1948 (Frederick Douglass); "Truth Goes to Washington," August 15, 1948 (Sojourner Truth); "Citizen Toussaint," October 3, 1948 (Toussaint L'Ouverture); "Investigator for Democracy," November 28, 1948 (Walter White); "Searcher for History," February 6, 1949 (W. E. B. Du Bois); "Woman with a Mission," April 10, 1949 (Ida B. Wells); "The Long Road," August 7, 1949 (Mary Church Terrell); "Black Hamlet," August 14, 21, 1949 (Henri Christophe); and "Father to Son," October 9, 1949 (Adam Clayton Powell Sr. and Jr.), *Destination Freedom* Collection, SCR.

65 For example, ibid., "Black Boy," March 20, 1949 (Richard Wright); "Poet in Bronzeville," September 18, 1949 (Gwendolyn Brooks); "Poet of Pine Hill," September 5, 1948 (James Weldon Johnson); "Pagan Poet," April 3, 1949 (Countee Cullen); and "Before I Sleep," April 17, 1949 (Paul Laurence Dunbar).

66 For example, ibid., "Dance Anthropologist," April 23, 1950 (Katherine Dunham); "Do Something, Be Somebody," March 6, 1948 (Canada Lee); "The Father of the Blues," September 12, 1948 (W. C. Handy); "Echoes of Harlem," November 7, 1948 (Duke Ellington); "Choir Girl from Philadelphia," December 18, 1948 (Marian Anderson); "The Chopin Murder Case," January 16, 1949 (Hazel Scott); and "Negro Cinderella," June 12, 1949 (Lena Horne).

67 For example, ibid., "Recorder of History," February 12, 1950 (Carter G. Woodson), and "Atlanta Thesis," March 5, 1950 (E. Franklin Frazier).

68 For example, ibid., "Arctic Biography," August 22, 1948 (Matthew Henson); "The Boy Who Was Traded for a Horse," October 17, 1948 (George Washington Carver); and "Transfusion," March 27, 1949 (Charles Drew).

69 Ibid., "Tales of Stackalee," July 17, 1949, and "The John Henry Story," July 24,

1949. Legendary sports heroes were also included. For example, ibid., "Little David," October 10, 1948 (Joe Louis); "The Rime of the Ancient Dodger," November 21, 1948 (Jackie Robinson); "The Ballad of Satchel Paige," May 15, 1949; and "Premonition of the Panther," March 12, 1950 (Sugar Ray Robinson).

70 Script and cassette recording, ibid., "Recorder of History," February 12, 1950 (Carter G. Woodson). See also MacDonald, "Radio's Black Heritage," 69.

71 Script and cassette recording, "Black Boy," *Destination Freedom,* March 20, 1949 (Richard Wright), *Destination Freedom* Collection, SCR. See also script, ibid., "One of Seventeen," November 14, 1948 (Mary McLeod Bethune), on how the "whole city came to the station to see her [off to school]. She was the first freedman's girl to explore education outside the county." The show also details Bethune's successful campaign to establish a college for blacks. Ibid., "Investigator for Democracy," November 28, 1948 (Walter White).

72 Ibid., "One of Seventeen," November 14, 1948 (Mary McLeod Bethune). The program concluded with a live speech by Bethune.

73 Script and cassette recording, ibid., "Recorder of History," February 12, 1950 (Carter G. Woodson).

74 Fictitious high school principal, quoted in ibid.

75 Richard Durham to Homer Heck, the show's director, June 27, 1948, quoted in Cordier, "History and Analysis of *Destination Freedom,*" 27. Durham wrote the letter following the table reading of the script for the first show, "The Knock-Kneed Man," on Crispus Attucks. Durham was angered by the director's attempt to mold the black characters into more subservient roles.

76 Script and cassette recording, "The Long Road," *Destination Freedom,* August 7, 1949 (Mary Church Terrell).

77 Ibid.

78 Scripts, ibid., "Execution Awaited: Prejudice, Part 2," October 2, 1949; "The Heart of George Cotton," August 8, 1948; "Anatomy of an Ordinance," June 5, 1949 (Rev. Archibald Carey); and "The Trumpet Talks," July 31, 1949 (Louis Armstrong).

79 Ibid., "Segregation, Incorporated," August 28, 1949.

80 Ibid., "Anatomy of an Ordinance," June 5, 1949 (Rev. Archibald Carey).

81 Ibid., "Woman with a Mission," April 10, 1949 (Ida B. Wells).

82 Ibid., "Investigator for Democracy," November 28, 1948 (Walter White).

83 Ibid., "Story of 1875," August 29, 1948 (Senator Charles Caldwell). Remarks from Booker T. Washington placed this view in a more immediate context: "The white man who begins by cheating a Negro usually ends by cheating a white man. The white man who begins by discriminating against a Negro soon discriminates against white men, too." Ibid., "Up from Slavery," March 13, 1949 (Booker T. Washington). See also the remark from Ida B. Wells: "[F]reedom that allowed the bigoted or the powerful to restrict the freedom of others was no freedom at all." Ibid., "Woman with a Mission," April 10, 1949 (Ida B. Wells).

84 Ibid., "The Story of 1875," August 28, 1948 (Senator Charles Caldwell).

85 Ibid., "The Liberators I," March 26, 1950 (William Lloyd Garrison).

86 MacDonald, *Richard Durham's "Destination Freedom,"* 6.

87 Ibid., 3.

88 Rampersad, *Life of Langston Hughes,* 153–54.

89 Richard Durham to Langston Hughes, November 17, 1949, box 52, folder, Dr–Dz, General Correspondence, Hughes Papers.

90 MacDonald, "Radio's Black Heritage," 70.

91 Richard Durham to Langston Hughes, October 8, 1951, box 52, folder, Dr–Dz, General Correspondence, Hughes Papers.

92 Horowitz, *"Negro and White, Unite and Fight!,"* 213–14, 235, 236. Thanks to David Montgomery for bringing Horowitz's book to my attention.

93 MacDonald, *Richard Durham's "Destination Freedom,"* 9; Clarice Durham to author, July 13, 1998, in author's possession; Neimark, "The Man Who Knows Muhammad Ali Best."

CONCLUSION

1 Reddick, "Educational Programs," 367–68, 386.

2 Canada Lee, "Radio and the Negro People," 1949, Canada Lee Papers, SCR.

BIBLIOGRAPHY

ARCHIVAL SOURCES

Atlanta, Georgia
Southern Education Foundation Archives, Robert W. Woodruff Library, Atlanta
    University Center

Birmingham, Alabama
Tutwiler Collection, Birmingham Public Library
Archives and History Division, Birmingham Southern College

Chicago, Illinois
Special Collections, University of Chicago Library
    William Benton Papers

Hyde Park, New York
Franklin D. Roosevelt Library
    Eleanor Roosevelt Papers
    Franklin D. Roosevelt Papers
        President's Official File
        President's Personal File
    Nathan Straus Papers

Madison, Wisconsin
Mass Communications Research Center, State Historical Society of Wisconsin,
    University of Wisconsin
    Erik Barnouw Papers
    Chet Huntley Papers
    NBC Collection
    WMCA Collection (unprocessed)

Minneapolis, Minnesota
Immigration History Research Center, University of Minnesota
    Rachel Davis DuBois Papers

New Haven, Connecticut
Yale University
    James Weldon Johnson Collection, Beinecke Rare Book Library
        Freedom's People Scripts
        Langston Hughes Papers
        Walter White Papers
        Richard Wright Papers
    Manuscripts and Archives, Sterling Library
        James Angell Papers
    Microfilm Collections, Manuscripts and Archives, Sterling Library
        W. E. B. Du Bois Papers
        Papers of the National Association for the Advancement of Colored

People, part 1, Meetings of the Board of Directors, 1909–50; part 9,
Discrimination in the United States Armed Forces, 1918–55, series A,
General Office Files on Armed Forces Affairs
Papers of the President's Committee on Civil Rights
Selected Documents on World War II from the Franklin D. Roosevelt
Library
Selected Records of the Office of War Information

New York, New York
Columbia University
Erik Barnouw Papers
Paul Lazarsfeld Papers
Oral History Collection
New York Public Library, Central Branch
America's Town Meeting of the Air Collection (unprocessed)
Performing Arts Library at Lincoln Center, New York Public Library
Americans All, Immigrants All Scripts
Schomburg Center for Research in Black Culture, New York Public Library
Audiovisual Division
Destination Freedom Collection (scripts and recordings)
New World A'Coming Collection (scripts and recordings)
Clipping Files
Canada Lee Papers
Radio Script Collection

Washington, D.C.
Library of Congress
Manuscripts Division
CBS Script Collection
Elmer Davis Papers
Archibald MacLeish Papers
National Urban League Papers
A. Philip Randolph Papers
Recorded Sound Division
NBC Manuscript Collection
Moorland-Spingarn Research Center, Howard University
Ambrose Caliver Papers (unprocessed)
Clipping Files
E. Franklin Frazier Papers
Alain Locke Papers
National Archives
Audiovisual Collections
Records of the Bureau of the Budget, Record Group 51
Records of the Civilian Aide to the Secretary of War (William Hastie–Truman
Gibson Papers), Record Group 107
Records of the Department of Interior, Record Group 48
Records of the Office of Education, Record Group 12

Records of the Office of Government Reports, Record Group 44
Records of the Office of War Information, Record Group 208
Smithsonian Institution
  Advertising History Collection
  Political History Collection

PUBLISHED SERIALS
America's Town Meeting of the Air, 1937–50
Education on the Air: Yearbook of the Institute for Education by Radio, Ohio State
  University, 1938–48
Opportunity, 1937–49
University of Chicago Round Table, 1937–50

PERSONAL INTERVIEWS
Peter Straus, November 1994, New York City (by telephone)
Ann Tanneyhill, November 28, 1994, Mashpee, Massachusetts

BOOKS, ARTICLES, AND DISSERTATIONS
Adamic, Louis. *My America, 1928–38*. New York: Harper & Brothers, 1938.
Allen, Robert C. *Speaking of Soap Operas*. Chapel Hill: University of North
  Carolina Press, 1985.
Anderson, Benedict. *Imagined Communities: Reflections on the Origins and Spread of
  Nationalism*. New York: Verso, 1991.
Anderson, Jervis. *A. Philip Randolph: A Biographical Portrait*. New York: Harcourt
  Brace Jovanovich, 1973.
Anderson, Karen. "Last Hired, First Fired: Black Women Workers during World
  War II." *Journal of American History* 69 (June 1992): 82–97.
Anderson, Mary. "Negro Women on the Production Front." *Opportunity* 21 (April
  1943): 37–38.
Archer, Leonard. *Black Images in American Theater*. New York: Pageant-Poseidon,
  1973.
Arnesen, Eric. "Following the Color Line of Labor: Black Workers and the
  Labor Movement before 1930." *Radical History Review* (Winter 1993): 53–87.
Barlow, William. "Commercial and Noncommercial Radio." In *Split Image: African
  Americans in the Mass Media,* ed. Jannette Dates and William Barlow, 175–250.
  Washington, D.C.: Howard University Press, 1990.
Barnouw, Erik. *Documentary: A History of the Non-Fiction Film*. New York: Oxford
  University Press, 1983.
———. *The Golden Web: 1933 to 1953*. Vol. 2 of *A History of Broadcasting in the United
  States*. New York: Oxford University Press, 1968.
———. *A Tower in Babel: 1933*. Vol. 1 of *A History of Broadcasting in the United States*.
  New York: Oxford University Press, 1966.
———, ed. *Radio Drama in Action: Twenty-Five Plays of a Changing World*. New
  York: Rinehart, 1945.
Barsam, Richard, ed. *Nonfiction Film Theory and Criticism*. New York: Dutton, 1976.
Baughman, James L. *The Republic of Mass Culture: Journalism, Filmmaking, and*

Broadcasting in America since 1941. Baltimore: Johns Hopkins University Press, 1992.

Bay, Mia. *The White Image in the Black Mind: African American Ideas about White People, 1830–1925*. New York: Oxford University Press, forthcoming.

Beasley, Maurine H. *Eleanor Roosevelt and the Media: A Public Quest for Self-Fulfillment*. Urbana: University of Illinois Press, 1987.

———, ed. *The White House Press Conferences of Eleanor Roosevelt*. New York: Garland, 1983.

Bennett, David. *Demagogues in the Depression: American Radicals and the Union Party, 1932–1936*. New Brunswick, N.J.: Rutgers University Press, 1969.

Berelson, Bernard. *Content Analysis in Communication Research*. Glencoe, Ill.: Free Press, 1952.

Bergreen, Lawrence. *Look Now, Pay Later: The Rise of Network Broadcasting*. New York: Doubleday, 1980.

Bernstein, Barton J., ed. *Politics and Policies of the Truman Administration*. Chicago: Quadrangle, 1970.

Berry, Mary Frances. *Black Resistance, White Law: A History of Constitutional Racism in America*. New York: Penguin, 1994.

———. *Military Necessity and Civil Rights Policy*. Port Washington, N.Y.: Kennikat, 1977.

Berry, Mary Frances, and John W. Blassingame. *Long Memory: The Black Experience in America*. New York: Oxford University Press, 1982.

Bilby, Kenneth. *The General: David Sarnoff and the Rise of the Communications Industry*. New York: Harper & Row, 1985.

Black Public Sphere Collective. *The Black Public Sphere: A Public Culture Book*. Chicago: University of Chicago Press, 1995.

Bliss, Edward, Jr. *Now the News: The Story of Broadcast Journalism*. New York: Columbia University Press, 1991.

Blum, John Morton. *V Was for Victory: Politics and American Culture during World War II*. New York: Harcourt Brace Jovanovich, 1976.

Bogle, Donald. *Slow Fade to Black: The Negro in American Film, 1900–1942*. New York: Oxford University Press, 1977.

———. *Toms, Coons, Mulattos, Mammies, and Bucks: An Interpretative History of Blacks in American Films*. New York: Viking, 1973.

Bourdieu, Pierre. *Language and Symbolic Power*. Cambridge: Harvard University Press, 1991.

Brinkley, Alan. *Voices of Protest: Huey Long, Father Coughlin, and the Great Depression*. New York: Vintage, 1983.

Brown, Earl. "The Negro Vote, 1944: A Forecast." *Harper's*, July 1944, 152–54.

Brown, Francis, and Joseph Roucek, ed. *Our Racial and National Minorities: Their History, Contributions, and Present Problems*. Englewood Cliffs, N.J.: Prentice-Hall, 1937.

Buchanan, Albert. *Black Americans in World War II*. Santa Barbara, Calif.: Clio, 1977.

Bunche, Ralph. *The Political Status of the Negro in the Age of FDR*. Chicago: University of Chicago Press, 1973.

Buni, Andrew. *Robert L. Vann of the "Pittsburgh Courier": Politics and Black Journalism.* Pittsburgh: University of Pittsburgh Press, 1974.

Caliver, Ambrose. "Developing Racial Tolerance in America." *Harvard Education Review* 11 (October 1941): 447–58.

———. "The Power Within: Commencement Address." *Tuskegee Messenger,* May–June 1936, 2–3, 21–23.

Cantril, Hadley. *The Invasion from Mars: A Study in the Psychology of Panic.* New York: Harper Torchbooks, 1940.

———, ed. *Public Opinion, 1935–1946.* Princeton: Princeton University Press, 1951.

Cantril, Hadley, and Gordon W. Allport. *The Psychology of Radio.* New York: Harper, 1935.

Capeci, Dominic, Jr. *Race Relations in Wartime Detroit: The Sojourner Truth Housing Controversy of 1942.* Philadelphia: Temple University Press, 1984.

Carey, James W. *Communication as Culture: Essays on Media and Society.* New York: Routledge, 1989.

Carlson, John Roy. *Under Cover: My Four Years in the Nazi Underworld of America.* New York: Dutton, 1943.

Carter, Elmer. "The Urban League." *Opportunity* 16 (December 1938): 355.

Chester, Edward W. *Radio, Television, and American Politics.* New York: Sheed and Ward, 1969.

Chomsky, Noam, ed. *The Cold War and the University.* New York: New Press, 1997.

Christian, Henry A. *Louis Adamic: A Checklist.* Kent, Ohio: Kent State University Press, 1971.

Cohen, Lizabeth. *Making a New Deal: Industrial Workers in Chicago, 1919–1939.* New York: Cambridge University Press, 1990.

Cordier, Hugh. "A History and Analysis of *Destination Freedom.*" Seminar paper, Northwestern University, 1949.

Corwin, Norman. *This Is War!* New York: Dodd, Mead, 1942.

Cripps, Thomas. *Black Film as Genre.* Bloomington: Indiana University Press, 1978.

———. *Making Movies Black: The Hollywood Message Movie from World War II to the Civil Rights Era.* New York: Oxford University Press, 1993.

———. *Slow Fade to Black: The Negro in American Film.* New York: Oxford University Press, 1973.

Cripps, Thomas, and David Culbert. "*The Negro Soldier* (1944): Film Propaganda in Black and White," in *Hollywood as Historian: American Film in a Cultural Context,* ed. Peter C. Rollins, 109–33. Lexington: University Press of Kentucky, 1983.

Crisell, Andrew. *Understanding Radio.* London: Routledge, 1994.

Crispin, George A. "Rachel Davis DuBois, Founder of the Group Conversation as an Adult Educational Facilitator for Reducing Intercultural Strife." D.Ed. dissertation, Temple University, 1987.

Culbert, David. "Education Unit in World War II: An Interview with Erik Barnouw." *Journal of Popular Culture* 12 (Fall 1978): 275–84.

———. *News for Everyman: Radio and Foreign Affairs in Thirties America.* Westport, Conn.: Greenwood Press, 1976.

Curran, James, ed. *Mass Communication and Society*. London: Edward Arnold, 1977.

Curran, James, and Michael Gurevitch, eds. *Mass Media and Society*. London: Edward Arnold, 1991.

Current Biography. New York: H. W. Wilson, 1945.

Czitrom, Daniel J. *Media and the American Mind: From Morse to McLuhan*. Chapel Hill: University of North Carolina Press, 1982.

Dalfiume, Richard M. *Desegregation of the U.S. Armed Forces: Fighting on Two Fronts, 1939–53*. Columbia: University of Missouri Press, 1969.

———. "The 'Forgotten Years' of the Negro Revolution." *Journal of American History* 60 (June 1968): 90–106.

Dates, Jannette, and William Barlow, eds. *Split Image: African Americans in the Mass Media*. Washington, D.C.: Howard University Press, 1990.

Davis, David Brion. *The Problem of Slavery in the Age of Revolution, 1770–1823*. Ithaca, N.Y.: Cornell University Press, 1975.

Davis, John. "The Negro in the United States Navy, Marine Corps, and Coastguard." *Journal of Negro Education* 11 (Summer 1943): 3.

Denning, Michael. "Common Ground: Theorizing Peoples in the Age of the CIO." Paper delivered at "Cultural Studies: Theory and Pluralism" conference, University of Melbourne, December 1992.

———. *The Cultural Front: The Laboring of American Culture in the Twentieth Century*. New York: Verso, 1997.

Dent, Gina, ed. *Black Popular Culture*. Seattle: Bay Press, 1992.

De Santis, Christopher C., ed. *Langston Hughes and the "Chicago Defender": Essays on Race, Politics, and Culture, 1942–62*. Urbana: University of Illinois Press, 1995.

Diner, Hasia R. *In the Almost Promised Land: American Jews and Blacks, 1915–1935*. Westport, Conn.: Greenwood Press, 1977.

Douglas, George H. *The Early Days of Radio Broadcasting*. Jefferson, N.C.: McFarland, 1987.

Douglas, Susan. *Inventing American Broadcasting, 1890–1920*. Baltimore: Johns Hopkins University Press, 1987.

Dower, John W. *War without Mercy: Race and Power in the Pacific War*. New York: Pantheon, 1986.

Drabeck, Bernard A., ed. *Archibald MacLeish: Reflections*. Amherst: University of Massachusetts Press, 1986.

Dryer, Sherman. *Radio in Wartime*. New York: Greenberg, 1942.

Duberman, Martin. *Paul Robeson*. New York: Knopf, 1988.

DuBois, Rachel Davis, and Corann Okorodudu. *All This and Something More: Pioneering in Intercultural Education*. Bryn Mawr, Pa.: Dorrance, 1984.

DuBois, Rachel Davis, and Rachel Schweppe. *The Jews in American Life*. New York: Nelson and Sons, 1935.

Du Bois, W. E. B. "Criteria of Negro Art." *Crisis,* October 1926, 294–97.

Dunning, John. *Tune in Yesterday: The Ultimate Encyclopedia of Old-Time Radio, 1925–1976*. Englewood Cliffs, N.J.: Prentice-Hall, 1976.

Early, Gerald. "American Education and the Postmodernist Impulse." *American Quarterly* 45 (June 1993): 220–30.

Edmerson, Estelle. "A Descriptive Study of the American Negro in United States

Professional Radio, 1922–1953." Master's thesis, Theater Arts, University of California at Los Angeles, 1954.

Egerton, John. *Speak Now against the Day: The Generation before the Civil Rights Movement in the South.* New York: Knopf, 1994.

Ely, Melvin. *The Adventures of Amos 'n' Andy: A Social History of an American Phenomenon.* New York: Free Press, 1991.

Everly, Philip K. *Music in the Air: America's Changing Tastes in Popular Music.* New York: Hastings House, 1982.

Fairclough, Adam. *Race and Democracy: The Civil Rights Struggle in Louisiana, 1915–1972.* Athens: University of Georgia Press, 1995.

Fang, Irving. *Those Radio Commentators!* Ames: Iowa State University Press, 1977.

Federal Security Agency, Office of Education. *Radio Script Catalog.* Washington, D.C.: Government Printing Office, 3d ed., 1946; 6th ed., 1950.

Fields, Barbara. "Slavery, Race, and Ideology in the United States of America." *New Left Review* 181 (May–June 1990): 95–118.

Finkle, Lee. *Forum for Protest: The Black Press during World War II.* Cranbury, N.J.: Associated University Presses, 1975.

Foley, Karen Sue. "The Political Blacklist in the Broadcast Industry: The Decade of the 1950s." Ph.D. dissertation, Ohio State University, 1972.

Fortunale, Peter, and Joshua E. Mills. *Radio in the Television Age.* Woodstock, N.Y.: Overlook Press, 1980.

Franklin, John Hope. *From Slavery to Freedom: A History of Negro Americans.* 5th ed. New York: Knopf, 1980.

Franklin, V. P. *Black Self-Determination: A Cultural History of African-American Resistance.* Chicago: Lawrence Hill, 1992.

Fredrickson, George M. *The Black Image in the White Mind: The Debate on Afro-American Character and Destiny, 1817–1914.* New York: Harper & Row, 1971.

Friedman, Saul. *No Haven for the Oppressed: United States Policy towards Jewish Refugees, 1938–1945.* Detroit: Wayne State University Press, 1967.

Fussell, Paul. *Wartime: Understanding and Behavior in the Second World War.* New York: Oxford University Press, 1989.

Gabler, Neal. *An Empire of Their Own: How the Jews Invented Hollywood.* New York: Crown, 1988.

Gates, Henry Louis. *Loose Canons: Notes on the Culture Wars.* New York: Oxford University Press, 1992.

Gerstle, Gary. *Working-Class Americanism: The Politics of Labor in a Textile City, 1914–1960.* New York: Cambridge University Press, 1989.

Gill, Gerald Robert. "Afro-American Opposition to the United States' Wars of the Twentieth Century: Dissent, Disinterest, and Discontent." Ph.D. dissertation, Howard University, 1985.

Gill, Glenda E. "Canada Lee: Black Actor in Non-Traditional Roles." *Journal of Popular Culture* 25 (Winter 1991): 79–89.

Gilmore, Glenda. *Gender and Jim Crow: Women and the Politics of White Supremacy in North Carolina, 1896–1920.* Chapel Hill: University of North Carolina Press, 1996.

Gitlin, Todd. *The Whole World Is Watching*. Berkeley: University of California Press, 1980.

Gleason, Philip. *Speaking of Diversity: Language and Ethnicity in Twentieth-Century America*. Baltimore: Johns Hopkins University Press, 1992.

Goings, Kenneth W. *"The NAACP Comes of Age": The Defeat of Judge John J. Parker*. Bloomington: Indiana University Press, 1990.

Gosnell, Harold F. "Obstacles to Domestic Pamphleteering by OWI in World War II." *Journalism Quarterly* 23 (December 1946): 364–69.

Graebner, William. *The Age of Doubt: American Thought and Culture in the 1940s*. Boston: Twayne, 1990.

Gray, Herman. *Watching Race: Television and the Struggle for "Blackness."* Minneapolis: University of Minnesota Press, 1995.

Greenfield, Thomas Allen. *Radio: A Reference Guide*. Westport, Conn.: Greenwood Press, 1989.

Griffin, Farah. *"Who Set You Flowin'?": The African-American Migration Narrative*. New York: Oxford University Press, 1995.

Gubar, Susan. *Racechanges: White Skin, Black Face in American Culture*. New York: Oxford University Press, 1997.

Guerrero, Ed. *Framing Blackness: The African American Image in Film*. Philadelphia: Temple University Press, 1993.

Guy-Sheftall, Beverly, ed. *Words of Fire: An Anthology of African-American Feminist Thought*. New York: New Press, 1995.

Guzman, Jesse Parkhurst, ed. *The Negro Yearbook: A Review of Events Affecting Negro Life, 1941–1946*. Tuskegee: Tuskegee Institute Press, 1947.

Halberstam, David. *The Powers That Be*. New York: Dell, 1980.

Hall, Jacquelyn Dowd, James Leloudis, Robert Korstad, Mary Murphy, Lu Ann Jones, and Christopher B. Daly. *Like a Family: The Making of a Southern Cotton Mill World*. Chapel Hill: University of North Carolina Press, 1987.

Hall, Stuart. "Reconstruction Work: Images of Post-War Black Settlement." In *Family Snaps: The Meanings of Domestic Photography,* ed. Jo Spence and Patricia Holland, 152–64. London: Virago, 1991.

Hansen, Burrell Fenton. "A Critical Evaluation of a Documentary Series of Radio Programs on Racial and Religious Prejudice." Ph.D. dissertation, University of Minnesota, 1953.

Harris, William H. *Keeping the Faith: A. Philip Randolph, Milton F. Webster, and the Brotherhood of Sleeping Car Porters, 1925–37*. Urbana: University of Illinois Press, 1977.

Hasse, John. *Beyond Category: The Life and Genius of Duke Ellington*. New York: Simon and Schuster, 1993.

Havig, Alan. "Beyond Nostalgia: American Radio as a Field of Study." *Journal of Popular Culture* 12 (Fall 1978): 218–27.

Heath, Gordon. *Deep Are the Roots: Memoirs of a Black Expatriate*. Amherst: University of Massachusetts Press, 1992.

Heide, Robert, and John Gilman. *Home Front America: Popular Culture of the World War II Era*. San Francisco: Chronicle, 1995.

Herzstein, Robert Edwin. *The War That Hitler Won: Goebbels and the Nazi Media Campaign.* New York: Paragon House, 1987.

Higginbotham, Evelyn Brooks. "African-American Women's History and the Metalanguage of Race." *Signs* 17 (Winter 1992): 251–75.

———. *Righteous Discontent: The Women's Movement in the Black Baptist Church, 1880–1920.* Cambridge: Harvard University Press, 1993.

Higham, John. "Multiculturalism and Universalism: A History and Critique." *American Quarterly* 45 (June 1993): 195–219.

———. *Send These to Me: Immigrants in Urban America.* Baltimore: Johns Hopkins University Press, 1984.

Hill, George H. *Black Media in America: A Resource Guide.* Boston: Hall, 1984.

Hill, Patricia Liggins. "Alain Locke on Black Folk Music." In *Alain Locke: Reflections on a Modern Renaissance Man,* ed. Russell J. Linnemann, 122–31. Baton Rouge: Louisiana State University Press, 1982.

Hill, Ruth Edmonds, ed. *The Black Women Oral History Project.* Vol. 9. Westport, Conn.: Meckler, 1991.

Hilmes, Michele. *Hollywood and Broadcasting: From Radio to Cable.* Urbana: University of Illinois Press, 1990.

———. "Invisible Men: *Amos 'n' Andy* and the Roots of Broadcast Discourse." *Critical Studies of Mass Communication* 10 (December 1993): 301–21.

———. *Radio Voices: American Broadcasting, 1922–1952.* Minneapolis: University of Minnesota Press, 1997.

Hine, Darlene Clark. "Black Migration to the Urban Midwest: The Gender Dimension, 1915–1945." In *The Great Migration in Historical Perspective: New Dimensions of Race, Class, and Gender,* ed. Joe William Trotter Jr., 127–46. Bloomington: Indiana University Press, 1991.

———. *Black Victory: The Rise and Fall of the White Primary in Texas.* Millwood, N.Y.: KTO Press, 1979.

———. *Black Women in White: Racial Conflict and Cooperation in the Nursing Profession, 1890–1950.* Bloomington: Indiana University Press, 1989.

———, ed. *The State of Afro-American History.* Baton Rouge: Louisiana State University Press, 1986.

Holloway, Jonathan. "Confronting the Veil: New Deal African American Intellectuals and the Evolution of a Radical Voice." Ph.D. dissertation, Yale University, 1995.

Holsinger, M. Paul, and Mary Anne Schofield, eds. *Visions of War: World War II in Popular Literature and Culture.* Bowling Green: Bowling Green State University Popular Press, 1992.

Holt, Thomas. "The Lonely Warrior: Ida B. Wells-Barnett and the Struggle for Black Leadership." In *Black Leaders of the Twentieth Century,* ed. John Hope Franklin and August Meier, 39–61. Urbana: University of Illinois Press, 1982.

Honey, Maureen. *Creating Rosie the Riveter: Class, Gender, and Propaganda during World War II.* Amherst: University of Massachusetts Press, 1984.

Honey, Michael K. *Southern Labor and Black Civil Rights: Organizing Memphis Workers.* Urbana: University of Illinois Press, 1993.

Horowitz, Roger. *"Negro and White, Unite and Fight!": A Social History of Industrial Unionism in Meatpacking, 1930–90*. Urbana: University of Illinois Press, 1997.

Hosley, David H. *As Good As Any: Foreign Correspondence on American Radio, 1930–1940*. Westport, Conn.: Greenwood Press, 1984.

Huggins, Nathan. "The Deforming Mirror of Truth: Slavery and the Master Narrative of American History." *Radical History Review* 49 (Winter 1991): 25–48.

———, ed. *Voices from the Harlem Renaissance*. New York: Oxford University Press, 1976.

———. *W. E. B. Du Bois: Writings*. New York: Library of America, 1986.

Hutchinson, Peyton Smith. "Marginal Man with a Marginal Mission: A Study of the Administrative Strategies of Ambrose Caliver, Black Administrator of Negro and Adult Education, the United States Office of Education, 1930–1962." Ph.D. dissertation, Michigan State University, 1975.

Ignatiev, Noel. *How the Irish Became White*. New York: Routledge, 1995.

Jackaway, Gwenyth L. *Media at War: Radio's Challenge to the Newspapers, 1924–1939*. Westport, Conn.: Praeger, 1995.

Jackson, Walter A. *Gunnar Myrdal and America's Conscience: Social Engineering and Racial Liberalism, 1938–1987*. Chapel Hill: University of North Carolina Press, 1990.

Janken, Kenneth Robert. *Rayford Logan and the Dilemma of the African American Intellectual*. Amherst: University of Massachusetts Press, 1993.

Johnson, Charles S. "The Present Status of Race Relations, with Particular Reference to the Negro." *Journal of Negro Education* 8 (July 1939): 323–35.

———. *To Stem This Tide: A Survey of Racial Tension Areas in the United States*. Boston: Pilgrim Press, 1943.

Jones, David Lloyd. "The United States Office of War Information and American Public Opinion during World War II, 1939–1945." Ph.D. dissertation, State University of New York at Binghamton, 1976.

Jones, Jacqueline. *Labor of Love, Labor of Sorrow: Black Women, Work, and the Family*. New York: Vintage, 1986.

Jones, Morris. *Americans All . . . Immigrants All: A Handbook for Listeners*. Washington, D.C.: Government Printing Office, Office of Education, 1939.

Kaltenborn, H. V. *Fifty Fabulous Years, 1900–1950*. New York: Putnam, 1950.

Kammen, Michael. *The Lively Arts: Gilbert Seldes and the Transformation of Cultural Criticism in the United States*. New York: Oxford University Press, 1996.

Kazal, Russell A. "Revisiting Assimilation: The Rise, Fall, and Reappraisal of a Concept in American Ethnic History." *American Historical Review* 100 (April 1995): 437–71.

Kelley, Robin D. G. *Hammer and Hoe: Alabama Communists during the Great Depression*. Chapel Hill: University of North Carolina Press, 1990.

———. "Notes on Deconstructing 'The Folk.'" *American Historical Review* 97 (December 1992): 1400–1408.

———. *Race Rebels: Culture, Politics, and the Black Working Class*. New York: Free Press, 1994.

———. "'We Are Not What We Seem': Rethinking Black Working-Class

Opposition in the Jim Crow South." *Journal of American History* 80 (June 1993): 75–112.

Kellogg, Peter J. "Northern Liberals and Black America: A History of White Attitudes, 1936–1952." Ph.D. dissertation, Northwestern University, 1971.

Kesselman, Louis Coleridge. *The Social Politics of FEPC: A Study in Reform Pressure Movements*. Chapel Hill: University of North Carolina Press, 1948.

Kirby, Jack Temple. *Media-Made Dixie: The South in the American Imagination*. Baton Rouge: Louisiana State University Press, 1978.

Kirby, John B. *Black Americans in the Roosevelt Era*. Knoxville: University of Tennessee Press, 1980.

Kivisto, Peter, and Dag Blanck, eds. *American Immigrants and Their Generations: Studies and Commentaries on the Hansen Thesis after Fifty Years*. Urbana: University of Illinois Press, 1990.

Kluger, Richard. *Simple Justice: The History of "Brown v. Board of Education" and Black America's Struggle for Equality*. New York: Vintage, 1977.

Kneebone, John T. *Southern Liberal Journalists and the Issue of Race, 1920–1944*. Chapel Hill: University of North Carolina Press, 1985.

Koppes, Clayton R., and Gregory D. Black. "Blacks, Loyalty, and Motion Picture Propaganda in World War II." *Journal of American History* 73 (September 1986): 383–406.

———. *Hollywood Goes to War: How Politics, Profits, and Propaganda Shaped World War II Movies*. New York: Free Press, 1987.

Korstad, Robert, and Nelson Lichtenstein. "Opportunities Found and Lost: Labor, Radicals, and the Early Civil Rights Movement." *Journal of American History* 75 (December 1988): 786–811.

Lash, John. "Educational Implications of the Negro College Radio Program." *Journal of Negro Education* 13 (Spring 1944): 162–68.

———. "The Negro and Radio." *Opportunity* 21 (October 1943): 158–61, 182–83.

Lawson, Steven. *Black Ballots: Voting Rights in the South, 1944–1969*. New York: Columbia University Press, 1976.

Lazarsfeld, Paul F., ed. *Radio Research, 1942–1943*. New York: Arno Press, 1944.

Lazarsfeld, Paul F., and Harry Field. *Radio and the Printed Page*. New York: Duell, Sloan and Pearce, 1940.

Lazarsfeld, Paul F., and Patricia L. Kendall. *Radio Listening in America: The People Look at Radio—Again*. New York: Prentice-Hall, 1948.

Lazarsfeld, Paul F., and Frank Stanton, eds. *Communications Research, 1948–1949*. New York: Harper, 1949.

———. *Radio Research, 1941*. New York: Duell, Sloan and Pearce, 1941.

Lee, Ulysses. *The Employment of Negro Troops*. Washington, D.C.: Government Printing Office, 1994.

Leiter, Robert D. *The Musicians and Petrillo*. New York: Bookman Associates, 1953.

Lemke-Santangelo, Gretchen. *Abiding Courage: African American Migrant Women and the East Bay Community*. Chapel Hill: University of North Carolina Press, 1996.

Leuchtenberg, William E. *Franklin D. Roosevelt and the New Deal, 1932–1940*. New York: Harper & Row, 1963.

Levine, Lawrence. "The Folklore of Industrial Society: Popular Culture and Its Audiences." *American Historical Review* 97 (December 1992): 1369–99.

———. *The Unpredictable Past: Explorations in American Cultural History.* New York: Oxford University Press, 1993.

Lewis, David Levering. "Parallels and Divergences: Assimilationist Strategies of Afro-American and Jewish Elites from 1910 to the Early 1930s." *Journal of American History* 71 (December 1984): 543–64.

Lewis, Earl. *In Their Own Interests: Race, Class, and Power in Twentieth-Century Norfolk, Virginia.* Berkeley: University of California Press, 1991.

Lewis, Thomas. *Empire of the Air: The Men Who Made Radio.* New York: Burlingame, 1991.

Linnemann, Russell J., ed. *Alain Locke: Reflections on a Modern Renaissance Man.* Baton Rouge: Louisiana State University Press, 1982.

Lipsitz, George. *Rainbow at Midnight: Labor and Culture in the 1940s.* Urbana: University of Illinois Press, 1994.

Liss, Joseph, ed. *Radio's Best Plays.* New York: Greenberg, 1947.

Livingstone, Sonia, and Peter Lunt. *Talk on Television: Audience Participation and Public Debate.* New York: Routledge, 1994.

Locke, Alain. "Art or Propaganda." In *Voices from the Harlem Renaissance,* ed. Nathan Huggins, 312–13. New York: Oxford University Press, 1976.

———. *The Negro and His Music.* Washington, D.C.: Associates in Negro Folk Education, 1936.

———. *Negro Art: Past and Present.* New York: Arno Press, 1936.

———. *The New Negro.* New York: Atheneum, 1925.

Locke, Alain, and Bernhard J. Stern. *When Peoples Meet: A Study in Race and Culture Contacts.* New York: Hinds, Hayden & Eldredge, 1946.

Logan, Rayford W., ed. *What the Negro Wants.* Chapel Hill: University of North Carolina Press, 1944.

Logan, Rayford W., and Michael R. Winston, eds. *Dictionary of American Negro Biography.* New York: Norton, 1982.

Lott, Eric. "Love and Theft: The Racial Unconscious of Blackface Minstrelsy." *Representations* 39 (Summer 1992): 23–50.

Lott, George E., Jr. "The Press Radio War of the 1930's." *Journal of Broadcasting* 14 (Summer 1970): 275–86.

Low, W. Augustus, and Virgil A. Clift, eds. *Encyclopedia of Black America.* New York: McGraw-Hill, 1981.

Lubell, Samuel. *The Future of American Politics.* New York: Harper & Row, 1952.

McChesney, Raymond. *Telecommunications, Mass Media, and Democracy: The Battle for the Control of U.S. Broadcasting, 1928–1935.* New York: Oxford University Press, 1993.

McCoy, Donald R., and Richard T. Ruetten. *Quest and Response: Minority Rights and the Truman Administration.* Lawrence: University Press of Kansas, 1973.

McCullough, David. *Truman.* New York: Simon and Schuster, 1992.

MacDonald, J. Fred. *Blacks and White TV: Afro-Americans in Television since 1948.* Chicago: Nelson-Hall, 1983.

———. *Don't Touch That Dial!: Radio Programming in American Life, 1920–1960.* Chicago: Nelson-Hall, 1979.

———. "Government Propaganda in Commercial Radio: The Case of Treasury Star Parade, 1942–1943." *Journal of Popular Culture* 12 (Fall 1978): 285–304.

———. "Radio and Television Studies and American Culture." *American Quarterly* 32, no. 3 (Bibliography Issue 1980): 301–17.

———. "Radio's Black Heritage: *Destination Freedom.*" *Phylon* 39, no. 1 (1978): 66–73.

———. *Richard Durham's "Destination Freedom": Scripts from Radio's Black Legacy, 1948–1950.* New York: Praeger, 1989.

McFadden, Margaret T. "'America's Boy Friend Who Can't Get a Date': Gender, Race, and the Cultural Work of the Jack Benny Program, 1932–1946." *Journal of American History* 80 (June 1993): 113–34.

MacGregor, Morris J. *Integration of the Armed Forces, 1940–1965.* Washington, D.C.: Government Printing Office, 1979.

Mackey, David R. *Drama on the Air.* Englewood Cliffs, N.J.: Prentice-Hall, 1951.

MacLeish, Archibald. *The American Story: Ten Broadcasts.* New York: Duell, Sloan and Pearce, 1944.

McLuhan, Marshall. *Understanding Media.* New York: McGraw-Hill, 1964.

McLuhan, Marshall, and Quentin Fiore. *The Medium Is the Message.* New York: Bantam, 1967.

McNeil, Geena Mae. "Charles Hamilton Houston: Social Engineer for Civil Rights." In *Black Leaders of the Twentieth Century,* ed. John Hope Franklin and August Meier, 221–40. Urbana: University of Illinois Press, 1982.

McQuire, Philip. *He, Too, Spoke for Democracy: Judge Hastie, World War II, and the Black Soldier.* Westport, Conn.: Greenwood Press, 1988.

McWilliams, Carey. *Brothers under the Skin.* Boston: Little, Brown, 1942.

Marsh, C. S., ed. *Educational Broadcasting, 1936: Proceedings of the First National Conference on Educational Broadcasting.* Chicago: University of Chicago Press, 1937.

Mayer, Milton. *Robert Maynard Hutchins: A Memoir.* Berkeley: University of California Press, 1993.

Meier, August, and John H. Bracey Jr. "The NAACP as a Reform Movement, 1909–1965: 'To Reach the Conscience of America.'" *Journal of Southern History* 59 (February 1993): 3–30.

Meier, August, and Elliott Rudwick. *Black History and the Historical Profession, 1915–1980.* Urbana: University of Illinois Press, 1986.

Merton, Robert K. *Mass Persuasion: The Social Psychology of a War Bond Drive.* New York: Oxford University Press, 1946.

Meyer, Leisa. *Creating GI Jane: Sexuality and Power in the Women's Army Corps during World War II.* New York: Columbia University Press, 1996.

Montalto, Nicholas. *A History of the Intercultural Education Movement.* New York: Garland, 1982.

Moon, Henry Lee. *Balance of Power: The Negro Vote.* Garden City, N.Y.: Doubleday, 1948.

Moore, Jesse Thomas, Jr. *A Search for Equality: The National Urban League, 1910–1961*. University Park: Pennsylvania State University Press, 1981.

Morgan, Chester M. *Redneck Liberal: Theodore G. Bilbo and the New Deal*. Baton Rouge: Louisiana State University Press, 1985.

Morley, David, and Kuan-Hsing Chen, eds. *Stuart Hall: Critical Dialogues in Cultural Studies*. New York: Routledge, 1996.

Morrison, Toni. *Playing in the Dark: Whiteness and the Literary Imagination*. New York: Vintage, 1993.

———, ed. *Race-ing Justice, En-gendering Power: Essays on Anita Hill, Clarence Thomas, and the Construction of Social Reality*. New York: Pantheon, 1992.

Morrison, Toni, and Claudia Brodsky Lacour, eds. *Birth of a Nation'hood: Gaze, Script, and Spectacle in the O. J. Simpson Case*. New York: Pantheon, 1997.

Morton, Patricia. *Disfigured Images: The Historical Assault on Afro-American Women*. Westport, Conn.: Praeger, 1991.

Mukerji, Chandra, and Michael Schudson, eds. *Rethinking Popular Culture: Contemporary Perspectives in Cultural Studies*. Berkeley: University of California Press, 1991.

Natanson, Nicholas. *The Black Image in the New Deal: The Politics of FSA Photography*. Knoxville: University of Tennessee Press, 1992.

Neimark, Paul. "The Man Who Knows Muhammad Ali Best." *Sepia,* May 1976, 22–30.

Newman, Mark. "Capturing the Fifteen Million Dollar Market: The Emergence of Black Oriented Radio." Ph.D. dissertation, Northwestern University, 1984.

Nimmo, Dan. *The Political Persuaders: The Techniques of Modern Election Campaigns*. Englewood Cliffs, N.J.: Prentice-Hall, 1970.

Norrell, Robert J. *Reaping the Whirlwind: The Civil Rights Movement in Tuskegee*. New York: Knopf, 1984.

Norton, Anne. *Republic of Signs: Liberal Theory and American Popular Culture*. Chicago: University of Chicago Press, 1993.

Novick, Peter. *That Noble Dream: The "Objectivity Question" and the American Historical Profession*. Cambridge: Cambridge University Press, 1988.

O'Kelly, Charlotte. "Black Newspapers and the Black Protest Movement: Their Historical Relationship, 1827–1945." *Phylon* 63 (Spring 1982): 1–14.

Omi, Michael, and Howard Winant. *Racial Formation in the United States: From the 1960s to the 1980s*. New York: Routledge, 1986.

Ottley, Roi. *Black Odyssey: The Story of the Negro in America*. New York: Scribner's Sons, 1948.

———. *The Lonely Warrior: The Life and Times of Robert S. Abbott*. Chicago: Henry Regnery, 1955.

———. *New World A-Coming*. New York: Arno Press, 1943.

Paley, William S. *As It Happened: A Memoir*. New York: Doubleday, 1979.

Paper, Lewis J. *Empire: William S. Paley and the Making of CBS*. New York: St. Martin's Press, 1987.

Parris, Guichard, and Lester Brooks. *Blacks in the City: A History of the National Urban League*. Boston: Little, Brown, 1971.

Patterson, James T. *Congressional Conservatism and the New Deal: The Growth of the*

Conservative Coalition in Congress, 1933–1939. Lexington: University Press of Kentucky, 1967.

Payne, Charles M. *I've Got the Light of Freedom: The Organizing Tradition and the Mississippi Freedom Struggle.* Berkeley: University of California Press, 1995.

Pells, Richard H. *Radical Visions and American Dreams: Culture and Social Thoughts in the Depression Years.* New York: Harper & Row, 1973.

Phelps, Shirelle, ed. *Who's Who among African Americans.* Detroit: Gale Research, 1998.

Plummer, Brenda Gayle. *Rising Wind: Black Americans and U.S. Foreign Affairs, 1935–1960.* Chapel Hill: University of North Carolina Press, 1996.

Polenberg, Richard. "The Good War?: A Reappraisal of How World War II Affected American Society." *Virginia Magazine of History and Biography* 100 (July 1992): 295–322.

———. *War and Society: The United States, 1941–45.* Philadelphia: Lippincott, 1972.

Public Papers of the Presidents: Harry S. Truman. Washington, D.C.: Government Printing Office, 1963.

Radway, Janice. *Reading the Romance: Women, Patriarchy, and Popular Literature.* Chapel Hill: University of North Carolina Press, 1984.

Rampersad, Arnold. *The Life of Langston Hughes,* vol. 2, *1941–1967: I Dream a World.* New York: Oxford University Press, 1988.

———, ed. *Richard Wright — Later Works: Black Boy (American Hunger), The Outsider.* New York: Library of America, 1991.

Reagon, Bernice. "World War II Reflected in Black Music: 'Uncle Sam Called Me.' " *Southern Exposure* 1, nos. 3 and 4 (1974): 170–84.

Reddick, L. D. "Educational Programs for the Improvement of Race Relations: Motion Pictures, Radio, the Press, and Libraries." *Journal of Negro Education* 13 (Summer 1944): 367–89.

Reed, Linda. *Simple Decency and Common Sense: The Southern Conference Movement, 1938–1963.* Bloomington: Indiana University Press, 1991.

Reed, Merl E. *Seedtime for the Modern Civil Rights Movement: The President's Committee on Fair Employment Practice, 1941–1946.* Baton Rouge: Louisiana State University Press, 1991.

Reinsch, J. Leonard. *Getting Elected: From Radio and Roosevelt to Television and Reagan.* New York: Hippocrene, 1988.

Riddle, Estelle Massey. "The Negro Nurse and the War." *Opportunity* 21 (April 1943): 44–45, 92.

Robinson, Armstead J., and Patricia Sullivan. *New Direction in Civil Rights Studies.* Charlottesville: University Press of Virginia, 1991.

Robinson, Thomas Porter. *Radio Networks and the Federal Government.* New York: Columbia University Press, 1943.

Roeder, George H., Jr. *The Censored War: American Visual Experience during World War II.* New Haven: Yale University Press, 1993.

Roediger, David. *The Wages of Whiteness: Race and the Making of the American Working Class.* New York: Verso, 1991.

Rogin, Michael. "Making America Home: Racial Masquerade and Ethnic

Assimilation in the Transition to Talking Pictures." *Journal of American History* 79 (December 1992): 1050–77.

Rosenman, Samuel, ed. *The Public Papers and Addresses of Franklin D. Roosevelt.* Vol. 2. New York: Macmillan, 1941.

Rubin, Joan Shelley. *The Making of Middlebrow Culture.* Chapel Hill: University of North Carolina Press, 1992.

Rupp, Leila J. *Mobilizing Women for War: Germany and American Propaganda, 1939–1945.* Princeton: Princeton University Press, 1978.

Sanchez, George J. *Becoming Mexican American: Ethnicity, Culture, and Identity in Chicano Los Angeles, 1900–1945.* New York: Oxford University Press, 1993.

Sanders, Mark A., ed. *A Son's Return: Selected Essays of Sterling A. Brown.* Boston: Northeastern University Press, 1996.

Sawyer, Mary R. "The Fraternal Council of Negro Churches, 1934–1964." *Church History* 59 (March 1990): 51–64.

Saxton, Alexander. *The Rise and Fall of the White Republic: Class, Politics, and White Culture in the Nineteenth Century.* London: Verso, 1990.

Sayre, Jeanette. *An Analysis of the Radiobroadcasting Activities of Federal Agencies.* Cambridge: Radiobroadcasting Research Project, Littauer Center, Harvard University, 1941.

Schlesinger, Arthur, Jr. *The Age of Roosevelt: Politics of Upheaval.* Boston: Houghton Mifflin, 1960.

Schwoch, James. *The American Radio Industry and Its Latin American Activities, 1900–1939.* Urbana: University of Illinois Press, 1990.

Scovronick, Nathan. "Broadcasting 'in the Public Interest': The N.B.C. Advisory Council, 1927–1941." Ph.D. dissertation, Rutgers University, 1973.

Seelye, Dorothea. "Broadcasting Tolerance to the Tolerant: *Americans All—Immigrants All.*" Master's thesis, American University, 1941.

Sekula, Allan. "On the Invention of Photographic Meaning." In *Thinking Photography,* ed. Victor Burgis, 84–109. London: MacMillan, 1982.

Seldes, Gilbert. *The Great Audience.* New York: Viking, 1950.

Shulman, Holly Cowan. *The Voice of America: Propaganda and Democracy, 1941–1945.* Madison: University of Wisconsin Press, 1990.

Siepman, Charles A. *Radio, Television, and Society.* New York: Oxford University Press, 1950.

Singal, Daniel Joseph. *The War Within: From Victorian to Modernist Thought in the South, 1919–1945.* Chapel Hill: University of North Carolina Press, 1982.

Sitkoff, Harvard. "Harry Truman and the Election of 1948: The Coming of Age of Civil Rights in American Politics." *Journal of Southern History* 37 (November 1971): 597–616.

———. *A New Deal for Blacks: The Emergence of Civil Rights as a National Issue.* New York: Oxford University Press, 1978.

———. "Racial Militancy and Interracial Violence in the Second World War." *Journal of American History* 58 (December 1971): 661–81.

Smith, Sally Bedell. *In All His Glory: The Life of William S. Paley—The Legendary Tycoon and His Brilliant Circle.* New York: Simon and Schuster, 1990.

Smulyan, Susan. *Selling Radio: The Commercialization of American Broadcasting, 1920–1934.* Washington, D.C.: Smithsonian Institution Press, 1994.

Snead, James. *White Screens, Black Images: Hollywood from the Dark Side.* New York: Routledge, 1994.

Snorgrass, J. William, and Gloria T. Woody, eds. *Blacks and Media: A Selected, Annotated Bibliography, 1962–1982.* Tallahassee: Florida A & M University Press, 1985.

Sollars, Werner. *Beyond Ethnicity: Consent and Descent in American Culture.* New York: Oxford University Press, 1986.

Southern, David W. *Gunnar Myrdal and Black-White Relations: The Use and Abuse of an American Dilemma, 1944–1969.* Baton Rouge: Louisiana State University Press, 1987.

Spaulding, Norman. "History of Black Oriented Radio in Chicago, 1929–1963." Ph.D. dissertation, University of Illinois–Champaign, 1981.

Spring, Joel. *Images of American Life: A History of Ideological Management in Schools, Movies, Radio, and Television.* Albany: State University of New York Press, 1992.

Steele, Richard W. "The Great Debate: Roosevelt, the Media, and the Coming of the War, 1940–1941." *Journal of American History* 71 (June 1984): 69–92.

———. *Propaganda in an Open Society: The Roosevelt Administration and the Media, 1933–1941.* Westport, Conn.: Greenwood Press, 1985.

Sterling, Christopher, and John Kittross. *Stay Tuned: A Concise History of American Broadcasting.* 2d ed. Belmont, Calif.: Wadsworth, 1990.

Stevenson, Nick. *Understanding Media Cultures: Social Theory and Mass Communication.* London: Sage, 1995.

Suggs, Henry Lewis. *P. B. Young, Newspaperman: Race, Politics, and Journalism in the New South, 1910–1962.* Charlottesville: University Press of Virginia, 1988.

———, ed. *The Black Press in the Middle West, 1865–1985.* Westport, Conn.: Greenwood Press, 1996.

———. *The Black Press in the South, 1965–1979.* Westport, Conn.: Greenwood Press, 1983.

Sullivan, Patricia. *Days of Hope: Race and Democracy in the New Deal Era.* Chapel Hill: University of North Carolina Press, 1996.

Summers, Harrison B., ed. *A Thirty Year History of Programs Carried on National Radio Networks in the United States, 1926–1956.* New York: Arno, 1971.

Tester, Keith. *Media, Culture, and Morality.* New York: Routledge, 1994.

Thatcher, Mary Anne. *Immigrants and the 1930s: Ethnicity and Alienage in the Depression and On-Coming War.* New York: Garland, 1990.

To Secure These Rights: The Report of the President's Committee on Civil Rights. New York: Simon and Schuster, 1947.

Trachtenberg, Alan. *Reading American Photographs: Images as History, Mathew Brady to Walker Evans.* New York: Hill and Wang, 1989.

Turner, John G. "Radio at Bennett College." *Opportunity* 21 (January 1943): 8–10, 29.

Turner, Patricia A. *Ceramic Uncles and Celluloid Mammies: Black Images and Their Influence on Culture.* New York: Anchor, 1994.

Tushnet, Mark V. *The NAACP's Legal Strategy against Segregated Education, 1925–1950*. Chapel Hill: University of North Carolina Press, 1987.

Tyler, Tracy F., ed. *Radio as a Cultural Agency: Proceedings of a National Conference on the Uses of Radio as a Cultural Agency in a Democracy*. Washington, D.C.: National Committee on Education by Radio, 1934.

U.S. Department of Commerce, Bureau of the Census. *Sixteenth Census of the United States, 1940: Housing*. Vol. 2, *General Characteristics: Part I: United States Summary*. Washington, D.C., 1943.

Von Eschen, Penny M. *Race against Empire: Black Americans and Anticolonialism, 1937–1957*. Ithaca, N.Y.: Cornell University Press, 1997.

Wacker, R. Fred. *Ethnicity, Pluralism, and Race: Race Relations Theory in America before Myrdal*. Westport, Conn.: Greenwood Press, 1983.

Waller, Judith. *Radio: The Fifth Estate*. Boston: Houghton Mifflin, 1950.

Ware, Gilbert. *William Hastie: Grace under Pressure*. New York: Oxford University Press, 1984.

Warren, Donald. *Radio Priest: Charles Coughlin, the Father of Hate Radio*. New York: Free Press, 1996.

Wasburn, Philo C. *Broadcasting Propaganda: International Radio Broadcasting and the Construction of Political Reality*. Westport, Conn.: Praeger, 1992.

Washburn, Patrick S. *A Question of Sedition: The Federal Government's Investigation of the Black Press during World War II*. New York: Oxford University Press, 1986.

Weinberg, Sydney. "What to Tell America: The Writers' Quarrel in the Office of War Information." *Journal of American History* 55 (June 1968): 73–89.

Weiss, Nancy. *Farewell to the Party of Lincoln: Black Politics in the Age of FDR*. Princeton: Princeton University Press, 1983.

———. *The National Urban League, 1910–1940*. New York: Oxford University Press, 1974.

White, Deborah Gray. *Arn't I a Woman?: Female Slaves in the Ante-Bellum South*. New York: Norton, 1985.

White, Walter. *A Man Called White*. New York: Arno Press, 1948.

———. *A Rising Wind*. Garden City, N.Y.: Doubleday, Doran, 1945.

Who's Who in Colored America. New York: Who's Who in Colored America Corp., 1927.

Widick, B. J. *Detroit: City of Race and Class Violence*. Detroit: Wayne State University Press, 1989.

Williams, Raymond. *Television: Technology and Cultural Form*. New York: Schocken, 1975.

Winfield, Betty Houchin. *FDR and the News Media*. Urbana: University of Illinois Press, 1990.

Winkler, Alan. "The Philadelphia Transit Strike of 1944." *Journal of American History* 59 (June 1972): 73–89.

———. *The Politics of Propaganda: The Office of War Information, 1942–1945*. New Haven: Yale University Press, 1978.

Wolseley, Roland E. *The Black Press, U.S.A.* Ames: Iowa State University Press, 1990.

Wright, Richard. *Twelve Million Black Voices*. New York: Thunder's Mouth Press, 1992.

Wyman, David S. *Paper Walls: America and the Refugee Crisis, 1939–1941*. New York: Pantheon, 1968.

Wynn, Neil. *The Afro-American and the Second World War*. London: Paul Elek, 1979.

Young, Donald. *American Minority Peoples: A Study of Racial and Cultural Conflicts in the United States*. New York: Harper & Brothers, 1932.

Young, James. *Black Writers of the Thirties*. Baton Rouge: Louisiana State University Press, 1973.

Zangrando, Robert L. *The NAACP Crusade against Lynching, 1909–1950*. Philadelphia: Temple University Press, 1980.

Zieger, Robert H. *American Workers, American Unions, 1920–1985*. Baltimore: Johns Hopkins University Press, 1986.

INDEX

ABC (American Broadcasting Company), 187, 189, 335 (nn. 103–5), 336 (n. 1), 339–40 (n. 46), 347 (n. 141), 348 (n. 4)

Adamic, Louis, 22, 58

African Americans: attitudes of, toward World War II, 66–67, 71, 89, 91, 106, 107–18, 130–31, 313–14 (n. 18), 314 (n. 21); as consumers, 12, 260, 287 (n. 29); education of, 38, 64–65, 85, 90, 102, 264, 302 (n. 3), 304 (n. 19); employment opportunities for, 66, 92–94, 158–68, 188–89; in federal government, 42, 64–66, 102, 136, 170, 296 (n. 63), 303 (n. 7), 316 (n. 34), 322 (n. 87); housing for, 91, 190, 253, 258, 266, 312 (n. 8), 348–49 (n. 5); migration of, to North and West, 66–67, 223, 260, 273; in military, 90–91, 122, 123–24, 130, 136–50; morale of, and OWI, 15, 107–35, 272–73, 312–13 (n. 9), 317 (n. 42); morale of, and War Department, 15, 107–8, 125, 135–48, 272–73; in 1930s–40s generally, 2–5, 58–59; photographic images of, 126, 129, 131, 134, 135, 144, 320 (n. 63), 322 (n. 85); political dissent of, against wars, 114–15; and politics of inclusion, 58–59, 60, 243; as "race rebels," 220; reentry of troops into civilian life, 148–52; relationship with media generally, 9–10, 43–44, 67–68; as soldiers in eighteenth and nineteenth centuries, 91–92, 94, 141–42, 324 (n. 104); as soldiers in World War I, 66, 71, 116, 124, 138–39, 171, 217; as soldiers in World War II, 90–91, 122, 123–24, 130, 139–40, 148–50, 185, 217, 324–25 (n. 111), 334 (n. 100); stereotypes of, 7–9, 11, 36, 43–44, 67–68, 140, 146, 169, 172, 190, 213, 214, 256–57, 325 (n. 116); violence

against, 90, 223, 266, 273, 312 (n. 8), 317 (n. 40), 324 (n. 104); War Department radio broadcast on role of, in army, 136–42. *See also* African American women; Desegregation; Racial equality; Segregation and discrimination; *and specific radio programs*

African American women: in *Destination Freedom*, 265, 353 (n. 66); in *Freedom's People*, 85; in "Heroines in Bronze," 157–58, 168–77, 180, 193; in military, 90, 116, 144, 171, 173–74, 256; in *New World A'Coming*, 256; stereotypes of, 8, 116, 169, 172, 219, 286–87 (n. 17), 330–31 (n. 48)

Agar, William, 168

AJC. *See* American Jewish Committee

Aldridge, Madeline, 173

Alexander, Will, 73, 317 (n. 42), 319 (n. 60)

Ali, Muhammad, 269

Allen, James, 132

All-white primaries, 223, 338 (n. 23)

American Civil Liberties Union, 165, 235, 344–45 (n. 111)

American Forum of the Air, 336 (n. 1)

American Jewish Committee (AJC), 46, 56

American Jewish Congress, 344 (n. 111)

Americans All, Immigrants All: accomplishments and significance of, 56–61, 272; Advisory Committee for, 26, 28, 46; African Americans as consultants to, 37–40; awards and critical acclaim for, 53, 300–301 (nn. 95, 97); controversy over title of, 26–27; description of, 14–15; dialect used in, 42; goals of, 21–22, 25–26, 28–29; as intercultural education, 34–35, 52–62, 272; "The Jews in America" episode, 45–52; mail responses to, 52–53, 298 (n. 87);

Dunbar, Paul Laurence, 86
Dunham, Katherine, 263, 265
Durham, Richard, 260–70, 275, 354 (n. 75)
Durlach, Theresa Mayer, 47–48

Early, Stephen, 97
Ebony magazine, 131, 261
Eisenhower, Dwight, 269
Eisenhower, Milton, 317 (n. 42)
Ellender, Allen, 233–35
Ellington, Duke, 138, 161
Embree, Edwin, 199
Emergency Committee of the Entertainment Industry, 177
Employment opportunities: for African Americans generally, 66, 92–94, 158–68, 188–89, 253, 329 (n. 34); in automobile industry, 161–62; in defense industry, 159, 165–68, 173; discussed on *Freedom's People*, 92–94; and FEPC, 71, 168, 204, 214–15, 236, 316 (n. 34), 324 (n. 106), 340 (n. 49); industrial workers, 286 (n. 14), 288 (n. 34); National Urban League's vocational opportunity campaign, 158–68, 327 (nn. 7–8)
English immigrants, 29–30, 292 (n. 24)
Equality. *See* Racial equality
Ethnicity, 27, 59–62, 302 (n. 112)
Ethnic stereotypes. *See* Racial and ethnic stereotypes
Ethridge, Mark, 207, 338 (n. 27)

Fair Employment Practices Commission (FEPC), 71, 168, 204, 214–15, 236, 316 (n. 34), 324 (n. 106), 340 (n. 49)
Farm Security Administration, 126, 321 (n. 82)
FBI. *See* Federal Bureau of Investigation
FCC. *See* Federal Communications Commission
Federal Bureau of Investigation (FBI), 305 (n. 28)

Federal Communications Commission (FCC), 305 (n. 31), 348 (n. 4)
Federal Council of Churches, 237, 347 (n. 149)
Federal Housing Authority, 249
Federal Theater Project, 143
FEPC. *See* Fair Employment Practices Commission
Fields, Barbara, 60
"Fighting Men," 325 (n. 111)
Films, 9–10, 108, 142–48, 172, 325 (nn. 113–14, 116)
Fisher, Clark, 214
Fisk University, 38, 64, 73, 151, 242
Fleming, John, 319 (n. 60)
Foster, A. L., 175
Franken, Jerry, 142
Frankfurter, Felix, 49
Fraternal Council of Negro Churches, 96, 310 (n. 87), 313 (n. 9)
Frazier, E. Franklin, 203, 204–6, 210, 221, 319 (n. 60), 343
Fredrickson, George, 293 (n. 28), 302 (n. 111)
Freedom's Journal, 9
Freedom's People: Advisory Committee for, 73–74, 81, 84, 88, 98; bibliographies and study guides for, 307 (n. 39); black artists and groups on, 72–73, 75; black women on, 85–86; Caliver's appearance on final show of, 98–100; Caliver's proposal for, 67–75, 104; depiction of class on, 74; description of, 15, 63–64, 69–70, 102–5; development of, 72–74, 138, 250; and Eleanor Roosevelt, 78, 83–84, 97–98, 307 (n. 40), 309 (n. 62), 311 (n. 92); end of, 100–102; final episode of, 94–101, 105, 225, 272, 311 (n. 92), 317 (n. 40); first episode of, 74–79; and Franklin Roosevelt, 78, 84, 97–100, 105, 225, 272, 309 (n. 62), 311 (n. 92), 317 (n. 40); funding for, 304 (n. 22); literature and the arts on, 84–86, 87; military service on, 91–92, 94, 141; music on,

74–78, 85, 88, 89, 95–96, 304 (n. 22),
305–6 (n. 32), 307 (nn. 37–38); narrator on, 74; and NBC, 68, 69–72,
81, 83, 87, 88, 101, 105, 167; phonograph recordings of, 98; political
theme of, 78, 94–100, 104–5; positive responses to, 79–83, 87, 93–94,
100–101; publicity for, 78–79, 80, 82,
83; religion on, 94–97; sciences on,
86–87; significance of, 102–5, 180,
269–70, 272; sign-off for, 75–76;
sign-on and opening lines of, 75; studio audience at broadcast of, 89;
treatment of segregation on, 88–90;
treatment of sports on, 88–90
French immigrants, 29–30, 292 (n. 24)

Garrison, Lloyd, 267
Garvey, Amy Jacques, 170
Gates, Henry Louis, 309 (n. 64)
Gershwin, George, 49
Gibson, Truman, 136–38, 141, 142,
146–52, 319–20 (n. 62), 322 (n. 94),
323 (n. 97), 326 (nn. 125–26), 334
(n. 100)
"Glass" radio script, 149–52, 189
Golden Gate Quartet, 73, 89
Gompers, Samuel, 49
Gosden, Freeman, 6, 286 (n. 12)
Gosnell, Harold, 318 (n. 53), 319
(n. 60)
Graham, Frank Porter, 122, 123
Granger, Lester, 126, 159, 165–68, 174,
185–87, 192, 330 (nn. 39, 41, 44), 336
(n. 112)
Graves, John Temple, 207, 210–12, 340
(nn. 48, 55)
Grayson, Mitchell, 350 (n. 15), 351
(n. 45)
Green Pastures, 96, 325 (n. 113)
Griffith, D. W., 10

Hale, Nathan, 269
Hall, Jacquelyn Dowd, et al., 286 (n. 14)
Hall, Stuart, 134
Halls of Congress, 352 (n. 48)

Hamilton, Colonel West, 92
Hammerstein, Oscar, 49
Handy, W. C., 75, 254, 307 (n. 38)
Harms, John, 338 (n. 37)
Hastie, William, 121–22, 125, 127,
136–37, 140, 142, 148, 315 (n. 30),
322 (nn. 87–88), 323 (nn. 101–2),
326 (n. 125)
Hays, Brooks, 242
Heath, Gordon, 258, 352 (n. 50)
Heisler, Stuart, 325 (n. 112)
Heller, Robert, 148
Henson, Matthew, 87
Here Comes Tomorrow, 261
Hernandez, Juan, 74, 96
"Heroines in Bronze," 157–58, 168–77,
180, 193
Herrick, Dwight, 168, 330 (nn. 39, 44)
Hilmes, Michele, 287 (n. 27)
Himes, Chester, 164–65
Hispanic Americans, 31–32, 200
Hitler, Adolf, 37, 50, 59, 110, 124–29,
131, 134–35, 143, 166, 196, 207, 302
(n. 110), 321 (n. 76)
Holiday, Billie, 254
Hoover, Herbert, 6, 302 (n. 3)
Horner, Henry, 289 (n. 6)
Houchins, Joseph, 73, 98
Houghteling, James, 46
Houseman, John, 143
Housing, 91, 190, 253, 258, 266,
312 (n. 8), 348–49 (n. 5)
Houston, Charles Hamilton, 227, 346
(n. 129)
Howard University, 65, 73, 122, 208–9,
339 (n. 42)
Huggins, Nathan, 60
Hughes, Langston: on America's Town
Meeting of the Air, 16, 210–15, 218,
221, 222, 237, 241, 340 (n. 51), 347
(n. 149); "Booker T. Washington in
Atlanta" by, 341 (n. 60); criticisms
of, 213–14; and Durham, 268;
"Freedom's Plow" by, 168, 181, 330
(n. 45); and Modern Minstrels, 268;
poetry of, on Freedom's People, 86; on

radio, 268; scripts on racial and religious tolerance by, 340 (n. 58)

Humphrey, Hubert, 229, 233–35, 240–41

Huntley, Chet, 180–81, 183

Ickes, Harold, 14, 21, 65

Imes, G. Lake, 121, 123, 317 (n. 48)

Imitation of Life, 172

Immigrants. See *Americans All, Immigrants All*

Institute for Education by Radio, 257, 267

Intercultural education, 22, 24, 27, 34–35, 44, 52–62, 272, 299–300 (n. 94)

Interior Department, 21, 54–55, 65, 288 (n. 36)

Intermarriage, 183, 205, 211, 218–19, 236, 237

International Workers Organization, 151–52

Irish immigrants, 30–31

"Is the Negro Oppressed?," 195–96

Italian immigrants, 30–31

Ives, Irving, 217

Jack Benny Show, 10, 103, 162

Japan, 113, 134, 144, 253, 314 (n. 21)

Japanese Americans, 32, 200, 210, 314 (n. 21)

Jernagin, Reverend W. H., 96–97

Jews, 45–52, 59, 110, 200, 237, 298 (nn. 81–82), 342 (n. 75). *See also* Anti-Semitism

"Jim Crow Is On the Run," 191

Johnson, Campbell, 98, 322 (n. 87)

Johnson, Charles S., 63, 73, 151, 242

Johnson, Guy, 73

Johnson, Jack, 88

Johnson, James Weldon, 40–41, 86, 95, 331 (n. 50)

Johnson, Lyndon, 315 (n. 32)

Johnson, Mordecai, 122–23, 208

Jubilee, 102

"Judgment Day," 142, 180

Kaltenborn, H. V., 27, 187

Kammen, Michael, 290 (n. 13), 294 (n. 46)

Kaplan, Rabbi Mordecai, 46

Kern, Jerome, 40

Ku Klux Klan, 219, 253, 264

La Follette, Charles, 341 (n. 63)

Landon, Alf, 288 (n. 33)

Lautier, Louis, 151

Lawson, Ed, 160–61, 329 (n. 37)

Lazarsfeld, Paul, 57, 320 (n. 64)

League of Nations, 256

Lee, Canada, 74, 142, 172, 251–52, 263, 275, 328 (n. 14), 351 (n. 45)

Leonard De Paur Chorus, 89, 307 (n. 37)

Lerner, Max, 237–38, 347 (n. 149)

"Let's Face the Race Question," 210–15

Levine, Lawrence, 286 (n. 14)

Lewis, William B., 119, 120, 122, 296 (n. 60)

Life magazine, 131, 322 (n. 85)

Lincoln, Abraham, 38, 95, 267

Lincoln University, 38, 121

Liss, Joseph, 123, 124

Locke, Alain: and *Americans All, Immigrants All*, 37–41, 43, 58, 65, 294 (n. 43), 296 (n. 60); and *America's Town Meeting of the Air*, 208, 209, 222; and *Freedom's People*, 15, 73, 74, 86, 88, 98, 104

Loeb, Charles, 336 (n. 112)

Logan, Rayford, 245

Long, Huey, 288 (n. 31)

Louis, Joe, 88–89, 138, 161–62, 313 (n. 9)

L'Ouverture, Toussaint, 263

Luce, Clare Boothe, 237, 347 (n. 149)

Lynching, 90, 207, 217, 266, 273, 312 (n. 8), 317 (n. 40). *See also* Antilynching campaign

McCloy, John, 151

McDougald, Elise, 170

McGill, Ralph, 200–203, 337 (n. 18), 338 (n. 27)

McIver, Marie, 87

McKay, Claude, 86

MacKaye, Milton, 321 (n. 75)

MacLeish, Archibald, 101, 109–11, 113, 115, 124, 143, 197, 208, 312–13 (nn. 8–9)

McWilliams, Carey, 190, 203, 204, 210–12, 340 (nn. 49, 55)

Malzac, Hugh, 252

March on Washington movement, 70–71, 117, 159, 166–67

Marion, Ira, 335 (n. 104)

Maynor, Dorothy, 73

Mays, Benjamin, 94–95

Media. *See* Black press; Films; Radio; Television

Mexican Americans, 31–32, 200

Military: African American soldiers in eighteenth- and nineteenth-century wars, 91–92, 94, 141–42, 324 (n. 104); African American soldiers in World War I, 66, 71, 116, 124, 138–39, 171, 217; African American soldiers in World War II, 90–91, 122, 123–24, 130, 139–40, 148–50, 185, 217, 324–25 (n. 111), 334 (n. 100); all-black military units, 142, 149, 323 (n. 102), 325 (n. 111), 326 (n. 125), 334 (n. 100); and "America's Negro Soldiers," 138–42; black women in, 90, 116, 144, 171, 173–74, 256; desegregation of, 227–28, 229; in *Freedom's People*, 91–92, 94, 141; and *Negro Soldier*, 108, 142–48; reentry of troops into civilian life, 148–52; segregation and discrimination in, 67, 70–71, 90–92, 94, 106, 111–15, 118, 136–38, 140, 165–66, 210–11, 226–27, 322–23 (nn. 88, 94, 102), 336 (n. 112); white resentment of African Americans in military uniforms, 140–41. *See also* Office of War Information; War Department; World War II

Miller, Dorie, 91, 213, 313 (n. 9)

"Minorities," 199–203

Minority Opinion, 189–92, 259, 336 (nn. 110, 112)

Minstrelsy, 7, 286 (n. 12)

Modern Minstrels, 268

Morgan v. Virginia, 338 (n. 23)

Morgenthau, Henry, 84

Morrison, Toni, 60

Morse, Wayne, 235

Moss, Carlton, 110–11, 143–47, 312–13 (n. 9), 325 (n. 112)

Movies. *See* Films

Muhammad, Elijah, 269

Muhammad Speaks, 269

Mulattoes, 218, 219

Muni, Paul, 330 (n. 45)

Murphy, George, 39–40, 43, 295 (n. 52)

Murray, Daniel, 9

Murrow, Edward R., 27, 291 (n. 20)

Music: on *Americans All, Immigrants All* African American episode, 40–41, 43; on *Freedom's People*, 74–78, 85, 88, 89, 95–96, 304 (n. 22), 305–6 (n. 32), 307 (nn. 37–38); on "Heroines in Bronze," 173, 331 (n. 53); in National Urban League's vocational opportunities campaign, 160–61, 328 (n. 12); on *Negro Soldier*, 144; on War Department radio program, 137–38

Mutual Radio Network, 121, 122, 167, 317–18 (nn. 47, 50), 325 (n. 111), 345–46 (n. 123)

My People, 121–24

Myrdal, Gunnar, 206, 227, 345 (n. 112)

NAACP. *See* National Association for the Advancement of Colored People

NAB. *See* National Association of Broadcasters

Nash, Phileo, 318 (n. 56), 319 (n. 60)

National Association for the Advancement of Colored People (NAACP):